Brain Slices

Brain Slices

Edited by
RAYMOND DINGLEDINE
University of North Carolina
Chapel Hill, North Carolina

Plenum Press • New York and London

Library of Congress Cataloging in Publication Data

Main entry under title:

Brain slices.

Bibliography: p.
Includes index.
1. Brain. 2. Electrophysiology—Technique. 3. Brain chemistry—Technique. 4. Microtomy. I. Dingledine, Raymond, 1948– [DNLM: 1. Brain—Physiology. WL 300 B814]
QP376.B734 1983 599′.01′88 83-22957
ISBN 0-306-41437-6

A Division of Plenum Publishing Corporation
233 Spring Street, New York, N.Y. 10013

Printed in the United States of America

Contributors

BRADLEY E. ALGER • Department of Physiology, University of Maryland School of Medicine, Baltimore, Maryland
PER ANDERSEN • Institute of Neurophysiology, University of Oslo, Oslo, Norway
MICHEL BAUDRY • Department of Psychobiology, University of California, Irvine, California
THOMAS H. BROWN • Department of Cellular Neurophysiology, Division of Neurosciences, City of Hope Research Institute, Duarte, California
BARRY W. CONNORS • Department of Neurology, Stanford University School of Medicine, Stanford, California
S. S. DHANJAL • Sobell Department of Neurophysiology, Institute of Neurology, The National Hospital, University of London, London, England
RAYMOND DINGLEDINE • Department of Pharmacology, University of North Carolina, Chapel Hill, North Carolina
JOHN GARTHWAITE • Department of Veterinary Physiology and Pharmacology, The University of Liverpool, Liverpool, England
A. GRINVALD • The Weizmann Institute of Science, Rehovot, Israel
MICHAEL J. GUTNICK • Unit of Physiology, Faculty of Health Sciences, Ben Gurion University of the Negev, Beer-Sheva, Israel
GLENN I. HATTON • Department of Psychology, and the Neuroscience Program, Michigan State University, East Lansing, Michigan

GRAEME HENDERSON • Department of Pharmacology, University of Cambridge, Cambridge, England

JØRN HOUNSGAARD • Department of Neurophysiology, Panum Institute, University of Copenhagen, Copenhagen, Denmark

DANIEL JOHNSTON • Program in Neuroscience, Section of Neurophysiology, Department of Neurology, Baylor College of Medicine, Houston, Texas

MARKUS KESSLER • Department of Psychobiology, University of California, Irvine, California

GREGORY L. KING • Department of Pharmacology, University of North Carolina, Chapel Hill, North Carolina

H. KITA • Department of Anatomy, College of Medicine, The University of Tennessee, Center for the Health Sciences, Memphis, Tennessee

S. T. KITAI • Department of Anatomy, College of Medicine, The University of Tennessee, Center for the Health Sciences, Memphis, Tennessee

PETER LIPTON • Department of Physiology, University of Wisconsin, Madison, Wisconsin

RODOLFO R. LLINÁS • Department of Physiology and Biophysics, New York University Medical Center, New York, New York

GARY LYNCH • Department of Psychobiology, University of California, Irvine, California

HENRY McILWAIN • Department of Biochemistry, St. Thomas's Hospital Medical School, London, England

CHARLES NICHOLSON • Department of Physiology and Biophysics, New York University Medical Center, New York, New York

ALAN NORTH • Neuropharmacology Laboratory, Department of Nutrition and Food Science, Massachusetts Institute of Technology, Cambridge, Massachusetts

PHILIP A. SCHWARTZKROIN • Department of Neurological Surgery, University of Washington, Seattle, Washington

T. A. SEARS • Sobell Department of Neurophysiology, Institute of Neurology, The National Hospital, University of London, London, England

M. SEGAL • The Weizmann Institute of Science, Rehovot, Israel

DENNIS A. TURNER • Department of Neurosurgery, VA Medical Center, Minneapolis, Minnesota

TIM S. WHITTINGHAM • Laboratory of Neurochemistry, National Institute of Neurological and Communicative Disorders and Stroke, National Institutes of Health, Bethesda, Maryland

JOHN WILLIAMS • Neuropharmacology Laboratory, Department of Nutrition and Food Science, Massachusetts Institute of Technology, Cambridge, Massachusetts

Preface

In little less than a decade brain slices have gained prominence among neurobiologists as appropriate tools to study cellular electrophysiological aspects of mammalian brain function. The purpose of this volume is to present in some detail several inquiries in the brain sciences that have benefited greatly by the use of brain slices. The book is directed primarily toward advanced students and researchers wishing to evaluate the impact these *in vitro* preparations of the mammalian brain are having on neurobiology.

The term *brain slice* has come to refer to thin (100–700 μm) sections of a brain region prepared from adult mammals and maintained for many hours *in vitro,* for either electrophysiological or biochemical studies. In addition to good accessibility, slices feature relatively intact synaptic connections that allow a variety of experiments not feasible with standard *in vivo* or tissue culture preparations. Certain electrophysiological studies once practical only with invertebrate models are becoming routine with mammalian brain slices. The ability to perform both biochemical and electrophysiological experiments on the same piece of CNS tissue provides additional bright prospects for future research. Although most of the electrophysiological studies have dealt with hippocampal slices, it should be evident from this book that slice methodology is not limited to the hippocampus.

The Appendix, "Brain Slice Methods," is a multiauthored treatment of the technical aspects of brain slice work, collected into one document. The procedures developed in many laboratories for the preparation of slices are described and compared. The comparison of slice with *in vivo* data from the viewpoints of electrophysiology, metabolic function, and histology receives a good deal of discussion. This has represented an opportunity for the authors to comment on their experiences, and those of others, in developing a range of techniques. The appendix is thus a broad-based working handbook that should be valuable to the prospective slicer as well as current practitioners.

This book will have served its purpose if it conveys some of the excitement generated by these preparations, and if it helps to establish the special promise slices have for advancing our understanding of the brain.

Raymond Dingledine

Chapel Hill, North Carolina

Contents

2. PASSIVE ELECTROTONIC STRUCTURE AND DENDRITIC PROPERTIES OF HIPPOCAMPAL NEURONS

DENNIS A. TURNER and PHILIP A. SCHWARTZKROIN

3. BIOPHYSICS AND MICROPHYSIOLOGY OF SYNAPTIC TRANSMISSION IN HIPPOCAMPUS

DANIEL JOHNSTON and THOMAS H. BROWN

4. HIPPOCAMPUS: SYNAPTIC PHARMACOLOGY
RAYMOND DINGLEDINE

5. ENERGY METABOLISM AND BRAIN SLICE FUNCTION
PETER LIPTON and TIM S. WHITTINGHAM

6. HIPPOCAMPUS: ELECTROPHYSIOLOGICAL STUDIES OF EPILEPTIFORM ACTIVITY *IN VITRO*
BRADLEY E. ALGER

8. OPTICAL MONITORING OF ELECTRICAL ACTIVITY: DETECTION OF SPATIOTEMPORAL PATTERNS OF ACTIVITY IN HIPPOCAMPAL SLICES BY VOLTAGE-SENSITIVE PROBES

A. GRINVALD and M. SEGAL

14. BRAIN SLICE WORK: SOME PROSPECTS

PER ANDERSEN

APPENDIX: BRAIN SLICE METHODS

BRADLEY E. ALGER, S. S. DHANJAL, RAYMOND DINGLEDINE, JOHN GARTHWAITE, GRAEME HENDERSON, GREGORY L. KING, PETER LIPTON, ALAN NORTH, PHILIP A. SCHWARTZKROIN, T. A. SEARS, M. SEGAL, TIM S. WHITTINGHAM, and JOHN WILLIAMS

Introduction
Cerebral Subsystems as Biological Entities

HENRY McILWAIN

Sixty years ago, preparing sections of the brain with investigative intent was carried out mainly by anatomists and neuropathologists in search of normal or abnormal structures visible to the eye or through a microscope. Having been obtained postmortem, the organ or its sections were not expected to show functional alterations and indeed were "fixed" with reactive chemicals to minimize endogenous changes.

That portions of the mammalian brain might *perform* in isolation was first demonstrated by those studying its energy metabolism. Variously ground and chopped tissues from different organs were known to respire, and it was Warburg (Krebs, 1981) who sought in the 1920s to prepare minimally altered pieces of tissue in order to examine, especially, the balance between aerobic and anaerobic utilization of substrates. The retina of small animals was the tissue that, in its natural condition, was sufficiently thin to be fully oxygenated from its outer surfaces by diffusion, and it thus became the prototypical tissue slice. In 1924, excised retina was compared with razor-cut sections from the brain of rats (Warburg *et al.*, 1924) in studies that had numerous successors in the following 30 years. My initial contributions were to show that adenosine triphosphate and phosphocreatine were resynthesized and maintained in such tissues when incubated in chemically defined

HENRY McILWAIN • Department of Biochemistry, St. Thomas's Hospital Medical School, London, SE1 7EH, England.

glucose saline solutions based on those of Ringer and of Krebs (Buchel and McIlwain, 1950; McIlwain, 1952).

Whether such tissues from guinea pig, rat, or man could respond to stimuli that simulated those normal to the brain was also, initially, given an answer by metabolic measurements (Anguiano and McIlwain, 1951; McIlwain, 1951, 1952, 1953; McIlwain *et al.*, 1952). Respiration and glycolysis of neocortical slices increased when they were stimulated electrically by relatively small voltage gradients; their inorganic phosphate increased and phosphocreatine fell. Such action could be conducted to a tissue slice by an incoming tract. Barbiturates and phenothiazines opposed such changes; there was specific antagonism to certain categories of electrical stimuli by particular anticonvulsants (Forda and McIlwain, 1953; Greengard and McIlwain, 1955; McIlwain, 1956; McIlwain and Greengard, 1957).

Then, with electrophysiological collaboration, resting membrane potentials were recorded in isolated, optimally maintained neocortical tissues (Li and McIlwain, 1957). The potentials were comparable in magnitude to those observed in the brain *in situ*; they were displaced by electrical stimuli and such displacement was opposed by low concentrations of a number of centrally acting agents including general depressants, local anaesthetics, phenothiazines, and butyrophenones (Hillman *et al*, 1963 McIlwain, 1964; Gibson and McIlwain, 1965). Such tissues also showed occasional action potentials in the absence of electrical stimulation. Moreover, by examining slices of piriform cortex with their prominent incoming lateral olfactory tract, it was relatively easy to observe action potentials extracellularly in response to a variety of stimuli applied to the tract, and also to study synaptic phenomena (Yamamoto and McIlwain, 1966).

Isolated tissues similarly prepared and maintained provided other routes to synaptic phenomena. Stimulation of piriform and neocortical slices in an *in vitro* superfusion system enabled the output of neurotransmitters to be measured concomitantly with measurement of the metabolic or electrical status of the tissue, when exposed to a variety of substrates and inhibitors (McIlwain and Snyder, 1970; Heller and McIlwain, 1973). Neurotransmitter mediators were also open to examination: measurement of the cyclic AMP (cAMP) content of tissues stimulated electrically in the presence of histamine and noradrenaline allowed the characterization of adenosine as a neurohumoral agent, and the study of its output on excitation (Kakiuchi *et al.*, 1969; Sattin and Rall, 1970; Pull and McIlwain, 1972).

At this point, the use of surviving, metabolically maintained tissues from the brain for electrophysiological and pharmacological studies was becoming the accepted, valued, and widely applied group of techniques that has prompted the preparation of this book. This naturally is very rewarding as is also the continuing critical appraisal of the methods that again is displayed in the following pages. The description "group of techniques" is necessary because the parts of the brain examined and the questions asked of them are diverse, as also are the experimental methods that succeed in their investigation. Indeed, this book illustrates well the manner in which cell-containing cerebral isolates (McIlwain, 1975) obtained by dissection, slicing, or chopping have contributed to knowledge of the great chemical and regional diversity of cerebral makeup. Brain slices are rarely studied as such; rather, a neocortical or hippocampal slice or a subdivision of such a slice becomes the relevant subsystem. It is now expected that any portion of the brain, prepared as an isolate while taking into account its structural units, should display important aspects of its functioning in the brain; or if such display is lacking, should lead to the examination of other isolates by other *in vitro* techniques in reaching further understanding.

That cerebral tissues were prepared for *in vitro* experiments as portions weighing some 10 to 100 mg was regarded initially as a prerequisite specified by the need to provide supplies of materials to them by diffusion, a process of limited range and rate. Now, however, the use of isolates of this size can be seen also as part of a process of structural and functional analysis of the brain, and as providing simpler and more readily investigated subsystems. Indeed, hippocampal isolates of 1 to 10 mg have been so employed. Isolates of this size comprise some 10^4 to 10^5 cells; and a large cerebral neuron may synapse with 10^3 to 10^5 other cells. Thus, the cell assembly provided in a typical isolate made for *in vitro* work may be the minimum size of unit needed for adequate display of the connectivity of a cerebral neuron in its adult environment. Attempts to use "isolated neurons" prepared from the brain have not achieved comparable success, and a reason for this can be seen in the physical strength and functional importance of synapses. Even supposing that technical developments yield a means of separating neural assemblies at their synaptic junctions, a great problem remains in disentangling the intertwined axons and dendrites. Thus, the minimum domain of a large cerebral neuron includes its contacts with some 10^3 to 10^5 other polarized cells, and isolates of 1 to 10 mg are not greatly above this minimum size.

Adequately prepared cerebral isolates of the types described in this book thus approach the status of biological entities, despite the trauma of their formation. The basis for this is cell autonomy; much is yet to be learned of the life of the individual cerebral neuron. Cell-containing cerebral isolates are often the appropriate system in which to investigate such autonomy, in fashions beyond those that at present preponderate. Thus, the interest in messenger substances can encompass not only neurotransmitters acting extracellularly, but also the second and third messengers of intracellular action (McIlwain, 1981a), and the processes of cytoplasmic transport that are so fundamentally connected with the working of the brain. The metaphorical flow of thought appears likely to have components activated by cytoplasmic flow of intracellular mediators (McIlwain, 1981b), a process itself susceptible to examination in cerebral isolates. The contributions to this book thus form a group of technical and interpretative advances that, if wisely applied, carry wide practical and theoretical implications to most of the neurosciences.

REFERENCES

Anguiano, G. and McIlwain, H., 1951, Convulsive agents and the phosphates of brain examined *in vitro*, *Br. J. Pharmacol.* **6**:448–453.

Buchel, L. and McIlwain, H., 1950, Narcotics and the inorganic and creatine phosphates of mammalian brain, *Br. J. Pharmacol.* **5**:465–473.

Forda, O. and McIlwain, H., 1953, Anticonvulsants on electrically stimulated metabolism of separated mammalian cerebral cortex, *Br. J. Pharmacol.* **8**:225–229.

Gibson, I. M. and McIlwain, H., 1965, Continuous recording of changes in membrane potential in mammalian cerebral tissues *in vitro*. Recovery after depolarization by added substances, *J. Physiol.* **176**:261–283.

Greengard, O. and McIlwain, H., 1955, Anticonvulsants and the metabolism of separated mammalian cerebral tissues, *Biochem. J.* **61**:61–68.

Heller, I. H. and McIlwain, H., 1973, Release of adenine derivatives from isolated subsystems of the guinea pig brain: Actions of electrical stimulation and of papaverine, *Brain Res.* **53**:105–116.

Hillman, H. H., Campbell, W. J., and McIlwain, H., 1963, Membrane potentials in isolated and electrically stimulated mammalian cerebral cortex. Effects of chlorpromazine, cocaine, phenobarbitone and protamine on the tissue's electrical and chemical responses to stimulation, *J. Neurochem.* **10**:325–339.

Kakiuchi, S., Rall, T. W., and McIlwain, H., 1969, The effect of electrical stimulation on the accumulation of adenosine 3:5-phosphate in isolated cerebral tissue, *J. Neurochem.* **16**:485–491.

Krebs, H. A., 1981, *Otto Warburg. Cell Physiologist, Biochemist and Eccentric*, Clarendon Press, Oxford.

Li, C-L. and McIlwain, H., 1957, Maintenance of resting membrane potentials in slices of mammalian cerebral cortex and other tissues *in vitro*, *J. Physiol.* **139**:178–190.

McIlwain, H., 1951, Metabolic response *in vitro* to electrical stimulation of sections of mammalian brain, *Biochem. J.* **49**:382–393.

McIlwain, H., 1952, Phosphates and nucleotides of the central nervous system, *Biochem. Soc. Symp.* **8**:27–43.

McIlwain, H., 1953, Substances which support respiration and metabolic response to electrical impulses in human cerebral tissues, *J. Neurol.Neurosurg. Psychiatry* **16**:257–266.

McIlwain, H., 1956, Electrical influences and the speed of chemical change in the brain, *Physiol. Rev.* **36:**355–375.

McIlwain, H., 1964, Actions of haloperidol, meperidine and related compounds on the excitability and ion content of isolated cerebral tissue, *Biochem. Pharmacol.* **13:**523–529.

McIlwain, H., 1975, Cerebral isolates and neurochemical discovery, *Biochem. Soc. Trans.* **3:**579–590.

McIlwain, H., 1981a, Brain: Intracellular and extracellular purinergic receptor-systems, in: *Purinergic Receptors* (G. Burnstock, ed.) Chapman and Hall, London, pp. 163–198.

McIlwain, H., 1981b, The flow of thought and the flow of substance in the brain, *Biol. Psychol.* **12:**147–169.

McIlwain, H. and Greengard, O., 1957, Excitants and depressants of the central nervous system, on isolated electrically stimulated cerebral tissues, *J. Neurochem.* **1:**348–357.

McIlwain, H. and Snyder, S. H., 1970, Stimulation of piriform and neocortical tissues in an *in vitro* flow-system: Metabolic properties and relase of putative neurotransmitters, *J. Neurochem.* **17:**521–530.

McIlwain, H., Ayres, P. J. W., and Forda, O., 1952, Metabolic response to electrical stimulation in separated portions of human cerebral tissues, *J. Ment. Sci.* **98:**265–272.

Pull, I. and McIlwain, H., 1972, Adenine derivatives as neurohumoral agents in the brain. Quantities liberated on excitation of superfused cerebral tissues, *Biochem. J.* **130:**975–981.

Sattin, A. and Rall, T. W., 1970, The effect of adenosine and adenine nucleotides on the cyclic adenosine 3′:5′-monophosphate content of guinea pig cerebral cortex slices, *Mol. Pharmacol.* **6:**13–23.

Warburg, D., Posener, K., and Negelein, E., 1924, Über den Stoffwechsel der Carcinomzelle, *Biochem. Z.* **152:**309.

Yamamoto, C. and McIlwain, H., 1966, Electrical activities in thin sections from the mammalian brain maintained in chemically defined media *in vitro, J. Neurochem.* **13:**1333–1343.

1

Comparative Electrobiology of Mammalian Central Neurons

RODOLFO R. LLINÁS

1. INTRODUCTION

In assessing the impact of the *in vitro* analysis of CNS function, especially in mammals, one issue has become clear; mammalian neurons are endowed with a large and intricate set of ionic conductances. The intricacy of these conductances relates not only to their ionic specificity, their voltage dependence, and their modulation by neurotransmitters and neuropeptides, but also to their location on the soma–dendritic regions of the neurons. This realization has forced all of us to reexamine the levels at which the rather involved interactions between neurons actually occur. For many years, most of the complexity demonstrable electrophysiologically in different regions of the nervous system was assumed to be produced by the synaptic interactions, i.e., the neuronal network. However, it is evident, following the development of the *in vitro* preparations, that much of the electrophysiology encountered in mammalian neurons does not derive necessarily from the attributes of networks but, rather, from the intrinsic electrical properties of the cells themselves. The functional implications of this conclusion are somewhat staggering, especially when one considers that the numbers of voltage-dependent ionic conductances present in a particular neuron may be

---

RODOLFO R. LLINÁS • Department of Physiology and Biophysics, New York University Medical Center, New York, New York 10016.

greater than ten, that the activation kinetics of such conductances may differ widely, and that their actions on neuronal integration may be equally diverse.

In writing a chapter on the comparative aspects of the electrophysiology of CNS neurons, my hope is to establish some trends from what may be considered a rather chaotic state of affairs. In this attempt, I feel that the simplest approach is to consider a "generalized" neuron and to superimpose upon it the electrophysiological findings obtained from the different types of neurons studied thus far *in vitro* in the hopes that an emerging pattern will become apparent. This approach, then, will be based on well-documented electrophysiological findings and will draw upon results from different preparations.

2. THE GENERALIZED NEURON

Let us assume, for purposes of discussion, a neuron consisting of the usual soma, axon, and dendritic tree. What voltage-dependent ionic conductances have been described regarding these different segments and how does their presence influence the integrative properties of the neuron?

2.1. *The Axon*

This is by far the most uniform of the elements to be considered. So far, research performed both *in vivo* and *in vitro* in mammalian neurons has demonstrated axonal conduction to be generated by Na-dependent action potentials that are, with only few exceptions (Fukuda and Kameyama, 1980), sensitive to tetrodotoxin (TTX). This voltage-dependent conductance, first modeled by Hodgkin and Huxley (1952a,b) for the squid giant axon and modified to fit the slightly different kinetics of other axons (cf., Cahalan, 1980), continues to be the universal mechanism for active axonal conduction except in some molluscans (Horn, 1977). The properties of these voltage-dependent conductances have been determined at both macroscopic current level as well as at the single channel level (Sigworth and Neher, 1980). Since a very good correlation has been determined between the ionic specificity, conductance, voltage dependence, and average open time of channels and the macroscopic ionic currents observed in voltage clamp, a direct reduction is possible from action potentials to single ionic channel properties. Although some of the properties relating to the voltage-dependent K conductance in mammalian myelinated axons are only now being understood, it seems clear that K channels are virtually absent in these axons (Chiu *et al.*, 1979), at least under conditions in which the myelin is intact. In short, one can assume that, until otherwise demonstrated, most axons that conduct spikes in the CNS do so by TTX-sensitive voltage-dependent Na conductances and that a K conductance may be associated in fact with this action potential only in some regions e.g., the initial segment of central neuron axons and their terminal arbor (Brigant and Mallart, 1982).

3. ELECTROPHYSIOLOGY OF THE NEURONAL SOMATA

3.1. The Na Conductances

3.1.1. The Fast Na Conductance. The simplicity and clarity of experimental design possible in relation to the electrophysiology of the axon is substantially less attainable when analyzing the excitable properties of the membrane of the neuronal body. Here, in addition to the Na conductance that extends into the soma as a continuation of the properties of the axon, other less well-understood conductances are present. The initial experiment on neuronal excitability of motoneurons already indicated over 20 years ago that the somata of motoneurons are capable of Na action potentials at least as determined by voltage-clamp studies and by analysis of extracellular field potentials (Terzuolo and Araki, 1961; Araki and Terzuolo, 1962; c.f. Barrett and Crill, 1980). Also, extracellular patch current recording supports the view that the somata of other central neurons generate active Na conductance capable of generating fast action potentials. Drawing upon our own work, it is evident that both Purkinje cell soma (Llinás and Sugimori, 1980a) and the inferior olivary cell somata (Llinás and Yarom, 1981a) demonstrate rapid inward currents that are TTX-sensitive and that disappear in the absence of extracellular Na.

3.1.2. The Noninactivating Na Conductance (g_{Na}). In many neurons, however, depolarization electrogenesis is not restricted to the activation of fast action potentials. Indeed, at least another voltage-dependent Na conductance must be considered at this point. This conductance, first encountered in Purkinje cells, is capable of generating graded and, on occasion, all-or-none plateau potentials lasting up to 15 sec in these cells. These plateau responses are produced by a noninactivating Na conductance located at the soma (Llinás and Sugimori, 1980a) and have a lower threshold for activation than the fast action potential. The time for onset is particularly slow, as indicated in Figure 1. A similar non-

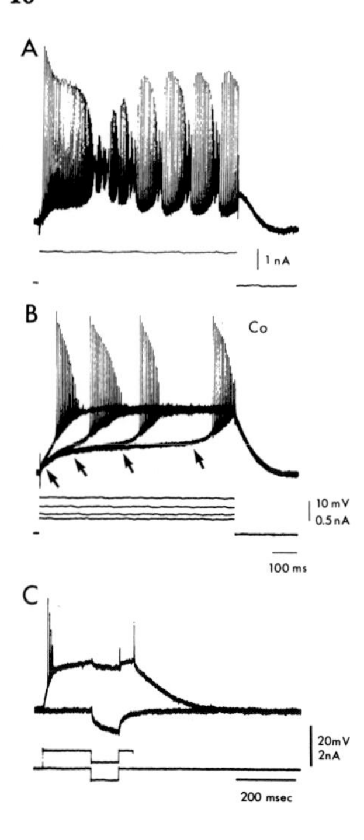

Figure 1. Noninactivating Na-dependent plateau potentials in guinea pig Purkinje cell. (A) Normal response of a Purkinje cell to a depolarizing current pulse. Activation consists of fast spikes superimposed on an oscillatory background potential produced by slow-Ca-dependent spiking. (B) Response after Ca conductance was blocked by the addition of Co to the bath. Direct stimulation elicits slow all-or-none depolarizing responses that generate fast action potentials and reach a plateau level. As the stimulus is increased, the onset moves to the left but the rate of rise and the final amplitude of the plateau are independent of stimulus amplitude in the range illustrated. The slow onset of the all-or-none responses is indicated by arrows. (C) Increased conductance during a plateau of this response is illustrated by the reduction of the voltage produced by a test hyperpolarizing pulse. Modified from Llinás and Sugimori, 1980a.

inactivating g_{Na} has been recently observed in neocortex (Stafstrom, *et al.*, 1982), thalamus (Jahnsen and Llinás, 1983a,b), and motoneurons (Schwindt and Crill, 1980), and a variance of this conductance may be the same as that present in the giant axon of squid (Matteson and Armstrong, 1982).

An interesting difference between this Na conductance and the one that generates the fast action potential is its sensitivity to some local anesthetics, in particular lidocaine. In concentrations of 10^{-5} M, lidocaine can produce a blockage of the fast action potentials leaving intact the noninactivating Na conductance (Sugimori and Llinás, 1980). This is of significance in the sense that the same local anesthetic has been shown to exert its anesthetic action by blocking Na channels in their inactive state (Cahalan, 1978).

Because the amplitude of the plateau potential is usually sufficiently large to activate voltage-dependent K conductances, a sizeable conductance increase is always associated with this potential (see Figure 1C). This seriously reduces the length constant, λ, and modifies the integrating properties of the neuron (Llinás and Sugimori, 1980a). Thus, following addition of Cs or tetraethylammonium to the bath, both of which are known to block g_K (Armstrong and Binstock, 1965; Hagiwara *et al.*, 1976), the amplitude of this plateau potential may approach sodium equilibrium potential (E_{Na}). This indicates that the lower value usually observed in the Purkinje cell (20 to 30 mV) is really produced by an equilibrium state between inward and outward currents. This current is ultimately responsible for the activation of action potentials when the depolarization of Purkinje cells is slow.

In addition to voltage-dependent Na, nine other ionic conductances are present in different somata of central neurons: seven for K, and two for Ca—the most complex electrophysiological properties probably being related to the K ion. I shall proceed first with discussion of the Ca conductances, which seem to be of two types: high and low threshold.

3.2. *Ca Conductances*

3.2.1. High-Threshold Ca Conductance. High-threshold Ca spikes recorded at somatic level were initially reported in Purkinje cells (Llinás and Hess, 1976; Llinás and Sugimori, 1980a); however, as will be discussed below, the location of this conductance is dendritic. Other examples of high threshold Ca spikes recorded at somatic level are the pyramidal cells in the hippocampus (Schwartzkroin and Slawsky, 1977; Wong and Prince, 1978), in dorsal root ganglia of rat (Fedulova *et al.*, 1981), and in rat spinal motoneurons (Fulton and Walton, 1981; Harada and Takahashi, 1983). This conductance is manifest by the ability to generate Ca-dependent action potentials in the absence of g_{Na}. The threshold for this "somatic" spike is actually much higher than that measured for the fast spike. All that is known about the actual distribution of the channels within the neuron is that under conditions where the soma is depolarized in the presence of TTX, action potentials can be generated that seem to have a lower threshold than those reported at dendritic level. In other neurons, such as the Purkinje cell, somatic Ca action potentials have not been found, although much research has been done in their pursuit. Basically two types of experiments have been designed to study this problem (Sugimori and Llinás, 1982):

1. Removal of Na and Ca from the bath and replacement of these ions by tris or choline leaves the neurons incapable of generating an inward current. At this stage, extracellular administration of Ba from an iontophoretic pipette can produce large action potentials when the pipette is positioned at the dendrites, but not at the soma of the neuron.

2. Recordings made using extracellular patch clamp technique of the nongigaseal variety indicate that Ca spikes at somatic level generate

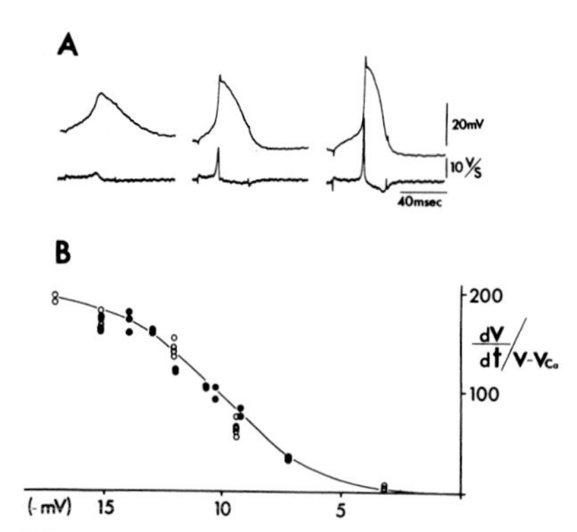

Figure 2. Deinactivation of the low-threshold Ca spike in an inferior olivary neuron. (A) Increase in rise time and amplitude of the somatic Ca spikes with hyperpolarization from rest after blockage of g_{Na} with TTX. The upper three records are the recorded spikes, the lower records the time derivative of the action potential obtained by electrical differentiation with a time constant of 5 μsec. (B) Maximum rate of rise (amplitude of the derivative corrected for driving force) as a function of membrane potential. Open and closed circles represent similar experiments in two different neurons. Abscissa indicates hyperpolarization from resting level (−60 mV); ordinate is the maximum rate of rise divided by driving force. Modified from Llinás and Yarom, 1981a.

only outward current, implying that no viable inward Ca current is present at the soma.

3.2.2. *Low-Threshold Ca Conductance.* In other neurons, such as the inferior olive (Llinás and Yarom, 1980, 1981a,b) and the thalamus (Llinás and Jahnsen, 1982), somatic Ca spikes have been observed following the hyperpolarization of the somatic membrane potential. Similar spikes have been reported in spinal sensory neurons (Murase and Randic, 1983). These Ca-dependent action potentials are shown to be produced when the membrane potential is hyperpolarized up to 75 mV at which level the action potential has its maximum amplitude and rate of rise (Figure 2). Under these circumstances, and if the membrane potential is held hyperpolarized, an action potential can be seen to produce a short-lasting spike that is lost in the absence of extracellular Ca or prevented by Ca blockers such as Cd or Co. Because such Ca-dependent spikes cannot be obtained from rest level, but are only observable following hyperpolarization, we named them low-threshold spikes (LTS). They seem to be located at the soma since, at least in the inferior olive (IO), where extracellular somatic recordings show that their activation is accompanied by a sizeable negative potential, indicating a rather powerful inward Ca current at that level. This LTS is produced by a voltage-dependent Ca conduction that is inactive at rest and deinactivated with membrane hyperpolarization, this deinactivation actually being not only voltage- but time-dependent. The presence of such a Ca conductance is of significance because it represents an important element in the generation and maintenance of the oscillatory firing seen in both the IO and thalamus (cf., Llinás, in press). As in the case of the somatic noninactivating Na component, they give the soma, and thus the axon, the ability to respond in a more autonomic manner than would be possible in their absence. Indeed, this autonomy seems to be at the basis of such phenomena as physiological tremor and the alpha rhythm (cf., Llinás, in press).

3.3. *The K Conductances*

The K conductance at somatic level may be categorized as voltage dependent and Ca dependent. This conductance, initially described in mollusks by Meech (1972), has been observed in many mammalian neurons (Krnjevic and Lisiewicz, 1972; McAfee and Yarowsky, 1979; Barrett *et al.*, 1980; Hotson and Prince, 1980; Llinás and Sugimori, 1980a,b; Llinás and Yarom, 1981a,b; Adams *et al.*, 1982a,b; MacDermott and Weight, 1982).

3.3.1. Ca-Dependent K Conductance ($g_{K[Ca]}$). In the case of the low-threshold Ca-dependent spike in the IO, this spike generates an afterhyperpolarization that is clearly Ca-dependent since substitution of extracellular Ca by Ba produces a longer potential not followed by the typical afterhyperpolarization normally observed. The inability of Ba to activate $g_{K[Ca]}$ has been demonstrated at many levels (cf., Meech, 1978).

3.3.2. Voltage-Dependent K Conductance. Six voltage-dependent K conductances have been described. The delayed (outward) rectifiers include the traditional I_K, first described by Hodgkin and Huxley (1952a,b), a new current found in sympathetic neurons (I_M), and one found in IO neurons. The remaining three currents are the two anomalous (or inward) rectifiers and a fast transient current (I_A).

3.3.2a. Delayed Rectification. At least three types of K conductance seem to be present in this type of voltage-dependent responses: (1) the usual delayed rectifier that is produced by the Hodgkin–Huxley type g_K that generates the falling phase of the fast action potential; (2) the so-called M current that has been described in sympathetic neurons (Brown and Adams, 1979, 1980; Constanti and Brown, 1981), in hippocampal pyramidal cells (Adams *et al.*, 1981; Halliwell and Adams, 1982), and in spinal cord neurons (Nowak and Macdonald, 1981). This is a noninactivating low-threshold current that is activated by depolarization and seems to be modulated by synaptic transmitter substance (Brown and Adams, 1979, 1980; MacDermott and Weight, 1980; Kobayashi *et al.*, 1981; Nowak and Macdonald, 1981; Adams *et al.*, 1982c). (3) Recently another delayed rectifier has been encountered in inferior olivary cells (Llinás and Yarom, 1980). Unlike the M current, it inactivates and is blocked by Cs. In addition, it has a Ca dependence. Although

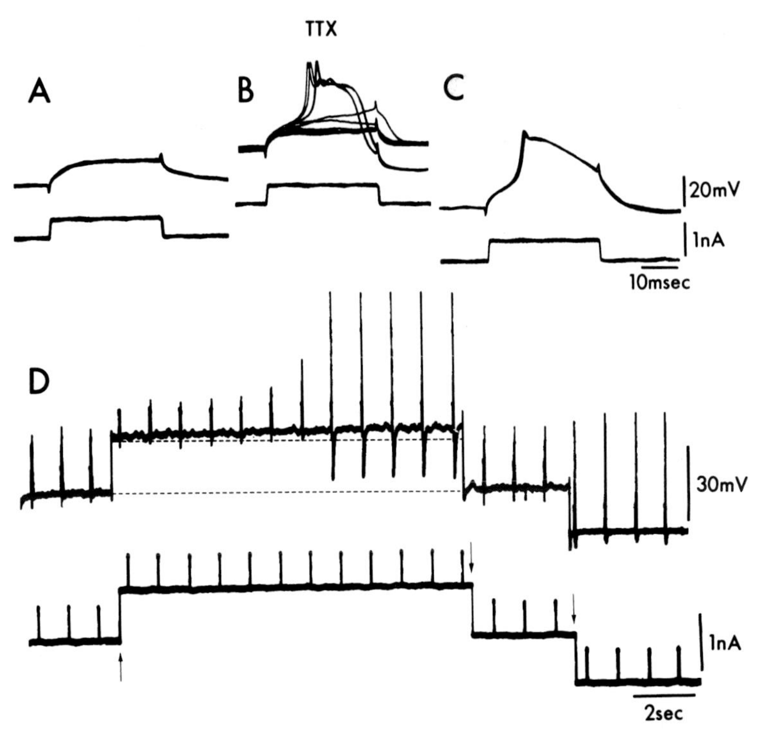

Figure 3. Inactivation of delayed rectification and the high-threshold Ca spike excitability in inferior olivary neurons. Subthreshold current pulses [lower record in (D)] at resting membrane level produce membrane depolarizations as seen at the beginning of record (D) (upper trace) and in record (A) at a fast sweep speed. A prolonged current pulse (upgoing arrow) is applied, which generates a large depolarization step that produces a reduction of the input resistance. Note the smaller-amplitude potential generated by the test current pulse. (B) This delayed rectification disappears with time, and finally full dendritic Ca spikes are generated by the last five current pulses. All 12 voltage reponses are superimposed at a faster sweep speed. At the end of the long polarization pulse (first downgoing arrow), the membrane resistance returns to control value. To the right, a prolonged direct-current hyperpolarization (second downgoing arrow) produces somatic Ca spikes, as shown at the fast sweep speed in (C). Modified from Llinás and Yarom, 1980.

the physiological implications of these conductances are not totally clear, at least in the inferior olive, this rectifier seems to have a significant role in neuronal integration since it slowly inactivates with time. Thus, if an inferior olivary cell is abruptly depolarized and the membrane potential is kept at a certain level, the input resistance progressively increases and the ability of the dendrite to communicate with the soma is facilitated (Llinás and Yarom, 1980; Figure 3). One may consider that this type of K conductance may play a rather significant role in plasticity and in long-term excitability changes of these neurons.

3.3.2b. Anomalous Rectification. Anomalous or inward rectification has been seen in many neurons in the CNS (Kandel and Tauc, 1966; Nelson and Frank, 1967; Purpura *et al.*, 1968; Schwartzkroin, 1975; Scholfield, 1978; Hotson *et al.*, 1979). As in other systems, anomalous rectification may be grouped into two categories: instantaneous and time-dependent.

(*i*) Instantaneous Component. A fast inward rectifying current has been observed in olfactory cortex (Scholfield, 1978; Constanti and Galvan, 1983) as well as in the inferior olive (R. Llinás and Y. Yarom, unpublished observations). This K channel appears to open when the membrane potential is hyperpolarized beyond a certain level. It is similar to current described in marine eggs (Hagiwara *et al.*, 1976; Ohmori, 1978; Fukushima, 1982) and was first observed in muscle (Katz, 1949; Adrian, 1969).

(*ii*) Time-Dependent Component. The instantaneous and the time-dependent components seem to be produced by different conductances. In some cases, the anomalous rectification resembles the f current in the heart (sinoartrial node) in which a step hyperpolarization is followed by a marked reduction of its amplitude as K conductances increase (Noma and Irisawa, 1976; Brown and Difrancisco, 1980; Yanagihara and Irisawa, 1980). Both type of anomalous rectification seem to modulate input resistance, and thus the integration properties of the neuron, in the sense of varying the degree of electrical coupling between soma and dendrites as the length constant, is reduced. This slow inward current has been described in hippocampal pyramidal cells (Wilson and Goldner, 1975; Adams and Halliwell, 1982; Halliwell and Adams, 1982) in which it was termed the I_Q by Adams and Halliwell (1982). This current may account for the observations of decreased slope conductance with hyperpolarization made *in vivo* in hippocampal cells (Purpura *et al.*, 1968) and pyramidal tract neurons (Takahashi *et al.*, 1967).

Yet another anomalous current has been reported in hippocampal pyramidal cells, but it is not identical to I_Q (Hotson *et al.*, 1979).

3.3.3. *The Fast Transient K Conductance (I_A).* The final variety of K current to be described has been seen at somatic level in some central neurons, such as spinal motoneurons (Barrett and Crill, 1974), vagal motoneurons (Yarom *et al.*, 1980), thalamic (Llinás and Jahnsen, 1982), hippocampal neurons (Gustafsson *et al.*, 1982), dorsal root ganglia (Adams *et al.*, 1982a), and cells in the central amygdala (R. Llinás, K. Walton, and A. Fernandez de Molina, unpublished observations). The presence of this current, termed I_A by Connor and Stevens (1971), is

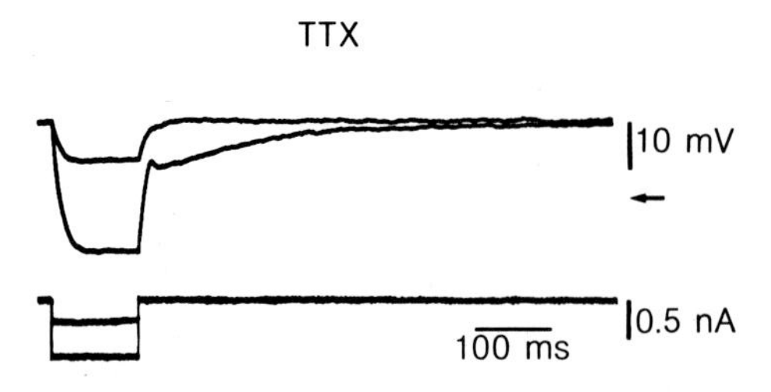

Figure 4. Examples of early K conductance in central neurons. Intracellular recording from central nucleus of the guinea pig amygdalar complex after addition of tetrodotoxin to the bath. Recording shows that direct hyperpolarization of the amygdala neuron, if large enough, is followed by a very slow return to base-line. This is not seen with a hyperpolarization of smaller amplitude. Lower trace indicates the amplitude and duration of the inward current pulse. Unpublished observations of R. Llinás, K. Walton, and A. Fernandez de Molina.

marked by a failure of the membrane potential to return immediately to baseline following hyperpolarization, but instead has a slow upward trajectory that may last for 50 to 300 msec (Figure 4). In many of these systems, and particularly in the vagal motoneuron (Yarom *et al*, 1980), this I_A current, prevents rebound excitation as the cell membrane level returns to baseline rather slowly.

In short, somatic regions of the cells are known to have at least the following conductances: inactivating Na conductance, noninactivating Na conductance, three types of delayed rectifier K conductance (g_K, the M current, and the inactivating type of conductance), two types of anomalous rectifiers (instantaneous and delayed), at least one variety of early K conductance, a $g_{K[Ca]}$, plus the inactivating and noninactivating Ca conductances.

The eleven different conductances that are present in nerve cells appear to be segregated among neurons such that they do not all appear in the same cell.

4. DENDRITIC ELECTROPHYSIOLOGY

The third aspect of our hypothetical neuron to be considered is the dendritic tree. Although intradendritic recordings were obtained *in vivo* in vertebrates more than 10 years ago (Llinás and Nicholson, 1971), the *in vitro* techniques have improved our understanding of the electroresponsive properties of this segment of the neuron. The fact that dendrites are electrophysiologically active, as suggested by electrophysiological evidence over the past 25 years. (Eccles *et al.*, 1958; Lorente de Nó and Condouris, 1959; Spencer and Kandel, 1961; Lux and Winter, 1968; Shapovalov and Grantyn, 1968; Kuno and Llinás, 1970; Czeh, 1972), was for the most part disregarded until rather recently.

Dendritic electrophysiology seems to be generated by membrane conductance changes to three ions: Na, K, and Ca. Sodium electroresponsiveness, at least at the base of the dendritic tree, has been suggested to be present in hippocampal pyramidal cells (Spencer and Kandel, 1961). Here, intracellular recordings indicated that an early prepotential fast enough to be a Na conductance could be recorded at somatic level in these neurons. These Na spikes have been confirmed in the *in vitro* preparation in these neurons by Wong *et al.* (1979). The time course and amplitude of these spikes suggest that, although they are close to the soma, they do not seem to be propagated to the soma in a continuous manner, but rather serve as an electrotonic coupler that may or may not trigger a somatic action potential.

More prominent than the Na currents are the Ca action potentials seen in neuronal dendrites. The Ca conductance, which was first encountered in the dendrites of Purkinje cells *in vivo* (Llinás and Hess, 1976; Llinás *et al.*, 1977), was then described *in vitro* in Purkinje cells (Llinás and Sugimori, 1978). Similar dendritic spikes have been encountered since then in other systems, e.g., hippocampal (Schwartzkroin and Slawsky, 1977; Wong and Prince, 1978), inferior olive (Yarom and Llinás, 1979), vagal motoneurons (Yarom *et al.*, 1980), spinal motoneurons (Schwindt and Crill, 1980; Walton and Fulton, 1981), spinal sensory neurons (Murase and Randic, 1983), and dorsal root ganglia (Matsuda *et al.*, 1978). At present, on-going studies in several types of cells in the neuraxis tend to suggest that Ca spiking, in fact, may be present in the dendrites of all CNS neurons. Because these spikes have been studied in most detail in Purkinje cells, I would like to describe them from that vantage point.

Intracellular recordings from Purkinje cell dendrites demonstrate two types of Ca-dependent action potentials: first, a plateau depolarization that may have a rather low threshold and that is generally accompanied by a prolonged Ca conductance. This Ca conductance may then be activated by direct stimulation of dendrites or by the activation of the climbing fiber. It may be observed either intracellularly as a transmembrane action potential or extracellularly as a large negative field having an all-or-none behavior. In addition to the above, Purkinje cell dendrites are capable of producing fast Ca-dependent action potentials that ride on these plateau depolarizations (Figure 5).

Whether these two phenomena represent variations of the distribution of Ca conductance or whether they represent two different varieties of Ca channels is not known at this time. Replacement of extracellular Ca by Ba produces in these cells, as in the inferior olivary neurons, prolonged action potentials that may last for many seconds.

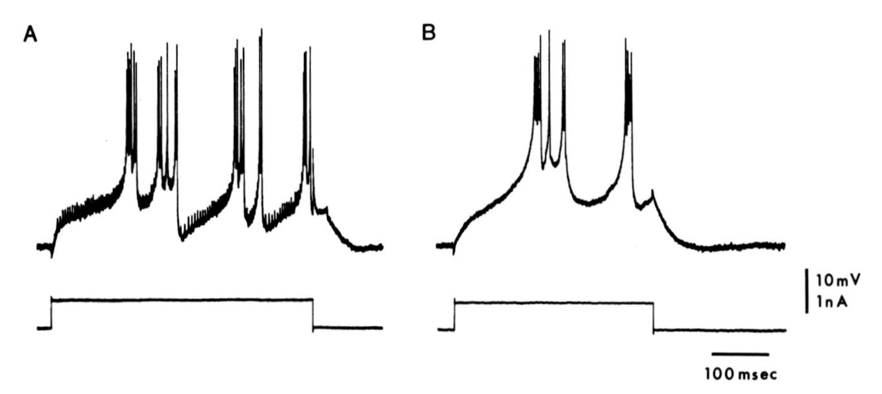

Figure 5. Ca-dependent spiking in Purkinje cells recorded intradendritically. (A) Direct depolarization of the dendrite generates a set of dendritic spike bursts interposed by small somatic Na-dependent potentials. (B) Addition of TTX to the bath has no effect on the voltage-dependent dendritic Ca spikes, but blocks the fast Na-dependent somatic firing. Modified from Llinás and Sugimori, 1980b.

This implies that the dendritic Ca channel demonstrates very little inactivation. The functional relevance of Ca dendritic activity to electrophysiology in general is clearly of great importance because Ca spikes not only serve to boost synaptic inputs to the soma but, in addition, they are followed by voltage-dependent K conductance changes, which in inferior olivary cells can be extraordinarily powerful (Llinás and Yarom, 1981a).

These dendritic Ca conductances may also serve to modulate the cell's integrative properties significantly. In the case of the inferior olive, the Ca dendritic response is the main protagonist determining the activities of these cells. Besides the clear effect on their electroresponsiveness, Ca entry in the dendrites far exceeds the level observed at either presynaptic terminals or probably at the soma, and thus a large set of biochemical changes may be assumed to be initiated by this Ca entry (cf., Llinás, 1979).

5. DISCUSSION

The above summary of comparative electroresponsiveness of mammalian central neurons indicates that the numbers and types of ionic conductance capable of being generated by the neuronal genome is clearly larger than those expressed in any particular cell. A more significant point is that neurons belonging to particular anatomical categories tend to have, beyond the morphological similarities that categorize them as belonging to a certain type, characteristics that are determined by their localization in the CNS. For instance, to our total surprise, in studying the mammalian thalamus, Jahnsen and I found the ionic properties to be very similar regardless of the sensory or motor function of the nucleus in which they are included (Llinás and Jahnsen, 1982). It would seem plausible that the neurons belonging to a particular set of nuclear subsets could have a kinship that makes them closer to each other than to other neurons in the neuraxis. However, this is probably not a generalizable statement. Indeed, nuclei such as the amygdalar complex demonstrate a large variety of electrophysiologically distinct neuronal elements. Similar statements may be made even for motoneurons which may vary as in the case of the vagal final common pathway as described above.

The obvious implications of grouping neuronal elements having similar electrophysiological properties is that the nervous system may be wired, to a certain extent, on the basis of particular electroresponsive resonance between neurons. Thus, it would not be surprising if neurons capable of oscillating at 10/sec, such as the cells of the IO, the thalamus, or the spinal cord, could be capable of performing a different type of network pattern transformation than the neurons not possessing such intrinsic properties. The functional role of these neuronal properties becomes evident when one considers such problems as the stages of consciousness in which alertness or different levels of sleep ordinarily implicate a resonant state between large bodies of cells in the CNS as indicated by the EEG waves which correlates with such states. Similar functional states may also be present in such different affective postures as rage and fear, and must ordinarily come to be by the the modulation of large numbers of neurons at many levels of the CNS.

From a more immediate point of reference, it is not surprising, given the large number of ionic conductances described, and the fact that many more are probably yet to be uncovered, that modern *in vitro* electrophysiology could appear to be, *prima facie,* in a rather chaotic state of affairs. However, the reality of the situation is that ultimately all of these conductances will coalesce to produce a more suitable algebra for understanding brain function in a global sense, this being the goal to which neuroscience ultimately aspires.

ACKNOWLEDGMENTS Research was supported by U.S. Public Health Service grant NS-13742 from the National Institute of Neurological and Communicative Disorders and Stroke.

6. REFERENCES

Adams, P. R. and Halliwell, J. V., 1982, A hyperpolarization-induced inward current in hippocampal pyramidal cells, *J. Physiol. (London)* **324:**62–63P.

Adams, P. R., Brown, D. A., and Halliwell, J. V., 1981, Cholinergic regulation of M-current in hippocampal pyramidal cells, *J. Physiol. (London)* **317:**29–30P.

Adams, P. R., Brown, D. A., and Constanti, A., 1982a, M-currents and other potassium currents in bullfrog sympathetic neurones, *J. Physiol. (London)* **330:**537–572.

Adams, P. R., Constanti, A., Brown, D. A., and Clark, R. B., 1982b, Intracellular calcium activates a fast voltage-sensitive potassium current in vertebrate sympathetic neurones, *Nature (London)* **296:**746–749.

Adams, P. R., Brown, D. A., and Constanti, A., 1982c, Pharmacological inhibition of the M-current, *J. Physiol. (London)* **332:**223–262.

Adrian, R. H., 1969, Rectification in muscle membrane, *Prog. Biophys Mol. Biol.* **19:**339–369.

Araki, T. and Terzuolo, C. A., 1962, Membrane currents in spinal motoneurons associated with the action potential and synaptic activity, *J. Neurophysiol.* **25:**772–789.

Armstrong, C. M. and Binstock, L., 1965, Anomalous rectification in the squid giant axon injected with tetraethylammonium chloride, *J. Gen. Physiol.* **48:**859–872.

Barrett, E. F., Barrett, J. N., and Crill, W. E., 1980, Voltage-sensitive outward currents in cat motoneurones, *J. Physiol. (London)* **304:**251–276.

Barrett, J. N. and Crill, W. E., 1974, Specific membrane properties of cat motoneurones, *J. Physiol. (London)* **239:**301–324.

Barrett, J. N. and Crill, W. E., 1980, Voltage clamp of motoneuron somata: Properties of the fast inward current, *J. Physiol. (London)* **304:**231–249.

Brigant, J. L. and Mallart, A., 1982, Presynaptic currents in mouse motor endings. *J. Physiol (London)* **333:**619–636.

Brown, H. F. and Adams, P. R., 1979, Muscarinic modification of voltage-sensitive currents in sympathetic neurones, *Soc. Neurosci. Abstr.* **5:**585.

Brown, H. F. and Adams, P. R., 1980, Muscarinic suppression of a novel voltage-sensitive K^+ current in a vertebrate neuron, *Nature (London)* **283:**673–679.

Brown, H. F. and Difrancisco, D., 1980, Voltage-clamp investigations of membrane currents underlying pace-maker activity in rabbit sino-atrial node, *J. Physiol. (London)* **308:**311–351.

Cahalan, M., 1978, Local anesthetic block of sodium channels in normal and pronase-treated squid giant axons, *Biophys. J.* **23:**285–311.

Cahalan, M., 1980, Molecular properties of sodium channels in excitable membranes, in: *The Cell Surface and Neuronal Function* (C. W. Cotman, G. Poste, and G. L. Nicolson, eds.), Elsevier North-Holland, Amsterdam, pp. 1–47.

Chiu, S. Y., Ritchie, J. M., Rogart, R. B., and Stagg, D., 1979, A quantitative description of membrane currents in rabbit myelinated nerve, *J. Physiol. (London)* **292:**149–166.

Connor, J. A. and Stevens, C. F., 1971, Inward and delayed outward membrane currents in isolated neural somata under voltage clamp, *J. Physiol. (London)* **213:**1–20.

Constanti, A. and Brown, D. A., 1981, M-currents in voltage-clamped mammalian sympathetic neurones, *Neurosci. Lett.* **24:**289–294.

Constanti, A. and Galvan, M., 1983, Fast inward-rectifying current accounts for anomalous rectification in olfactory cortex neurones, *J. Physiol. (London)* **385:**153–178.

Czeh, G., 1972, The role of dendritic events in the initiation of monosynaptic spikes in frog motoneurons, *Brain Res.* **39:**505–509.

Eccles, J. C., Libet, B., and Young, R. R., 1958, The behavior of chromatolysed motoneurons studied by intracellular recording, *J. Physiol.* (*London*) **143:**11–40.

Fedulova, S. A., Kostyuk, P. G., and Veselovsky, N. S., 1981, Calcium channels in the somatic membrane of the rat dorsal ganglion, effects of cAMP, *Brain Res.* **214:**210–214.

Fukuda, J. and Kameyama, M., 1980, Tetrodotoxin-sensitive and tetrodotoxinresistant sodium channels in tissue-cultured spinal ganglion neurons from adult mammals, *Brain Res.* **182:**191–197.

Fukushima, Y., 1982, Blocking kinetics of the anomalous potassium rectifier of tunicate egg studied by single channel recording, *J. Physiol.* (*London*) **331:**311–331.

Fulton, B. and Walton, K., 1981, Calcium-dependent spikes in neonatal rat spinal cord *in vitro, J. Physiol.* (*London*) **317:**25–26P.

Gustafsson, B., Galvan, M., Grafe, P., and Wigstrom, H., 1982, A transient outward current in a mammalian central neurone blocked by 4-amino-pyridine, *Nature* (*London*) **299:**252–254.

Hagiwara, S., Miyazaki, S., and Rosenthal, N. P., 1976, Potassium current and the effect of cesium on this current during anomalous rectification of the egg cell membrane of a starfish, *J. Gen. Physiol.* **67:**621–628.

Halliwell, J. V. and Adams, P. R., 1982, Voltage clamp analysis of muscarinic excitation in hippocampal neurons, *Brain Res.* **250:**71–92.

Harada, V. and Takahashi, T., 1983, The calcium component of the action potential in spinal motoneuron of the rat, *J. Physiol.* (*London*) **335:**89–100.

Hodgkin, A. L. and Huxley, A. F., 1952a, The dual effect of membrane potential on sodium conductance in the giant axon of *Loligo, J. Physiol.* (*London*) **116:**497–506.

Hodgkin, A. L. and Huxley, A. F., 1952b, A quantitative description of membrane current and its application to conduction and excitation in nerve, *J. Physiol.* (*London*) **117:**500–544.

Horn, R., 1977, Tetrodotoxin-resistant divalent action potentials in an axon of *Aplysia, Brain Res.* **133:**177–182.

Hotson, J. R. and Prince, D. A., 1980, A calcium-activated hyperpolarization follows repetitive firing in hippocampal neurons, *J. Neurophysiol.* **43:**409–419.

Hotson, J. R., Prince, D. A., and Schwartzkroin, P. A., 1979, Anomalous inward rectification in hippocampal neurons, *J. Neurophysiol.* **42:**889–895.

Jahnsen, J. and Llinás, R., Electrophysiological properties of mammalian thalamic neurones: An *in vitro* study, *J. Physiol.*, in press.

Jahnsen, J. and Llinás, R., Ionic basis for the electrical activation and the oscillatory properties of thalamic neurons *in vitro, J. Physiol,* in press.

Kandel, E. R. and Tauc, L., 1966, Anomalous rectification in the metacerebral giant cells and its consequence for synaptic transmission, *J. Physiol.* (*London*) **183:**287–304.

Katz, B., 1949, Les constantes éléctriques de la membrane du muscle, *Arch. Sci. Physiol.* **3:**285–299.

Kobayashi, H., Hashiguchi, T., Tosaka, T., and Mochida, S., 1981, Muscarinic antagonism of a persistent outward current in sympathetic neurons of rabbits and its partial contribution to the generation of the slow EPSP, *Neurosci. Lett.* **6**(suppl.)**:**S64.

Krnjevic, K. and Lisiewicz, A., 1972, Injections of calcium ions into spinal motoneurones, *J. Physiol.* (*London*) **225:**363–390.

Kuno, M. and Llinás, R., 1970, Enhancement of synaptic transmission by dendritic potentials in chromatolysed motoneurones of the cat, *J. Physiol.* (*London*) **210:**807–821.

Llinás, R., 1979, The role of calcium in neuronal function, in: *The Neurosciences: Fourth Study Program* (F. O. Schmitt and F. G. Worden, eds.), M.I.T. Press, Cambridge, Mass. pp. 555–571.

Llinás, R., Rebound excitation as the physiological basis for tremor: A biophysical study of the oscillatory properties of mammalian central neurons *in vitro*, in: *International Neurological Symposia: Tremor* (R. Capildeo and L. J. Findley, eds.), Macmillan, London, in press.

Llinás, R. and Hess, R., 1976, Tetrodotoxin-resistant dendritic spikes in avian Purkinje cell, *Proc. Nat. Acad. Sci. USA* **73**:2520–2523.

Llinás, R. and Jahnsen, H., 1982, Electrophysiology of mammalian thalamic neurones *in vitro*, *Nature* (*London*) **297**:406–408.

Llinás, R. and Nicholson, C., 1971, Electrophysiological properties of dendrites and somata in alligator Purkinje cells, *J. Neurophysiol.* **34**:534–551.

Llinás, R. and Sugimori, M., 1978, Dendritic calcium spiking in mammalian Purkinje cells: *In vitro* study of its function and development, *Soc. Neurosci. Abstr.* **4**:66.

Llinás, R. and Sugimori, M., 1980a, Electrophysiological properties of *in vitro* Purkinje cell somata in mammalian cerebellar slices, *J. Physiol.* (*London*) **305**:171–195.

Llinás, R. and Sugimori, M., 1980b, Electrophysiological properties of *in vitro* Purkinje cell dendrites in mammalian cerebellar slices, *J. Physiol.* (*London*) **305**:197–213.

Llinás, R. and Yarom, Y., 1980, Electrophysiological properties of mammalian inferior olivary cells *in vitro*, in: *The Inferior Olivary Nucleus: Anatomy and Physiology* (J. Courville, C. de Montigny, and Y. Lamarre, eds.), Raven Press, New York, pp. 379–388.

Llinás, R. and Yarom, Y., 1981a, Properties and distribution of ionic conductances generating electroresponsiveness of mammalian inferior olivary neurones *in vitro*, *J. Physiol.* (*London*) **315**:569–584.

Llinás, R. and Yarom, Y., 1981b, Electrophysiology of mammalian inferior olivary neurones *in vitro*. Different types of voltage-dependent ionic conductances, *J. Physiol.* (*London*) **315**:549–567.

Llinás, R., Sugimori, M., and Walton, K., 1977, Calcium dendritic spikes in the mammalian Purkinje cells, *Soc. Neurosci. Abstr.* **3**:58.

Lorente de Nó, R. and Condouris, G. A., 1959, Decremental conduction in peripheral nerve. Integration of stimuli in the neuron, *Proc. Natl. Acad. Sci. USA* **45**:592–617.

Lux, H. D. and Winter, P., 1968, Studies on EPSPs in normal and retrograde recting facial motoneurones, *Proc. Int. Union Physiol. Sci.* **7**:818.

MacDermott, A. B. and Weight, F. F., 1980, The pharmacological blockade of potassium conductance in voltage-clamped bullfrog sympathetic neurons, *Fed. Proc.* **39**:2074.

MacDermott, A. B. and Weight, F. F., 1982, Action potential repolarization may involve a transient, Ca^{2+}-sensitive outward current in a vertebrate neurone, *Nature* (*London*) **300**:185–188.

Matsuda, Y., Yoshida, S., and Yonezawa, T., 1977, Tetrodotoxin sensitivity and Ca component of action potentials of mouse dorsal root ganglion cells cultures *in vitro*, *Brain Res.* **154**:69–82.

Matteson, D. R. and Armstrong, C. M., 1982, Evidence for a population of sleepy sodium channels in squid axon at low temperature, *J. Gen. Physiol.* **79**:739–758.

McAfee, D. A. and Yarowsky, P. J., 1979, Calcium-dependent potentials in the mammalian sympathetic neurone, *J. Physiol.* (*London*) **290**:507–523.

Meech, R. W., 1972, Intracellular calcium injection causes increased potassium conductance in aplysia nerve cells, *Comp. Biochem. Physiol.* **42A**:493.

Meech, R. W., 1978, Calcium-dependent potassium activation in nervous tissues, *Am. Rev. Biophys. Bioeng.* **7**:1–18.

Murase, K. and Randic, M., 1983, Electrophysiological properties of rat spinal dorsal horn neurones *in vitro*: Calcium-dependent action potentials, *J. Physiol.* (*London*) **334**:141–153.

Nelson, P. G. and Frank, K., 1967, Anomalous rectification in cat spinal motoneurons and effects of polarizing currents on excitatory postsynaptic potentials, *J. Neurophysiol.* **30:**1097–1113.

Noma, A. and Irisawa, H., 1976, Membrane currents in the rabbit sino atrial node cell as studied by double microelectrode method, *Pflügers Arch.* **364:**45–52.

Nowak, L. M. and Macdonald, R. L., 1981, DL-muscarine decreases a potassium conductance to depolarize mammalian spinal cord neurons in cell culture, *Neurosci. Abstr.* **7:**725.

Ohmori, H., 1978, Inactivation kinetics and steady-state noise in the anomalous rectifier of tunicate cell membranes, *J. Physiol. (London)* **281:**77–99.

Purpura, D. P., Prevelic, N., and Santini, M., 1968, Hyperpolarizing increase in membrane conductance in hippocampal neurons, *Brain Res.* **7:**310–312.

Scholfield, C. N., 1978, Electrical properties of neurones in the olfactory cortex slice *in vitro, J. Physiol. (London)* **275:**547–557.

Schwartzkroin, P. A., 1975, Characteristics of CA1 neurons recorded intracellularly in the hippocampal slice, *Brain Res.* **85:**423–435.

Schwartzkroin, P. A. and Slawsky, M., 1977, Probable calcium spikes in hippocampal neurons, *Brain Res.* **135:**157–161.

Schwindt, P. C. and Crill, W. E., 1980, Properties of a persistent inward current in normal and TEA-injected motoneurons, *J. Neurophysiol.* **43:**1700–1724.

Shapovalov, A. I. and Grantyn, A. A., 1968, A suprasegmental synaptic influence on chromatolysed motor neurones, *Biofizika* **3:**260–269.

Sigworth, F. J. and Neher, E., 1980, Single Na channel currents observed in cultured rat muscle cells, *Nature (London)* **287:**447–449.

Spencer, W. A. and Kandel, E. R., 1961, Electrophysiology of hippocampal neurons. IV. Fast preprotentials, *J. Neurophysiol.* **24:**272–285.

Stafstrom, C. E., Schwindt, P. C., and Crill, W. E., 1982, Negative slope conductance due to a persistent subthreshold sodium current in cat neocortical neurons *in vitro, Brain Res.* **236:**221–226.

Sugimori, M. and Llinás, R., 1980, Lidocaine differentially blocks fast and slowly inactivating sodium conductance in Purkinje cells: An *in vitro* study in guinea pig cerebellum using iontophoretic glutamic acid, *Soc. Neurosci. Abstr.* **6:**468.

Sugimori, M. and Llinás, R., 1982, Role of dendritic electroresponsiveness in neuronal integration: *In vitro* study of mammalian Purkinje cells, *Soc. Neurosci. Abstr.* **8:**446.

Takahashi, K., Kubota, K., and Masatake, U., 1967, Recurrent facilitation in cat pyramidal tract cells, *J. Neurophysiol.* **30:**22–34.

Terzuolo, C. A. and Araki, T., 1961, An analysis of intra- versus extracellular potential changes associated with activity of single spinal motoneurons, *Ann. N. Y. Acad. Sci.* **94:**547–558.

Walton, K. and Fulton, B., 1981, Role of calcium conductance in neonatal motoneurons of isolated rat spinal cord, *Soc. Neurosci. Abstr.* **7:**246.

Wilson, W. A. and Goldner, M. M., 1975, Voltage clamping with a single microelectrode, *J. Neurobiol.* **6:**411–422.

Wong, R. K. S. and Prince, D. A., 1978, Participation of calcium spikes during intrinsic burst firing in hippocampal neurons, *Brain Res.* **159:**385–390.

Wong, R. K. S., Prince, D. A., and Basbaum, A. I., 1979, Intradendritic recordings from hippocampal neurons, *Proc. Natl. Acad. Sci. USA* **76:**986–990.

Yanagihara, K. and Irisawa, H., 1980, Inward current activated during hyperpolarization in the rabbit sino atrial node, *Pflügers Arch.* **385:**11–19.

Yarom, Y. and Llinás, R., 1979, Electrophysiological properties of mammalian inferior olive neuron in *in vitro* brain stem slices and *in vitro* whole brain stem, *Soc. Neurosci. Abstr.* **5**:109.

Yarom, Y., Sugimori, M., and Llinás, R., 1980, Inactivating fast potassium conductance in vagal motoneurons in guinea pigs: An *in vitro* study, *Soc. Neurosci. Abstr.* **6**:198.

2

Passive Electrotonic Structure and Dendritic Properties of Hippocampal Neurons

DENNIS A. TURNER and PHILIP A. SCHWARTZKROIN

1. INTRODUCTION

1.1. *Why Develop Models of Neurons?*

Most central nervous system (CNS) neurons possess a considerable spatial dispersion of structures specialized for postsynaptic reception. Variability in dendritic structure, as well as differing distributions of spines (Scheibel and Scheibel, 1968) and other synaptic specializations, confer unique capabilities on each cell type. Neurons in the hippocampus demonstrate a complexity of neuronal shape (Figure 1), and dispersion of spine synapses that are typical particularly of cortical tissue (Minkwitz, 1976; Wenzel *et al.*, 1981). The various classes of hippocampal neurons also possess a distribution of nonsynaptic ionic conductances that can modify signal transfer from input sites to the summation or recording site (Jack *et al.*, 1975; Rall, 1977), and thus complicate the interpretation of synaptic inputs. Such signal modification may occur according to both passive cable attenuation and nonlinear forms of distortion, neither of which are usually subject to intuition. Quantitative models of neurons

---

DENNIS A. TURNER • Department of Neurosurgery, VA Medical Center, Minneapolis, Minnesota 55417. PHILIP A. SCHWARTZKROIN • Department of Neurological Surgery, University of Washington, Seattle, Washington 98195.

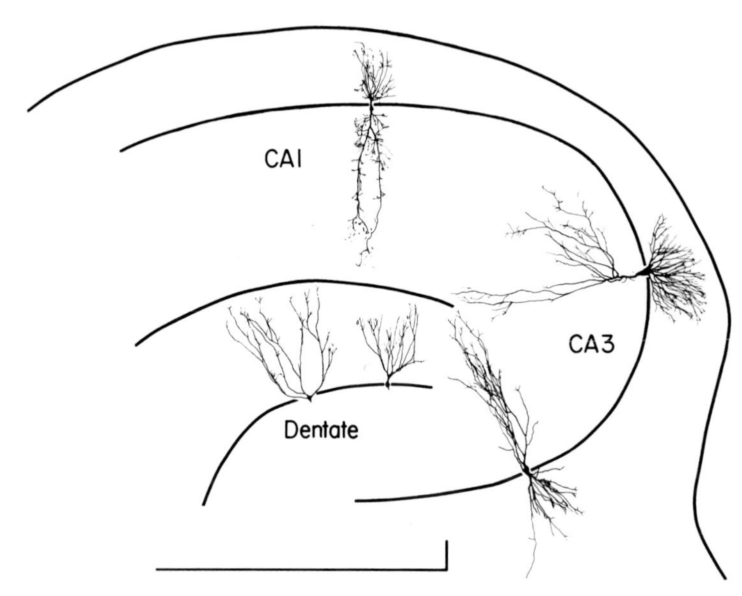

Figure 1. This montage provides representative examples of intracellularly-stained hippocampal neurons, following injection of HRP into cells recorded *in vitro* [CA1 cell from Turner and Schwartzkroin (1980), CA3 and dentate cells from Turner and Schwartzkroin (1983)]. The different size and extent of the dendritic trees are well illustrated. The scale shows 1 mm (horizontal) and 100 μm (vertical).

have been developed both to enhance our ability to extrapolate from a recorded signal back to the original synaptic signal, and also to help us understand the signal's importance and role in the process of neuronal integration. Complex processing and multiple interactions can be approached in a model in a way that cannot be done experimentally.

A set of theoretical models of dendritic signal processing has been derived by Rall (1977) and elaborated by others (Jack *et al.*, 1971, 1975, 1981; Norman, 1972). Because these mathematical treatments deal with the passive electrotonic transfer of signals within complex neuronal geometry, the resulting electrical description has been termed *electrotonic structure*. In general, these models are based on linear differential equations for current flow across membranes, and specifically account for the spatial separation of input and measuring sites. The basic analytical unit may be either discrete and lumped (compartmental models) or continuous (cable models), or mixed. Most of these models have been applied to hippocampal neurons, in attempts to reproduce the observed effects (at physiologically recordable sites or on a neuron's output) of events occurring at remote locations—dendritic synaptic inputs, other distal ionic conductance changes, electrical coupling, etc. This form of evaluation leads to a functional analysis of dendritic and neuron structure, as opposed to a morphometric (i.e., purely anatomical) description. The electrotonic evaluation incorporates spatial features of the neuron, as well as temporal features of the input signal, into the analysis of signal modification and attenuation (or amplification).

The most general electrical model transforms a single neuron into an equivalent circuit or network (Carnevale and Johnston, 1982). This approach provides an empirical description of the input impedance (the AC or transient resistance) at any point in a generalized network of soma and dendrites, without prior knowledge of the electrical pathways involved. Both the cable and compartmental models, on the other hand, require *a priori* assumptions about the structure of the neuron under analysis; these assumptions then lead to suggestions about the relation between electrotonic and morphometric structure of the neuron in question (Rall, 1977). Each of the electrotonic models furnishes a different portrayal of a neuron, depending on the nature of the assumptions regarding its structure. A general goal of all these models, however, has been the electrical characterization of how inputs at distal sites (which are not accessible to direct recording) are "seen" by the cell's somatic integrating site. These electrotonic models, then, attempt to describe the electrical processing that occurs within a single neuron.

The interaction of signals between neurons may also be characterized, though usually by models that abstract single neurons into computing elements (Perkel *et al.*, 1981; Leung, 1982). These circuit models vary as a function of the number and complexity of cellular components included. The neuronal net that is simulated may correspond to a smaller or larger number of cells, often ranging from two to several hundred. Such interneuronal modeling has been specifically applied to the hippocampus (Traub and Wong, 1981; Leung, 1982) in attempts to describe possible mechanisms of both normal and epileptic activities. More generally, these neuron models have been implemented in order to place our understanding of physiological observations on a firmer theoretical basis (Perkel *et al.*, 1981).

Predictions based on these theoretical models can, and should, be experimentally testable. The interplay between a theoretical understanding and the complexity of actual, direct observation may serve to sharpen both the insight into neuronal mechanisms and the limitations, assets, and basic parameters of the model. This process of attempting direct comparison of a model and its "real" counterpart in the experimental domain provides a substantial yet necessary challenge to both physiologist and theoretician. Data from various laboratories, obtained using a variety of techniques, can be integrated and tested in appropriate models. Successful model predictions (i.e., model and experimental data in agreement) may serve as building blocks for more complex (and realistic) views of a neuronal system. Model predictions "disproved" by experimental results can lead to generation of revised or new hypotheses.

1.2. Hippocampal Slices as a Substrate for Theoretical Models

The hippocampus illustrates many of the general features of synapse dispersion found throughout the cortex, both in the variability of neuron structure and the complexity of interneuronal circuitry. Since this region has been rather well defined anatomically and physiologically, it may serve as a convenient "model" for telencephalic function and structure. Further, the hippocampus can now be studied using the *in vitro* slice technique, a method which provides considerable advantage and convenience for the electrophysiologist. Hippocampal slices are, to be sure, one level of abstraction beyond the hippocampus, and are thus a "model within a model." However, modeling of both single-cell electrotonic structure and of small neuronal nets can be facilitated by the *in vitro* slice approach, provided that the experimenter remains aware of the limitations imposed by this further level of abstraction (Schwartzkroin, 1981). With each subsequent shell of "modeling" of a system, a set of assumptions may limit the immediate relevance of the model to the original system. For some analyses, the difficulty in experimentally validating these assumptions may decrease the general usefulness of the "model."

However, there are many advantages of the *in vitro* technique for the study of dendritic and neuronal models. These include the greater mechanical stability of *in vitro* over *in vivo* recording, allowing better quality, longer term intracellular penetrations; visual identification and manipulation of both input pathways and output pathways, facilitating the discrete input/output analyses needed to study neuronal integration; limitation and control of external influences on the neuronal populations; the general absence of spontaneous activity, allowing the more accurate measurement of "resting" membrane properties (without anesthetic); and significantly easier recovery and processing of stained cells (Figure 1). This chapter focuses on the characterization of electrotonic structure, dendritic properties, and interactions of neurons in the hippocampus. Most of the theoretical modeling is based on data derived from intracellular recordings from hippocampal neurons *in vitro*. Both the methods and the insights may be extended to enhance our understanding of cortical neurons in general, and to illustrate typical features of dendritic and interneuronal processing.

2. ELECTROTONIC AND CIRCUITRY MODELS—AN OVERVIEW OF METHODS

2.1. *Network Model*

The general network model (Jack *et al.*, 1975; Carnevale and Johnston, 1982) assesses the electrical interaction between two arbitrary neuronal sites and involves no assumed constraints regarding the morphological structure of a neuron. This model lumps an entire neuron into an equivalent passive electrical network, which can include both impedances and capacitances. Due to its general nature, this model does not include any specific electrotonic parameters and cannot describe electrotonic structure. However, it does make possible the physiological evaluation of signal transfer between a recording and an arbitrary input site, without any prior assumptions regarding the path that the signal must travel. Potentially, this model may be useful in evaluating the existence and functional significance of remote sites of electrical coupling between neurons, as well as of remote chemical synapses. Such an assessment is based on the predictable manner in which current injection by recording electrode influences the input signal. Though no experimental reports of the application of this model are yet available, it is possible to evaluate the range of expected network attenuation factors with segmental cable calculations (Turner and Schwartzkroin, 1983).

2.2. *Continuous Cable Models*

The continuous cable models (Figure 2) have been derived from the cable theory of axons (Rall, 1959, 1977; Norman, 1972; Jack *et al.*, 1975). As depicted in Figure 2A, a cable segment is assumed to be uniform in cross-section and to be adequately described as a function of length along the cable. A neuron may be approximated by either one overall cable segment or by a series of such cylindrical segments. The application of the infinite cable equations to dendrites involves several strict assumptions: neglect of extracellular resistivity, an isopotential lumped soma, and the portrayal of dendritic branches as smooth cable segments. These cable models include a mixture of components: a lumped compartment for a soma and continuous electrical decay along the cable segments.

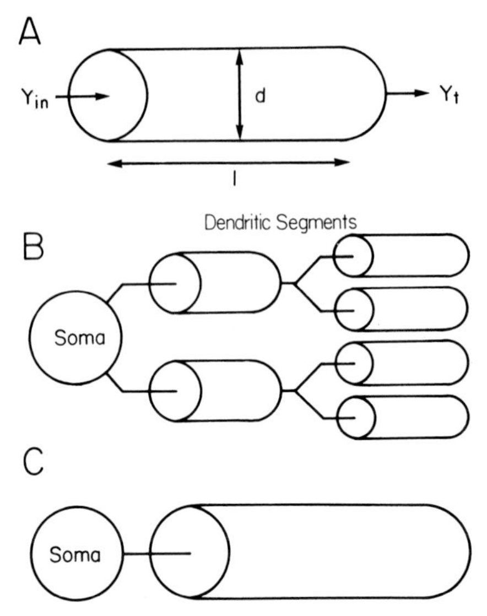

Figure 2. (A) A simplified cable segment for general calculations, showing the geometric measures of length (l) and diameter (d). The boundary conditions are represented here by a normalized admittance (AC conductance) at either end, depending upon orientation of the segment; Y_{in} is the input admittance and Y_t is the terminating admittance. (B) A schematic of the general branching cable model, showing a number of individual segments making up each major dendrite. The input admittances are summed at the junction of several segments, and together form the new terminating admittance of the next most proximal segment. Each neuron may be approximated by a series of these cable segments (in the range of 300 to 600 segments per neuron). (C) An illustration of the equivalent cylinder model. Given certain assumptions about the branching pattern of a cell, the equivalent cylinder model can reduce a complex dendritic tree [such as in (B)] to a single cable segment, usually with a specific, lumped soma attached to it.

The electrotonic length of the model cell is derived from each cable segment's length normalized by its length constant. There are several variations on the continuous cable model that have been employed in the analysis of hippocampal neurons.

2.2.1. *Segmental Branching Cable Model.* The anatomically-derived branching cable model (Rall, 1959; Barrett and Crill, 1974; Turner and Schwartzkroin, 1980, 1983; Turner, 1982) assumes no constraints on dendritic branching structure. It represents the neuron by a number of individual cable segments, the dimensions of which are based on anatomical reconstructions (to any degree of precision desired) of the cell. This segmental representation (Figure 2B) is assumed to depict an entire dendritic tree (Figure 1) as a connected ensemble of smooth, cylindrical cable pieces, each possessing a set of boundary conditions (Norman, 1972). Like the general network model, either a steady-state or transient input impedance can be calculated for any site. Unlike the network model, this electrical description employs the cable equations and specific dendritic structure to calculate signal transfer between any two sites of a neuron. Morphological features such as dendritic spines may be included in the evaluation, and single inputs onto either dendritic shafts or spines may be simulated. Disadvantages of this form of the continuous cable model include the relatively slow computing speed associated with the complex structure of the neuron under study, and the considerable difficulty in constructing an accurate segmental representation of a neuron. However, because of both its general scope and intimate relation to a neuron's dendritic morphology, this segmental evaluation may be quite powerful in simulating single synaptic inputs.

2.2.2. Equivalent Cylinder Cable Model. The equivalent cylinder cable model (Rall, 1969; Jack *et al.*, 1971, 1981; Iansek and Redman, 1973; Brown *et al.*, 1981; Durand *et al.*, 1983; Johnston, 1981) assumes that a neuron can be represented as a single equivalent cable and lumped soma (Figure 2C). This contraction of an otherwise complicated dendritic branching system into a single cable segment allows a tremendous simplification of the cable calculations. An analysis using this model involves the measurement of physiologically observed transient responses at the soma (either to step or to brief pulse inputs). The time constants and coefficients measured in response to such inputs can be used to derive a single equivalent electrotonic length of the neuron. It is then possible to simulate synaptic inputs on a single major dendrite and to predict the subsequent electrical transfer of these inputs to the soma (Jack *et al.*, 1981). However, because the individual dendritic branches are lumped in this model, specific dendritic inputs cannot be stimulated. This model is also useful in the assessment of tonic conductance changes in the dendritic periphery, as might be induced by drugs (Carlen and Durand, 1981), though it appears to be relatively insensitive to the actual location of the conductance change. This model requires a major assumption in addition to those mentioned for the general cable model—that a neuron can be adequately represented as an equivalent cylinder; this hypothesis has been subject to criticism for both motoneurons (Barrett and Crill, 1974) and hippocampal neurons (Turner and Schwartzkroin, 1980, 1983).

2.3. Compartmental Models

The compartmental models (Rall, 1967; Perkel and Mulloney, 1978; Traub and Llinás, 1979; Perkel *et al.*, 1981) portray a neuron's dendritic structure as a variable number of isopotential compartments (approximately 5 to 100; Figure 3), each with a set of lumped electrical parameters. Each compartment includes a number of voltage-dependent membrane conductances and communicates with adjacent regions through a connecting resistance. Electrotonic distance can be calculated from the connecting resistance, which is analogous to the internal resistivity of a dendritic trunk. Both distal synaptic inputs (Rall, 1967) and spatially-

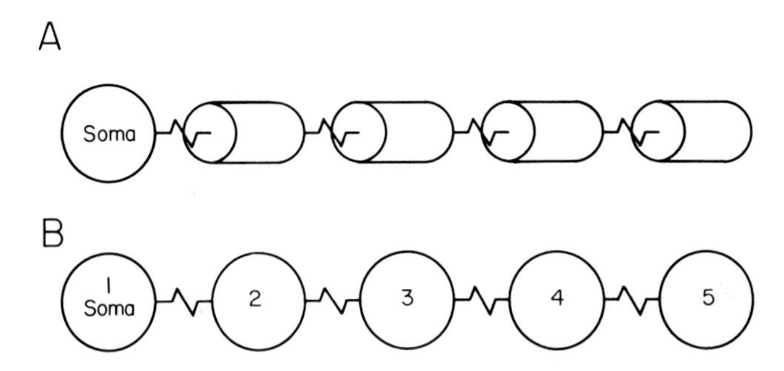

Figure 3. (A) Representation of an equivalent cylinder cable (Figure 2C) with several isopotential compartments. The area for each cylindrical segment is used to derive the total lumped conductance and admittance for each compartment. The compartments are linked by equivalent resistors, which represent the internal resistivity of the cell's dendrites. This compartment model may be normalized to the dimensions of an equivalent cylinder neuron (see Traub and Llinás, 1979). (B) Usual compartmental designation, with the circles representing isopotential regions possessing lumped parameters and the resistor representing linkage between compartments (after Rall, 1967).

distributed nonsynaptic ionic conductances (Traub and Llinás, 1979) have been considered in detail using this form of model. However, the model parameters and compartment localization may not correlate well with anatomically measurable neuronal dimensions. Except in a general sense, it may be difficult to apply findings from this model to anatomically discrete sites. However, there are significant advantages to this model in terms of decreased computation time; this model is especially appropriate when a continuous decrement of signal voltage (as along a cable segment) is not required to simulate observed phenomena. Thus, faste events can be incorporated into model simulations, including action potentials (Traub and Llinás, 1979) and other "active" currents.

2.4. *Neuronal Circuit Models*

Models of small neuron nets (Perkel and Mulloney, 1978; Traub and Wong, 1981, 1983; Leung, 1982; Wong and Traub, 1983) abstract a neuron into a single computing element, within a larger scheme of interneuronal integration (Figure 4). These models provide for an analysis of the interrelationships (both excitatory and inhibitory) between neurons, and can vary in complexity, depending on the number of elements included. Simpler models might portray an interaction between only two elements and attempt to determine whether they share a common synaptic input or interaction (Perkel *et al.*, 1981). More complex models

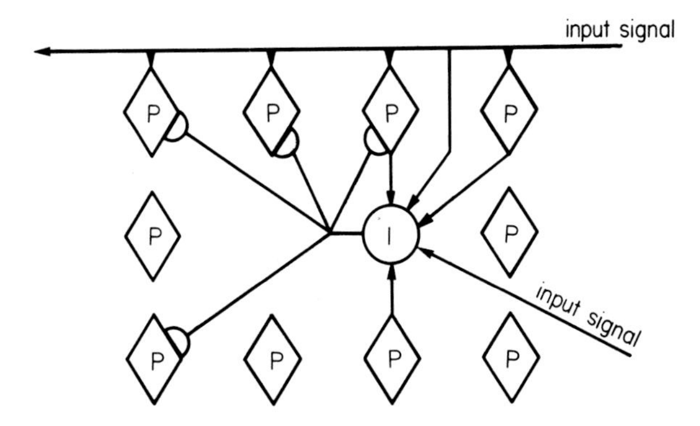

Figure 4. A schematic of a multineuronal circuit model, with each diamond representing a principal cell (P) of the circuit and the circle portraying a local circuit neuron (an "interneuron") (I). This particular diagram outlines one possible scheme for inhibition, with the arrows indicating excitatory inputs and the semicircle endings portraying inhibitory influences. Each neuron may be included as a single computing element, which sums all of the various inputs and produces a single output that is then fed back into the circuit. This model illustrates a circuit incorporating both feedback and feed-forward inhibition.

can include larger cell networks of 100 to 500 elements (Traub and Wong, 1981, 1983; Leung, 1982; Wong and Traub, 1983) and might attempt to explain how the interaction of those elements result in the characteristic behavior of small regions of the CNS. These models do not generally account for the electrotonic structure of a particular computing element, but rather represent intraneuronal integration as a simple weighted sum of many inputs or as a general transfer function. However, each computing element can perform rather complex weighting functions, and intrinsic cell properties (such as afterhyperpolarizations) can be included (Traub and Wong, 1981). Because the circuitry included in the scheme of interactions usually needs to be simplified to its "essential" components (an arbitrary decision at best) as compared to the corresponding neuroanatomical data, the model results are rather selective.

3. MODELS CONSTRUCTED WITH DATA FROM WORK ON HIPPOCAMPAL SLICES

3.1. *Continuous Cable Models*

3.1.1. Segmental Branching Cable Model. The anatomically-derived branching cable model has been employed in analyses of the major classes of neurons recorded in the *in vitro* hippocampus (Turner and Schwartzkroin, 1980, 1983; Turner, 1982). These physiological/anatomical studies on the same cell involved an electrophysiological characterization of input resistances and time constants with subsequent intracellular dye injection and anatomic reconstruction on the cell (Figure 1). From these data, a cable evaluation of the approximated dendritic structure was carried out. This model, assuming a uniform passive membrane, could demonstrate many features of dendritic processing that are critical in the prediction of dendritic synaptic efficacy. Derived electrical values (for both steady-state and transient analyses using the branching cable model) are presented in Table I. These include estimates for dendrite-to-soma conductance ratio (ρ), and for equivalent–cylinder (E–C) electrotonic length (L); this latter estimate can be compared to the segmental L terminations for the same cell to test the adequacy of an equivalent–cylinder simplification of the cell dendrites.

Relative estimates for steady-state synaptic efficacy in a dentate granule cell and CA3 pyramidal cell are illustrated in Figure 5. One index of comparison for different synaptic sites on the same neuron is the steady-state *orthograde charge transfer.* This value is equivalent to the charge transferred from a dendritic site to the soma, for a particular steady-state current input at that dendritic site. The solid line in Figure 5 illustrates the theoretical prediction of this charge transfer value for an E–C (Rinzel and Rall, 1974) of variable electrotonic length, while the individual points indicate calculations for actual dendritic sites on these dentate and CA3 cells. The individual site values lie reasonably close to the theoretical E–C predictions and both appear to be a function of summated electrotonic length to a given site. As found theoretically in the network model (Carnevale and Johnston, 1982), this orthograde charge transfer value was calculated to be identical to *retrograde voltage transfer,* which represents the steady-state effectiveness of a voltage input at the soma as seen at each dendritic site. It is evident from Figure 5 that even the most distal sites in these hippocampal neurons still transfer approximately 20% of dendritically injected current to the soma, indicating a significant effect attributable to passive electrotonic transfer.

The steady-state values for orthograde voltage transfer (from a dendritic site to the soma) did not follow the E–C predictions, but rather appeared to be inversely proportional to the local input resistance (R_D). Both the R_D value and the steady-state voltage transfer values varied considerably; for sites at the same electrotonic distance from the soma, these values could differ by a factor of 10. This finding of the segmental model is in contrast with the E–C model (Rinzel and Rall, 1974), and to values calculated for motoneurons (Barrett and Crill, 1974). The seg-

Table I. Comparision of Hippocampal Electrical Parameters[a]

Hippocampal cell type	R_N[b] (MΩ)	Time constant[c] (msec)	Time constant ratio[d]	Dentrite-to-soma conductance ratio[e] (ρ) Anat	Phys	Electrotonic length[e] (L) Anat	Phys
Dentate granule neurons							
Brown *et al.* (1981)	38.4	11.6	8.1	—	1.5	—	0.94
	3.0	1.5	0.2		0.2		0.08
Durand *et al.* (1983)	58.6	16.2	—	—	7.6	—	1.13
	8.6	1.4			2.3		0.06
Turner (1982)	53.5	12.3	10.4	9.3	3.4	1.02	0.96
	3.8	1.4	2.7	1.9	0.8	0.05	0.17
CA1 pyramidal cells							
Turner *et al.* (1980)	27.8	10.9	—	3.5	—	0.98	—
	2.1	1.0		0.4		0.08	
Brown *et al.* (1981)	42.6	15.2	8.2	—	1.5	—	0.90
	3.6	1.2	0.7		0.2		0.08
Turner (1982)	30.0	17.9	9.2	6.7	2.3	0.69	0.88
	5.5	2.0	1.4	1.6	0.6	0.08	0.05
CA3 pyramidal cells							
Johnston (1981)	—	23.6	8.4	—	2.4	—	0.96
		1.9	0.7		0.3		0.06
Brown *et al.* (1981)	39.0	19.3	7.8	—	1.2	—	0.93
	3.2	1.8	1.2		0.4		0.09
Turner (1982)	37.1	22.3	9.0	17.5	1.9	0.61	0.91
	3.3	4.5	2.2	1.2	0.6	0.10	0.12

[a] Values are expressed as mean ± standard error of the mean, and have been abstracted from the studies given on the left of the table.
[b] R_N represents the measured input resistance, as estimated by injecting small hyperpolarizing current pulses through an electrode in the cell soma.
[c] *Time constant* represents the longest exponential decay from the current pulse and is the largest of the time constant series for a cell.
[d] *Time constant ratio* is the ratio of the first two-time constants extracted from the physiologically observed transients.
[e] For both the *dendrite-to-soma conductance ratio* (ρ) and *electrotonic length* (L), the anatomically-derived value (Anat) reflects the mean of individual segmental terminations, while the physiologically observed value (Phys) has been calculated from the time constant ratio and corresponding coefficients, according to the methods described by Brown *et al.* (1981) and Johnston (1981).

mental branching model results indicate that deviations from the E–C assumption (Rall, 1959, 1969), which are considerable in the hippocampal pyramidal cells (Turner and Schwartzkroin, 1980, 1983), tend to be more prominent for the R_D and voltage transfer than for current transfer. The discrepancy between voltage and current transfer values may be significant in understanding the inadequacy of the E–C lumping procedure (and the E–C model in general) for hippocampal neurons.

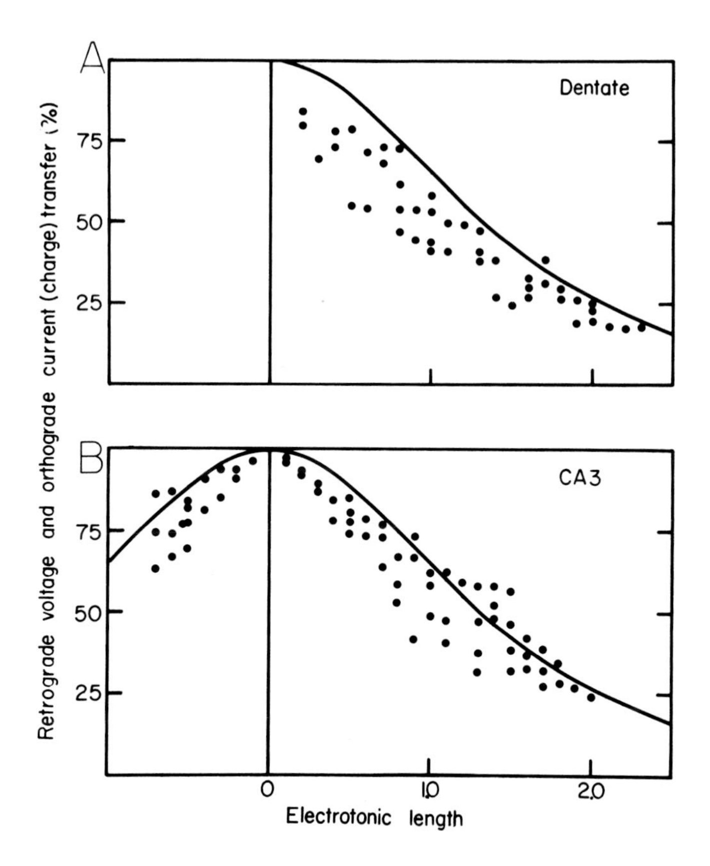

Figure 5. (A) A plot of electrical transfer values, from the steady-state segmental branching model (Turner and Schwartzkroin, 1983) for a dentate granule cell. The heavy line indicates the predicted retrograde voltage transfer (from the soma to the dendritic terminus) and orthograde current (or charge) transfer (from the dendritic terminus to the soma) for an equivalent cylinder with a single input to a terminal segment (Rinzel and Rall, 1974). The dots represent the calculated electrical transfer for a number of simulated, steady-state dendritic spine inputs at different points in the dendritic tree. As predicted by the network model (Carnevale and Johnston, 1982), these two different values, retrograde voltage transfer and orthograde current transfer, were essentially identical for a given site, and hence are presented together. As can be seen from the plot, even the most distal dendritic site transfers about 20% of a steady-state input current to the soma, indicating a potentially important influence of these synapses. Proximal dendritic sites transfer an even larger amount of current to the soma. (B) A similar plot, but for a representative CA3 pyramidal cell (the findings are also typical for CA1 pyramidal cells). Spine inputs to the basilar dendrites are to the left of the bar and inputs to apical dendrites to the right of the bar. These electrotonic transfer figures cluster around the predicted equivalent cylinder values derived for a terminus of the same electrotonic length as the terminal segment. As was the case for the dentate granule cells, even distal sites transfer 20 to 25% of injected current (steady-state) to the soma.

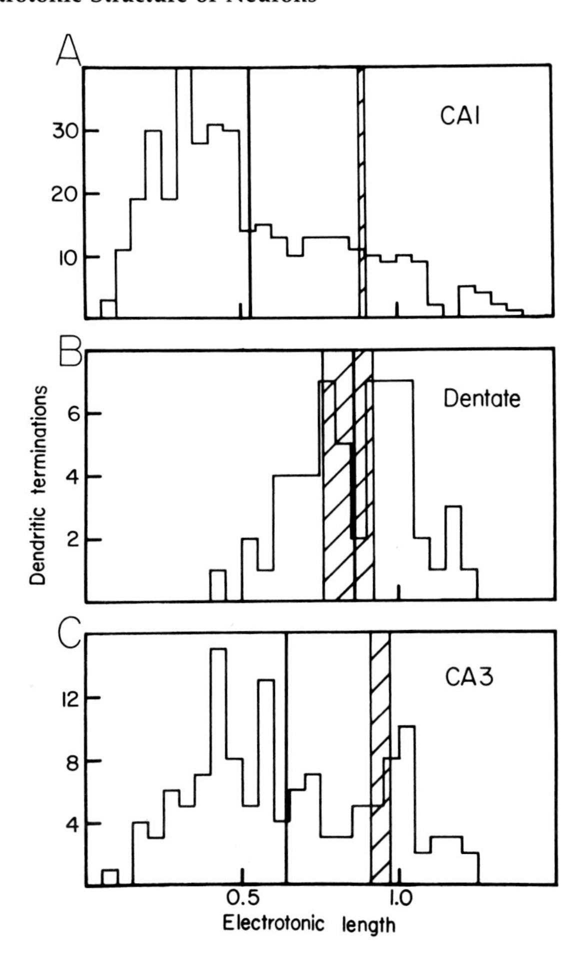

Figure 6. (A) Data from a typical CA1 pyramidal cell (Turner, 1982), illustrating the relation between individual segmental terminations (numbers of terminations on the central axis) and the electrotonic length to that site of termination (horizontal axis). The dark vertical bar indicates the mean electrotonic length of these individual terminations (equivalent to "Anat" in Table I). The hatched bars represent two separate estimates for the equivalent cylinder electrotonic length (L) of the neuron; one derived from observed electrophysiological transients, and the other from charging responses calculated by the transient segmental model (Turner, 1982). In CA1 cells, the anatomically derived mean L was always lower than the equivalent cylinder L estimates. (B) A similar histogram of individual dendritic terminations for a dentate granule neuron. In this cell class, the two equivalent cylinder predictions for L usually enclosed the anatomically derived mean value. There were relatively few nonuniform electrotonic terminations, indicating that the granule cells approached the equivalent cylinder assumption quite closely. (C) A representative plot of individual segmental terminations versus electrotonic length for a CA3 pyramidal cell, which demonstrates a finding similar to that found for CA1 pyramids. The mean terminating value appeared to be lower than the equivalent cylinder estimates, partly because of the bias imparted by numerous short basilar dendritic terminations.

Figure 6 portrays a comparison of these different estimates for overall L for intracellularly-stained hippocampal neurons (Turner, 1982). The E–C values for electrotonic length (the hatched bars) appear to be close to the mean of the segmental terminating electrotonic lengths (i.e., the average of the electrotonic length to the tip of each dendritic branch) for the granule cells. However, the CA1 and CA3 pyramidal cell values for L tend to diverge from the E–C values; such divergence has also been demonstrated in calculations for motoneurons (Barrett and Crill, 1974). As an overall average (Turner, 1982, Table I), the E–C values did not significantly deviate from the segmental averages, both computing to be approximately 0.9 length constants. This similarity was reflected in the comparison of somatic step transients, where there appeared to be no significant difference between the observed physiological step transients and those computed from either the E–C model (Brown *et al.*, 1981) or the segmental model (Turner, 1982). Thus, the neuronal deviations from the E–C assumption appeared to make little difference for the analysis of somatic transients. One implication of this finding is that somatic recordings are relatively insensitive to the exact dendritic structure of a neuron.

The very numerous small dendritic spines (Scheibel and Scheibel, 1968; Minkwitz, 1976; Wenzel *et al*, 1981) on cortical neurons have only rarely been subject to functional analysis, although several hypotheses have been suggested regarding their function (Diamond *et al*, 1971; Rall, 1974; Fifkova and Van Harreveld, 1977). With the segmental cable model, dendritic spines have been evaluated using "average" electron-microscopic dimensions (Turner, 1982; Turner and Schwartzkroin, 1983). In this analysis, incorporation of spines altered electrotonic properties of hippocampal cells only minimally. Further, in studying signal transfer across the spine to the dendrite, no significant electrical effect was calculated, presumably because of the spines' short length (0.2 to 0.4 μm). These preliminary calculations did not, however, take into account the great variability and diversity of spine dimensions nor the possible nonuniformity of internal and membrane spine properties. Rather, only minimum and average values for spine length and spine neck diameter were considered. The absence of electrical function of the thin spine neck found in this model contrasts sharply with the theoretical possibilities, wherein significant attenuation of electrical signals has been postulated (Diamond *et al*, 1971; Rall, 1974; Fifkova and Van Harreveld, 1977). This unexpected model finding is apparently attributable to the extremely high local input resistance at the base of the spine neck; this R_D was an order of magnitude higher than the resistance of even the longest, narrowest spine neck considered in the model. These findings appear to be valid for simulations of both steady-state and transient inputs to dendritic spines (Turner, 1982; Schwartzkroin and Turner, 1983), and prod us to consider other possible roles of dendritic spines in neuronal function (e.g., structural or metabolic roles).

Figure 7 illustrates two simulations of a transient input, in the form of a time-dependent conductance change (Turner, 1982). Figure 7A indicates the time course of this smooth conductance change applied to the soma and the direct response "seen" at the soma. Figure 7B portrays the same simulated conductance input (note change in voltage scales) applied to a dendritic spine on a dentate granule cell, and the indirect or transferred response "recorded" at the soma. As indicated by theoretical analyses of waveform shaping (Jack *et al*, 1971; Rinzel and Rall, 1974), the soma response is much slower, and more gradual in appearance than the spine input. This slowing appears to be dependent mainly on the electrotonic length of the transfer path from input spine to source, indicating the "leakiness" of the cable segments to high frequencies (Jack *et al.*, 1975). A set of such indirect soma responses to spine inputs were averaged ($n = 72$), to calculate the efficacy of an average "quantal" synaptic input. For these smooth waveform inputs, an integrated conductance transient of 12×10^{-10} S was necessary in order to obtain a 100 μV average peak response at the soma (Turner, 1982). This value lies very close to that observed physiologically (McNaughton *et al.*, 1981).

3.1.2. Long Pulse Equivalent Cylinder Model. In one series of electrophysiology/modeling experiments, the equivalent cylinder model has been used to analyze the time constants in the charging and decaying response to step inputs (Brown *et al.*, 1981; Johnston, 1981) in *in vitro* hippocampal neurons. This analysis of step inputs involves the separation of the various time constants which constitutes a neuron's charging response to current injection (Rall, 1969). Graphical differentation of the charging response (Rall, 1977) has allowed these authors to calculate both the time constants of charging and their respective coefficients (Brown *et al.*, 1981; Johnston, 1981). These studies have indicated that the major classes of hippocampal neurons (CA1 and CA3 pyramidal cell and dentate granule cells) appear rather uniform in their electrotonic structure (Table I). The average electrotonic length was computed to be about 0.9 length constants for all of these cells, in spite of significant differences in geometric structure among the three cell types. This finding argues strongly for the concept that even distal dendritic synapses in hippocampal cells can have significant effects at a somatic integrating site; due to the compact electrical structure of these neurons, any "av-

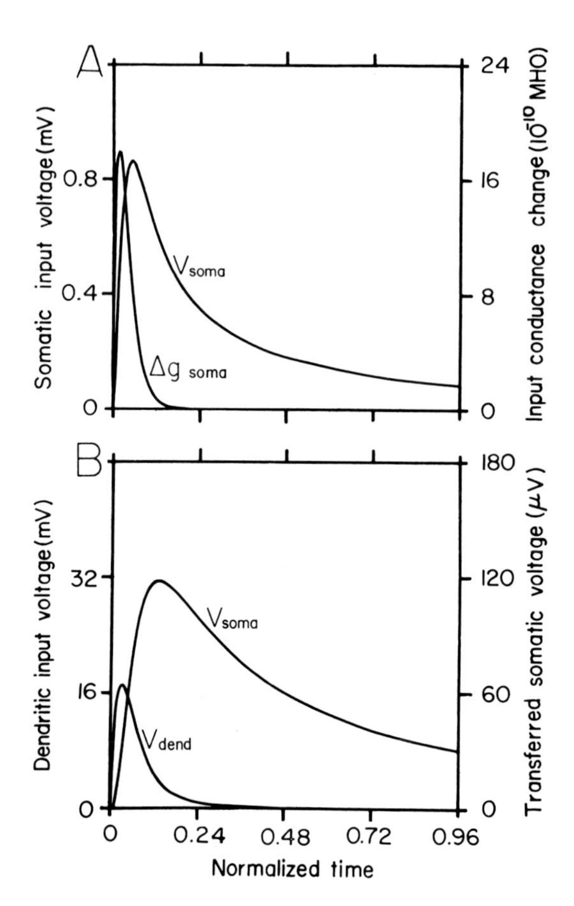

Figure 7. (A) A model plot demonstrating the effects of a transient conductance change [fast input transient of Rinzel and Rall (1974), $\alpha = 50$, equilibrium potential = 55 mV], applied directly to the soma of a dentate granule cell. The faster curve shows the time course of the conductance change (Δg_{soma}) while the slower curve illustrates the direct soma voltage response (V_{soma}) to such a Δg. This conductance input injected an equivalent input charge of 6.7×10^{-14} C into the soma, with the voltage response peaking at approximately 875 μV. (B) The same conductance change input, but applied directly to a dendritic spine. The input impedance at this site was approximately 40 times that at the soma. The fast curve shows the dendritic voltage response (V_{dend}) to the conductance change, while the slower trace is the transferred soma response (V_{soma}) after passive conduction to the soma. Note the difference in scale for the two curves. The soma peak is approximately 120 μV, with a rise time (10 to 90% amplitude) of 0.18 τ and a halfwidth of 0.52 τ. Using this modeling technique, the waveform responses at the soma may be analyzed for many different dendritic input sites. Also, a given input may be analyzed when injected at a dendritic shaft site or at a dendritic spine. When such a comparison was made, no systematic alteration of the input signal or output waveform resulted when input was at a dendritic spine as opposed to a dendritic shaft. This result challenges previous predictions and theories (Diamond *et al.*, 1971, Rall, 1974 and Scheibel and Scheibel, 1968) regarding the possible electrical roles of dendritic spines.

erage" synapse would transfer approximately 40 to 45% of charge to the soma.

Reconstruction of the transient responses of these neurons from the model showed a generally good correlation between the measured responses and those predicted by the E–C cable model. A specific analysis of dendritic structure thus does not appear to be essential in the prediction of responses to somatic inputs, though the effect of dendritic loading does need to be explicitly considered.

3.1.3. Brief Pulse Equivalent Cylinder Model. In other experiments, the equivalent cylinder model has been employed in the analysis of brief pulse inputs applied at the soma and the subsequent decay of somatic voltage (Iansek and Redman, 1973; Durand *et al*, 1983). One such study (Durand *et al*, 1983) focused on the electrotonic structure of *in vitro* dentate granule neurons (Table I). Values of average electrotonic length from this investigation are only slightly different from those derived using the long step input techniques and emphasize the same point—namely, the electrical compactness of hippocampal neurons. The use of almost instantaneous short pulses greatly facilitates time constant analysis of the voltage response, since the tedious work of differentiating long step responses can be replaced by automatic computer analysis of short pulse decay. Further, it avoids nonlinearities that are associated with longer pulses. Results of both E–C methods provide an averaged, overall representation of a neuron's dendritic structure. However, short pulse analysis based on the E–C model has recently indicated a need to incorporate a somatic current shunt into neuronal models developed from intracellular recording and current injection. Such a shunt, attributable to electrode-induced injury, results in faster decay constants than predicted by the E–C model (Durand *et al*, 1983).

3.2. The Compartmental Model

The compartmental model has been employed in an analysis of the spatial dispersion of dendritic ionic conductances (Traub and Llinás, 1979), largely using data obtained from hippocampal neurons *in vitro* (Schwartzkroin, 1975, 1977). With a differential distribution of regenerating conductances (mainly Na^+ and Ca^+), the necessary ionic conditions for repetitive firing and bursting were defined for CA1 and CA3 pyramidal neurons. The description of dendritic conductances and their locations given by this model has provided a theoretical basis for many of the observed phenomena seen in hippocampal neurons, such as Ca^+ spikes, fast prepotentials, afterpotentials, and the characteristic relation between current injection and cell firing frequency. An important result of this model was a definition of the location of conductances that would simulate such cell behaviors as burst discharge. Consistent with recent experimental findings, this model showed the critical involvement of calcium and slow potassium conductances in producing and timing epileptiform bursting. The model has also reinforced the recent focus on dendrites as important sites of integration and potential generation, since the model properly predicts cell activity when specific conductances are localized in the dendrites.

3.3. *Neuronal Circuitry Model*

The behavior of a small neuronal population in the *in vitro* hippocampus has recently been studied with comparative experimental and modeling data (Traub and Wong, 1981, 1983; Wong and Traub, 1983). Though each neuron was included in the model as a single compartment element, many of the experimentally observed features of both normal and epileptic (produced by penicillin) hippocampus were reproduced. In addition, interesting recurrent phenomena of relatively long duration (1 to 2 sec) could be explained, such as the genesis of penicillin-induced bursting and synchronization of the neuron population. Thus, though this model considerably simplified the intrinsic properties of each cell (e.g., including only selected ionic conductances), and described each neuronal element as a simple summing point for a small number of inputs, the model could predict complex population phenomena successfully.

In another study analyzing a network of CA1 hippocampal neurons (Leung, 1982), the overall transfer function of the network was emphasized. This model led to the prediction of population patterns of activity (i.e., field potentials), including the theta rhythm. This model, and other attempts to describe rhythmic or oscillatory discharge in hippocampus (Mates and Horowitz, 1976) have emphasized a key role for recurrent inhibitory circuitry.

3.4. *Overview of Hippocampal Slice Data*

Characteristic values for various neuronal properties (input resistance, time constant, time constant ratios, and electrotonic length) predicted by different electrotonic models were gratifying similar for neurons studied *in vitro* (Table I). The input resistances of different hippocampal neuronal types were surprisingly similar; time constants and membrane resistance varied proportionally to each other and differed among the three hippocampal cell groups. All hippocampal cells showed a rather uniform, compact L (average overall L of 0.9), which is significantly shorter than the value found for motoneurons (average overall L of 1.5). This average L value deviates significantly from segmental L values in pyramidal cells. However, soma recordings do not appear to distinguish the "real" dendritic tree (i.e., the complexity of individual segments) from an equivalent cylinder simplification.

In contrast to the agreement of the various models regarding electrotonic length in hippocampal neurons (Table I), predicted values for dendrite-to-soma conductance ratio (ρ) appear discordant. This discrepancy may be due both to biological variability, and to the inherently difficult task of estimating this value (Rall, 1969, 1977; Jack *et al.*, 1971). In addition, physiological criteria for a soma region (i.e., an area of relative isopotentiality) are obviously different from anatomical criteria for separating soma and dendrite. Given the high variability of ρ values both within and between electronic models, it is questionable what information this measure currently provides.

The uniform L values among hippocampal neurons contrast with the distinct variability in geometric extent among the three classes of cells (Turner and Schwartzkroin, 1983). The CA3 pyramidal cells have about four times the surface area of the dentate granule cells, and have time constants 3 to 4 times longer, but their electrotonic lengths are similar. These surface area and time constant values varied among neuron types, but both were proportional to specific membrane resistance. It appeared, then, that specific membrane resistance changed in order to keep the total or average electrotonic length values constant. This alteration in membrane resistance may be one mechanism that "compensates" for variation in dendritic morphology, allowing inputs onto distal dendrites of extensive dendritic trees to have significant effects on cell integration.

4. DISCUSSION

4.1. *Insights from the Models of Electrotonic Structure*

Almost all excitatory synaptic inputs are onto the dendrites (and spines) of hippocampal neurons, but physiological recordings directly from these sites are virtually impossible. Thus, if information about synapse-to-soma relationships is important for our understanding of cell function, it is necessary to *predict* actual electrical events in dendrites on the basis of somatic recordings. This predictive task is a primary function of electrotonic models, and the motivation for trying to understand the electrotonic structure of neurons. The types of synaptic inputs that one wishes to simulate, and their location, will determine the choice of model. For example, for the evaluation of single dendritic (or dendritic spine) input, the segmental branching cable model (Turner, 1982; Turner and Schwartzkroin, 1983) may be preferred, since it provides a detailed structural analysis that is not possible with other models. Analysis of inputs from a region of a major dendrite may be facilitated by the brief pulse equivalent cylinder model (Jack *et al.*, 1981), which can easily define electrical transfer of mass synaptic input to the soma. The compartmental model, too, may define dendritic events (Rall, 1967; Traub and Llinás, 1979) (though it suffers from a lack of anatomical clarity as to the location of the input), and has been particularly useful in incorporating the triggering of regenerative dendritic events into a model of cell function. In different situations, then, each model has been helpful in defining the process of neuronal integration.

Since dendrites clearly affect how the soma responds to somatic inputs, an evaluation of neuronal electrotonic structure can also shed light on cell responses to somatic inputs. Models that lump an extensive dendritic tree into a simplified equivalent cylinder provide an adequate picture of this electrotonic structure (Brown *et al.*, 1981; Carlen and Durand, 1981; Johnston, 1981). Simulations of somatic inputs using the E–C models may be useful in gaining an understanding of synaptic inputs located close to the soma (as in the case of mossy fiber inputs onto CA3 neurons) (Yamamoto, 1982), and for evaluating the efficacy of voltage clamp techniques applied to hippocampal neurons (Johnston and Brown, 1983).

Given information about neuronal electrotonic structure provided by these models, one may then address the questions of whether the dendrites are merely structural features (to gain more synaptic inputs from a broader field?) or whether they play a significant role in shaping, transforming and weighting the many inputs to a neuron. One report has indicated that little electrical processing or differential weighting may be performed at a dendritic level in motoneurons (Jack *et al.*, 1981) for all dendritic inputs appeared to possess similar transfer efficacy to the soma. In contrast, model studies of dendritic transfer in hippocampal neurons (Turner and Schwartzkroin, 1980, 1983; Turner, 1982) suggest that laminar segregation of inputs onto different levels of the dendrite can provide one means by which various afferent pathways may be "scaled" or adjusted in their relative efficacy. Further physiological studies of synaptic efficacy from discrete dendritic regions are needed to elucidate the actual role of dendritic electrical processing, and to test the adequacy of the passive electrotonic models (Andersen *et al.*, 1980; Barnes and McNaughton, 1980).

A question related to that of dendritic processing is whether (or how) dendritic spines may function electrophysiologically. Current model studies investigating the electrophysiological role of spines (Turner, 1982; Turner and Schwartzkroin, 1983) provide a preliminary indication of limited spine importance for a cell's electrotonic structure. Perhaps these dendritic appendages possess an electrical function not yet elucidated. They may, however, play a primarily structural role without direct electrical significance. Thus, spines may enhance the number of presynaptic connections to a cell (Swindale, 1981), or be of metabolic importance (Gray, 1982), or provide a morphological substrate for synaptic plasticity (Fifkova and Van Harreveld, 1977; Crick, 1982).

4.2. *Models of Cell Circuitry*

However complicated the integration and electrical processing within a single neuron might be, the complex interactions between neurons are keys to our understanding of the actual functioning of the CNS. Clearly, complex population phenomena can be reproduced in models that consider neurons as simple computing elements that are interconnected in specified patterns. It is certainly important to include single cell properties in larger and more complex models of cell interaction. However, the development of models describing *network* processing is a present challenge to the electrophysiologist and modeling theoretician. The initial efforts at network evaluation (Perkel *et al.*, 1981; Traub and Wong, 1981, 1983; Leung, 1982) synthesize much of our present information about local circuits, but clearly represent only a beginning step. Future attempts at network modeling will hopefully include the complex single neuron electrical properties as well as more realistic views of the complex connectivities within local circuits.

4.3. *Contributions of the Slice Preparation to Model Studies*

The slice technique considerably facilitates both physiological and anatomical analysis of single neuron properties and small cortical circuits (MacVicar and Dudek, 1980; Knowles and Schwartzkroin, 1981). The stability of intracellular penetrations and the feasibility of impaling small cells are both enhanced over *in vivo* cortical preparation (Schwartzkroin, 1975, 1977); resultant recordings are of high quality, thus providing useful information about cell properties for modeling work. The reduced inputs and low spontaneous activity of the simplified "deafferented" slice neurons allow the measurement of "basal" resting membrane parameters in the sense that activity is less biased by tonic inputs than in *in vivo* preparations (Turner and Schwartzkroin, 1980). Although this characteristic of slices is useful in obtaining baseline neuronal data, it also presents a somewhat false picture of *in situ* neurons that normally function with a bias provided by both tonic and phasic inputs.

Another important feature of slices is the improved accessibility of the tissue, both in visualization of specific laminae (corresponding to identifiable recording site and input pathways), and also in facilitation of pharmacologic manipulation of the tissue. Thus, specific afferent inputs or conductances can be blocked by mechanical means (e.g., "microsurgery") or by use of channel or receptor blockers (Andersen *et al.*, 1980; Yamamoto, 1982; Wong and Traub, 1983). Also, because of the improved accessibility, neurons that have been intracellularly stained are much easier to recover and process, as compared to the *in vivo* requirements for extensive searching and serial sectioning (MacVicar and Dudek, 1982; Turner, 1982; Turner and Schwartzkroin, 1983). This slice simplification, of course, also may limit the varieties of neurons encountered in the tissue so that the investigator must be aware of the biases of the selection process.

4.4. *Experimental Directions and Testable Predictions*

The emphasis of these model studies thus far has been on the understanding of electrical processing during synaptic integration, both at the single neuron and circuitry levels. Each kind of model, given its inherent assumptions, produces a set of predictions regarding the behavior of a cell or population of cells. These predictions, such as the purely passive electrotonic transfer of synaptic signals from dendrite to soma, need to be carefully tested at the experimental level. For example, model predictions about the voltage and conductance characteristics of quantal synaptic events can now be approached using the *in vitro* slice technique (McNaughton, *et al.*, 1981; Yamamoto, 1982). Investigations into the relative "weighting" of different afferent inputs, as a function of synapse location in the dendritic tree, can also be analyzed (Andersen *et al.*, 1980). Such studies could begin to include interaction of synaptic events with voltage-dependent membrane phenomena that have the potential for input amplification (Wong and Prince, 1978, 1979). Both synaptic and regenerative membrane events have been modeled for hippocampal neurons, but the integration of synaptic inputs (at different sites on the cell surface and with different types of conductances), and their effects on local intrinsic phenomena are still to be explored (Lux and Schubert, 1975; Barnes and McNaughton, 1980). Similarly, we can now study at both experimental and theoretical levels, the interaction of single cell events with extracellular changes (extracellular ionic concentration changes, alterations and variations in extracellular space with respect to current flow) or with other cellular events in a local circuit.

The productivity of theoretical modeling work is dependent on the degree to which the modeler and experimentalist establish a real interaction. That such collaboration can be fruitful is amply illustrated by recent studies based on data from experiments on hippocampal neurons *in vitro*. Models have been used to show that hippocampal neurons (1) are electrotonically short, (2) have substantial current transfer, but poor voltage transfer, from dendritic synapse to soma, (3) have surprisingly similar electrotonic structures given their diverse morphologies, (4) may use membrane resistivity changes as a mechanism for determining synaptic efficacy at a given dendritic site, (5) show little dependence on spines for processing synaptic input, and (6) can account for some population rhythms by virtue of their intrinsic properties. From modeling insights, we might predict that (1) distal synapses on hippocampal cells have significant influences on cell output, (2) truncation or distortion of dendrites, such as occurs in many neurological abnormalities, should have significant effects on cell processing capabilities, (3) rearrangement of membrane conductance (densities, location) due to cellular trauma can alter cellular potentials, and (4) changes in single cell properties can alter complex population phenomena independently of changes in connectivity. These model-based predictions can be formulated in a qualitative and specific manner, and are experimentally testable.

5. REFERENCES

Andersen, P., Silfvenius, H., Sundberg, F. H., and Sveen, O., 1980, A comparison of distal and proximal dendritic synapses on CA1 pyramids in guinea pig hippocampal slices *in vitro*, *J. Physiol. (London)* **307**:273–299.

Barnes, C. A. and McNaughton, B. L., 1980, Physiological compensation for loss of afferent synapses in rat hippocampal granule cells during senescence, *J. Physiol. (London)* **309**:473–485.

Barrett, J. N. and Crill, W. E., 1974, Influence of dendritic location and membrane properties on the effectiveness of synapses on cat motoneurons, *J. Physiol. (London)* **239**:325–345.

Brown, T. H., Fricke, R. A., and Perkel, D. H., 1981, Passive electrical constants in three classes of hippocampal neurons, *J. Neurophysiol.* **46**:812–827.

Carlen, P. L. and Durand, D., 1981, Modelling the postsynaptic location and magnitude of tonic conductance changes resulting from neurotransmitters or drugs, *Neuroscience* **6**:839–846.

Carnevale, N. T. and Johnston, D., 1982, Electrophysiological characterization of remote chemical synapses, *J. Neurophysiol.* **47**:606–621.

Crick, F., 1982, Do dendritic spines twitch? *Trends Neurosci.* **5**:44–46.

Diamond, J., Gray, E. G., and Yasargil, G. M., 1971, The function of the dendritic spine: An hypothesis, in: *Excitatory Synaptic Mechanisms* (P. Andersen and K. Jansen, eds.), Universitetsforlaget, Oslo, pp. 213–222.

Durand, D., Carlen, P. L., Gurevich, N., Ho, A., and Kunov, H., 1983, Electrotonic parameters of rat dentate granule cells measured using short current pulses and HRP staining, *J. Neurophysiol.*, in press.

Fifkova, E. and Van Harreveld, A., 1977, Long-lasting morphological changes in dendritic spines of dentate granular cells, following stimulation of the entorhinal area, *J. Neurocytol.* **6**:211–230.

Gray, E. G., 1982, Rehabilitating the dendritic spine, *Trends Neurosci.* **5**:5–6.

Iansek, R. and Redman, S. J., 1973, An analysis of the cable properties of spinal motoneurones using a brief intracellular current pulse, *J. Physiol. (London)* **234**:613–636.

Jack, J. J. B., Miller, S., Porter, R., and Redman, S. J., 1971, The time course of minimal excitatory postsynaptic potentials evoked in spinal motoneurones by group IA afferent fibers, *J. Physiol. (London)* **215**:353–380.

Jack, J. J. B., Noble, D., and Tsien, R. W., 1975, *Electric Current Flow in Excitable Cells*, Clarendon Press, Oxford.

Jack, J. J. B., Redman, S. J., and Wong, K., 1981, The components of synaptic potentials evoked in cat spinal motoneurones by impulses in single group IA afferents, *J. Physiol. (London)* **321**:65–96.

Johnston, D., 1981, Passive cable properties of hippocampal CA3 pyramidal neurons, *Cell. Mol. Neurobiol.* **1**:41–55.

Johnston, D. and Brown, T. H., 1983, Interpretation of voltage-clamp measurements in hippocampal neurons, *J. Neurophysiol.* **50**:464–486.

Knowles, W. D., and Schwartzkroin, P. A., 1981, Local circuit synaptic interactions in hippocampal brain slices, *J. Neurosci.* **1**:318–322.

Leung, L. S., 1982, Nonlinear feedback model of neuronal populations in the hippocampal CA1 region, *J. Neurophysiol.* **47**:845–868.

Lux, H. D. and Schubert, P., 1975, Some aspects of the electroanatomy of dendrites, in: *Advances in Neurology*, Vol. 12 (G. W. Kreutzberg, ed.), Raven Press, New York, pp. 29–44.

MacVicar, B. A. and Dudek, F. E., 1980, Local synaptic circuits in rat hippocampus: Interactions between pyramidal cells, *Brain Res.* **184**:220–223.

MacVicar, B. A. and Dudek, F. E., 1982, Electrotonic coupling between granule cells of the rat dentate gyrus: Physiological and anatomical evidence, *J. Neurophysiol.* **47**:579–592.

Mates, J. W. B. and Horowitz, J. M., 1976, Instability in a hippocampal neuronal network, *Compt. Programs Biomed.* **6**:74–84.

McNaughton, B. L., Barnes, C. A., and Andersen, P., 1981, Synaptic efficacy and EPSP summation in granule cells of rat fascia dentata studied *in vitro*, *J. Neurophysiol.* **46**:952–966.

Minkwitz, H-G., 1976, Zur Entwicklung der Neuronenstruktur des Hippocampus während der prä und postnatalen Ontogenese der Albinoratte. III. Mitteilung: Morphometrische Erfassung der ontogenetischen Veränderungen in Dendritenstruktur und Spine Besatz an Pyramiden-Neuronen (CA1) des Hippocampus, *J. Hirnforsch.* **17**:255–275.

Norman, R. S., 1972, Cable theory for finite length dendritic cylinders with initial and boundary conditions, *Biophys. J.* **12**:25–45.

Perkel, D. H. and Mulloney, B., 1978, Electrotonic properties of neurons: Steady-state compartmental model, *J. Neurophysiol.* **41**:621–639.

Perkel, D. H., Mulloney, B., and Budelli, R. W., 1981, Quantitative methods for predicting neuronal behavior, *Neuroscience* **6**:823–837.

Rall, W., 1959, Branching dendritic trees and motoneuron membrane resistivity, *Exp. Neurol.* **1**:491–527.

Rall, W., 1967, Distinguishing theoretical synaptic potentials computed for different soma-dendritic distributions of synaptic input, *J. Neurophysiol.* **30**:1138–1168.

Rall, W., 1969, Time constants and electrotonic length of membrane cylinders and neurons, *Biophys. J.* **9**:1483–1508.

Rall, W., 1974, Dendritic spines, synaptic potency and neuronal plasticity, in: *Cellular Mechanisms Subserving Changes in Neuronal Activity* (C. D. Woody, K. A. Brown, T. J. Crow, Jr., and J. D. Knispel, eds.), Brain Information Service, Los Angeles, pp. 13–21.

Rall, W., 1977, Core conductor theory and cable properties of neurons, in: *Handbook of Physiology.* Section I. The Nervous System, *Volume 1 Cellular Biology of Neurons* (E. R. Kandel, ed.), Williams & Wilkins, Bethesda, pp. 39–98.

Rinzel, J. and Rall, W., 1974, Transient response in a dendritic neuron model for current injected at one branch, *Biophys. J.* **14**:759–790.

Scheibel, M. E. and Scheibel, A. B., 1968, On the nature of dendritic spines—Report of a workshop, *Comm. Behav. Biol.* **I**(A):231–265.

Schwartzkroin, P. A., 1975, Characteristics of CA1 neurons recorded intracellularly in the hippocampal '*in vitro*' slice preparation, *Brain Res.* **85**:423–436.

Schwartzkroin, P. A., 1977, Further characteristics of hippocampal CA1 cells *in vitro, Brain Res.* **128**:53–68.

Schwartzkroin, P. A., 1981, To slice or not to slice, in: *Electrophysiology of Isolated Mammalian CNS Preparations* (G. A. Kerkut and H. V. Wheal, eds.), Academic Press, London, pp. 15–50.

Swindale, N. V., 1981, Dendritic spines only connect, *Trends Nuerosci.* **4**:240–241.

Traub, R. D. and Llinás, R., 1979, Hippocampal pyramidal cells: Significance of dendritic ionic conductances for neuronal function and epileptogenesis, *J. Neurophysiol.* **42**:476–496.

Traub, R. D. and Wong, R. K. S., 1981, Penicillin-induced epileptiform activity in the hippocampal slice: A model of synchronization of CA3 pyramidal cell bursting, *Neuroscience* **6**:223–230.

Traub, R. D. and Wong, R. K. S., 1983, Synchronized burst discharge in the disinhibited hippocampal slice. II. Model of the cellular mechanism, *J. Neurophysiol.* **49**:459–471.

Turner D. A., 1982, Soma and dendritic spine transients in intracellularly-stained hippocampal neurons, *Neurosci. Abstr.* **8**:945.

Turner, D. A. and Schwartzkroin, P. A., 1980, Steady-state electrotonic analysis of intracellularly-stained hippocampal neurons, *J. Neurophysiol.* **44**:184–199.

Turner, D. A. and Schwartzkroin, P. A., 1983, Electrical characteristics of dendrites and dendritic spines in intracellularly-stained CA3 and dentate neurons, *J. Neuroscience,* in press.

Wenzel, J., Stender, G., and Duwe, G., 1981, The development of the neuronal structure of the fascia dentata of the rat. Neurohistologic, morphometric, ultrastructural and experimental investigations, *J. Hirnforsch.* **22**:629–683.

Wong, R. K. S. and Prince, D. A., 1978, Participation of calcium spikes during intrinsic burst firing in hippocampal neurons, *Brain Res.* **159**:385–390.

Wong, R. K. S. and Prince, D. A., 1979, Dendritic mechanisms underlying penicillin-induced epileptiform activity, *Science* **204**:1228–1231.

Wong, R. K. S. and Traub, R. D., 1983, Synchronized burst discharge in the disinhibited hippocampal slice. I. Initiation in CA2–CA3 region, *J. Neurophysiol.* **49**:442–458.

Yamamoto, C., 1982, Quantal analysis of excitatory postsynaptic potentials induced in hippocampal neurons by activation of granule cells, *Exp. Brain Res.* **46**:170–176.

3

Biophysics and Microphysiology of Synaptic Transmission in Hippocampus

DANIEL JOHNSTON and THOMAS H. BROWN

1. MOTIVATION FOR STUDYING THE BIOPHYSICS AND MICROPHYSIOLOGY OF CORTICAL SYNAPSES

Most of what is known about the microphysiology and biophysics of synaptic transmission has derived from the study of just three preparations: the vertebrate and arthropod neuromuscular junctions and the squid giant synapse (Katz, 1969; Martin, 1977; Takeuchi, 1977). Although the general facts and laws gleaned from these three classical preparations probably apply also to cortical synapses, this inference remains an article of faith. Furthermore, there are structural and functional specializations in the cortex that cannot readily be addressed in the three classical preparations.

Examples of problems resulting from such specializations include the role of dendrites and their spines in synaptic interactions and integration (Jack *et al.*, 1975; Rall, 1977; Crick, 1982); the effect on synaptic information transfer of spatial inhomogenieties in the active and passive membrane conductances of the dendritic and somatic membrane (Traub

DANIEL JOHNSTON • Program in Neuroscience, Section of Neurophysiology, Department of Neurology, Baylor College of Medicine, Houston, Texas 77030. THOMAS H. BROWN • Department of Cellular Neurophysiology, Division of Neurosciences, City of Hope Research Institute, Duarte, California 91010.

and Llinás, 1979; Wong *et al.*, 1979; Benardo *et al.*, 1982); and the mechanisms underlying interesting forms of use-dependent synaptic plasticity, such as homosynaptic and associative long-term potentiation (Bliss and Gardner-Medwin, 1973; Bliss and Lomo, 1973; Bliss, 1979; Bliss and Dolphin, 1982; Levy and Desmond, 1983; Brown and Barrionuevo, submitted). In addition, there are neurological disorders, such as epilepsy, that probably can only be fully understood by direct study of the commonly affected synaptic networks (Johnston and Brown, 1983a).

Considerations such as these clearly underscore the current need for quantitative studies of the biophysics and microphysiology of synaptic transmission between cortical neurons. To fulfill this need, the first step is obviously to select and develop a suitable mammalian cortical synaptic preparation. Such a preparation ideally would incorporate some of the useful features of the three classical preparations and would also enable the experimenter to investigate some of the possibly unique properties of synapses in the mammalian brain. In the following sections, we will describe the use of the *in vitro* hippocampal slice preparation for the study of the biophysics and microphysiology of synaptic transmission at the mossy-fiber synapses onto CA3 pyramidal neurons. This system offers many of the technical advantages found in the three classical synaptic preparations.

2. CRITERIA FOR SELECTING A SUITABLE CORTICAL SYNAPTIC PREPARATION

In thinking about developing a mammalian cortical preparation that would be suitable for studying the biophysics and microphysiology of synaptic transmission, we considered seven criteria to be of paramount importance. These criteria are listed and described below.

2.1. *Identifiable Neurons and Synapses*

In order to compare results among experiments, it is extremely important to be able to record consistently from the same neuron or at least from the same class of neuron. In addition, the synapses under study should be ultrastructurally distinguishable from other synapses; that is, the exact anatomical location of the synapses under study should be amenable to investigation. This would permit the electrotonic localization of the synapses and the evaluation of errors associated with measuring their biophysical properties. Furthermore, structure–function relationships unique to the particular synapse under study could be explored.

2.2. *Stable Intracellular Recordings*

Long-term intracellular recordings are usually required for experiments in which detailed measurements of synaptic function are to be made. This is usually not possible with *in vivo* preparations because of their intrinsic instabilities caused by cardiorespiratory functions. *In vitro* preparations offer excellent stability for intracellular recordings and are preferred for this reason.

2.3. *Minimal Diffusional Barriers*

It is important that the experimenter have direct control of the extracellular milieu. Changes in the chemical composition of the extracellular fluid often must be made rapidly and completely during an experiment—sometimes while maintaining an intracellular recording. Rapid and complete exchange of solution requires direct access to the bulk medium as well as minimal diffusional barriers between the bulk medium and the extracellular space adjacent to the neuron under study. Many *in vitro* preparations satisfy this requirement.

2.4. *Monosynaptic Connection from Stimulus Site*

Monosynaptic stimulation of the synapses under study is necessary for detailed analyses of synaptic function. Polysynaptic inputs with their intervening interneurons would add numerous unknown parameters and would greatly complicate the analysis of unitary synaptic events. When monosynaptic inputs are normally concomitant with polysynaptic inputs, the latter must be amenable to separation from the monosynaptic response by pharmacological or electrical means.

2.5. *Minimal Electrotonic Distance between Subsynaptic Membrane and Recording Site*

The quantitative analysis of synaptic function requires that the potential at the subsynaptic membrane be known and accurately controlled. Few, if any, synaptic preparations are completely isopotential with the recording microelectrodes throughout the frequency spectrum of the synaptic current waveform. It is important, therefore, that the degree of nonisopotentiality be small and well characterized so that errors resulting from the nonisopotentiality of the synapses can be evaluated.

2.6. *Measurement of Single Quantal Events*

One of the two most powerful neurophysiological tools for analyzing synaptic function is quantal analysis. A proper quantal analysis requires, ideally, the ability to resolve clearly single quantal events above the background recording noise levels.

2.7. *Application of Voltage-Clamp Techniques*

The other neurophysiological tool that has proven indispensable for quantitative analytical studies of synaptic function is the voltage clamp. Voltage-clamp techniques and quantal analyses have provided the key insights into our current understanding of the biophysics and microphysiology of synaptic transmission in the three classical synaptic preparations.

2.8. *A Synapse That Satisfies These Criteria in the Hippocampal Slice Preparation*

The mossy-fiber-to-CA3 pyramidal neuron excitatory synapse in the hippocampal slice preparation satisfies all seven of the criteria listed above (Johnston and Brown, 1983b; Brown and Johnston, 1983). Pyramidal neurons of the CA3 region can be identified easily in the *in vitro* slice. The slice preparation offers excellent stability and minimal diffusional barriers. The mossy-fiber excitatory synapse can be stimulated monosynaptically, and the normal concomitant inhibitory input can be separated electrically or blocked with any of several pharmacological agents (Section 4). Mossy-fiber synapses are electrotonically close to the somatic recording site, and voltage-clamp techniques have been successfully applied to both voltage-dependent and synaptic conductance changes in CA3 neurons (Sections 3 and 4). Finally, both excitatory and inhibitory quantal events can be resolved in CA3 neurons, and the prospects for performing a meaningful quantal analysis of evoked release appear good (Sections 5 and 6).

3. DEVELOPMENT OF VOLTAGE-CLAMP TECHNIQUES FOR APPLICATION TO HIPPOCAMPAL SYNAPSES

The importance of applying voltage-clamp techniques to cortical neurons cannot be overemphasized. However, when one considers the special problems associated with voltage clamping a small, morphologically complex neuron embedded somewhere deep within a slab of tissue, the task appears hopeless. The most formidable problem is inadequate space clamp and the evaluation of the errors in the data resulting therefrom. This issue is addressed in Section 3.2 and is dealt with in detail elsewhere (Johnston and Brown, 1983b).

3.1. *Techniques for Voltage Clamping Small Cortical Neurons*

3.1.1. Double-Microelectrode Voltage Clamp. The traditional method for voltage clamping whole neurons utilizes two microelectrodes: one for voltage recording, and the other for passing current. This method has been applied successfully to large invertebrate cells for over 20 years (Hagiwara and Saito, 1959). Double-microelectrode voltage clamping has also been applied to spinal motoneurons (Frank *et al.*, 1959; Araki and Terzuolo, 1962). However, not until the relatively recent work of Crill and associates (Barrett and Crill, 1980; Barrett *et al.*, 1980; Schwindt and Crill, 1977, 1980a,b,c, 1981) has significant information regarding the properties of spinal motoneurons been provided by this method.

The double-microelectrode voltage-clamp method presents special problems when applied to small cortical neurons. First, the individual neurons are usually not visible. In hippocampal slices, although the cell-body regions are clearly visible through a dissecting microscope, *individual neurons* are difficult to distinguish clearly even with special contrast-enhancing optics. In principle, however, clear visualization of single neurons may be accomplished in the hippocampus by using very thin slices, such as recently shown by Llinás and Sugimori (1980) in cerebellum. Second, surface neurons may be damaged by slicing. Therefore, the healthy cells are probably at least one cell layer beneath the surface of a slice. Manipulation of two independent electrodes through tissue to the surface of an embedded cell may be difficult. Two electrodes fused together could overcome this problem but would add complexity to an experiment by increasing the time associated with electrode preparation. Third, two electrodes separated by only a few microns, such as would be necessary to impale a small neuron, would have significant capacity coupling that would require special shielding between the electrodes and would significantly reduce the loop gain of the clamp. Fourth, the smaller cortical neurons do not tolerate well impalement by two independent microelectrodes, and the yield of healthy cells from such experiments is exceedingly low.

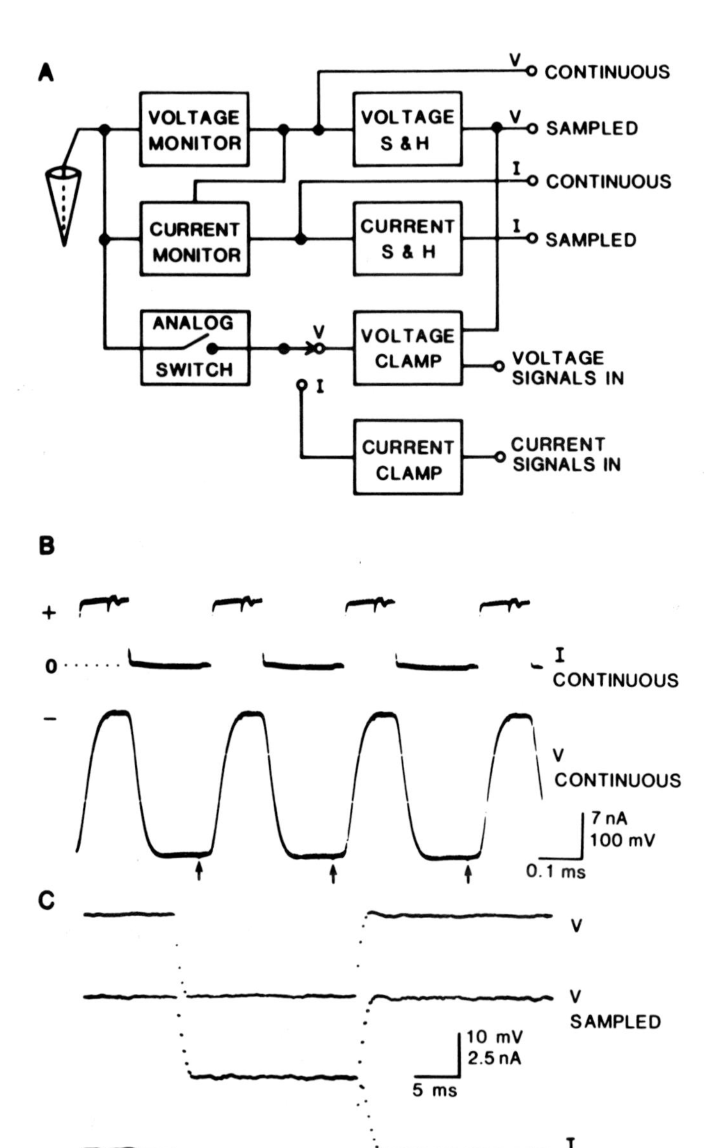
A
V CONTINUOUS
VOLTAGE MONITOR
VOLTAGE S & H
V SAMPLED
I CONTINUOUS
CURRENT MONITOR
CURRENT S & H
I SAMPLED
ANALOG SWITCH
V
I
VOLTAGE CLAMP
VOLTAGE SIGNALS IN
CURRENT CLAMP
CURRENT SIGNALS IN
B
+
0
−
I CONTINUOUS
V CONTINUOUS
7 nA
100 mV
0.1 ms
C
V
V SAMPLED
10 mV
2.5 nA
5 ms
I SAMPLED

3.1.2. *Single-Microelectrode Voltage Clamp*

3.1.2a. Advantages of Using the Single-Microelectrode Voltage Clamp for Hippocampal Neurons. In theory, the single-microelectrode voltage clamp (SEC), which uses a time-share circuit (Wilson and Goldner, 1975; Johnston *et al.*, 1980; Johnston, 1981; Johnston and Brown, 1981; Brown and Johnston, 1983, Johnston and Brown, 1983b), can circumvent some of the problems inherent in the double-microelectrode clamp (see Figure 1). For example, less damage to cells should occur with impalement by only a single microelectrode. Although electrode design is exceedingly important with the SEC, routine electrode preparation is less time-consuming than with pairs of glued electrodes that require special shielding. Probably the most significant advantage of the SEC over traditional double-microelectrode clamping, however, is the greater yield of healthy cells. The yield of healthy cells amenable to voltage clamping with the SEC should be only slightly less than the yield of healthy cells obtained from using standard intracellular recordings with bridge-balance amplifiers. The somewhat lower yield arises from the need to use microelectrodes with slightly lower resistances (see below).

3.1.2b. Practical Considerations for Use of the Single-Microelectrode Voltage Clamp. The most important limitation in the use of the SEC is the microelectrode characteristics. The electrodes should have low resistances, and the overall input to the headstage should have extremely low capacity to ground. The optimum range for electrode resistances is about 10 to 50 MΩ for hippocampal neurons (Johnston *et al.*, 1980; Johnston, 1981; Johnston and Brown, 1981; Brown and John-

Figure 1. The time-share system used for both current- and voltage-clamp experiments. (A) Simplified block diagram for the single-electrode-clamp (SEC) device. (B) Illustration of the unsampled (continuous) voltage and current monitored from the SEC. The upper trace is the 3 kHz rectangular (40% duty cycle) wave current signal that is passed through the microelectrode into the cell. The lower trace is the potential developed across the electrode resistance and the cell membrane. Note that the voltage decays fully prior to the sample point, indicated by the arrows. The success of the system relies on the fact that the voltage across the electrode resistance decays rapidly (within about 100 μsec) relative to the decay of voltage across the cell membrane (which has a time constant of 20 to 50 msec). (C) Illustration of the sampled voltages and currents during a voltage-step command to a model cell. The model cell consists of a parallel resistor and capacitor (20 MΩ, 0.001 μF) connected in series to a standard 30-MΩ micropipette. The upper trace is the potential across the model cell, as measured with a standard intracellular preamplifier (Transidyne MPA-6). The middle trace is the sampled potential from the SEC. The bottom trace is the sampled current from the SEC. Note that the sampled potential measured with the SEC is very similar to the actual potential. From Brown and Johnston, 1983.

ston, 1983). The microelectrodes should be capable of passing 10 to 30 nA without rectification. It is our experience that electrodes with resistances much higher than 50 MΩ severely limit the performance characteristics of the SEC. The electrode characteristics should be assessed prior to and throughout the experiment by monitoring the unsampled electrode potential with a dedicated monitoring oscilloscope (see Figure 1).

The need for low input capacity requires shielding of as much of the microelectrode as possible. The shielding should be driven at the same potential as the output of the unity gain amplifier in the headstage. Also, the electrode tip should be minimally submerged in the experimental saline; otherwise, the capacitance to ground becomes too large. Coating the electrode tip with an insulator such as Q-dope, Sylgard, or wax has been reported to reduce tip capacitance significantly (Cornwall and Thomas, 1981).

With appropriate electrode design and proper shielding, the SEC is capable of stepping the membrane potential near the tip of the electrode ± 30 mV within 1 or 2 msec (Figure 1) (Brown and Johnston, 1983; Johnston and Brown, 1983b). The SEC should also be capable of holding the potential constant throughout membrane-conductance changes. This latter requirement is difficult to achieve for large and rapidly changing conductances. Typically, the membrane potential can be well controlled during ordinary synaptic events (Brown and Johnston, 1983). The frequency response of the SEC system used for our hippocampal recordings was approximately 550 Hz (Johnston and Brown, 1983b) (Figure 2).

3.2. *The Problem of Space Clamp*

The most serious restriction on biophysical analyses of voltage-dependent and synaptic conductances in cortical neurons is the lack of adequate space clamp. A perfect voltage clamp means that the voltage across the membrane has a temporally constant and known value. A perfect space clamp is achieved when a perfect voltage clamp is applied to that portion of membrane through which current is measured. Only under these conditions will the measured current be proportional to membrane conductance (Carnevale and Johnston, 1982; Johnston and Brown, 1983b). Few, if any, experimental preparations provide perfect voltage- and space-clamp conditions. Several invertebrate systems, such as squid giant axon and isolated spherical neurons from molluscan ganglia, approximate the ideal situation, but most other preparations deviate significantly from the theoretically perfect space clamp.

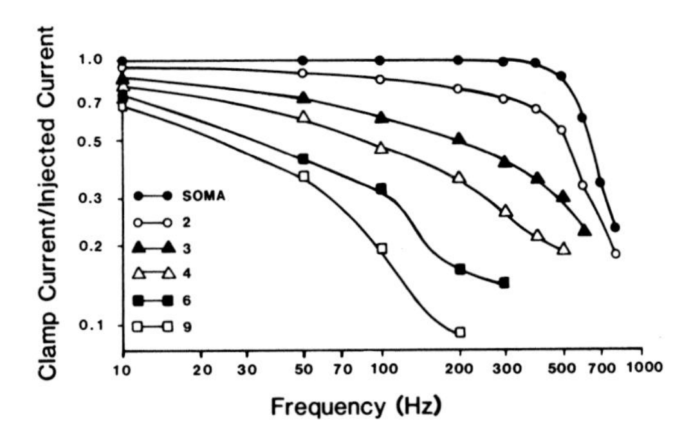

Figure 2. Current-attenuation ratios in an equivalent-cylinder compartmental model that mimics the electrotonic structure of an average CA3 pyramidal neuron. A time-share single-electrode voltage clamp was applied to the soma of this analog compartmental model through a standard 30-MΩ microelectrode. Next, a sinusoidal conductance change, in series with a battery, was placed in parallel with different compartments of the model. The current injected by the conductance change was measured at different frequencies, along with the voltage-clamp current in the soma. A log–log plot of current attenuation versus frequency (Bode plot) is shown for the conductance change located on several different compartments. The uppermost curve, where the conductance change was located on compartment 1, illustrates the frequency response of our single-electrode voltage clamp. (From Johnston and Brown, 1983b).

Cortical neurons, with their complex dendritic arborizations, are particularly unfavorable for good voltage- and space-clamp conditions. It is important, therefore, to assess quantitatively the deviation from ideal conditions for the physiological events of interest. When this is done, it quickly becomes apparent that the range of questions that can reasonably be answered using these techniques in intact cortical neurons is rather restricted. For example, it would be imprudent to inquire into the kinetics of the sodium current underlying the action potential in a cortical neuron; but the kinetics of certain synaptic events or of some of the slower voltage-dependent conductances that may be proximal to the cell soma might be amenable to analysis.

3.2.1. Theoretical Considerations. The most general approach to determining electrical coupling between a membrane conductance change and a recording site would make no assumptions regarding the electrotonic structure of the neuron. Such an approach was recently outlined by Carnevale and Johnston (1982), where a neuron with a synaptic input was represented as a two-port system. One of the strengths of this approach is that general formulations can be derived for current, voltage, and charge spread in a neuron. The result is an increased insight into the ways in which the electrotonic structure of a neuron can influence the measurement of physiological signals. One of the weaknesses of the approach, however, is that the parameters of the model, the knowledge of which would allow for accurate determination of coupling parameters, are difficult to measure experimentally.

The traditional alternative approach is to assume that the cell can be reasonably well represented by a relatively simple electrotonic model (such as one of the various equivalent dendritic cylinder models of Rall) and then estimate the parameters of the model using well-characterized methods. This approach is quite useful and has been outlined in detail by Rall and others during the past 20 years (for review, see Jack *et al.*, 1975; Rall, 1977).

3.2.2. *Electrotonic Structure of Hippocampal Neurons.* A simple Rall model (lumped soma attached to a finite, nontapering, equivalent cylinder) provides a reasonable representation of the electrotonic structure of hippocampal neurons for certain purposes (Brown *et al.*, 1981; Johnston and Brown, 1983b). Although it is possible that more complex Rall models (such as the incorporation of a tapering equivalent dendritic cylinder—see Turner and Schwartzkroin, submitted) could provide greater accuracy, for addressing a number of key questions the simplest form of the Rall model appears to be adequate. Two independent studies have provided data for the parameters of this simple version of the Rall model for CA3 pyramidal neurons (Brown *et al.*, 1981; Johnston, 1981). The best estimate of these parameters suggests an electrotonic length of 0.9 for the dendrites and a dendritic-to-somatic conductance ratio of about 1.5. Using these parameters, we performed a number of analog simulations to assess the performance of the SEC in measuring local and nonlocal conductance changes in these neurons (Johnston and Brown, 1983b).

3.2.3. *Errors Associated with Voltage Clamping Hippocampal Neurons.* A frequency analysis of the SEC connected (via a standard microelectrode immersed in saline) to the soma of the model cell indicated that the cable properties of the dendrites are rate-limiting for accurately measuring nonlocal conductance changes (Johnston and Brown, 1983b; see Figure 2). The bandwidth (−3 dB point) of physiological signals that can be measured by a voltage clamp in the soma is 300, 55, 23, 13, and 8 Hz for events generated, respectively, at electrotonic distances of 0.06, 0.18, 0.30, 0.54, and 0.9 from the cell soma (Johnston and Brown, 1983b). Practically, this means that for physiological signals generated much beyond an electrotonic distance of about 0.1 from the soma, a voltage clamp applied to the soma will give a very distorted representation of both the amplitude and kinetics of the actual conductance change. This is true regardless of the type of voltage-clamp device one employs. Simulations of this type are useful for assessing specific errors associated with voltage clamping events in hippocampal neurons, but only if the electrotonic location of the relevant conductance changes can be estimated.

3.2.4. Electrotonic Localization of Mossy-Fiber Synapses. We have used three methods to estimate the electrotonic distance of the mossy-fiber synapses from the somata of CA3 pyramidal neurons, and all three have yielded similar values (Johnston and Brown, 1983b; Brown and Johnston, 1982). The mossy-fiber synaptic boutons are large and readily identifiable (Figure 3) in both light and electron microscopic studies (Blackstad and Kjaerheim, 1961; Hamlyn, 1961, 1962; Haug, 1967). Anatomical studies have indicated that the mossy-fiber synapses terminate on the proximal portion of the apical dendrites (Lorente de Nó, 1934; Blackstad and Kjaerheim, 1961; Blackstad *et al.*, 1970; Amaral and Dent, 1981; Johnston and Brown, 1983b; Figure 3).

The first approach we used to estimate the electrotonic location of the mossy fibers involved measurement of the average diameter of the apical dendrites of CA3 neurons, measurement of the anatomical distance (from the soma) to the mossy-fiber synapses, and then calculation of the fraction of a length constant from the soma to the mossy-fiber endings. The results of this approach yielded an electrotonic distance of 0.07 for the *most distal* mossy-fiber synapses.

The second approach was based on the calculated attenuation factors from the somata to the mossy-fiber synapses, using compartmental models of reconstructed CA3 neurons. The results of this approach yielded an average electrotonic distance of 0.06 for the *most distal* mossy-fiber synapses.

The third approach utilized the theoretical studies of Rall (Rall, 1967; Rall *et al.*, 1967), which showed that the ratio of rise time to halfwidth (the "shape index") of synaptic potentials measured in the soma varies in a predictable manner as a function of the electrotonic distance of the synapses from the cell soma. Shape-index plots for mossy-fiber evoked synaptic potentials, in which the mixed inhibitory component (see Section 4) was either absent or blocked pharmacologically, indicated that the synapses were at electrotonic distances of less than 0.06 from the cell soma.

We conclude from these studies that mossy-fiber synapses are at electrotonic distances of less than 0.1 from the somata of CA3 neurons. This estimate of electrotonic distance permits calculation of the errors associated with the measurement of various parameters of synaptic function. For example, in considering the *most distal* synapses, the meas-

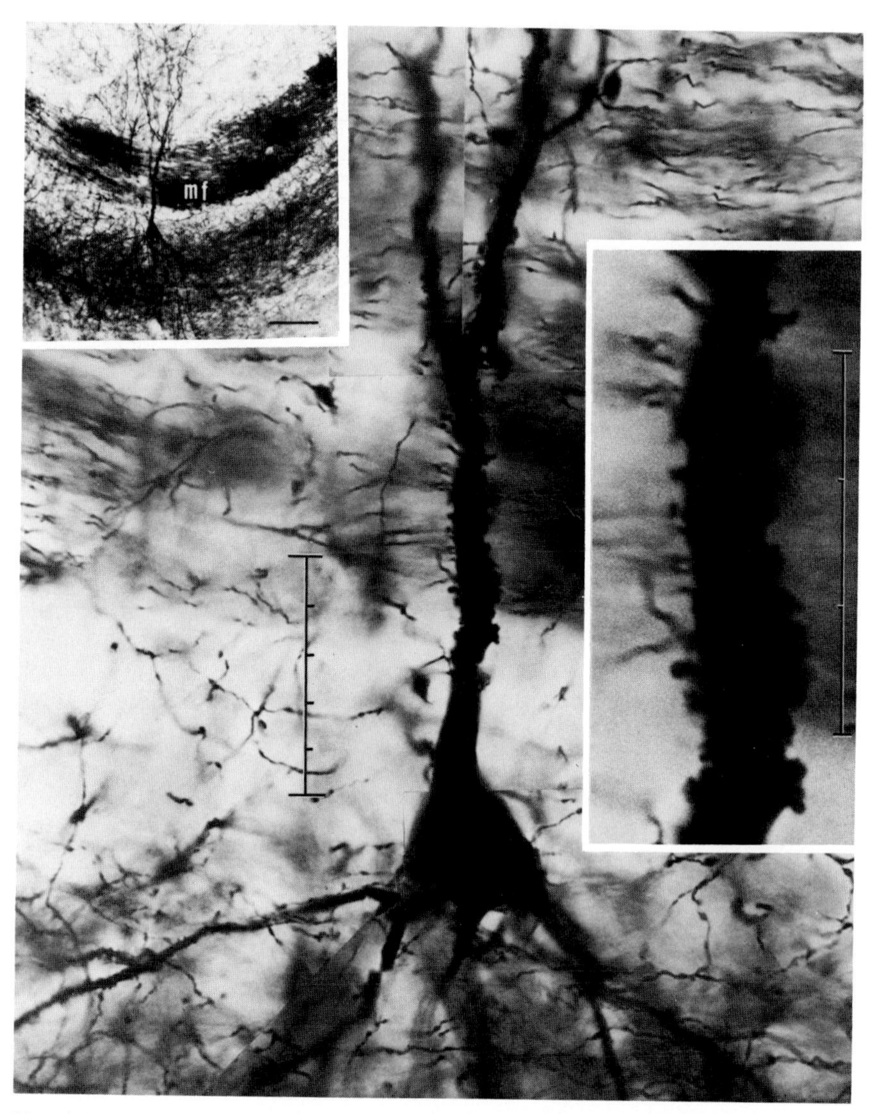

Figure 3. Photomicrograph of a transverse section through the hippocampus that was stained using the rapid Golgi method. Upper left: low-power view of a CA3 neuron, indicating its location within the regio inferior. The dark band passing through the apical region (*stratum lacunosum*) represents numerous Golgi-stained mossy-fiber synaptic expansions (scale, 100 μm). Center: a higher-power view of the same CA3 pyramidal neuron. Several large mossy-fiber synaptic expansions can be seen within the plane of focus. The dendritic thorny excrescences (the very characteristic large postsynaptic spines on which the mossy-fiber synapses are located) are also apparent (scale, 50 μm). Lower right: higher magnification of a portion of the apical dendrite of the same neuron, illustrating the thorny excrescences more clearly (scale, 30 μm). From Johnston and Brown, 1983b.

urement of reversal potential and net charge injected by the mossy-fiber synapses should be quite accurate; the measurement of decay kinetics of the synaptic current should be less accurate, but still within 10 to 15% of the true values; while the measurement of the amplitude and the kinetics of the synaptic conductance could conceivably be in error by as much as 20% (Johnston and Brown, 1983b).

4. CURRENT- AND VOLTAGE-CLAMP STUDIES OF EVOKED SYNAPTIC EVENTS IN HIPPOCAMPAL NEURONS

4.1. *Analysis of Mossy-Fiber Evoked Synaptic Potentials*

Stimulation of the dentate granule cells usually elicits a complex synaptic waveform in CA3 pyramidal neurons. Often the complexity of the synaptic response cannot be appreciated fully until the membrane potential of the pyramidal neuron is changed from its resting level. When the postsynaptic cell is depolarized, the synaptic potential usually becomes clearly biphasic, consisting of an early depolarizing phase and a late hyperpolarizing phase (Figure 4). When the postsynaptic cell is depolarized by as little as 20 to 30 mV from normal resting potential, the synaptic waveform can sometimes appear to be entirely hyperpolarizing, and a false reversal potential for the depolarizing phase could be obtained (Figure 4).

The early excitatory component of the synaptic waveform represents the monosynaptic response, while the late component is likely due to feedforward or recurrent inhibition, and it can readily be blocked (although not necessarily specifically) by any of several pharmacological agents (picrotoxin, bicuculline, penicillin, and others). Following blockade of the late component, the early component can be shown to have a reversal potential of about 0 mV (Lebeda *et al.*, 1982; Johnston and Brown, 1983a; Brown and Johnston, 1983). The false reversal potential obtained with the mixed response is likely due to temporal overlap of the early and late components, the late component being much larger than the early component at depolarized potentials. The temporal overlap between the conductance changes associated with the early and late components cannot be determined from current-clamp measurements. Therefore, the properties of the early component cannot be investigated without proper separation from the late component.

Using pharmacological and/or ionic manipulations, the two phases of the complex synaptic waveform resulting from stimulation of the mossy fibers can be separated (Brown and Johnston, 1983). The early

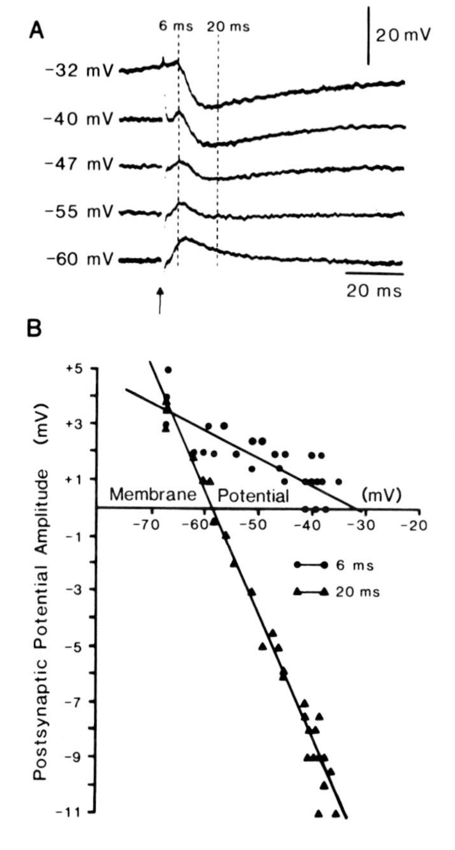

Figure 4. Synaptic potential waveform evoked by granule cell stimulation. The intracellular recording was from a CA3 neuron into which we had previously injected Cs^+. The membrane potential was changed by passing steady outward current while stimulating the granule cells at 0.5 Hz. (A) Changes in the synaptic potential amplitude and waveform are illustrated at several different membrane potentials. The dotted lines indicate 6 msec and 20 msec latencies from the beginning of the stimulus artifact (indicated by the arrow), at which time measurements were made. (B) The measurements made at latencies of 6 msec (circles) and 20 msec (triangles) are plotted as a function of the membrane potential. The solid lines represent the least-squares linear regression fits to the data points. Extrapolation of the line through the 6-msec data points would greatly underestimate the true reversal potential of the early excitatory component due to overlapping excitatory and inhibitory conductances (see text for details). From Brown and Johnston, 1983.

or excitatory component reverses at a potential close to 0 mV, as mentioned above, while the late or inhibitory component reverses at more negative potentials. For technical reasons, an accurate determination of the reversal potential of the inhibitory component has not been made, but the range is probably between about −45 and −65 mV (Brown and Johnston, 1983). The demonstration of an actual reversal of the entire excitatory component is significant because it effectively rules out any appreciable contribution to the waveform by an electrical synapse.

4.2. *Analysis of Mossy-Fiber Evoked Synaptic Currents*

Under voltage-clamp conditions, the biphasic nature of the synaptic currents evoked by stimulating the mossy fibers is quite clear (Figure 5). In many cells, the early currents can be reversed completely (Figure

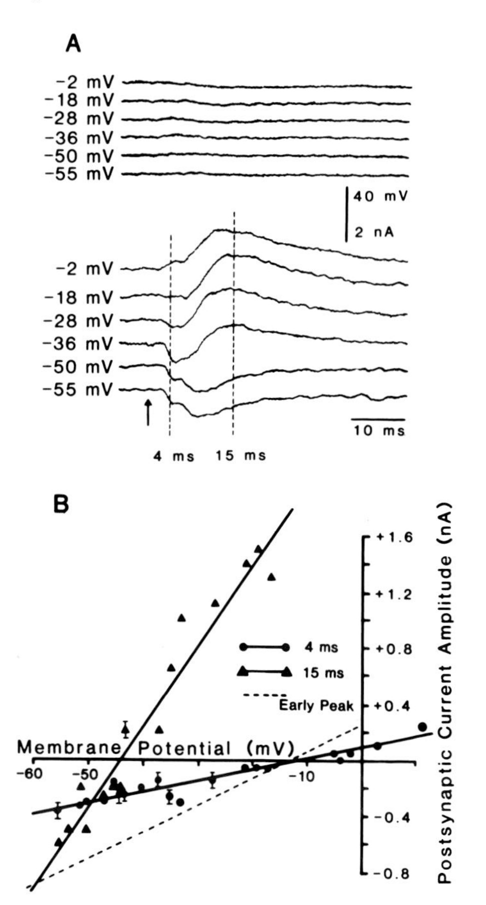

Figure 5. Reversal of the early synaptic currents. (A) Upper traces are membrane potential; lower traces are voltage-clamp current. The dotted lines indicate the times (4 and 15 msec) from the stimulus artifact onset (arrow) at which measurements were made. An actual reversal of the early current can be seen at the most depolarized holding potential (−2 mV). (B) Plots of the synaptic currents, measured at 4 and 15 msec, as a function of the membrane potential. The solid lines are the best fits to these data points; the broken line is the best fit to measurements made at the first peak in the current waveform. For clarity, the data for the latter measurements are not illustrated. When three or more measurements were made, the standard error bars are illustrated. From Brown and Johnston, 1983.

5). At the reversal potential for the early currents, the time course of the late currents can be determined. Measurement of the early and late currents as a function of the holding potential yields estimates of the underlying conductance changes. Both the excitatory and inhibitory components appear to result from a classical conductance-increase mechanism typical of many chemical synapses. The beginning of the conductance increase associated with the inhibitory component is delayed about 1 to 5 msec from the onset of the early excitatory conductance increase. For the excitatory component, the typical peak conductance increase was about 20 nS, while the peak conductance increase associated with the inhibitory component was about five times larger (see Brown and Johnston, 1983).

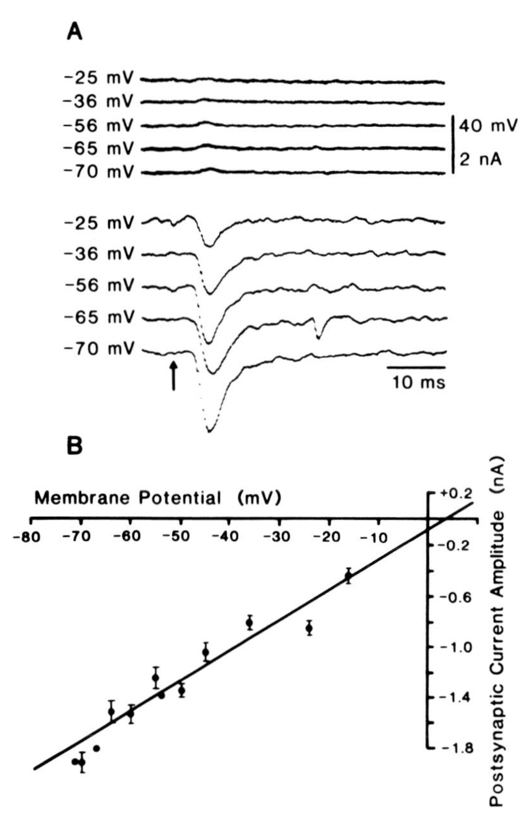

Figure 6. Evoked synaptic current apparently lacking an inhibitory component. (A) Upper traces are membrane potentials; lower traces are voltage-clamp currents. In the current trace obtained at a holding potential of −65 mV, a small spontaneous synaptic current is evident following the experimentally evoked current. The arrow indicates the onset of the stimulus artifact. (B) Plot of peak currents as a function of the holding potential. Each data point is the mean of two or more measurements. When three or more measurements were used to calculate the mean, the standard errors are indicated by the vertical bars. From Brown and Johnston, 1983.

Under fortuitous conditions, a purely excitatory current waveform can be obtained in CA3 neurons (see Figure 6). The excitatory current was shown to peak within about 2 msec and to decay exponentially with a time constant of 3 to 4 msec. The current waveform $I(t)$ can be fit closely by an alpha function (cf., Jack *et al.*, 1975)

$$I(t) = \kappa t \exp(-\alpha t) \tag{1}$$

using (in this case) $\alpha = 500/\text{sec}$ (see Brown and Johnston, 1983). Under the more usual circumstances, when the inhibitory component is present, it is necessary to use pharmacological agents to study the kinetics of the excitatory component (Figure 7). A rigorous examination of the excitatory conductance change in the presence of inhibitory blockers has not yet been performed, primarily because of the presently unknown effects of these agents on the channels associated with the excitatory component.

In summary, both current- and voltage-clamp studies of the mossy-fiber evoked synaptic input have provided consistent results. Both types of studies suggest that the excitatory and inhibitory components behave like classical synapses, in that a clear reversal potential can be demonstrated. In addition, the voltage-clamp results provided information about the temporal overlap of the two components of the usual synaptic response and demonstrated that both components result from a conductance-increase mechanism. It is apparent that reversal-potential measurements made (under current clamp) on the mixed response can be grossly in error. Also, the presence of a mixed response has important implications for efforts to perform a quantal analysis at these synapses (see Section 6.1). The large value of the inhibitory conductance increase, relative to that of the excitatory component, seems to be important for the normal functioning of the hippocampus (see Section 7.3).

5. CURRENT- AND VOLTAGE-CLAMP STUDIES OF SPONTANEOUS MINIATURE SYNAPTIC EVENTS IN HIPPOCAMPAL NEURONS

5.1. *The Quantum Hypothesis*

According to the quantum hypothesis (del Castillo and Katz, 1954; Katz, 1969), neurotransmitter substances are discharged from synaptic endings in the form of integral numbers of multimolecular packets or quanta. Quanta are also released spontaneously (in the absence of presynaptic nerve impulses), giving rise in the postsynaptic cell to the random occurrence of brief voltage perturbations termed *spontaneous miniature synaptic potentials*. Evoked synaptic potentials are proposed to result from the nearly synchronous release of some integral number of quanta, produced by the arrival of an action potential in the presynaptic nerve terminal. The average size of an evoked potential $\overline{V}$ is determined by two factors: the average number of quanta released, termed the mean quantal content m; and the mean amplitude of the postsynaptic response associated with a single quantal event, termed the mean quantal size $\bar{q}$.

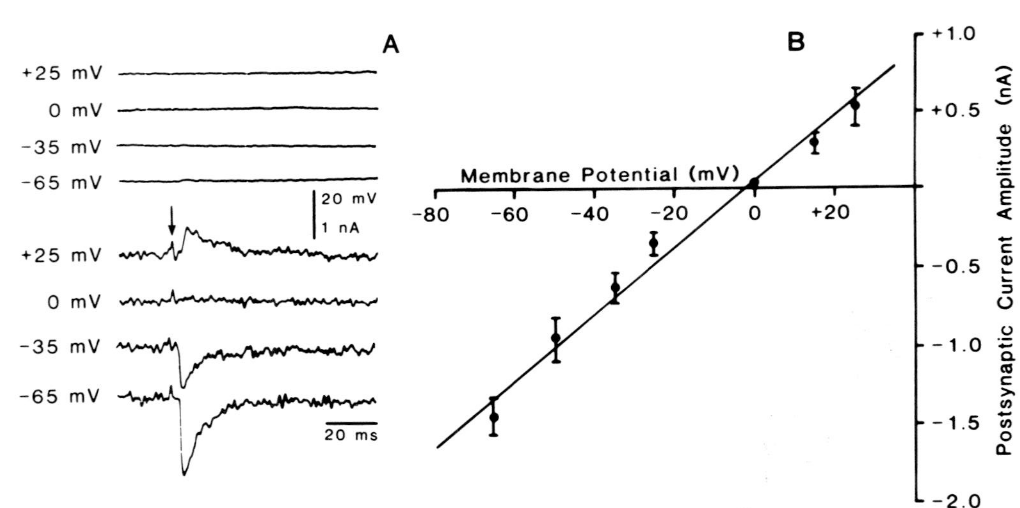

Figure 7. Reversal of an apparently pure excitatory synaptic current. The slice was bathed in 10 μM picrotoxin to block the inhibitory synaptic component, and recordings were done with a CsC1-filled micropipette. (A) Voltage-clamp records at four different holding potentials, indicated at the left of the traces. Upper traces show the membrane potential during the synaptic input. Note that the potential is well controlled. Lower traces are the corresponding clamp currents. Arrow indicates the stimulus artifact. The currents are clearly reversed at a holding potential of +25 mV. (B) Plot of the mean (± SE) peak synaptic current as a function of the membrane potential. Each mean is based on 4 to 14 trials. Averaging of the synaptic currents measured at each membrane potential reduced the scatter of data points about the regression line. The regression line gave a reversal potential of −1.9 mV and a conductance increase of 20.6 nS. From Brown and Johnston, 1983.

Thus, ignoring nonlinear summation of quantal responses (Martin, 1976; Stevens, 1976; McLachlan and Martin, 1981), we see that

$$\overline{V} = m\overline{q} \tag{2}$$

The quantum hypothesis and the techniques of quantal analysis have proven to be extraordinarily powerful tools for understanding the biophysics and microphysiology of synaptic transmission—especially in understanding the mechanisms responsible for *changes* in synaptic efficacy. In particular, the techniques of quantal analysis permit one to determine whether changes in synaptic strength are due to alterations in m or $\overline{q}$. Quantal analysis of several short-term forms of use-dependent increases in synaptic efficacy has revealed that they are due to increases in m and that they are therefore presynaptic in origin (Barrett and Magleby, 1976).

5.2. *Discovery of Spontaneous Miniature Synaptic Potentials in Hippocampal Neurons*

One must agree with Krnjevic's (1980) conclusion that there has been little convincing evidence that evoked transmitter release in the cortex is quantal in nature. His conclusion was based in part on the observation that previous studies had failed to show spontaneous quantal release in cortical neurons. Indeed, the occurrence of spontaneous miniature synaptic potentials is commonly taken as presumptive evidence for the quantal nature of synaptic transmission.

We were therefore very interested in knowing whether spontaneous miniature synaptic potentials could be detected in hippocampal neurons. The hippocampal slice preparation proved to be quite useful for such studies for two reasons. First, the slice affords the experimenter convenient control over the composition of the extracellular solution. By bathing the slices in solutions containing tetrodotoxin (TTX) and $MnCl_2$, it is possible to block all regenerative membrane responses and evoked (nerve-impulse-dependent) transmitter release. Thus, any remaining synaptic activity can be convincingly argued to represent nerve-impulse-independent transmitter release. Second, the technical advantages of the hippocampal slice preparation result in improved recordings over what can be conveniently obtained *in vivo*. The resultant more negative resting potentials and greater input resistances increase the amplitude of excitatory synaptic potentials, thereby increasing the chances of resolving spontaneous miniature potentials above the recording noise levels.

Intracellular recordings from well-impaled hippocampal neurons in the CA3 region typically reveal an incessant barrage of synaptic activity (Figure 8A). Of course, some of the synaptic activity could represent evoked (nerve-impulse-dependent) release. However, when the slices are bathed in solutions containing TTX and $MnCl_2$, we continue to see a high frequency of spontaneous synaptic activity (Figure 8B). The timing of occurrence of these spontaneous miniature synaptic potentials conforms approximately to a Poisson process, in that the intervals between successive events are approximately exponentially distributed (Figure 8C), and the number of events occurring in successive nonoverlapping time intervals is predictable from a Poisson law (Figure 8D).

5.3. *Current- and Voltage-Clamp Studies of Single Quantal Events*

In a subsequent series of experiments, we have begun to examine the properties of spontaneous miniature synaptic events under current- and voltage-clamp conditions using the SEC described in Section 3. The slices were continually perfused with normal saline to which was always added 1 μg/ml TTX and 2 to 5 mM $MnCl_2$ (substituted for $CaCl_2$) for the purpose of blocking all regenerative spiking activity and evoked (nerve-impulse-dependent) synaptic events. In many of these experiments, we used micropipettes that were filled with Cs_2SO_4. Injection of Cs^+ into the postsynaptic cells enabled us to change the membrane potential to extreme voltages (see Johnston *et al.*, 1980; Johnston and Brown, 1981, 1983a; Brown and Johnston, 1983).

Under current-clamp conditions, when the membrane potential was adjusted to voltages in the range of −125 to −50 mV (by passing inward or outward current via the microelectrode), the miniature potentials were always depolarizing. The amplitudes of the miniature potentials were sometimes as large as 12 mV when the cells were hyperpolarized. The smallest reliably detectable events were about 1 mV in amplitude. Some relatively large miniature potentials are shown in Figure 9A. When this same cell was depolarized to −5 mV, the miniature potentials were all hyperpolarizing, as shown in Figure 9B. The transition is shown in Figure 10, which illustrates typical records from the same cell at each of six different membrane potentials, ranging from −80 mV to +10 mV. As the cell was depolarized to −55 mV, the miniature potentials became smaller, and a larger proportion of the events appeared to be buried in the background noise. At a membrane potential of about −30 mV, the events were still smaller and were difficult to resolve above the noise.

Note, however, that when the cell was depolarized to −20 mV, the miniature potentials appeared to be hyperpolarizing. With further de-

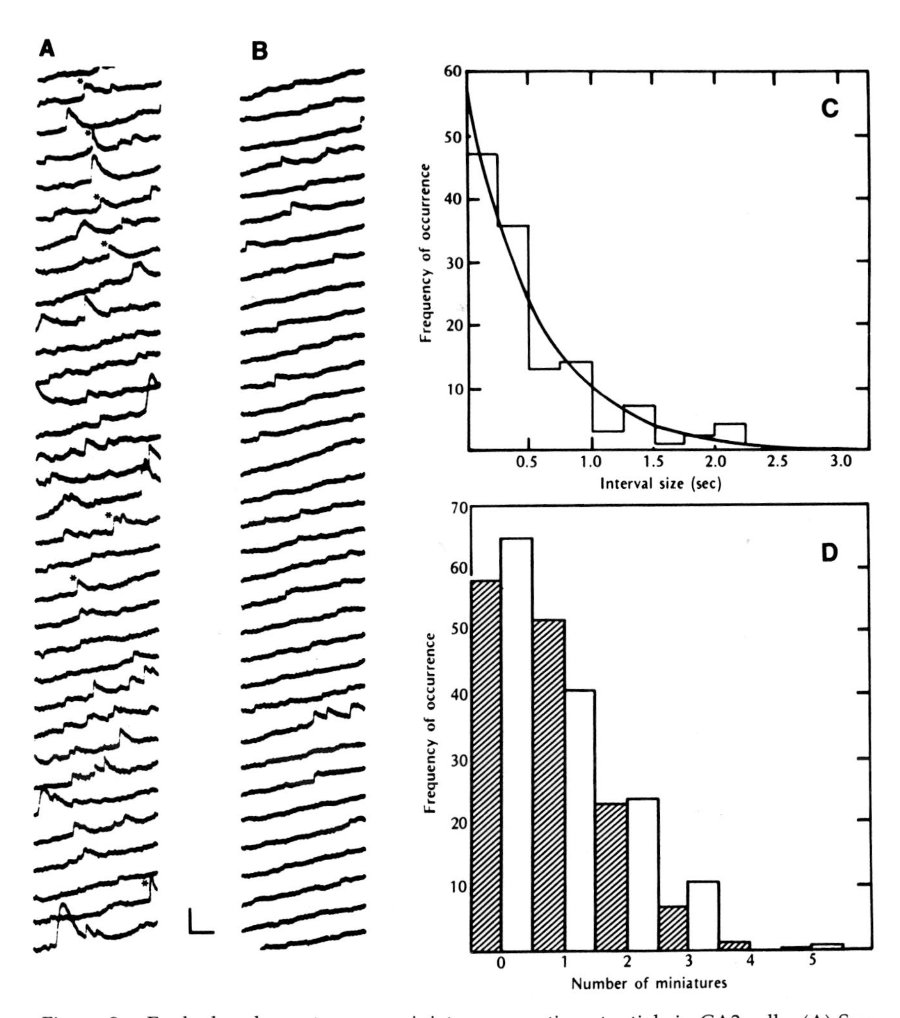

Figure 8. Evoked and spontaneous miniature synaptic potentials in CA3 cells. (A) Synaptic activity in a cell that was bathed in normal saline (lacking TTX and $MnC1_2$). Some of these synaptic potentials were evoked by stimulating the granule cells just above threshold at a point in time indicated by the asterisks. (B) Spontaneous miniature synaptic potentials recorded in another cell 3 hr after perfusion with the standard experimental saline (containing TTX plus $MnC1_2$). Calibration: 5 mV and 50 msec. (C) Frequency distribution of intervals between successive miniature potentials. The expected frequency distribution of intervals, based on an exponential density function, is plotted (solid curved line). (D) Frequency distribution of the number of miniature potentials occurring in successive 0.50-sec time bins (open histogram). The expected frequency distribution, based on a Poisson probability function, is also shown (hatched histogram). From Brown *et al.*, 1979.

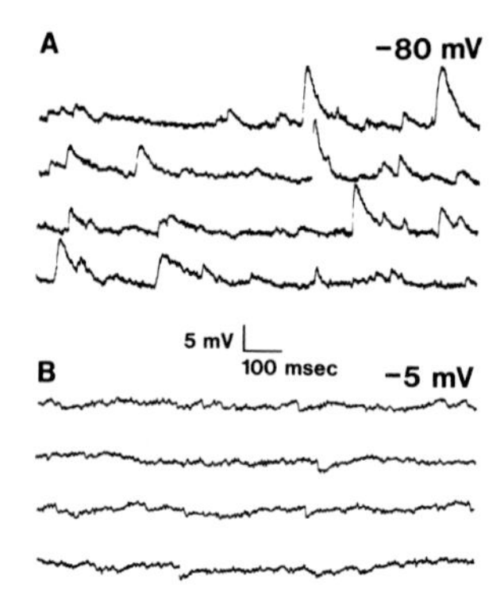

Figure 9. Examples of spontaneous miniature potentials at two different membrane potentials. The slice was bathed in the standard experimental saline (containing TTX and $MnCl_2$). (A) Relatively large amplitude depolarizing miniatures were evident at a membrane potential of −80 mV. (B) At a membrane potential of −5 mV, the miniature potentials were all hyperpolarizing. From Brown and Johnston, unpublished.

polarization to −10 mV and then +10 mV, the hyperpolarizing events became larger and more readily resolved above the background noise. These hyperpolarizing events were observed in all 45 cells that we examined under these conditions. Their rate of occurrence was commonly greater than 5 Hz, but the rate was difficult to determine accurately due to the relatively poor signal-to-noise ratio. In some cells, it was possible to detect the joint occurrence of both hyperpolarizing and depolarizing events at the same membrane potential (−45 to −30 mV), as illustrated in Figure 11.

In principle, the hyperpolarizing events might be either reversed excitatory miniatures or else a second population of inhibitory miniature potentials. To distinguish between these two possibilities, we performed parallel experiments (on an additional set of 49 cells) in which we added to the previously used experimental saline one of several agents known to block synaptic inhibition in the hippocampus. These blocking agents included bicuculline (10 μM), picrotoxin (10 μM), and penicillin (3.4 mM).

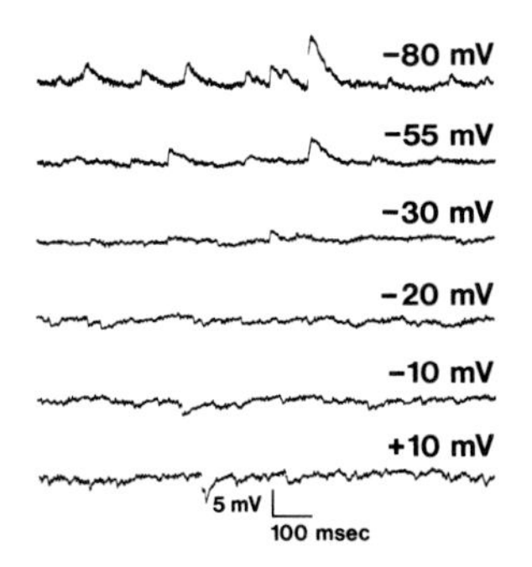

Figure 10. The transition in the amplitude and polarity of the miniature potentials, as a function of the membrane potential, is illustrated for the same cell shown in Figure 9. As the membrane was depolarized from −80 mV to −30 mV, both the amplitude and frequency of depolarizing events declined. At membrane potentials from −20 to +10 mV, the miniatures are all hyperpolarizing, increasing in amplitude and apparent frequency with depolarization. From Brown and Johnston, unpublished.

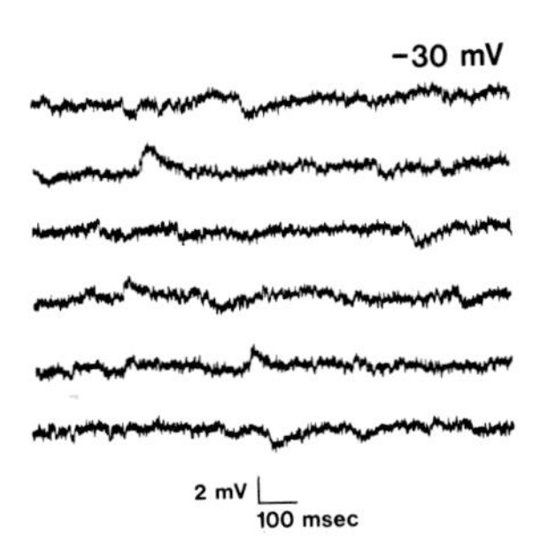

Figure 11. Illustration of the joint occurrence of both hyperpolarizing and depolarizing miniature potentials at the same membrane voltage (−30 mV). The slice was bathed in the standard experimental saline. This was observed in several cells at membrane potentials between −45 and −30 mV. From Brown and Johnston, unpublished.

In the presence of the inhibitory blockers, we saw the usual depolarizing miniatures when the membrane potentials were held in the range of −125 to −50 mV. As before, the amplitude of these events decreased as the cells were depolarized. However, as we continued to depolarize the cells, there was no hint of hyperpolarizing potentials. Instead, when the cells were depolarized to values close to 0 mV, the voltage traces were quite flat (Figure 12A). This result suggested that the hyperpolarizing potentials that we normally observe at depolarized potentials are inhibitory miniatures rather than reversed excitatory miniatures. The effect of inhibitory blocking agents was reversible; following a change to the standard experimental saline (containing TTX and $MnCl_2$ but no inhibitory blocking agents), hyperpolarizing events always reappeared when the cells were maintained at depolarized potentials.

To determine whether it would be possible to reverse the excitatory miniatures, we attempted to depolarize some of these cells to extreme potentials. An example of a recording from a cell that we were able to depolarize to +75 mV is shown in Figure 12B. This is the same cell that

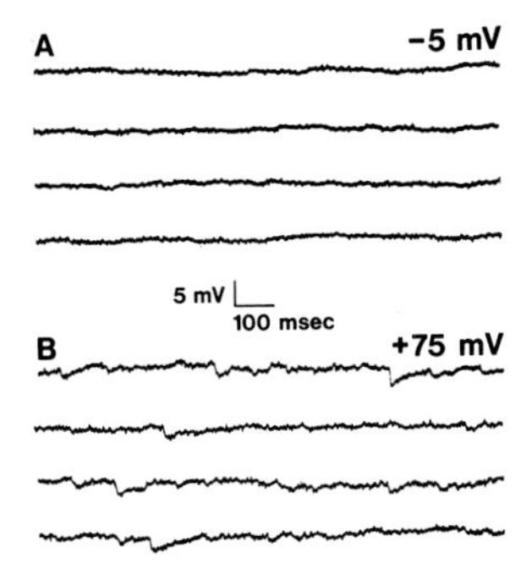

Figure 12. Recordings from an exceptional cell that we were able to depolarize to extreme potentials. The slice was bathed in the standard experimental saline plus penicillin. (A) When the membrane was maintained at potentials between −110 mV and −50 mV, we saw the usual depolarizing miniature potentials. However, at −5 mV the traces were quite flat, as indicated. The hyperpolarizing events that are normally evident at this potential were entirely blocked. (B) When the cell was depolarized to +75 mV, we were able to see hyperpolarizing miniatures. These events may be reversed excitatory miniatures. From Brown and Johnston, unpublished.

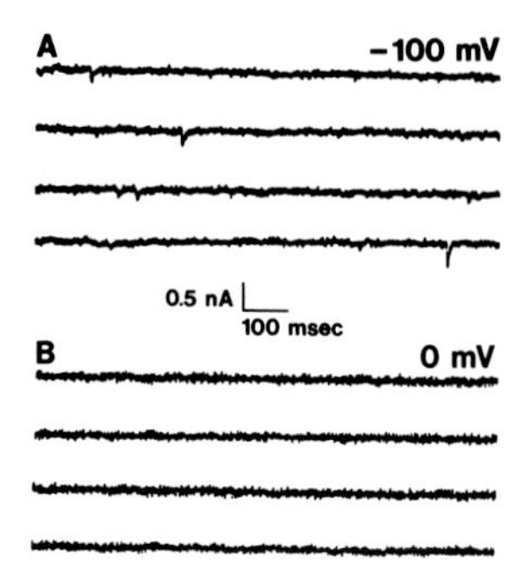

Figure 13. Records of miniature currents from a slice bathed in the standard experimental saline plus penicillin. (A) At a membrane potential of −100 mV, the usual inward currents were present. (B) When the cell was depolarized to 0 mV, the typical outward currents were abolished. Instead, the voltage trace became quite flat. By contrast, in the absence of inhibitory blocking agents, outward currents were routinely observed at this potential. From Brown and Johnston, unpublished.

was illustrated in Figure 12A. As indicated, when the cell was depolarized to +75 mV, there were clear hyperpolarizing events.

Under voltage-clamp conditions, when the membranes were held at potentials between −125 and −50 mV, the miniature currents were always inward, and the peak amplitudes rarely exceeded 0.5 nA. Within this voltage range, the amplitude of the currents decreased as the cells were depolarized. An example of some miniature currents recorded in a cell that was voltage clamped at -100 mV is shown in Figure 13A. These records were from a slice that was bathed in saline containing picrotoxin. As indicated in Figure 13B, when the cell was depolarized to 0 mV, outward currents were not observed. However, in slices that were not bathed in saline containing inhibitory blocking agents, outward currents were routinely observed in cells that were voltage clamped at potentials near 0 mV.

These results suggest that we can resolve, under both current- and voltage-clamp conditions, two classes of spontaneous quantal events—excitatory and inhibitory—and that the latter can be blocked pharmacologically with any of several agents. Unfortunately, in these early preliminary studies, the signal-to-noise ratios under both current- and voltage-clamp conditions were rather poor. We have subsequently developed methods to improve greatly the signal-to-noise ratio under both current- and voltage-clamp conditions. It should now be possible to perform more quantitative and analytical studies of the quantal nature of synaptic transmission in hippocampal neurons (see below).

6. IMPLICATIONS FOR PERFORMING A QUANTAL ANALYSIS OF EVOKED RELEASE

There are four traditional intracellular methods for performing a quantal analysis of evoked release—the method of failures, the variance method, the so-called "direct" method, and techniques based on deconvolution or Fourier-type analysis (see Martin, 1977; Jack *et al.*, 1981; McNaughton *et al.*, 1981). The results presented in the previous sections have important implications for efforts to apply these methods to hippocampal synapses.

6.1. *The Problem of Mixed Synaptic Responses*

As we demonstrated in Section 4, the usual experimentally evoked synaptic response consists of a sequence of two temporally overlapping conductance increases. The observed synaptic potential is the result of an early, monosynaptic, excitatory response and a delayed, disynaptic or polysynaptic, inhibitory response. Obviously, no form of quantal analysis based on response fluctuations (the variance method or deconvolution or Fourier-type techniques) can be meaningfully applied to such an aggregate response. The use of these techniques to analyze the monosynaptic response requires that the normal concomitant inhibitory component be eliminated if one is to obtain a meaningful estimate of m and q. As we have shown, the inhibitory component can be readily blocked using a variety of techniques.

6.2. *Some Advantages of Voltage-Clamp Analysis*

The fact that we can apply voltage-clamp techniques to certain hippocampal synapses is significant for several of the methods of quantal analysis, particularly those based on response fluctuations. First, voltage clamping eliminates the problem of nonlinear quantal summation, for which adequate correction procedures are difficult to apply in practice (Martin, 1976; Stevens, 1976; McLachlan and Martin, 1981). Second, voltage clamping eliminates the problem of nonohmic postsynaptic membrane responses. Such nonohmic membrane responses can be particularly troublesome in current-clamp studies of synaptic response fluctuations in hippocampal neurons. Third, any inhibitory contaminants in the postsynaptic response are clearly revealed under voltage-clamp conditions. Fourth, by performing the analysis on the integral of the synaptic currents (to obtain the net charge transfer) the signal-to-noise ratio becomes very favorable (see below), and the problem of quantal latency fluctuations (Williams and Bowen, 1974) is eliminated. Fifth, charge measured by a voltage clamp in the soma can be an accurate measure of the charge injected by the synapse (Carnevale and Johnston, 1982). In particular, errors arising from the electrotonic distances of the mossy-fiber synapses from the recording microelectrode should be very small when performing the analysis on charge measurements (Brown and Johnston, 1983; Johnston and Brown, 1983b).

6.3. *The Importance of a Suitable Signal-to-Noise Ratio*

Having a suitable signal-to-noise ratio for single quantal events is important regardless of the method of quantal analysis. The exact requirements depend on the particular method. To use the method of failures, one must be able to distinguish clearly single quantal releases above the background noise levels. The requirements for the variance method are possibly less stringent, depending on the required accuracy. Until now, we felt that the signal-to-noise ratio was not adequate. However, the success we have had recently in further improving the signal-to-noise ratio has encouraged us to begin performing a quantal analysis of evoked release.

6.4. *A Caveat Concerning Release and Quantal Size Statistics*

Some of the considerations discussed above suggest the feasibility of performing a meaningful quantal analysis of evoked release in hippocampal neurons. However, we hasten to point out that there remain numerous pitfalls in applying any of the traditional methods of quantal analysis to cortical neurons, where the true probability release function may not be a simple Poisson or binomial law (see Brown *et al.*, 1976; Hatt and Smith, 1976; Barton and Cohen, 1977; Zucker, 1977; McLachlan, 1978; Perkel and Feldman, 1979; Bennett and Lavidis, 1979; Jack *et al.*, 1981) and the quantal size density distribution (see McLachlan, 1978) for the synapses of interest has not yet been well characterized. A complete discussion of these issues will be considered elsewhere.

7. SIGNIFICANCE FOR SELECTED PROBLEMS IN CORTICAL PHYSIOLOGY

The work described here was motivated by the belief that there has been a pressing need to develop the capability of studying the biophysics and microphysiology of cortical synapses. The implications of such an approach are probably best appreciated with a few concrete examples. In the following subsections, we have therefore selected three problems in cortical neurophysiology that we find interesting and that now appear amenable to quantitative analysis.

7.1. *Mechanism of Long-Term Synaptic Potentiation*

There has been considerable interest in homosynaptic and associative long-term synaptic potentiation (LTP) as possible synaptic substrates for information storage in the nervous system (see Chung, 1977; Bliss, 1979; Bliss and Dolphin, 1982; Brown and McAfee, 1982; Brown and Barrionuevo, submitted; Levy and Desmond, 1983). Homosynaptic LTP is an enchanced synaptic efficacy that can be produced by brief tetanic stimulation of a synaptic input and that can last for hours, days, or weeks. Indeed, some feel it may be permanent. Associative LTP is a possibly related long-term enhanced synaptic efficacy that occurs in one (weak) synaptic input only if it is tetanically stimulated in conjunction with nearly concurrent stimulation of a second (strong) synaptic input. In our intracellular recordings performed on the hippocampal slice, we find that both homosynaptic and associative LTP can last for hours, often with little or no decrement.

Until now, most of the research on homosynaptic and associative LTP utilized extracellular recordings, so that almost nothing is known about the underlying neurophysiological mechanisms. Voltage-clamp studies will be needed to determine the cause of the increased synaptic efficacy that we observe intracellularly (Figure 14). For example, are the net inward synaptic currents increased and/or are there changes in the postsynaptic active or passive membrane properties? If the synaptic currents are larger, is this due to a change in the conductance for the excitatory and/or inhibitory components of the synaptic response? Alternatively, is the reversal potential for the excitatory and/or inhibitory component altered? Clearly, these possibilities are not necessarily mutually exclusive.

Quantal analysis could help distinguish among certain hypotheses for the underlying mechanisms. As indicated earlier, a number of use-dependent increases in synaptic efficacy—facilitation, augmentation, and posttetanic potentiation—have now been shown by quantal analysis to be presynaptic in origin (Barrett and Magleby, 1976). In the case of each of these relatively short-term increases in synaptic efficacy, the mechanism has involved a greater amount of transmitter release, due to an increase in m (with no change in $\bar{q}$). According to one hypothesis (see Bliss and Dolphin, 1982), LTP is also due to an increase in the amount of transmitter release. However, an alternative hypothesis (Baudry and Lynch, 1980) has proposed that LTP is due to an increase in the number of postsynaptic neurotransmitter receptors. In this case, we would predict an increase in $\bar{q}$ with no change in m. Of course, it is possible that both hypotheses are correct, in which case we could see an increase in both m and $\bar{q}$.

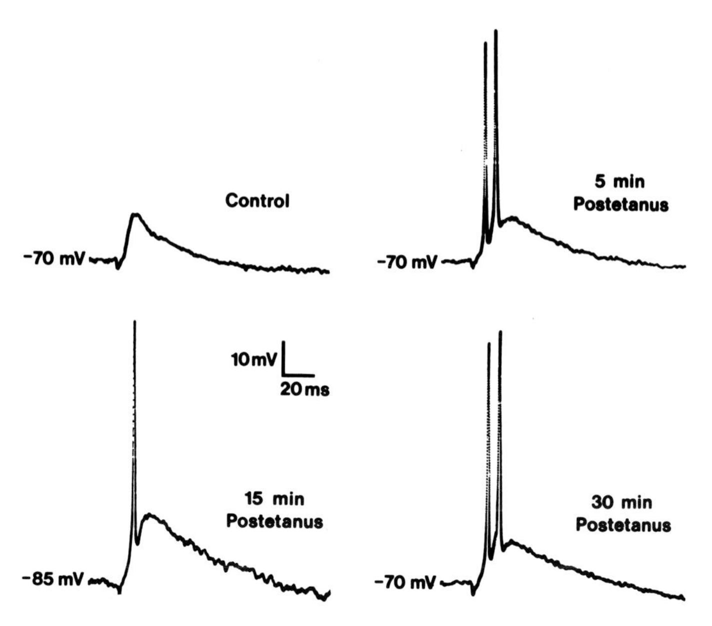

Figure 14. Illustration of homosynaptic LTP in a CA3 pyramidal neuron. Upper left: Representative synaptic potential recorded prior to the induction of LTP. The synaptic response is subthreshold. Upper right: A typical example of the synaptic response taken 5 min after tetanic stimulation (100 Hz for 1 sec) of the afferent pathway. Note that the synaptic response is now suprathreshold, eliciting two action potentials. Lower right: 30 min later the synaptic response is still suprathreshold and appears to have shown little or no decrement. Lower left: It will be noted that the LTP that we observe here is not a subtle effect, in that when the cell is hyperpolarized by a full 15 mV, the synaptic response is still suprathreshold for eliciting a single action potential. The action potential amplitudes are greatly attenuated due to filtering. From Brown and Johnston, unpublished.

7.2. *Role of Dendrites and Their Spines in Synaptic Information Transfer and Integration*

The complex morphology of cortical dendrites unquestionably plays an important role in the overall information processing of these cells (see Rall and Rinzel, 1973; Barrett and Crill, 1974; Rinzel and Rall, 1974; Jack *et al.*, 1975; Redman, 1976; Rall, 1977; Shepherd, 1979; Eaton, 1980; Brown *et al.*, 1981). Understanding the role of this complex geometry has, historically, been an intellectually intriguing problem. Even the fine details of the morphology, such as the presence of dendritic spines, have long invited theoretical speculations (Scheibel and Scheibel, 1968; Valverde and Ruiz-Marcos, 1969; Diamond *et al.*, 1971; Rall, 1974, 1978; Swindale, 1981; Crick, 1982). According to one popular notion, changes in spine shape could (via a reduction in the axial spine resistance) increase synaptic efficacy and thereby serve as an anatomical substrate for learning and memory (Rall, 1974, 1978; Jack *et al.*, 1975; Crick, 1982). Spine-shape changes have now been reported to accompany LTP and a variety of learning or social experiences (see Van Harreveld and Fifkova, 1975; Fifkova and Van Harreveld, 1977; Coss and Globus, 1978, 1979; Coss *et al.*, 1980; Lee *et al.*, 1980; Berard *et al.*, 1981; Burgess and Coss, 1980, 1981; Desmond and Levy, 1981; Fifkova and Anderson, 1981; Anderson and Fifkova, 1982; Coss and Brandon, 1982; Fifkova and Delay, 1982).

Although the mathematical theory required to address these cable-theory problems has been available for some time, the physiological data required for performing realistic computer simulations were missing. To perform such simulations, one must know the synaptic conductance waveform and its reversal potential (Rall, 1967; Jack and Redman, 1971; Johnston and Brown, 1983b; Segev and Parnas, 1983). The application of voltage-clamp techniques to evoked as well as single quantal synaptic events in hippocampal neurons has provided this key piece of information, which is necessary for performing theoretical calculations of the expected role of dendrites and their spines in synaptic information transfer and integration. Such information, when incorporated into realistic simulations, may provide a basis for evaluating competing hypotheses (for example, on the role of spines), and the simulations are certain to provide insights that will help direct further experimental research.

A related problem, which was raised earlier (Section 1), concerns the effect on synaptic signaling of spatial inhomogeneities in the passive and active conductances of the dendritic and somatic membrane. Obtaining the pertinent experimental data appears, at first, to be an insurmountable problem. However, the recent successful application of patch-clamp techniques to selected portions of hippocampal membranes (Gray *et al.*, 1982) has opened up several new experimental opportunities. This technique obviously affords extremely precise spatial resolution of membrane properties.

7.3. *Currents Underlying Epileptiform Discharges*

Epilepsy is a potentially debilitating disease that affects up to 2% of the population. Understanding the underlying mechanisms could lead to more rational treatment and, in some cases, perhaps even prevention of the onset of the disease. One of the more extensively studied animal models of epilepsy involves topical application to the cortex of convulsant agents such an penicillin, bicuculline or picrotoxin (Ayala *et al.*, 1973; Prince, 1978; Johnston and Brown, 1983a). Following this treatment, periodic interictal epileptiform discharges are recorded in the electrocorticogram. The intracellular correlate of this interictal discharge is a burst, which has been termed a paroxysmal depolarizing shift (PDS) (Matsumoto and Ajmone Marsan, 1964). The PDS is characterized by a sudden 20 to 50 mV depolarization of the membrane potential, lasting 100 msec or longer (Ayala *et al.*, 1973).

The mechanism underlying the PDS has been extensively studied because it is thought to bear directly on our understanding of the cellular basis of epilepsy (Dichter and Spencer, 1969; Ayala *et al.*, 1973; Prince, 1978; Johnston and Brown, 1983a). The most influential hypothesis regarding the mechanism of PDS generation maintains that the PDS is a *network-driven burst* that is sustained by an underlying giant compound excitatory postsynaptic potential (EPSP) (Ayala *et al.*, 1973). The network synchronization was proposed to result from an imbalance between recurrent excitation and inhibition. A recently proposed alternative hypothesis maintains that the PDS is simply an *endogenous burst* triggered by synaptic events of the usual amplitude (Prince, 1978; Schwartzkroin and Wyler, 1980).

Using the biophysical techniques that we devised for the *in vitro* hippocampal slice, it was possible to test four quantitative predictions (described in Johnston and Brown, 1981, 1983a; Lebeda *et al.*, submitted) of the giant EPSP hypothesis. The results were unequivocal. All four predictions of the giant EPSP hypothesis were confirmed, and the results were not reconcilable with the alternative hypothesis. Some of these same techniques were subsequently applied (Gutnick *et al.*, 1982) to evaluate the PDS in neocortex (the hippocampus is archicortex, evolutionarily intermediate between paleocortex and neocortex). Again, the results confirmed the predictions of the giant EPSP hypothesis.

Based on what is known about network-driven bursting in general and about hippocampal physiology in particular, it is not difficult to provide a plausible working hypothesis (for details, see Johnston and Brown, 1983a) for the origin of interictal discharges in hippocampus. In the CA3 subfield of the hippocampus, reciprocal excitatory synapses are thought to exist (Lebowitz *et al.*, 1971; MacVicar and Dudek, 1980). In addition, there are powerful recurrent or feedforward inhibitory connections, which are blocked by any of several convulsant agents (Brown and Johnston, 1980; Dingledine and Gjerstad, 1980; Lebeda *et al.*, 1982, submitted; Johnston and Brown, 1980, 1983a). Blockade of the normal concomitant inhibitory synaptic conductances would leave a self-excitatory (positive feedback) network among the CA3 neurons. The neuronal activity in such a network of hippocampal neurons is easily shown to tend toward synchronization (see Traub and Wong, 1982). These considerations underscore the importance of the normal powerful inhibitory conductances (described in Section 4) for the proper functioning of the hippocampus and probably other regions of cortex in which there are networks with strong recurrent excitation.

7.4. *Summary and Conclusions*

Cortical synaptic networks display some interesting and possibly even unique features. We have therefore felt that it was worth the time investment to develop the capability of applying to these synapses the same powerful electrophysiological methods that have proven so valuable in studies of the three best understood synapses. The particular synaptic preparation that we have selected—the *in vitro* hippocampal mossy-fiber synapse—now appears amenable to voltage-clamp and quantal analysis. Application of these techniques to mossy-fiber synapses has already provided some important insights, and we can reasonably expect rapid progress to be made in understanding some of the intriguing characteristics of these synapses.

ACKNOWLEDGMENTS. This work was supported by NIH grants NS11535, NS15772, NS18295, and NS18861 and McKnight Foundation Scholar's and Development Awards.

8. REFERENCES

Amaral, D. G. and Dent, J. A., 1981, Development of the mossy fibers of the dentate gyrus. I. A light and electron microscopic study of the mossy fibers and their expansions, *J. Comp. Neurol.* **195:**51–86.

Anderson, C. L. and Fifkova, E., 1982, Morphological changes in the dentate molecular layer accompanying long-term potentiation, *Soc. Neurosci. Abstr.* **8:**279.

Araki, T. and Terzuolo, C. A., 1962, Membrane currents in spinal motoneurons associated with the action potential and synaptic activity, *J Neurophysiol.* **25:**772–789.

Ayala, G. F., Dichter, M., Gumnit, R. J., Matsumoto, H., and Spencer, W. A., 1973, Genesis of epileptic interictal spikes. New knowledge of cortical feedback systems suggests a neurophysiological explanation of brief paroxysms, *Brain Res.* **52:**1–17.

Barrett, E. F. and Magleby, K. L., 1976, Physiology of the cholinergic transmission, in: *Biology of Cholinergic Function,* (A. M. Goldberg and I. Hanin, eds.), Raven Press, New York, pp. 29–100.

Barrett, J. N. and Crill, W. E., 1974, Influence of dendritic location and membrane properties on the effectiveness of synapses on cat motoneurones, *J. Physiol. (London)* **239:**325–345.

Barrett, J. N. and Crill, W. E., 1980, Voltage clamp of cat motoneurone somata: Properties of the fast inward current, *J. Physiol. (London)* **304:**231–249.

Barrett, E. F., Barrett, J. N., and Crill, W. E., 1980, Voltage-sensitive outward currents in cat motoneurones, *J. Physiol. (London)* **304:**251–276.

Barton, S. B. and Cohen, I. S. 1977, Are transmitter release statistics meaningful? *Nature (London)* **268:**267–268.

Baudry, M. and Lynch, G., 1980, Hypothesis regarding the cellular mechanism responsible for long-term synaptic potentiation in the hippocampus, *Exp. Neurol.* **68:**202–204.

Benardo, L. S., Masukawa, L. M., and Prince, D. A., 1982, Electrophysiology of isolated hippocampal pyramidal dendrites, *J. Neurosci.* **2:**1614–1622.

Bennett, M. R. and Lavidis, N. A., 1979, The effect of calcium ions on the secretion of quanta evoked by an impulse at nerve terminal release sites, *J. Gen. Physiol.* **74:**429–456.

Berard, D. R., Burgess, J. W., and Coss, R. G., 1981, Plasticity of dendritic spine formation: A state-dependent stochastic process, *Int. J. Neurosci.* **13:**93–98.

Blackstad, T. W. and Kjaerheim, A., 1961, Special axo-dendritic synapses in the hippocampal cortex: Electron and light microscopic studies on the layer of mossy fibers, *J. Comp. Neurol.* **117:**133–159.

Blackstad, T. W., Brink, K., Hem, J., and Jeune, B., 1970, Distribution of hippocampal mossy fibers in the rat. An experimental study with silver impregnation methods, *J. Comp. Neurol.* **138:**433–450.

Bliss, T. V. P., 1979, Synaptic plasticity in the hippocampus, *Trends Neurosci.* **2:**42–45.

Bliss, T. V. P. and Dolphin, A. C., 1982, What is the mechanism of long-term potentiation in the hippocampus? *Trends Neurosci.***5:**289–290.

Bliss, T. V. P. and Gardner-Medwin, A. R., 1973, Long-lasting potentiation of synaptic transmission in the dentate area of the unanaesthetized rabbit following stimulation of the perforant path, *J. Physiol. (London)* **232:**357–374.

Bliss, T. V. P. and Lomo, T., 1973, Long-lasting potentiation of synaptic transmission in the dentate area of the anaesthetized rabbit following stimulation of the perforant path, *J. Physiol. (London)* **232:**331–356.

Brown, T. H. and Barrionuevo, G. Associative long-term synaptic potentiation in hippocampal slices, submitted.

Brown, T.H. and Johnston, D., 1980, Two classes of miniature synaptic potentials in CA3 hippocampal neurons, *Soc. Neurosci. Abstr.* **164:**10.

Brown, T.H. and Johnston, D., 1982, Electrotronic localization of hippocampal mossy fiber synapses, *Soc. Neurosci. Abstr.* **8:**380.

Brown, T. H. and Johnston, D., 1983, Voltage-clamp analysis of mossy fiber synaptic input to hippocampal neurons, *J. Neurophysiol.* **50:**487–507.

Brown, T. H. and McAfee, D. A., 1982, Long-term synaptic potentiation in superior cervical ganglion, *Science* **215:**1411–1413.

Brown, T. H., Perkel, D. H., and Feldman, M. W., 1976, Evoked neurotransmitter release: Statistical effects of nonuniformity and nonstationarity, *Proc. Natl. Acad. Sci. USA* **73:**2913–2917.

Brown, T. H., Wong, R. K. S., and Prince, D. A., 1979, Spontaneous miniature synaptic potentials in hippocampal neurons, *Brain Res.* **174:**194–199.

Brown, T. H., Fricke, R. A., and Perkel, D. H., 1981, Passive electrical constants in three classes of hippocampal neurons, *J. Neurophysiol.* **46:**812–827.

Burgess, J. W. and Coss, R. G., 1980, Crowded jewel fish show changes in dendritic spine density and spine morphology, *Neurosci. Lett.* **17:**277–281.

Burgess, J. W. and Coss, R. G., 1981, Short-term juvenile crowding arrests the developmental formation of dendritic spines on tectal interneurons in jewel fish, *Dev. Psychobiol.* **14:**389–396.

Carnevale, N. T. and Johnston, D., 1982, Electrophysiological characterization of remote chemical synapses, *J. Neurophysiol.* **47**:606–621.

Chung, S-H., 1977, Synaptic memory in the hippocampus. *Nature* (*London*) **266**:677–678.

Cornwall, M. C. and Thomas, M. V., 1981, Glass microelectrode tip capacitance: Its measurement and a method for its reduction, *J. Neurosci. Meth.* **3**:225–232.

Coss, R. G. and Brandon, J. G., 1982, Rapid changes in dendritic spine morphology during the honeybee's first orientation flight, in: *The Biology of Social Insects,* (M. D. Breed, C. D. Michener, and H. E. Evans, eds.), Westview Press, Boulder, Colorado, pp. 338–342.

Coss, R. G. and Globus, A., 1978, Spine stems on tectal interneurons in jewel fish are shortened by social stimulation, *Science* **200**:787–790.

Coss, R. G. and Globus, A., 1979, Social experience affects the development of dendritic spines and branches on tectal interneurons in the jewel fish, *Dev. Psychobiol.* **12**:347–358.

Coss, R. G., Brandon, J. G., and Globus, A., 1980, Changes in morphology of dendritic spines on honeybee calycal interneurons associated with cumulative nursing and foraging experiences, *Brain Res.* **192**:49–59.

Crick, F., 1982, Do dendritic spines twitch? *Trends Neurosci.* **5**:44–46.

del Castillo, J. and Katz, B. 1954, Quantal components of the end-plate potential, *J. Physiol.* (*London*) **124**:560–573.

Desmond, N. L. and Levy, W. B., 1981, Ultrastructural and numerical alterations in dendritic spines as a consequence of long-term potentiation, *Anat. Rec.* **199**:68.

Diamond, J., Gray, E. G., and Yasargil, G. M., 1971, The function of the dendritic spine: An hypothesis, in: *Excitatory Synaptic Mechanism* (P. Andersen and K. Jansen eds.), Universitetsforlaget, Oslo, pp. 213–222.

Dichter, M. and Spencer, W. A., 1969, Penicillin-induced interictal discharges from the cat hippocampus. I. Characteristics and topographical features, *J. Neurophysiol.* **32**:649–662.

Dingledine, R. and Gjerstad, L., 1980, Reduced inhibition during epileptiform activity in the *in vitro* hippocampal slice, *J. Physiol.* (*London*) **305**:297–313.

Eaton, D., 1980, How are the membrane properties of individual neurons related to information processing in neural circuits? in: *Information Processing in the Nervous System* (H. M. Pinsker and W. D. Willis, Jr., eds.), Raven Press, New York, pp. 39–57.

Fifkova, E. and Anderson, C., 1981, Stimulation-induced changes in dimensions of stalks of dendritic spines in the dentate molecular layer, *Exp. Neurol.* **74**:621–627.

Fifkova, E. and Delay, R. J., 1982, Cytoplasmic actin in dendritic spines as a possible mediator of synaptic plasticity, *Soc. Neurosci. Abstr.* **8**:279.

Fifkova, E. and Van Harreveld, A., 1977, Long-lasting morphological changes in dendritic spines of dentate granular cells, following stimulation of the entorhinal area, *J. Neurocytol.* **6**:211–230.

Frank, K., Fuortes, M. G. F., and Nelson, P. G., 1959, Voltage clamp of motoneuron soma, *Science* **130**:38–39.

Gray, R., Kellaway, J., and Johnston, D., 1982, Electrical properties of acutely isolated hippocampal neurons, *Physiologist* **25**:221.

Gutnick, M. J., Connors, B. W., and Prince, D. A., 1982, Mechanisms of cortical epileptogenesis *in vitro, J. Neurophysiol.* **48**:1321–1335.

Hagiwara, S. and Saito, N., 1959, Membrane potential change and membrane current in supramedullary nerve cell of puffer, *J. Neurophysiol.* **22**:204–221.

Hamlyn, L. H., 1961, Electron microscopy of mossy fibre endings in Ammon's horn, *Nature* (*London*) **190**:645–648.

Hamlyn, L. H., 1962, The fine structure of the mossy fibre endings in the hippocampus of the rabbit. *J. Anat.* **96:**112–126.

Hatt, H. and Smith, D. O., 1976, Nonuniform probabilities of quantal release at the crayfish neuromuscular junction, *J. Physiol.* (*London*) **259:**395–404.

Haug, F.-M. S., 1967, Electron microscopic localization of zinc in hippocampal mossy fiber synapses by a modified sulfide silver procedure, *Histochemie* **8:**355–368.

Jack, J. J. B. and Redman, S. J., 1971, The propagation of transient potentials in some linear cable structures, *J. Physiol.* (*London*) **215:**283–320.

Jack, J. J. B., Noble, D., and Tsien, R. W., 1975, *Electric Current Flow in Excitable Cells,* Oxford University Press, London.

Jack, J. J. B., Redman, S. J., and Wong, K., 1981, The components of synaptic potentials evoked in spinal motoneurones by impulses in a single group Ia afferents, *J. Physiol.* (*London*) **321:**65–96.

Johnston, D., 1981, Passive cable properties of hippocampal CA3 pyramidal neurons, *Cell. Mol. Neurobiol.* **1:**41–55.

Johnston, D. and Brown, T. H., 1980, Miniature inhibitory and excitatory synaptic potentials in hippocampal neurons, *Fed. Proc.* **39:**2071.

Johnston, D. and Brown, T. H., 1981, Giant synaptic potential hypothesis for epileptiform activity, *Science* **211:**294–297.

Johnston, D. and Brown, T. H., 1983a, Mechanism of neuronal burst generation, in: *Electrophysiology of Epilepsy* (P. A. Schwartzkroin and H. V. Wheal, eds.), Academic Press, New York, in press.

Johnston, D. and Brown, T. H., 1983b, Interpretation of voltage-clamp measurements in hippocampal neurons, *J. Neurophysiol.* **50:**464–486.

Johnston, D., Hablitz, J. J., and Wilson, W. A., 1980, Voltage clamp discloses slow inward current in hippocampal burst-firing neurones, *Nature* (*London*) **286:**391–393.

Katz, B., 1969, *The Release of Neuronal Transmitter Substances,* Charles A. Thomas, Springfield, Illinois.

Krnjevic, K., 1980, Neurobiology. General principles related to epilepsy, in: *Antiepileptic Drugs: Mechanisms of Action* (G. H. Glaser, J. K. Penry, and D. M. Woodbury, eds.) Raven Press, New York, pp. 127–154.

Lebeda, F. J., Hablitz, J. J., and Johnston, D., 1982, Antagonism of GABA-mediated responses by *d*-tubocurarine in hippocampal neurons, *J. Neurophysiol.* **48:**622–632.

Lebeda, F. J., Brown, T. H., and Johnston, D., Synaptic mechanisms underlying epileptiform discharges in hippocampal neurons, submitted.

Lebovitz, R. M., Dichter, M., and Spencer, W. A., 1971, Recurrent excitation in the CA3 region of cat hippocampus, *Int. J. Neurosci.* **2:**99–108.

Lee, K. S., Schottler, F., Oliver, M., and Lynch, G., 1980, Brief bursts of high-frequency stimulation produce two types of structural changes in rat hippocampus, *J. Neurophysiol.* **44:**247–258.

Levy, W. B. and Desmond, N., 1983, The rules of elemental synaptic plasticity, in: *Synaptic Modification, Neuron Selectivity and Nervous System Organization* (W. B. Levy, J. Anderson, and S. Lehmkuhle, eds.), Lawrence Erlbaum Association, Hillsdale, New Jersey, in press.

Llinás, R. and Sugimori, M., 1980, Electrophysiological properties of *in vitro* Purkinje cell somata in mammalian cerebellar slices, *J. Physiol.* (*London*) **305:**171–195.

Lorente de Nó, R., 1934, Studies on the structure of the cerebral cortex. II. Continuation of the study of the ammonic system, *J. Psychol. Neurol.* **46:**113–117.

MacVicar, B. A. and Dudek, F. E., 1980, Local synaptic circuits in rat hippocampus: Interactions between pyramidal cells, *Brain Res.* **184:**220–223.

Martin, A. R., 1976, The effect of membrane capacitance on nonlinear summation of synaptic potentials, *J. Theor. Biol.* **59**:179–187.
Martin, A. R., 1977, Junctional transmission. II. Presynaptic mechanisms, in: *Handbook of Physiology*, Section I: *The Nervous System* (J. M. Brookhart and V. B. Mountcastle, eds.), American Physiological Society, Bethesda, Maryland, pp. 329–355.
Matsumoto, H. and Ajmone Marsan, C., 1964, Cortical cellular phenomena in experimental epilepsy: Interictal manifestation, *Exp. Neurol.* **9**:286–304.
McLachlan, E. M., 1978, The statistics of transmitter release at chemical synapses, in: *International Review of Physiology. Neurophysiology III*, Volume 17, (R. Porter, ed.), University Park Press, Baltimore, pp. 49–117.
McLachlan, E. M. and Martin, A. R., 1981, Non-linear summation of end-plate potentials in the frog and mouse, *J. Physiol.* (*London*) **311**:307–324.
McNaughton, B. L., Barnes, C. A., and Andersen, P., 1981, Synaptic efficacy and EPSP summation in granule cells of rat facia dentata studied *in vitro*, *J. Neurophysiol.* **46**:952–966.
Perkel, D. H. and Feldman, M. W., 1979, Neurotransmitter release statistics: Moment estimates for inhomogenous Bernoulli trials, *J. Math. Biol.* **7**:31–40.
Prince, D. A., 1978, Neurophysiology of epilepsy, *Annu. Rev. Neurosci.* **1**:395–415.
Rall, W., 1967, Distinguishing theoretical synaptic potentials computed for different soma-dendritic distributions of synaptic input, *J. Neurophysiol.* **30**:1138–1168.
Rall, W., 1974, Dendritic spines, synaptic potency and neuronal plasticity, in: *Cellular Mechanisms Subserving Changes in Neuronal Activity* (C. D. Woody, K. A. Brown, T. J. Crow, Jr., and J. D. Knispel, eds.), Brain Information Service, Los Angeles, pp. 13–21.
Rall, W., 1977, Core conductor theory and cable properties of neurons, in: *Handbook of Physiology*, Section I: *The Nervous System* (J. M. Brookhart and V. B. Mountcastle, eds.), American Physiological Society, Bethesda, Maryland, pp. 39–97.
Rall, W., 1978, Dendritic spines and synaptic potency, in: *Studies in Neurophysiology*, presented to A. K. McIntyre, R. Porter, ed., Cambridge University Press, pp. 203–209.
Rall, W. and Rinzel, J., 1973, Branch input resistance and steady attenuation for input to one branch of a dendritic neuron model, *Biophys. J.* **13**:648–688.
Rall, W., Burke, R. E., Smith, T. G., Nelson, P. G., and Frank, K., 1967, Dendritic location of synapses and possible mechanisms for the monosynaptic EPSP in motoneurons. *J. Neurophysiol.* **30**:1169–1193.
Redman, S. J., 1976, A quantitative approach to integrative function of dendrites, in: *International Review of Physiology. Neurophysiology II*, Volume 10, (R. Porter, ed.), University Park Press, Baltimore, pp. 1–35.
Rinzel, J. and Rall, W., 1974, Transient response in a dendritic neuron model for current injected at one branch, *Biophys. J.* **14**:759–790.
Scheibel, M. E. and Scheibel, A. B., 1968, On the nature of dendritic spines—Report on a workshop, *Comm. Behav. Bio.* **I**(A):231–265.
Schwartzkroin, P. A. and Wyler, A. R., 1980, Mechanisms underlying epileptiform burst discharge, *Ann. Neurol.* **7**:95–107.
Schwindt, P. C. and Crill, W. E., 1977, A persistent negative resistance in cat lumbar motoneurons, *Brain Res.* **120**:173–178.
Schwindt, P. C. and Crill, W. E., 1980, Role of a persistent inward current in motoneuron bursting during spinal seizures, *J. Neurophysiol.* **43**:1296–1318.
Schwindt, P. C. and Crill, W. E., 1980, Properties of a persistent inward current in normal and TEA-injected motoneurons, *J. Neurophysiol.* **43**:1700–1724.
Schwindt, P. C. and Crill, W. E., 1980, The effects of barium on cat spinal motoneurons studied by voltage clamp, *J. Neurophysiol.* **44**:827–846.

Schwindt, P. C. and Crill, W. E., 1981, Differential effects of TEA and cations on outward ionic currents of cat motoneurons, *J. Neurophysiol.* **46:**1–16.

Segev, I. and Parnas, I., 1983, Synaptic integration mechanisms. Theoretical and experimental investigation of temporal postsynaptic interactions between excitatory and inhibitory inputs, *Biophys J.* **41:**41–50.

Shepherd, G. M., 1979, *The Synaptic Organization of the Brain,* 2nd ed., Oxford University Press, New York.

Stevens, C. F., 1976, A comment on Martin's relation, *Biophys. J.* **16:**891–895.

Swindale, N. V., 1981, Dendritic spines only connect, *Trends Neurosci.* **4:**240–241.

Takeuchi, A., 1977, Junctional transmission. I. Postsynaptic mechanisms, in: *Handbook of Physiology,* Section I: *The Nervous System,* Volume 1, Part 1 (J. M. Brookhart and V. B. Mountcastle, eds.), American Physiological Society, Bethesda, Maryland, pp. 295–328.

Traub, R. D. and Llinás, R., 1979, Hippocampal pyramidal cells: Significance of dendritic ionic conductances for neuronal function and epileptogenesis, *J. Neurophysiol.* **42:**476–496.

Traub, R. D. and Wong, R. K. S., 1982, Cellular mechanism of neuronal synchronization in epilepsy, *Science* **216:**745–747.

Turner, D. A. and Schwartzkroin, P. A., Electrical characteristics of dendrites and dendritic spines in intracellularly-stained CA3 and dentate hippocampal neurons, submitted.

Valverde, F. and Ruiz-Marcos, A., 1969, Dendritic spines in the visual cortex of the mouse: Introduction to a mathematical model, *Exp. Brain Res.* **8:**269–283.

Van Harreveld, A. and Fifkova, E., 1975, Swelling of dendritic spines in the fascia dentata after stimulation of the perforant fibers as a mechanism of posttetanic potentiation, *Exp. Neurol.* **49:**736–749.

Williams, J. D. and Bowen, J. M., 1974, Effects of quantal unit latency on statistics of Poisson and binomial neurotransmitter release mechanisms, *J. Theor. Biol.* **43:**151–165.

Wilson, W. A. and Goldner, M. M., 1975, Voltage clamping with a single microelectrode, *J. Neurobiol.* **6:**411–422.

Wong, R. K. S., Prince, D. A., and Basbaum, A. I., 1979, Intradendritic recordings in hippocampal neurons, *Proc. Natl. Acad. Sci. USA* **76:**986–990.

Zucker, R. S., 1977, Synaptic plasticity at crayfish neuromuscular junctions, in: *Identified Neurons and Behavior of Arthropods* (G. Hoyle, ed.), Plenum Press, New York, pp. 49–65.

4

Hippocampus

Synaptic Pharmacology

RAYMOND DINGLEDINE

1. INTRODUCTION

Central nervous system pharmacology is in the midst of a new stage of development. With the advent of sensitive and specific radioreceptor binding assays for studying drug–receptor interactions, rapid advances have been made in the biochemical investigation of drug and transmitter mechanisms. Our knowledge of the physiological actions of neuroactive drugs has also progressed in recent years, due in large part to the extensive use of the iontophoretic technique coupled with extra- and intracellular recording from cells in the intact brain. Although such *in vivo* studies provide an important and necessary foundation for any serious investigation of the effects of a drug on the nervous system, detailed information about the site and mode of action of drugs is very difficult to obtain. *In vivo* electrophysiological studies of neurons in the mammalian brain generally suffer from two difficulties that prohibit detailed analysis: (1) insufficient mechanical stability to permit long-lasting intracellular recordings on a routine basis and (2) an inability to know the equilibrium concentration of drugs in the interstitial space. These drawbacks can be overcome with the use of brain slices, which forms the basis of a strong rationale for the utility of these preparations in neu-

---

RAYMOND DINGLEDINE • Department of Pharmacology, University of North Carolina, Chapel Hill, North Carolina 27514.

ropharmacological investigations. Brain slices, like all preparations, suffer from disadvantages that must be borne in mind when interpreting data. Such technical aspects of brain slice preparations are discussed in some detail in the Appendix.

The purpose of this chapter is to illustrate a range of neuropharmacological studies that have taken advantage of the *in vitro* nature of hippocampal slices and to discuss in more detail some of our own studies that have been carried out over the past 5 years. Before considering the electrophysiological aspects of drug action in the hippocampus, a brief survey of the cellular localization of endogenous neuroactive agents will be presented in order to provide an anatomical framework for these studies.

2. LOCALIZATION OF TRANSMITTERS AND ENDOGENOUS NEUROACTIVE AGENTS IN THE HIPPOCAMPAL FORMATION

Figure 1 depicts highly schematized drawings of a transverse section through the hippocampus. The most likely transmitter in several of the major intrinsic pathways is shown, along with the distribution of extrinsic afferents and neuroactive peptides. Anatomical precision is intentionally omitted in order to highlight the chemical nature of hippocampal pathways.

2.1. *Transmitter Candidates*

2.1.1. Intrinsic Excitatory Pathways. A striking organizational feature of the hippocampal formation is the laminar arrangement of its excitatory pathways, as demonstrated both anatomically and electrophysiologically (reviewed by Andersen, 1975). Recent evidence supports the long-held notion that acidic amino acids serve as neurotransmitters for several of these excitatory pathways. Biochemical studies of the release of glutamate, and electrophysiological studies with acidic amino acid antagonists, strongly support glutamate as a major transmitter of the lateral perforant path (White *et al.*, 1977; Koerner and Cotman, 1981). The excitatory transmitters used by the Schaffer/commissural afferents to the CA1 pyramidal cells are less certain, but may include glutamate, aspartate, or related compounds (Nadler *et al.*, 1976; White *et al.*, 1979; Wieraszko and Lynch, 1979; Lanthorn and Cotman, 1981; Koerner and Cotman, 1982). Excitatory transmission between granule cell mossy fibers and CA3 pyramidal cells is resistant to known acidic amino acid

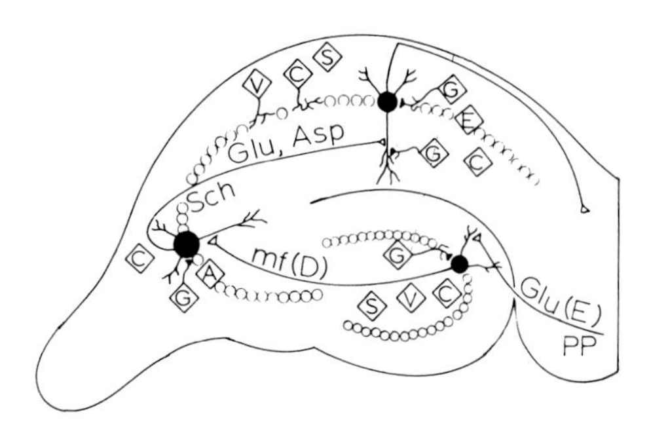

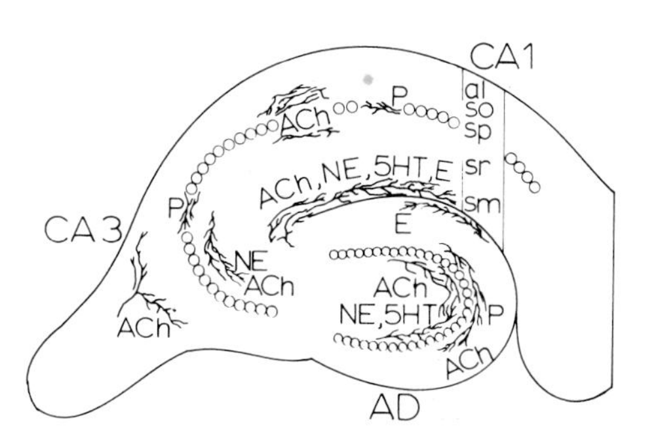

Figure 1. Schematic drawings of hippocampal circuitry, emphasizing the neurochemistry of the pathways. Abbreviations: CA1 and CA3, hippocampal subfields. AD, area dentata. Layers: al, alveus; so, stratum (s.) oriens; sp, s. pyramidale; sr, s. radiatum; sm, s. lacunosum moleculare. Interneurons: G, GABA; E, met- or leu-enkephalin; C, cholecystokinin; S, somatostatin; V, vasoactive intestinal polypeptide; A, angiotensin II. Intrinsic pathways: pp, perforant path; mf, mossy fibers; Sch, Schaffer collaterals; Glu, glutamate; Asp, aspartate; D, dynorphin. Extrinsic afferents: ACh, acetylcholine; NE, norepinephrine; 5HT, serotonin; P, substance P. The principle neurons (pyramidal cells, granule cells) are shown in black.

antagonists (White *et al.*, 1979), and the mossy fibers do not appear to release selectively either glutamate or aspartate (Nadler *et al.*, 1978). Since neither monoamines nor acetylcholine have been found by histochemical methods in this pathway (although it does contain dynorphin—see below), the principle excitatory transmitter of the mossy fibers remains unknown.

2.1.2. *Intrinsic Inhibitory Pathways.* A wealth of information indicates that the basket cells described by Lorente de Nó (1934), which invest the pyramidal and granule cell bodies with a plexus of synapses, use GABA as their principle transmitter (Andersen, 1975; Storm-Mathisen, 1977). Basket cells are activated by pyramidal or granule cell axons and participate in the classical recurrent inhibitory pathway of the hippocampus (Andersen *et al.*, 1964). A variety of circumstantial evidence suggests that dendritic inhibition mediated by GABA also exists in the hippocampus. The axons of certain basket cells ramify within the dendritic zones (Lorente de Nó, 1934). GABA receptors, as well as the GABA synthetic enzyme glutamic acid decarboxylase, are distributed throughout all layers of the hippocampus (Storm-Mathisen, 1977; Ribak *et al.*, 1978; Andersen *et al.*, 1980). Inhibitory postsynaptic potentials (IPSPs) are occasionally observed that are not accompanied by a measurable conductance increase at the soma, suggestive of a remote synaptic origin (Dingledine and Langmoen, 1980). Compelling physiological evidence for this suggestion was provided by Alger and Nicoll (1979, 1982), who demonstrated evoked inhibitory synaptic potentials in CA1 pyramidal neurons that could be blocked by focal dendritic application of GABA antagonists or tetrodotoxin (TTX). Similarly, Silfvenius *et al.* (1980) showed that focal application of the GABA antagonist, penicillin, into the dendritic region of pyramidal cells induced epileptiform bursting suggestive of dendritic disinhibition. Although the evidence for the existance of dendritic inhibitory synapses (at least in the CA1 region) is building, the neurons mediating these inhibitory potentials have yet to be identified anatomically.

2.1.3. Extrinsic Modulatory Pathways. The precise distribution of the hippocampal cholinergic input, which originates in the medial septum and diagonal band, is only partly understood. This is due to discrepancies among the staining patterns of the various markers used to trace this pathway. There is general agreement, however, that the most dense innervation of cholinergic fibers occurs within the stratum (s.) oriens of CA3 and the hilus of the fascia dentata, with less dense innervation of the inner portion of the dentate molecular layer (Storm-Mathisen, 1977; Lynch *et al.*, 1978; Crutcher *et al.*, 1981; Kimura *et al.*, 1981). In addition, the distribution of both cholinesterase and choline acetyltransferase suggests a cholinergic innervation of the juxtapyramidal region, expecially in CA1 (Storm-Mathisen, 1977; Lynch *et al.*, 1978; Kimura *et al.*, 1981), although this projection could not be demonstrated by orthograde transport of horseradish peroxidase (HRP) (Crutcher *et al.*, 1981). Entirely unresolved is the identity of the cellular targets of the septohippocampal pathway.

The noradrenergic innervation of the hippocampal formation is thought to arise solely from the locus coeruleus. Noradrenergic axons are found in a rather dense plexus within the hilus of the dentate gyrus, and also in s. radiatum of CA3 and s. lacunosum-moleculare of CA1 (Lindvall and Bjorkland, 1974). The serotonergic projection to the hippocampus arises from the medial and dorsal raphe nuclei, innervating approximately the same areas served by noradrenergic fibers (Moore and Halaris, 1975; Azmitia and Segal, 1979). However, both noradrenergic and serotonergic fibers appear to be distributed diffusely throughout the rest of the hippocampus. As with the cholinergic projection, the identity of the target cells for monoamine inputs is unknown.

16 1833$$4A01

2.2. Neuroactive Peptides

Immunohistochemical techniques hold great promise for the mapping of specific neuronal systems. These techniques have recently begun to be applied to hippocampal neurons and pathways, which have been found to contain a number of neuroactive peptides. Chief among these may be the opioid peptides. Gall *et al.* (1981) described enkephalinlike immunoreactivity within fibers of the lateral perforant path and s. lacunosum-moleculare of CA1, and scattered enkephalin-containing neuronal perikarya throughout s. pyramidale of the CA1 region. The morphology of the stained cells was unlike that of typical pyramidal neurons, suggesting that pyramidal basket cells or another type of interneuron was involved. The granule cells and mossy fiber system were also heavily stained, although this staining was later shown to be due mostly, if not exclusively, to dynorphin (McGinty *et al.*, 1983).

Immunoreactive staining for cholecystokinin (CCK) has been reported in a population of interneurons lying within s. oriens of CA1; these cells were postulated to be a type of basket cell since their processes extended into s. pyramidale (Greenwood *et al.*, 1981; Handelmann *et al.*, 1981). Additional CCK-positive cells were found scattered throughout s. radiatum of CA1 and within the polymorphic zone of the dentate hilus. By the use of an immunofluorescent double label staining technique, a separate population of somatostatin-containing interneuronlike cells was shown by the same authors to reside deep within s. oriens of CA1 (Greenwood and Winstead, 1981). Somatostatin-staining neurons were also found within the dentate hilus.

The distribution of several other peptides within the hippocampus has been mapped immunohistochemically. Thus, vasoactive intestinal polypeptide (VIP) is reported to have approximately the same distribution as CCK immunoreactivity (Loren *et al.*, 1979). Substance-P-like immunoreactive fibers have been demonstrated within s. pyramidale of CA1 (Vincent *et al.*, 1981), and antisera to angiotensin II have been reported to stain a population of pyramidal-like cells in CA1 and CA3 (Haas *et al.*, 1980). These neurochemical findings add new significance to the classical picture of hippocampal interneurons. The original Golgi studies of Lorente de Nó (1934) delineated many subtypes of interneuron within the hippocampus, and it now appears that multiple populations of interneuron can be recognized by selective staining. It will be important to determine whether the "excitatory" peptides CCK (Dodd and Kelly, 1981), somatostatin (Dodd and Kelly, 1979), VIP (Dodd *et al.*, 1979), and angiotensin (Haas *et al.*, 1982) coexist with GABA in a sub-

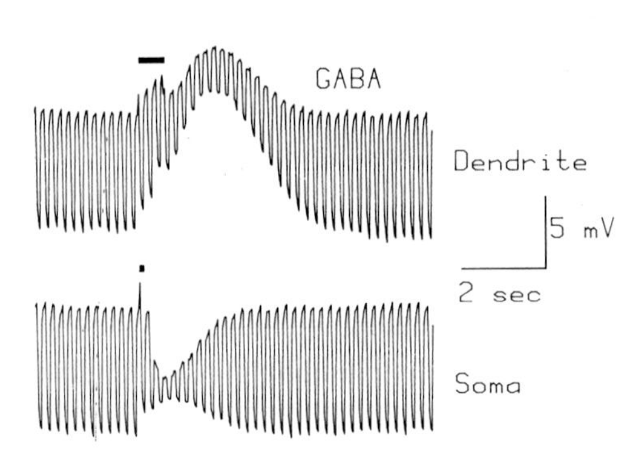

Figure 2. Effect of somatic and dendritic applications of GABA. GABA was iontophoresed into the cell body layer (bottom; 190-nA, 160-msec pulse), and into the layer of apical dendrites (top, 190 nA, 600 msec). During each drug application a train of hyperpolarizing current pulses (−0.33 nA, 90 msec) was injected across the membrane to monitor its input resistance. Tetrodotoxin (1 μM) was present to eliminate action potentials. (S. Korn and R. Dingledine, unpublished observation.)

population of interneurons, and, if so, under what circumstances, if any, they are released.

3. CELLULAR ACTIONS OF NEUROACTIVE DRUGS IN HIPPOCAMPAL SLICES

3.1. *GABA*

A number of investigators have been intrigued by the inhibitory actions of GABA on hippocampal neurons, especially in light of the role of disinhibition in experimental models of epilepsy and synaptic plasticity. When GABA is applied iontophoretically into the region of the pyramidal cell bodies, where the recurrent inhibitory synapses are most dense, a hyperpolarization with increased conductance is typically observed. The physiological significance of this GABA effect is widely accepted, since it mimics the action of synaptically released GABA at recurrent inhibitory (somatic) synapses and probably also at dendritic inhibitory synapses (Andersen *et al.*, 1980; Alger and Nicoll, 1982). Andersen *et al.* (1978a) were the first to report that, in contrast to results obtained with somatic applications, focal application of GABA into the dendritic layers of the slice usually depolarized pyramidal cells. The two spatially separate actions of GABA acting on the same cell are shown in Figure 2. This observation has since been extended and repeated in several laboratories (Alger and Nicoll, 1979, 1982; Andersen *et al.*, 1980; Thalmann *et al.*, 1981). Hyperpolarizing responses are not exclusively limited to the cell body, however, but can also be demonstrated on dendritic membranes (Alger and Nicoll, 1982; Wong and Watkins, 1982).

The hyperpolarizing GABA response is thought to be due predominantly to a rise in chloride conductance, although rigorous evidence for this statement is rather meager. The reversal potential is −65 to −70 mV in 133 mM chloride (Andersen *et al.*, 1980). Reduction of extracellular chloride concentration has been shown to shift the reversal potential of the hyperpolarizing response in the depolarizing direction (Thalmann *et al.*, 1981). Conversely, elevation of intracellular chloride by leakage from KCl-filled micropipettes converts the hyperpolarization into a large depolarization. It has been our experience that hyperpolarizing responses to GABA are much more robust when potassium methylsulfate is used to fill the micropipettes rather than potassium acetate. This observation is probably related to the report that acetate depresses GABA-mediated hyperpolarizing IPSPs in hippocampal pyramids recorded *in vivo* (Eccles *et al.*, 1977).

The ionic mechanism of the depolarizing GABA response is uncertain. The reversal potential is estimated to be −40 to −50 mV in 133 mM chloride (Andersen *et al.*, 1980). A chloride component is suggested by the finding that low chloride solutions increase the response (S. Korn and R. Dingledine, unpublished observations) and shift the reversal potential in a depolarizing direction (Thalmann *et al.*, 1981). This mechanism, however, would imply a reversed intracellular chloride gradient in soma and dendrites, which is incompatible with the demonstration of dendritic hyperpolarizing responses (Alger and Nicoll, 1982; Wong and Watkins, 1982). The depolarization is often reduced in low sodium solutions (S. Korn and R. Dingledine, unpublished observations), which suggests that an inward sodium current may be responsible for the depolarization; the large conductance change, then, could result from the characteristic (inward) chloride flux associated with GABA responses. Quantitative study of this issue is hampered both by the small size of the pyramidal cells, which may facilitate redistribution of intracellular ions during imposed changes in extracellular ions, and by the remote location of the majority of depolarizing receptors from the cell body, the usual site of impalement.

Pharmacological studies of the two effects suggest that different receptors may be involved (Alter and Nicoll, 1982). Although both effects can be blocked by bicuculline, the depolarizing action of GABA appears to be more sensitive. Likewise, the depolarizing action of GABA is reported to be potentiated more by pentobarbitone than is the hyperpolarizing response. Finally, the GABA analog, 4,5,6,7-tetrahydroisoxazolo [5,4-*c*]pyridin-3-ol (THIP) appears to preferentially activate the hyperpolarizing receptors. It is unlikely that the depolarizing response is due to release of transmitter from excitatory afferents synapsing onto pyramidal cell dendrites as both responses were resistant to blocking synaptic transmission with low Ca^{2+}/high Mg^{2+} solutions (Andersen *et al.*, 1980; Wong and Watkins, 1982).

Hyperpolarizing GABA responses, both *in vivo* and *in vitro*, show rather striking plasticity that can take two (possibly related) forms. The first is a marked fade in the evoked changes in conductance and especially potential that can occur during relatively prolonged (several second) GABA applications (Ben-Ari *et al.*, 1981; Wong and Watkins, 1982). Second, for tens of seconds after tetanic stimulation of orthodromic inputs, hyperpolarizing GABA responses can actually be converted to depolarizations (Wong and Watkins, 1982). The fade in the potential and conductance changes should probably be considered separately. Failure to maintain a high GABA-evoked conductance during prolonged GABA applications could be due to receptor desensitization, accumulation of intracellular chloride with subsequent inactivation of chloride conductance (cf. Gold and Martin, 1982), or possibly a slow uptake process forcing re-equilibration to a lower GABA concentration. The fading of GABA-evoked hyperpolarizations is probably caused by a shift in the chloride Nernst potential secondary to intracellular accumulation or extracellular depletion of the ion, and/or diffusion of GABA to distant depolarizing receptors. A rapid depolarizing shift in the chloride Nernst potential may largely explain the inversion of sign seen by Wong and Watkins (1982) after tetanic stimulation; the cells transiently depolarized during the stimulus train, which would allow the chloride Nernst potential to become established at a level depolarizing to the resting potential.

The fade phenomenon is important to understand as it may be related to two major aspects of hippocampal physiology—epilepsy and short-term potentiation following repetitive orthodromic stimulation. IPSPs in the *in vivo* hippocampus are especially labile in the face of repetitive activation of inhibitory fibers (Ben-Ari *et al.*, 1979); the reduction in IPSP–conductance was shown by Ben-Ari *et al.* to be accompanied by a reduction in the responses of pyramidal cells to iontophoretic application of GABA, implying a postsynaptic mechanism. The significance of reduced inhibition for epilepsy was recently emphasized by studies of the epileptogenic action of penicillin. Penicillin's effect was shown to be due in part to its ability to block the synaptic inhibitory action of GABA on pyramidal cells; this action causes the conversion of excitatory postsynaptic potential and inhibitory postsynaptic potential (EPSP–IPSP) sequences to nearly pure EPSPs, which presumably become capable of triggering dendritic inward currents responsible for bursting (Dingledine and Gjerstad, 1979, 1980; Wong and Prince, 1979;

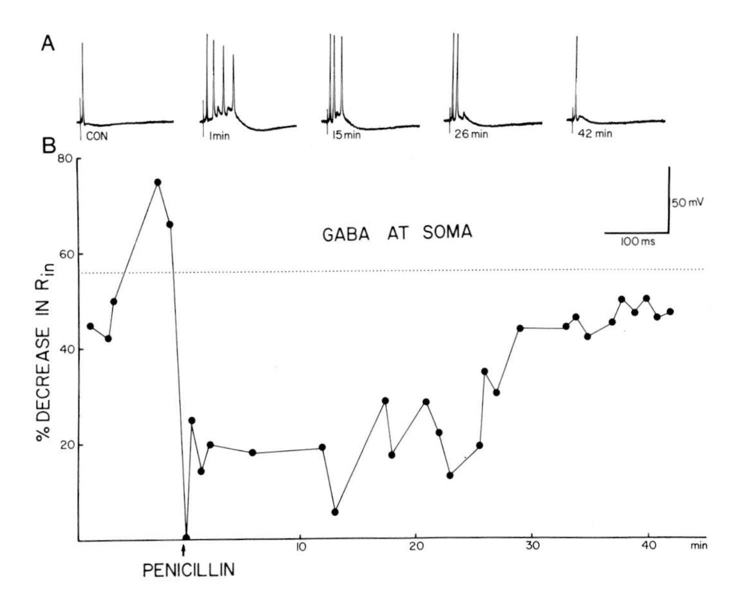

Figure 3. Temporal relation between the epileptiform effect and the GABA-blocking effect of penicillin. (A) Single sweeps show the response of a pyramidal cell to an orthodromic stimulus before (CON) and at various times after application of a droplet of ACSF containing penicillin (170 mM). The time below each sweep refers to the time on the abscissa of Figure 3B. (B) The maximum reduction in input resistance produced by GABA (60 nA, 6 sec) ejected near the soma of the impaled cell is plotted against time. The penicillin droplet was added at the time indicated on the abscissa. Note the similar rates of recovery for both effects of penicillin. From Dingledine and Gjerstad, 1980.

Figures 3, 4). Figure 3 shows that the timecourse of the epileptogenic effect of focally applied penicillin matches that of the GABA-blocking effect. Figure 4 demonstrates the gradual loss of the recurrent IPSP, coupled with the gradual change of the orthodromic EPSP–IPSP sequence, following a droplet of penicillin.

3.2. *Acetylcholine*

Acetylcholine (ACh) appears to have at least two distinct muscarinic actions in the hippocampus: a direct, rather slowly developing depolarizing action on pyramidal cells, and a rapid presynaptic inhibitory effect on both excitatory and inhibitory afferents to pyramidal cells. Both effects have been seen in *in vivo* recordings. The depolarizing effect largely accounts for the slow excitatory action of ACh on hippocampal neurons (Biscoe and Straughan, 1966), while the presynaptic effect may

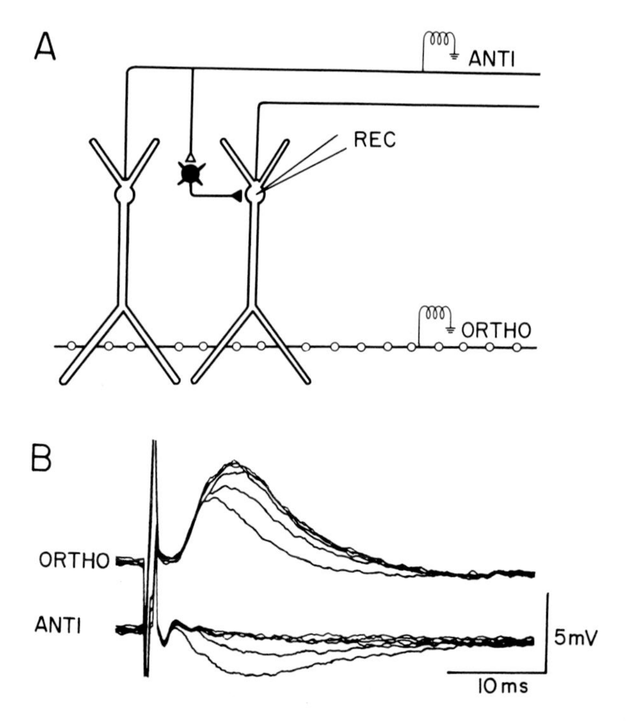

Figure 4. Unmasking of prolonged EPSP during IPSP blockade by penicillin. (A) Schematic diagram of experimental arrangement. Subthreshold antidromic (ANTI) and orthodromic (ORTHO) stimuli were delivered alternately to a CA1 pyramidal cell that had been impaled by a microelectrode (REC). Stimuli were at 0.75 Hz. The ANTI stimulus, via a recurrently activated inhibitory interneuron, evoked a pure IPSP, while the ORTHO stimulus produced a mixed EPSP–IPSP. (B) Families of responses to the two stimulating lines shown in Figure 4A, just before and after the application of a penicillin droplet (17 mM). Each trace is the average of 10 consecutive stimuli. The lowest trace of each of the two responses was obtained in the 30-sec control period before penicillin, the middle trace 0 to 30 sec after penicillin, and the upper traces in the four consecutive 30-sec periods thereafter. The initial downward blip following the stimulus artifact in the ANTI responses represents the antidromic field potential. From Dingledine and Gjerstad, 1979.

appear principally as a disinhibition (Ben-Ari *et al.*, 1981). In addition to these two effects, ACh can sometimes produce a rapid hyperpolarization that has properties consistent with an increase in potassium conductance (Segal, 1982).

The depolarizing effect of ACh is accompanied by a slowly developing rise in input resistance (Dodd *et al.*, 1981; Benardo and Prince, 1982a), which often becomes apparent only after the depolarization is well under way or has plateaued (Dodd *et al.*, 1981; Segal, 1982). The rise in input resistance is not blocked by TTX, manganese, or cadmium (Benardo and Prince, 1982b; Haas, 1982; Halliwell and Adams, 1982), making it unlikely to be caused by an indirect release of excitatory transmitters. Several studies suggested that this direct effect of ACh was due mainly to reduction of resting potassium conductance (Dodd *et al.*, 1981; Benardo and Prince, 1982b), in accordance with the findings of Krnjević *et al.* (1971) on the muscarinic action of ACh on neocortical neurons. This hypothesis was confirmed in a voltage-clamp study by Halliwell and Adams (1982), who described a voltage-dependent potassium current, the M-current, that was active over the range −70 to −40 mV and was reduced by muscarinic agonists. The ACh-evoked depolarization can be reduced by manganese or TTX (Benardo and Prince, 1982b), suggesting that it is caused by voltage-sensitive calcium and sodium currents that are unleashed by blocking potassium conductance.

One property of the direct ACh effect may be of special significance for understanding the physiology of the septohippocampal system. ACh tends to promote burst firing in pyramidal cells, whether triggered by brief depolarizing current pulses (Figure 5) or by orthodromic stimuli. This action, which is readily understood in terms of reducing voltage-sensitive potassium currents, would be expected to exaggerate the responses of pyramidal cells to excitatory inputs. The depolarizing action of ACh, and the facilitation of excitatory responses, are observed most readily when ACh is applied near the somatic, as opposed to the dendritic, region of the pyramidal cell (Valentino and Dingledine, 1981). Benardo and Prince (1982a) emphasized the long-lasting nature of the direct effect of ACh on pyramidal cells. Because of its persistence, and its ability to facilitate the effects of other excitatory afferents, the direct depolarizing action of ACh on pyramidal neurons can be thought of as modulatory in nature.

In addition to its slow depolarizing action, ACh exerts a rather rapid muscarinic inhibitory effect upon both excitatory and inhibitory afferents to pyramidal cells (Figure 6). This effect of ACh was first discovered by Yamamoto and Kawai (1967), and later studied in more detail by Hounsgaard (1978) and Valentino and Dingledine (1981). Focal dendritic applications of ACh reduce the field EPSP as well as the intracellular EPSP. Several lines of evidence lead to the conclusion that this effect of ACh does not represent a form of remote postsynaptic inhibition, but rather is caused by a decrease in the amount of transmitter released per afferent stimulus. First, dendritic ACh applications reduce local EPSPs without altering membrane potential or input resistance of pyramidal cells. Second, responses to brief dendritic applications of L-glutamate are unaffected when the evoked EPSP is dramatically inhibited. Third, the

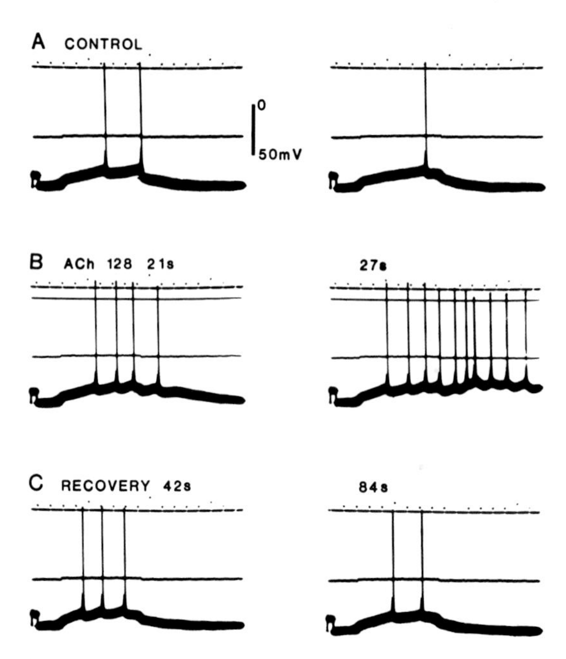

Figure 5. Burst firing evoked by ACh in a pyramidal neuron. (A) In control, standard depolarizing current pulses delivered through the microelectrode elicit 1 or 2 action potentials at long latency. The current monitor is shown in the middle trace, and a 20-msec timebase in the upper trace. (B) At 21 and 27 sec after starting ACh iontophoresis, the current pulse elicits repetitive firing that lasts beyond the duration of the current pulse. (C) Recovery at 42 and 84 sec after terminating the ACh application. From J. Dodd, R. Dingledine, and J. S. Kelly, unpublished observations.

timecourse of the reduced EPSP is unchanged. A local fall in dendritic input resistance sufficient to reduce the EPSP would be expected to reduce the membrane time constant in the subsynaptic region and thus increase both rise and fall times of the inhibited EPSP. This point was verified on an analog model of a dendritic neuron having an electrotonic structure similar to that of pyramidal cells (Figure 7C). However, the change in timecourse of the ACh-inhibited EPSP expected from this model was not observed (Figure 7A,B). Thus, the EPSP reduction appears to be mediated simply by a fall in transmitter release. Focal application of ACh into s. pyramidale also reduces the amplitude of somatic IPSPs (Figure 6C). This effect occurs well before the characteristic depolarization and rise in input resistance develop, which is consistent with presynaptic inhibition of GABAergic transmission.

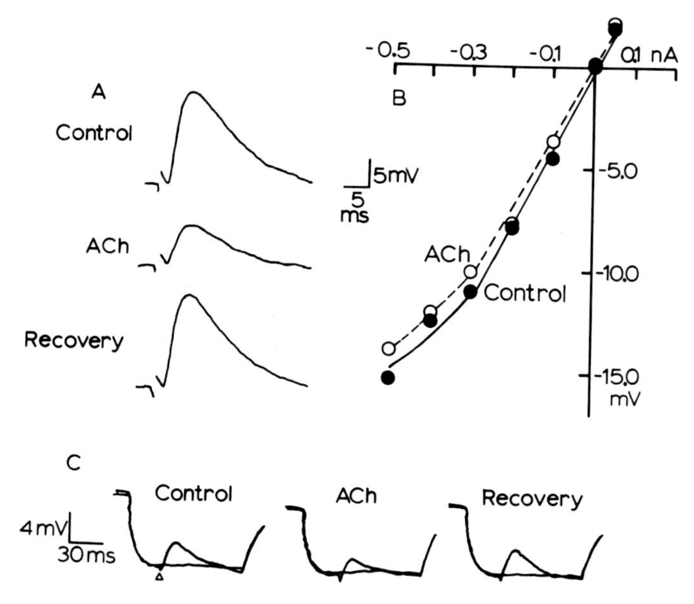

Figure 6. Depression by ACh of excitatory and inhibitory synaptic potentials. (A) An EPSP recorded from a CA1 pyramidal cell before (Control), 2 to 6 sec after (ACh) and following recovery from an iontophoretic dose of ACh (100 nA, 8 sec) into the layer of activated apical dendritic synapses. Each trace is the average of four sweeps. (B) Current–voltage plot of the same cell. Input slope resistance was 37 MΩ in control and 35 MΩ during ACh. (C) A depolarizing recurrent IPSP evoked by alveus stimulation during a hyperpolarizing current pulse prior to (Control), 3 sec after (ACh), and 52 sec after (Recovery) the start of a 15-sec iontophoretic application of ACh (100 nA) into the cell layer. The small open triangle denotes the stimulus artifact. Adapted from Valentino and Dingledine, 1981.

How might the presynaptic inhibitory action of ACh come about? The fact that spatially separate EPSPs and IPSPs could both be reduced by ACh makes it unlikely that the effect is related to any particular transmitter. Possibilities include direct depolarization of terminals (perhaps by reducing potassium conductance), local potassium accumulation leading to terminal depolarization, or interference with stimulus secretion coupling by blocking a voltage-sensitive inward calcium current in the nerve terminal. The finding that the size of the presynaptic fiber volley is unaffected at the time of peak EPSP reduction (Valentino and Dingledine, 1981; but see Hounsgaard, 1978) would appear to rule against the first two mechanisms, if it is accepted that the volley represents a valid measure of the strength of excitatory input to the dendrites (Andersen *et al.*, 1978b).

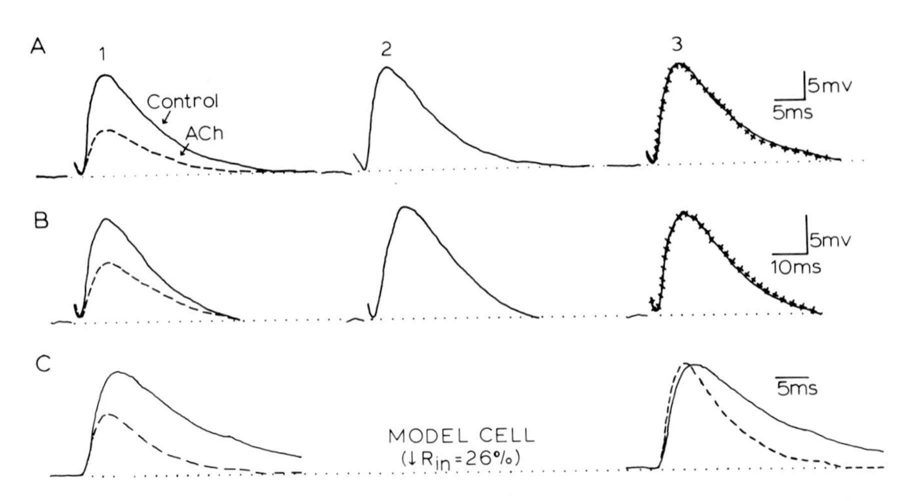

Figure 7. Effect of dendritically applied ACh on the timecourse of the EPSP in two CA1 pyramidal cells (A,B). (A,B) In 1, an EPSP is shown before (solid line) and after (dashed line) focal application of ACh into the apical dendritic tree. 2, Recovery of the EPSP following ACh application. 3, The EPSPs recorded in 1 were normalized so that the peak heights were the same. The solid line represents the control while the x's represent the EPSP recorded during ACh. Each trace is the average of four sweeps. (C) A local conductance shunt quickens the timecourse of an "EPSP" generated in an analog (11-compartment) cable model of a pyramidal cell. A remote EPSP was produced by placing a conductance in series with a battery in the ninth compartment, while recording from compartment 1. A remote conductance shunt produced near the site of EPSP generation (compartment 8) increased both rising and falling phases of the electrotonic potential. The cable had an electrotonic length of 1.1 length constants and a time constant of 20 msec. Adapted from Valentino and Dingledine, 1981.

3.3. *Opioid Peptides*

The cellular actions of opioids in the hippocampus have received a good deal of attention since it was demonstrated that intraventricular injection of these drugs can evoke electrographic seizure activity in limbic areas (Urca *et al.*, 1977; Henricksen *et al.*, 1978). *In vivo* recordings revealed that enkephalins elevated the firing rate of pyramidal cells, an action that was attributed to the inhibition of adjacent inhibitory interneurons (Zieglgänsberger *et al.*, 1979). These observations on the hippocampus *in vivo* were followed by numerous studies carried out with hippocampal slices in an effort to determine the mode of action of opioids in this limbic structure.

A characteristic action of opioid peptides in the hippocampal slice is to enhance the response of CA1 pyramidal neurons to excitatory syn-

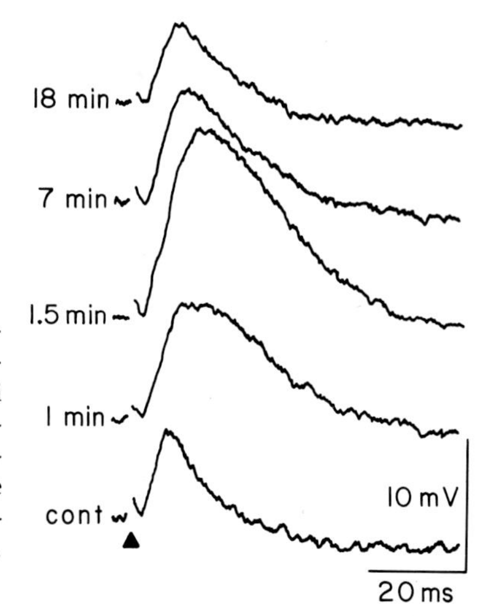

Figure 8. Potentiation of an EPSP by [D-ala^2,D-leu^5]-enkephalin (DADL). Each trace is the average of four responses to s. radiatum stimuli in control (cont) and at various times after application of a droplet of 10 μM DADL. The arrowhead marks the stimulus artifact. The membrane potential (−70 mV) and input resistance (15 MΩ) were unchanged by DADL. From Dingledine, 1981.

aptic activation. This has been shown in field potential recordings of evoked population spikes (Corrigall and Linseman, 1980; Dunwiddie *et al.*, 1980; Dingledine, 1981; Lynch *et al.*, 1981) and in intracellular recordings of evoked EPSPs (Dingledine, 1981; Robinson and Deadwyler, 1981; Masukawa and Prince, 1982; Figure 8). A number of possible modes of action by which opioids might produce this effect have been ruled out. Opioid peptides do not depolarize pyramidal cells, increase input resistance, or lower spike threshold (Nicoll *et al.*, 1980; Dingledine, 1981; Masukawa and prince, 1982). Two observations indicate that excitatory transmitter release is unaffected by opioids. First, dendritically recorded field EPSPs are unchanged (Corrigall and Linseman, 1980; Dunwiddie *et al.*, 1980; Dingledine, 1981; Lynch *et al.*, 1981; but see Haas and Ryall, 1980). Second, iontophoretic mapping experiments indicate that the site of action of the peptide [D-ala^2, D-leu^5]-enkephalin (DADL) is in or near s. pyramidale, quite far from the sites of excitatory transmitter release in the dendrites (Dingledine, 1981). Opioids do not appear to facilitate the activation of dendritic calcium or sodium currents, since neither calcium spikes nor depolarizations evoked by focal dendritic application of excitatory amino acids are enhanced by opioid peptides (Haas and Gähwiler, 1980; Dingledine, 1981). How, then, do opioids facilitate evoked EPSPs?

The original suggestion of a disinhibitory mechanism has received support from a number of laboratories, and it is worthwhile examining this hypothesis in some detail. It is presently accepted that both somatic and dendritic GABA-mediated postsynaptic inhibitions exist in the hippocampus. Somatic inhibitory synapses are predominantly if not exclusively made by the basket cells described by Lorente de Nó (1934). Dendritic synaptic inhibition is less well defined in terms of its cells of origin and functional roles, although it is clear that hyperpolarizing IPSPs can be elicited that are mediated by dendritic GABA receptors (Alger and Nicoll, 1979, 1982). Dendritic, and probably also somatic, inhibitory synapses can be activated by orthodromic stimulation in the slice. The low threshold and short latency of orthodromic IPSPs indicates that feedforward as well as feedback pathways are involved. The detailed studies of Alger and Nicoll (1982) demonstrate clearly that feedforward and feedback IPSPs are mediated by separate (although probably overlapping) sets of inhibitory synapses, and the rich morphological diversity of hippocampal interneurons is consistent with this argument.

Three reports are of special significance for the disinhibition hypothesis. Lee *et al.* (1980) described a population of nonpyramidal cells whose stimulus-evoked burst discharges could be suppressed by iontophoresis of D-ala-met-enkephalinamide. This population of cells had many characteristics of inhibitory interneurons (Schwartzkroin and Mathers, 1978; Knowles and Schwartzkroin, 1981), with the exception that the evoked burst firing was only observed to occur well after the pyramidal cell population spike. Suppression of interneuron firing by opioids could then account for potentiation of evoked pyramidal cell spiking if, as supposed by Lee *et al.*, short latency, opioid sensitive interneuron spikes were masked by the large field potentials and thus not observed. Nicoll *et al.* (1980) provided the first direct evidence that opioids could block evoked IPSPs and reduce the frequency of spontaneous IPSPs. Their recordings were made in the presence of 100 μM pentobarbital to potentiate IPSPs. Since spontaneous IPSPs could also be blocked by TTX, they presumably resulted from spontaneous discharge of inhibitory interneurons. These data supported the hypothesis that opioids could suppress the firing of a population of inhibitory interneurons.

Can a reduction of IPSPs explain the ability of opioids to potentiate evoked EPSPs and population spikes? Certainly the prolongation of the EPSP by DADL shown in Figure 8 is suggestive of reduced orthodromic inhibition. Other evidence indicates, however, that feedforward and feedback IPSPs are not *necessarily* blocked by opioids in concentrations sufficient to potentiate both intracellular EPSPs and population spikes (Haas and Ryall, 1980; Dingledine, 1981; Figure 9). This is in striking contrast to the action of the GABA-blocking agent, penicillin, that invariably and dramatically blocks IPSPs at concentrations sufficient to

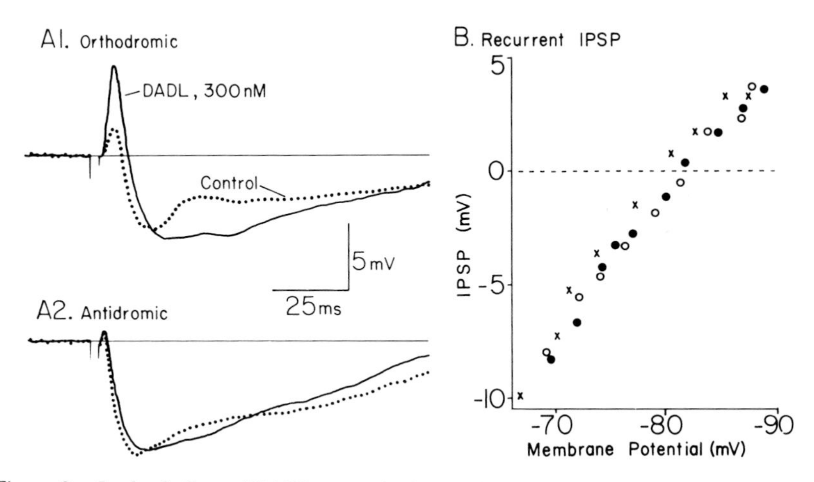

Figure 9. Lack of effect of DADL on evoked IPSPs. (A) Responses to orthodromic (1) and antidromic (2) stimuli before (Control) and after perfusing with 300 nM DADL. The EPSP is enhanced with no discernable reduction of either orthodromic or antidromic IPSPs. Each trace is the average of four sweeps. Subsequent perfusion with 500 nM naloxone returned the orthodromic response to control size (not shown). (B) The voltage dependence of the recurrent IPSP in control (solid circles), DADL (open circles) and DADL plus naloxone (Xs). Note no effect of DADL on the recurrent IPSP. Adapted from Dingledine, 1981.

produce burst firing or prolong intracellularly recorded EPSPs (Dingledine and Gjerstad, 1980; Figure 4). There are other differences between the actions of penicillin and opioids in the hippocampal slice. The protracted intense burst firing so characteristic of the penicillin-treated slice is not observed with opioids, which elicit a much milder facilitation of orthodromic discharge. Additionally, opioids produce a characteristic large shift to the left in the input-output (IO) curve formed by plotting population spike amplitude as a function of the field EPSP (Dingledine, 1981), whereas penicillin (and also pentobarbital) causes a much smaller and less consistent effect on the IO curve (A. A. Roth and R. Dingledine, unpublished observations). In Figure 10, a large shift in the IO curve produced by DADL was shown not to be accompanied by a reduction in the strength of recurrent inhibition, while a relatively mild shift of the IO curve by penicillin was accompanied by the expected loss of inhibition. Thus, opioids and penicillin differ dramatically in their effects on orthodromic activation of pyramidal cells, indicating that different sets of inhibitory synapses may be affected.

All of the above data can be reconciled if the principal or most sensitive site of action of opioids is on a restricted population of inhi-

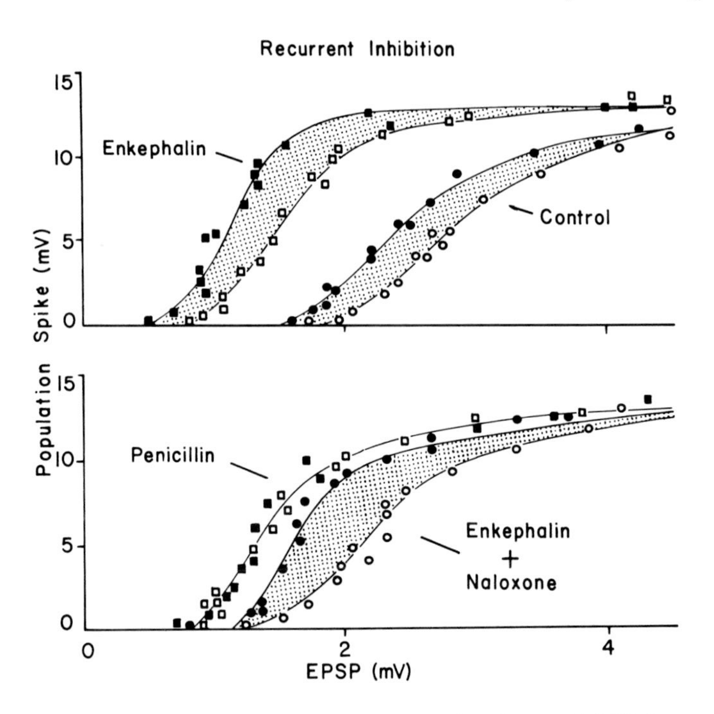

Figure 10. Reduction of recurrent inhibition by penicillin, but not DADL. Input–output curves of population spike amplitude as a function of field EPSP were plotted either without (■,●) or with (□,○), a conditioning antidromic stimulus delivered 25 msec prior to the orthodromic stimulus. The antidromic volley activated a recurrent inhibitory circuit that caused a small shift to the right of the input–output curve. The amount of this shift (shaded area) was taken as a measure of the intensity of recurrent inhibition. The top panel shows that the characteristic large shift to the left in the input–output curve caused by perfusion with 500 nM DADL is not accompanied by a reduction in recurrent inhibition. The bottom panel shows partial recovery from the effect of DADL after perfusion with DADL plus 500 nM naloxone, and the expected obliteration of recurrent inhibition produced by subsequent perfusion with 0.4 mM penicillin. This low concentration of penicillin caused only a small shift in the input–output curve. From Dingledine, 1981.

bitory GABAergic interneurons. This interneuron population would not include the classical basket cell. To comply with published data (Lee *et al.*, 1980; Dingledine, 1981), the cell bodies should lie in or near the pyramidal cell layer, while the inhibitory synapses would be made exclusively or predominantly on pyramidal cell dendrites. Thus, in contrast to the blocking action of penicillin on all postsynaptic GABAergic synapses in the CA1 region, opioids would appear to act primarily on a dendritic subset of inhibitory synapses.

Table I. Relative Potencies for Excitatory Action of Opioids in Hippocampus[a]

	EC50 (nM)	Potency ratio
DADL[b]	68	1.00
DSThr[b]	330	0.21
β-endorphin	450	0.15
Morphiceptin	1600	0.040
Morphine	3000	0.020
Ethylketocyclazocine	>10000	<0.007

[a] The concentration required to produce a half-maximal shift in the input–output curve was estimated by interpolation from dose–response curves. Thirteen to fifty-five slices were used for each dose–response curve, except for ethylketocyclazocine (N = 8 slices).
[b] Abbreviations: DADL: [D-ala^2,D-leu^5]-enkephalin; DSThr: Tyr-D-Ser-Gly-Phe-Leu-Thr. DADL and DSThr are selective δ-receptor agonists, while morphiceptin is a selective μ-receptor agonist. Ethylketocyclazocine is considered to act preferentially at κ opioid receptors (from Valentino and Dingledine, 1982).

Both μ and δ opioid receptors have been demonstrated in the hippocampus in binding assays (Chang *et al.*, 1979). Recent findings indicate that agonists selective for either μ or δ receptors produce similar, if not identical, facilitatory effects on pyramidal cell activation (Valentino and Dingledine, 1982; Bostock *et al.*, 1983). A method was developed for quantitating the shift to the left in the IO curve produced by opioids, and full-dose response curves were constructed for morphine and several peptides. All opioids produced a similar maximum effect. From the relative potencies (Table I) it appears that δ receptors play a prominant role in the opioid effect. However, the potency of the μ-specific agonist, morphiceptin, was 1.6 μM, far too low to activate δ receptors. Indeed, the half-maximally effective concentration of morphiceptin was found to match its dissociation constant from μ receptors in hippocampus when the binding assay was carried out under physiological conditions (Bostock *et al.*, 1983). It thus appears that both receptor subtypes may subserve similar functions in hippocampus. Surprisingly, the κ agonist ethylketocyclazocine was inactive (see also Gähwiler and Maurer, 1981), and the potency of morphine was low, even though μ receptors are present. Thus, the pharmacological profile of opioids in hippocampus does not appear to fit the receptor classifications developed for peripheral preparations.

3.4. *Excitatory Amino Acids*

It is now quite clear that multiple receptors exist for glutamate and other acidic amino acids (Watkins and Evans, 1981). Most of the evidence comes from studies in spinal cord, although recent pharmacological studies support the notion of a similar (but not identical—see Koerner and Cotman, 1982) family of receptors in hippocampus (Collingridge *et al.*, 1982a). The most well-characterized acidic amino acid receptor is activated strongly by *N*-methyl-D-aspartate, and is called the NMDA receptor; L-glutamate is a weak agonist at this site, whereas L-aspartate is somewhat stronger. A comparison of the ionic mechanisms underlying the depolarizing action of various acidic amino acids may help to clarify the physiological roles served by each receptor subtype. To this end, Hablitz and Langmoen (1982) showed that the reversal potential for glutamate-evoked depolarization matched that of the EPSP and was shifted in a negative direction in low sodium solution.

In our laboratory, the actions of L-glutamate and *N*-methyl-DL-aspartate (NMA), an NMDA receptor agonist, were compared on CA1 pyramidal cells (Dingledine, 1982, 1983a,b). Interestingly, the effects of NMA and glutamate, though both depolarizing, were strikingly different. NMA evoked calcium spikes and at low iontophoretic doses produced an apparent increase in input resistance, whereas the most common effect of glutamate was a simple depolarization with conventional resistance decrease. The effects of NMA, but not those of glutamate, were prevented by calcium channel blockers (Figure 11) and the NMA depolarization was potentiated by potassium channel blockers. The conductance change produced by NMA was voltage dependent and could be abolished at sufficiently hyperpolarized potentials. The results of both pharmacologic and ionic manipulations indicated that NMA can directly activate a voltage-dependent calcium conductance in pyramidal cells. The apparent rise in input resistance could be explained by NMA inducing a region of negative slope conductance. Our work, and that of Hablitz and Langmoen (1982), indicate that NMA and glutamate have predominantly different modes of action: NMA appears to preferentially activate calcium conductance, whereas glutamate mostly activates sodium and potassium conductances.

Electrophysiological evidence has thus been provided for multiple receptors for acidic amino acids on pyramidal neurons. Similar results have been obtained by MacDonald and Wojtowicz (1982) on cultured mouse spinal cord cells. Synaptic excitation at a variety of CNS synapses is reported to be sensitive to NMDA antagonists (Watkins and Evans, 1981). In particular, in the hippocampal CA1 region NMDA receptor antagonists may prevent long-term potentiation (LTP) produced by brief tetanic stimuli (Collingridge *et al.*, 1982b) but not conventional synaptic transmission at low frequency (Koerner and Cotman, 1982). If the natural transmitter released by tetanic stimulation does act on NMDA re-

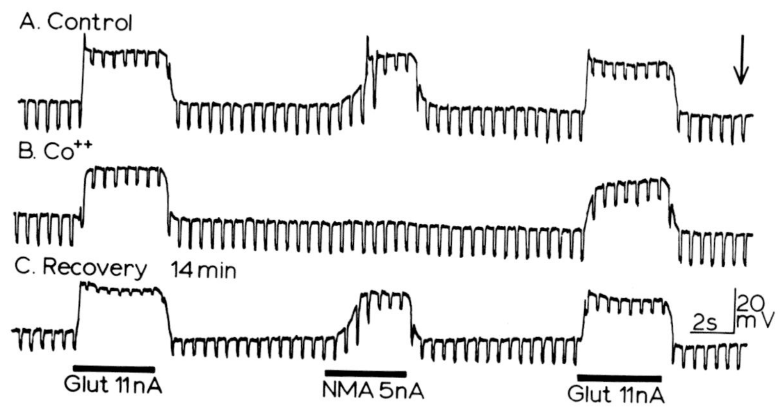

Figure 11. Cobalt selectively blocks *N*-methyl-DL-aspartate (NMA) evoked conductance changes. (A) NMA and L-glutamate (Glut) were alternately iontophoresed from a two-barrelled pipette at regular intervals as illustrated. At the arrow, a droplet of 10-mM Co^{2+} in ACSF was applied to the surface of the slice near the iontophoretic pipette. The trace in (B) began 14 sec after Co^{2+} application. The NMA response was completely blocked, while the glutamate response was unaffected (with the exception of an initial small calcium spike). (C) Recovery developed gradually. Tetrodotoxin (1 μM) was present to eliminate sodium action potentials. From Dingledine, 1983a.

ceptors, the above studies suggest that a voltage-dependent calcium conductance would be chemically activated during the tetanus.

4. CONCLUSIONS AND FUTURE DIRECTIONS

The hippocampal slice, due to its mechanical stability and controllable environment, has proved a useful preparation to examine the actions of drugs on neurons in this brain region. It is encouraging that, with a few notable exceptions, findings obtained in different laboratories have been mutually supportive. The majority of studies to date, including ours, have addressed some aspect of the chemical regulation of neuronal excitability, especially with regard to burst firing in pyramidal cells. This research approach is straightforward and readily linked to a group of major neurologic diseases, the hippocampal epilepsies. It can be anticipated that, with expanding knowledge of the cellular distribution of neuroactive peptides, more subtle aspects of hippocampal pharmacology will receive attention in the future.

It will be increasingly important to integrate more fully the findings obtained from pharmacological and physiological studies in the hippocampus in order to appreciate the functional implications of the presence and distribution of our favorite drug receptors. Inevitably, this will require some turning back to the intact brain to gain a satisfactory understanding. For example, detailed physiological studies from intact animals will be unavoidable to gain a satisfactory understanding of the roles the two effects of ACh play in mediating the actions of the septohippocampal pathway, and to learn the behavioral consequences of the synaptic mechanisms underlying LTP.

The brain slice can also provide a meaningful bridge between *in vivo* functional studies and the more mechanistic studies of receptor-linked ion channels that are possible in tissue culture. By carrying out concurrent studies of drug action on hippocampal physiology in the slice, and of receptor-activated individual ion channels in hippocampal cultures, one can cautiously begin to understand the linkage between drug binding and ultimate physiological response, and to reach a richer appreciation of the drug effect itself. In any event, the hippocampal slice should remain a workhorse preparation for the neuropharmacologist for many years to come.

ACKNOWLEDGMENTS. I thank Dr. Gregory King and Stephen Korn for helpful comments on the manuscript. Research reported here was supported in part by NIDA grant DA-02360, NIH grant NS-17771, and a Sloan Foundation Fellowship.

5. REFERENCES

Alger, B. E. and Nicoll, R. A., 1979, GABA-mediated biphasic inhibitory responses in hippocampus, *Nature,* (*London*) **281:**315–317.

Alger, B. E. and Nicoll, R. A., 1982, Feed-forward dendritic inhibition in rat hippocampal pyramidal cells studied *in vitro, J. Physiol.* (*London*) **328:**105–123.

Andersen, P., 1975, Organization of hippocampal neurons and their connections, in: *The Hippocampus,* Volume 1 (R. L. Isaacson and K. H. Pribram, eds.),Plenum Press, New York, pp. 155–175.

Andersen, P., Bie, B., Ganes, T., and Mosfeldt Laursen, A., 1978a, Two mechanisms for effects of GABA on hippocampal pyramidal cells, in: *Iontophoresis and Transmitter Mechanisms in the Mammalian Central Nervous System* (R. Ryall and J. S. Kelly, eds.), Elsevier N. Holland Biomedical Press, pp. 178–181.

Andersen, P., Silfvenius, H., Sundberg, S. H., Sveen, O., and Wigström, H., 1978b, Functional characteristics of unmyelinated fibres in the hippocampal cortex, *Brain Res.* **144:**11–18.

Andersen, P., Eccles, J. C., and Løyning, Y., 1964, Pathway of postsynaptic inhibition in the hippocampus, *J. Neurophysiol.* **27:**608–619.

Andersen, P., Dingledine, R., Gjerstad, L., Langmoen, I. A., and Mosfeldt-Laursen, A., 1980, Two different responses of hippocampal pyramidal cells to application of gamma-amino butyric acid, *J. Physiol.* (*London*) **305:**279–296.

Azmitia, E. C. and Segal, M., 1979, An autoradiographic analysis of the differential ascending projections of the dorsal and median raphe nuclei in the rat, *J. Comp. Neurol.* **179:**641–668.

Benardo, L. S. and Prince, D. A., 1982a, Cholinergic excitation of mammalian hippocampal pyramidal cells, *Brain Res.* **249:**315–331.

Benardo, L. S. and Prince, D. A., 1982b, Ionic mechanisms of cholinergic excitation in mammalian hippocampal pyramidal cells, *Brain Res.* **249:**333–344.

Ben-Ari, Y., Krnjevic, K., Reiffenstein, R., and Ropert, N., 1981, Intracellular observations on disinhibitory action of acetylcholine in hippocampus, *Neuroscience* **6:**2475–2484.

Biscoe, T. J. and Straughan, D. W., 1966, Micro-electrophoretic studies of neurones in the cat hippocampus, *J. Physiol.* (*London*) **183:**341–359.

Bostock, E., Dingledine, R., Xu, G., and Chang, K.-J., Epileptiform effect of a mu-opioid receptor agonist, morphiceptin, in the hippocampal slice, submitted.

Chang, K.-J., Cooper, B. R., Hazum, E., and Cuatrecasas, P., 1979, Multiple opiate receptors: Different regional distribution in the brain and differential binding of opiates and opioid peptides, *Mol. Pharmacol.* **16:**91–104.

Collingridge, G. L., Kehl, S. J., and McLennan, H., 1982a, Antagonism of amino acid induced excitation of CA1 hippocampal neurones *in vitro, J. Physiol.* (*London*) **322:**52P.

Collingridge, G. L., Kehl, S. J., and McLennan, H., 1982b, The effect of excitatory amino acid agonists and antagonists on the Schaffer collateral input to rat CA1 hippocampal neurones, *J. Physiol.* (*London*) **322:**53P.

Corrigall, W. A. and Linseman, M. A., 1980, A specific effect of morphine on evoked activity in the rat hippocampal slice, *Brain Res.* **192:**227–238.

Crutcher, K. A., Madison, R., and Davis, J. N., 1981, A study of the rat septohippocampal pathway using anterograde transport of horseradish peroxidase, *Neuroscience* **6:**1961–1973.

Dingledine, R., 1981, Possible mechanisms of enkephalin action on hippocampal CA1 pyramidal neurons, *J. Neurosci.* **1:**1022–1035.

Dingledine, R., 1982, Amino acid activated calcium conductance in hippocampal pyramidal cells, *Soc. Neurosci. Abstr.* **8:**796.

Dingledine, R., 1983a, *N*-Methyl-aspartate activates voltage dependent calcium conductance in rat hippocampal pyramidal neurones, *J. Physiol.* (*London*) in press.

Dingledine, R., 1983b, Excitatory amino acids: Modes of action on hippocampal pyramidal cells, *Fed. Proc.*, **42:**2281–2285.

Dingledine, R. and Gjerstad, L., 1979, Penicillin blocks hippocampal IPSPs, unmasking prolonged EPSPs, *Brain Res.* **168:**205–209.

Dingledine, R. and Gjerstad, L., 1980, Reduced inhibition during epileptiform activity in the *in vitro* hippocampal slice, *J. Physiol.* (*London*) **305:**297–313.

Dingledine, R. and Langmoen, I. A., 1980, Conductance changes and inhibitory actions of hippocampal recurrent IPSPs, *Brain Res.* **185:**277–287.

Dodd, J. and Kelly, J. S., 1979, Is somatostatin an excitatory transmitter in the hippocampus? *Nature* (*London*) **273:**674–675.

Dodd, J. and Kelly, J. S., 1981, The actions of cholecystokinin and related peptides on pyramidal neurones of the mammalian hippocampus, *Brain Res.* **205:**337–350.

Dodd, J., Kelly, J. S., and Said, S. I., 1979, Excitation of CA1 neurones of the rat hippocampus by the octacosapeptide, vasoactive intestinal polypeptide (VIP), *Br. J. Pharmacol.* **66:**125P.

Dodd, J., Dingledine, R., and Kelly, J. S., 1981, The excitatory action of acetylcholine on hippocampal neurones of the guinea pig and rat maintained *in vitro, Brain Res.* **207:**109–127.

Dunwiddie, T., Mueller, A., Palmer, M., Steward, J., and Hoffer, B., 1980, Electrophysiological interactions of enkephalins with neuronal circuitry in the rat hippocampus. I. Effects on pyramidal cell activity, *Brain Res.* **184:**311–330.

Eccles, J. C., Nicoll, R. A., Oshima, T., and Rubia, F. J., 1977, The anionic permeability of the postsynaptic membrane of hippocampal pyramidal cells, *Proc. R. Soc. Lond.* [*Biol*] **198:**345–361.

Gähwiler, B. H. and Maurer, R., 1981, Involvement of μ-receptors in the opioid-induced generation of bursting discharges in hippocampal pyramidal cells, *Regulatory Peptides* **2:**91–96.

Gall, C., Brecha, N., Karten, H. J., and Chang, K.-J., 1981, Localization of enkephalin-like immunoreactivity to identified axonal and neuronal populations of the rat hippocampus, *J. Comp. Neurol.* **198:**335–350.

Gold, M. R. and Martin, A. R., 1982, Intracellular Cl^- accumulation reduces Cl^- conductance in inhibitory synaptic channels, *Nature* **299:**828–830.

Greenwood, R. S. and Winstead, K. K., 1981, Immunocytochemical double labeling for cholecystokinin and somatostatin in the rat hippocampus, *Soc. Neurosci. Abstr.* **7:**100.

Greenwood, R. S., Godar, S. E., Reaves, T. A., and Hayward, J. N., 1981, Cholecystokinin in hippocampal pathways, *J. Comp. Neurol.* **203:**335–350.

Haas, H. L., 1982, Cholinergic disinhibition in hippocampal slices of rat, *Brain Res.* **233:**200–204.

Haas, H. L. and Gähwiler, B. H., 1980, Do enkephalins directly affect calcium-spikes in hippocampal pyramidal cells? *Neurosci. Lett.*, **19:**89–92.

Haas, H. L. and Ryall, R. W., 1980, Is excitation by enkephalins of hippocampal neurons in the rat due to presynaptic facilitation or to disinhibition? *J. Physiol.* (*London*) **308:**315–330.

Haas, H. L., Felix, D., Celio, M. R., and Inagami, T., 1980, Angiotensin II in the hippocampus. A histochemical and electrophysiological study, *Experientia* **36:**1394–1395.

Haas, H. L., Felix, D., and Davis, M. D., 1982, Angiotensin excites hippocampal pyramidal cells by two mechanisms, *Cell. Mol. Neurobiol.* **2:**21–32.

Hablitz, J. J. and Langmoen, I. A., 1982, Excitation of hippocampal pyramidal cells by glutamate in the guinea-pig and rat, *J. Physiol.* (*London*) **325:**317–331.

Halliwell, J. V. and Adams, P. R., 1982, Voltage clamp analysis of muscarinic excitation in hippocampal neurons, *Brain Res.* **250:**71–92.

Handelmann G., Meyer, D. K., Beinfeld, M. C., and Oertel, W. H., 1981, CCK-containing terminals in the hippocampus are derived from intrinsic neurons: An immunohistochemical and radioimmunological study, *Brain Res.* **224:**180–184.

Henricksen S. J., Bloom, F. E., McCoy, F., Ling, N., and Guillemin, R., 1978, β-endorphin induces non-convulsive limbic seizures, *Proc. Nat'l. Acad. Sci. USA* **75:**5221–5225.

Hounsgaard, J., 1978, Presynaptic inhibitory action of acetylcholine in area CA1 of the hippocampus, *Exp. Neurol.* **62:**787–797.

Kimura, H., McGeer, P. L., Peng, J. H., and McGeer, E. G., 1981, The central cholinergic system studied by choline acetyltransferase immunohistochemistry in the cat, *J. Comp. Neurol.* **200:**151–200.

Knowles, W. D. and Schwartzkroin, P. A., 1981, Local circuit synaptic interactions in hippocampal brain slices, *J. Neurosci.* **1:**318–322.

Koerner, J. F. and Cotman, C. W., 1981, Micromolar L-2-amino-4-phosphonobutyric acid selectively inhibits perforant path synapses from lateral entorhinal cortex, *Brain Res.* **216:**192–198.

Koerner, J. F. and Cotman, C. W., 1982, Response of Schaffer collateral-CA1 pyramidal cell synapses of the hippocampus to analogues of acidic amino acids, *Brain Res.* **251:**105–115.

Krnjević, K., Pumain, R., and Renaud, L., 1971,The mechanism of excitation by acetylcholine in the cerebral cortex, *J. Physiol.* (*London*) **215:**247–268.

Lanthorn, T. H. and Cotman, C. W., 1981, Baclofen selectively inhibits excitatory synaptic transmission in the hippocampus, *Brain Res.* **225:**171–178.

Lee, H. K., Dunwiddie, T., and Hoffer, B., 1980, Electrophysiological interactions of enkephalins with neuronal circuitry in the rat hippocampus. II. Effects on interneuron excitability, *Brain Res.* **184:**331–342.

Lindvall, O. and Bjorkland, A., 1974, The organization of the ascending catecholamine neurone systems in the rat brain, as revealed by the glyoxylic acid fluorescence method, *Acta Physiol. Scand.* **73**(suppl 412):1–48.

Loren, I., Emson, P. C., Fahrenkrug, J., Bjorklund, A., Alumets, J., Hakänson, R., and Sundler, F., 1979, Distribution of vasoactive intestinal polypeptide in the rat and mouse brain, *Neuroscience* **4:**1953–1976.

Lorente de Nó, R., 1934, Studies on the structure of the cerebral cortex. II. Continuation of the study of the ammonic system, *J. Psychol. Neurol.* (*Leipzig*) **46:**113–177.

Lynch, G., Rose, G., and Gall, C., 1978, Anatomical and functional aspects of the septohippocampal projections, in: *Functions of the Septo-Hippocampal System,* Ciba Foundation Symposium no. 58, Elsevier-North Holland, Amsterdam, pp 5–20.

Lynch, G. S., Jensen, R. A., McGaugh, J. L., Davila, K., and Oliver, M. W., 1981, Effects of enkephalin, morphine and naloxone on the electrical activity of the *in vitro* hippocampal slice preparation, *Exp. Neurol.* **71:**527–540.

MacDonald, J. F. and Wojtowicz, J. M., 1982, The effects of L-glutamate and its analogues upon the membrane conductance of central murine neurones in culture, *Canad. J. Physiol. Pharmacol.* **60:**282–296.

Masukawa, L. M. and Prince, D. A., 1982, Enkephalin inhibition of inhibitory input to CA1 and CA3 pyramidal neurons in the hippocampus, *Brain Res.* **249:**271–280.

McGinty, J. F., Henricksen, S. J., Goldstein, A., Terenius, L., and Bloom, F. E., 1983, Dynorphin is contained within hippocampal mossy fibers: Immunochemical alterations after kainic acid administration and colchicine-induced neurotoxicity, *Proc. Natl. Acad. Sci. USA* **80:**589–593.

Moore, R. Y. and Halaris, A. E., 1975, Hippocampal innervation by serotonin neurons of the midbrain raphe in the rat, *J. Comp. Neurol.* **152:**163–174.

Nadler, J. V., Vaca, K. W., White, W. F., Lynch, G. S., and Cotman, C. W., 1976, Aspartate and glutamate as possible transmitters of excitatory hippocampal afferents, *Nature* (*London*) **260:**538–540.

Nadler, J. V., White, W. F., Vaca, K. W., Perry, B. W., and Cotman, C. W., 1978, Biochemical correlates of transmission mediated by glutamate and aspartate, *J. Neurochem.* **31:**147–155.

Nicoll, R. A., Alger, B. E., and Jahr, C. E., 1980, Enkephalin blocks inhibitory pathways in the vertebrate CNS, *Nature* (*London*) **287:**22–25.

Ribak, C. E., Vaughn, J. E., and Saito, K., 1978, Immunocytochemical localization of glutamic acid decarboxylase in neuronal sonata following colchicine inhibition of axonal transport, *Brain Res.* **140:**315–332.

Robinson, J. H. and Deadwyler, S. A., 1981, Intracellular correlates of morphine excitation in the hippocampal slice preparation, *Brain Res.* **224:**375–387.

Schwartzkroin, P. A. and Mathers, L. H., 1978, Physiological and morphological identification of a non-pyramidal hippocampal cell type, *Brain Res.* **157:**1–10.

Segal, M., 1982, Multiple actions of acetylcholine at a muscarinic receptor studied in the rat hippocampal slice, *Brain Res.* **246:**77–87.

Silfvenis, H., Olofsson, S., and Ridderheim, P.-A., 1980, Induced epileptiform activity evoked from dendrites of hippocampal neurones, *Acta Physiol. Scand.* **108:**109–111.

Storm-Mathisen, J., 1977, Localization of transmitter candidates in the brain: The hippocampal formation as a model, *Prog. Neurobiol.* **8:**119–181.

Thalmann, R. H., Peck, E. J., and Ayala, G. F., 1981, Biphasic response of hippocampal pyramidal neurons to GABA, *Neurosci. Lett.* **21:**319–324.

Urca, G., Frenk, H., Liebeskind, J. C., and Taylor, A. N., 1977, Morphine and enkephalin: Analgesic and epileptic properties, *Science* **197:**83–86.

Valentino, R. J. and Dingledine, R., 1981, Presynaptic inhibitory effect of acetylcholine in the hippocampus, *J. Neurosci.* **1:**787–792.

Valentino, R. J. and Dingledine, R., 1982, Pharmacological characterization of opioid effects in the rat hippocampal slice, *J. Pharmacol. Exp. Ther.* **223:**502–509.

Vincent, S. R., Kimura, H., and McGeer, E. G., 1981, Organization of substance P fibers within the hippocampal formation demonstrated with a biotin–avidin immunoperoxidase technique, *J. Comp. Neurol.* **199:**113–123.

Watkins, J. C. and Evans, R. H., 1981, Excitatory amino acid transmitters, *Annu. Rev. Pharmacol. Toxicol.* **21:**165–204.

White, W. F., Nadler, J. V., Hamberger, A., Cotman, C. W., and Cummins, J. T., 1977, Glutamate as a transmitter of hippocampal perforant path, *Nature* (*London*) **270:**356–357.

White, W. F., Nadler, J. V., and Cotman, C. W., 1979, The effect of acidic amino acid antagonists on synaptic transmission in the hippocampal formation *in vitro, Brain Res.* **164:**177–194.

Wieraszko, A. and Lynch, G., 1979, Stimulation-dependent release of possible transmitter substances from hippocampal slices studied with localized perfusion, *Brain Res.* **160:**372–376.

Wong, R. K. S. and Prince, D. A., 1979, Dendritic mechanisms underlying penicillin-induced epileptiform activity, *Science* **204:**1228–1231.

Wong, R. K. S. and Watkins, D. J., 1982, Cellular factors influencing GABA response in hippocampal pyramidal cells, *J. Neurophysiol.* **48:**938–951.

Yamamoto, C. and Kawai, N., 1967, Presynaptic action of acetylcholine in thin sections from the guinea pig dentate gyrus *in vitro, Exp. Neurol.* **19:**176–187.

Zieglgänsberger, W., French, E., Siggins, G., and Bloom, F., 1979, Opioid peptides may excite hippocampal pyramidal neurons by inhibiting adjacent inhibitory interneurons, *Science* **205:**415–417.

5

Energy Metabolism and Brain Slice Function

PETER LIPTON and TIM S. WHITTINGHAM

1. INTRODUCTION

The brain slice preparation has been used in a wide variety of investigations since its development in the 1930s by Quastel and by Elliot (Elliot and Wolfe, 1962). Prior to the 1970s, most of this work centered on slice metabolism, with a particular focus on the metabolic consequences of electrical activity (McIlwain and Bachelard, 1971). Although these studies were fundamental to the whole development of the brain slice as a useful preparation, they had a serious drawback in that normal electrophysiological responses could not be obtained from these cortical preparations. Thus, electrical activity was mimicked by profound membrane depolarizations, produced either by high-frequency electrical stimulation or by large changes in extracellular K^+ concentrations. More recently, the development of the olfactory and the hippocampal slice preparations (Yamamoto and Kurokawa, 1970; Skrede and Westgaard, 1971) has opened the door to much more sophisticated studies of the relationship between neural activity and energy metabolism than was possible with the previous cortical slice preparation. Thus, it is now possible to correlate metabolic and electrophysiological changes in different con-

PETER LIPTON • Department of Physiology, University of Wisconsin, Madison, Wisconsin 53706. TIM S. WHITTINGHAM • Laboratory of Neurochemistry, National Institute of Neurological and Communicative Disorders and Stroke, National Institutes of Health, Bethesda, Maryland 20205.

ditions and thereby determine mechanisms by which neural transmission affects metabolism and vice-versa (Yamamoto and Kurokawa, 1970). In the second half of this chapter, we shall discuss the effects of altered energy metabolism on neural transmission. We shall focus on our own work in which we have tried to determine the mechanisms involved in the rapid inhibition of transmission that occurs in the brain when the supply of oxygen is removed. Although this represents an application of the slice technique to an important physiological problem, there is a more universal question concerning slice energy metabolism; namely, to what extent does the energy metabolism and ion balance of the slice resemble that of the tissue *in situ*? The answer to this question may affect the interpretation of a wide variety of studies on slices, including electrophysiological and pharmacological investigations. The early studies of McIlwain and others demonstrated that energy metabolism in cortical slices was severely compromised with respect to *in situ* tissue (summarized in McIlwain and Bachelard, 1971). Our studies on the hippocampal slice demonstrate that this preparation suffers from similar deficits (Lipton and Heimbach, 1977; Lipton and Whittingham, 1979). Thus, the large number of electrophysiological studies that are being carried out on the hippocampal slice (and the same considerations should apply to other tissue slices) are being carried out on tissue that is quite different from the *in situ* tissue in terms of some very fundamental parameters. In the first half of the chapter, we summarize the differences in energy-related parameters between the slice and *in situ* tissue, and attempt to account for these differences. The purpose is to provide the scientist who uses tissue slices with information and analysis that may help relate results obtained *in vitro* to events occuring *in situ*. Clearly, this is an important relationship to understand if results from slice preparations are to be applied to *in situ* situations.

2. INTEGRITY OF THE SLICE PREPARATION

2.1. *Comparison of Energy-Related Parameters in Slices and in Situ*

2.1.1. Respiration. The rate of oxygen consumption in brain slices is about 40% of its value *in situ*. Thus, slices of guinea pig cortex in phosphate buffer respire at a rate of about 1 μmole O_2/g fresh tissue per min (McIlwain, 1953) while they respire at a slightly higher rate of about 1.3 μmole O_2/g per min in a bicarbonate buffer (Rolleston and Newsholme, 1967). Rat cortex slices respire at a somewhat higher rate of about 1.6 μmole/g per min (Benjamin and Verjee, 1980) and the rate appears to be independent of which anion is serving as the major buffer base (Elliot and Wolfe, 1962). Although there are no values reported for the respiration of guinea pig brain *in situ*, there are a multitude of reported measurements of rat brain respiratory rates. For rats anesthetized with nitrous oxide, the cortical respiratory rate is about 4 μmole O_2/g fresh tissue per min (Bertman *et al.*, 1979), while the value is slightly higher in unanesthetized animals (Ghajar *et al.*, 1982).

It has been suggested that the low slice respiration may be due to the absence of spontaneous neural activity in this preparation (McIlwain and Bachelard, 1971). Thus, both barbiturate anesthetics (Siesjo, 1978) and coma (Sokoloff, 1971) effect an approximate 50% reduction in respiration of the intact brain, bringing it to a value close to that observed for the slices, and anesthetics have very little effect on slice respiration (McIlwain, 1953). In addition, the fraction of respiration that is dependent upon active Na^+/K^+ transport is similar in the two preparations. In the absence of Ca^{2+}, about 50% of slice respiration is blocked by ouabain (Whittam, 1962), while in the perfused anesthetized canine brain, about 40% of the respiration is blocked by a combination of lidocaine and ouabain (Astrup *et al.*, 1981). Further support comes from the observation that auditory deprivation rapidly reduces the glucose consumption of auditory cortex by about 50% (Sokoloff, 1981).

Although reduced spontaneous activity may account for at least some of the lowered respiration, attempts to draw this conclusion by subjecting the slice to intense electrical stimulation do not seem valid. Thus, while high-frequency pulses do increase tissue respiration twofold, intracellular ion balance is grossly disturbed (Keesey *et al.*, 1965) and tissue PCr/ATP falls significantly. Similar changes are observed in intact brain tissue during seizure activity (Meldrum and Nilsson, 1976), suggesting that high-frequency stimulation of brain slices really represents a pathological paradigm rather than a model for normal spontaneous activity. The very large effect of anesthetics and sensory deprivation on brain respiration *in situ* is not understood. It is not clear that the effect is due solely to decreased ion movements and exocytotic activity in the tissue.

2.1.2. *Aerobic Glycolysis.* There is very little aerobic lactate production *in vivo*. Thus, measured rates of glucose consumption and respiration suggest that a maximum of 5% of metabolized glucose forms lactic acid (Hawkins *et al.*, 1971; Norberg and Siesjo, 1975). For example, in unanesthetized rats, an oxygen consumption of 4.6 μmole/g per min is accompanied by a glucose utilization of 0.83 μmole/g per min (Ghajar *et al.*, 1982). In brain slices, the situation is quite different. Between one-third and two-thirds of metabolized glucose forms lactic acid in a situ-

Table I. Energy-Related Parameters in Vivo and in Vitro for Guinea Pig Hippocampus[a]

	In vivo[b]	*In vitro*[c]	Change *in vitro* (%)
ATP	29.1	13.0	−55
ADP	3.0	2.1	−30
AMP	0.4	0.5	+25
Total adenylates	31.2	15.5	−50
Phosphocreatine	56.0	32.2	−43
Creatine	59.7	27.1	+191
Lactate	8.9	17.1	−68
PCr/ATP	1.92	2.46	+137
Energy charge	0.946	0.903	−4.5

[a] All values, except ratios, in nM/mg protein.
[b] Animals' brains were funnel-frozen during anesthesia with 4 mg/100 g Surital.
[c] Slices incubated for 6 hr in standard Krebs buffer.

ation where glucose is essentially the only metabolizable substrate. Thus, in guinea pig cortex slices, glucose is consumed at a rate of 0.34 μmole/g tissue per min while lactate is produced at a rate of 0.2 μmole/g per min (Rolleston and Newsholme, 1967). In rat cortex slices, a rate of respiration equal to 1.7 μmole/g per min is accompanied by a rate of lactate production of 0.8 μmole/g per min (Benjamin and Verjee, 1980).

The mismatching of glucose utilization to respiration is characteristic of strongly stimulated brain tissue *in vivo*. Such data suggest that the capacity of mitochondria is normally less than that of the glycolytic pathway and, for an as yet unidentified reason, this mismatch is being expressed in brain slices. A straightforward explanation is that an anoxic core in the slice is causing the increase in glycolysis; however, as will be seen later, this is probably not the case. It is possible that the high intracellular pH (see below) strongly activates the glycolytic pathway (Wu and Davis, 1981). Alternatively, in some regions of the slice, the mitochondria may be severely damaged while the glycolytic system is still functioning.

2.1.3. *High-Energy Phosphates.* Table I compares the *in vivo* and *in vitro* levels of the adenylates and phosphocreatine (PCr), as well as creatine and lactate, from hippocampal tissue. There is an approximate 50% decrease in the *in vitro* content of most of these compounds, the notable exceptions being AMP and lactate. Both of these compounds are elevated in the slice. The PCr to ATP ratio appears to increase for tissue slices even though total adenylate (ATP + ADP + AMP) and total creatine (PCr + creatine) contents both decrease by 50%. The levels of ATP and PCr for *in situ* hippocampal tissue are slightly higher than levels in cortex but, in general, the metabolic profiles of the two regions are quite similar (Folbergrova *et al.*, 1981). The *in situ* values shown here are quite similar to those reported by Folbergrova *et al.* (1981) for rat hippocampus.

2.1.4. Energy Charge. The energy charge (EC) is a measure of the fraction of tissue adenylate content which is readily available in the form of high-energy phosphate bonds. It is calculated as: EC = (ATP + 0.5 ADP)/(ATP + ADP + AMP). *In vivo* EC is approximately 0.94 to 0.95 in rat cerebral cortex (Salford *et al.*, 1973; MacMillan, 1975; Folbergrova *et al.*, 1981) and hippocampus from rat (Folbergrova *et al.*, 1981) and guinea pig (T. S. Whittingham, unpublished observations). This compares to a steady-state energy charge of about 0.90 in the hippocampal slice from guinea pig, incubated for 4 to 8 hr (Table I). Thus, there is apparently a slight fall in energy charge for *in vitro* tissue.

2.1.5. Intracellular pH. To date, intracellular pH (pH_i) in most mammalian tissues, including brain, has only been estimated indirectly by calculating the H^+ ion concentration from measured values of reactants in an equilibrium reaction involving hydrogen ions (Roos and Boron, 1981). The two most reliable methods for doing this are to use the weak acid, DMO, and the CO_2–HCO_3 equilibrium (Siesjo, 1978). The results lead to calculations of pH_i values in the *in situ* rat brain of between 6.94 and 7.05 (MacMillan and Siesjo, 1972; Lai *et al.*, 1973; Arieff *et al.*, 1976). Siesjo and co-workers have demonstrated that the reaction catalyzed by creatine kinase can also be used to calculate pH_i (MacMillan and Siesjo, 1972; Siesjo *et al.*, 1972). It has the great advantage of providing the ability to estimate rapid (1 min) changes in pH. However, if the *in vitro* value for the equilibrium constant is used in the calculation, the resultant pH is very alkaline, at 8.1 (Kobayashi *et al.*, 1977). MacMillan and Siesjo (1972) found that a value for the equilibrium constant of $1.41 \times 10^8\ M^{-1}$, rather than the biochemically more correct value of $1.66 \times 10^9\ M^{-1}$ (Veech *et al.*, 1979), was needed to bring the "creatine kinase pH" into agreement with the values calculated by the other two methods. Although the absolute value of the creatine kinase pH is, then, not reliable, it does provide good estimates of changes in pH_i when brain tissue is perturbed by altered CO_2 or by hypoxia (MacMillan and Siesjo, 1972; Siesjo *et al.*, 1972; Norberg *et al.*, 1975). Thus, the equilibrium constant does not seem to be strongly affected by altered metabolic states of the tissue. Using the value of the creatine kinase equilibrium constant calculated for the *in situ* brain (MacMillan and Siesjo, 1972), we have calculated pH_i values for rat and guinea pig slices of between 7.34 and 7.49. The DMO equilibrium technique produced similar values in these slices (Kass and Lipton, 1982; P. Lipton and T. S. Whittingham, unpublished observations). Thus, measurements in our laboratories show a significant alkaline shift in slice pH_i. Other studies, however, determined a pH_i of 6.98 for rat cortical slices using the DMO method (Hertz *et al.*, 1970).

As with the *in situ* tissue, there is good evidence that the creatine kinase pH yields reliable estimates of rapid changes during altered metabolic states *in vitro*. Thus, we calculate a 0.4 unit acidification of the hippocampal slice during 4 min of anoxia, and a larger shift during 10 min of anoxia (Kass and Lipton, 1982). The magnitudes of these changes are very close to the values recorded *in vivo* by other techniques (Kaasik *et al.*, 1970; Ljunggren *et al.*, 1974). In addition, when we exposed hippocampal slices to 18% CO_2, intracellular pH showed a rapid 0.4 unit acidification (P. Lipton and T. S. Whittingham, unpublished observations). This value is very close to the effect of a similar increase in CO_2 *in vivo* calculated by other methods (Siesjo *et al.*, 1972).

The reason for the alkaline shift in slices is not apparent, considering that it occurs in tissue that exhibits an increased rate of aerobic glycolysis. One possible explanation is that there is a significant shift in some slice buffer base concentrations, such as bicarbonate ion. Alternatively, the transport characteristics of active H^+ extrusion across the plasmalemma or other cellular membrane may be altered.

2.1.6. Intracellular Ion and Water Contents. Cell Na^+ and K^+ levels are strongly dependent upon energy metabolism and are important determinants of cell function. Along with cell water content, they are good indicators of the health of a preparation. Unfortunately, it is difficult to use the literature to compare *in situ* and slice values for these parameters. This difficulty stems partly from the different ways in which results are expressed (i.e., ions/protein; ions/dry weight; ions/fresh tissue weight; ions/tissue weight after 60 min incubation). It also stems from the uncertainty associated with determinations of extracellular water spaces and the fact that different extracellular markers are used by different workers (Katzmann and Pappius, 1973; Cohen, 1974). Furthermore, while a purported extracellular marker such as inulin may be restricted to one free compartment *in situ*, it may have access to additional compartments in the slice (Cohen, 1974; Amtorp, 1979). In this section, we shall rationalize the data from several laboratories so that values of these parameters in slices and *in situ* can be meaningfully compared.

Figure 1 shows the distribution of tissue water both *in situ* and in the slice. The water spaces are referred to on a basis of tissue dry weight

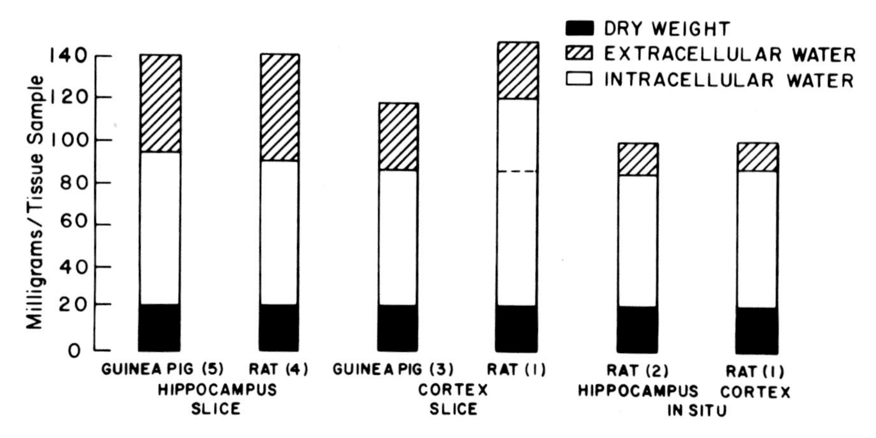

Figure 1. Water compartments in brain slices and *in situ* brain. The bars show the amounts of intracellular and extracellular water associated with 20 mg of dry weight. The different groups have been appropriately recalculated to this reference point. All slice data have been calculated on the assumption that 20% of total inulin is not in an extracellular space. Results in (3) were reworked as this assumption was not made in that case. *In situ* data do not require this correction because of the small size of the inulin space. (1) Amtorp (1979). (2) Rees *et al.* (1982). (3) Keesey *et al.* (1965). (4) Kass and Lipton (1982). (5) Lipton and Robacker (unpublished results).

and, in one case (Amtorp, 1979), this has been done by our using a dry weight/protein ratio of 2.17, a value we obtain for fresh hippocampal tissue from both the rat and guinea pig. Both *in situ* (Lund-Anderson, 1974; Amtorp, 1979) and in the slice (Lipton and Heimbach, 1978), about 20% of the tissue-associated inulin is in a very slowly exchanging compartment that is not considered to be extracellular. While other workers have not allowed for this, we have amended their data to derive new values for both fluid spaces and for ion concentrations. Such manipulations are indicated in the legend as are other treatments of literature data which were done to permit a more ready comparison between values in different preparations. The figure demonstrates that there is very little difference in apparent cell volume between the slices and the *in situ* preparation (intracellular volume is well measured as intracellular water/dry weight, as there is little loss of dry weight during 2 hr of slice incubation). In the rat cortex, where the cell volume does appear greatly enlarged (Amtorp, 1979), this is largely because the calculated extracellular water does not include some of the space in which inulin is actually distributed. Thus, extracellular water is underestimated and the intracellular space accordingly overestimated. The "volume" under the dashed line in this bar is considered by Amtorp to be a closer approximation to the true intracellular space. It is clearly similar to that in the other preparations.

Although these data seem to suggest that the cells in the slice are not swollen, in fact they probably are. Thus, a significant portion of the dry weight of the tissue is probably not associated with the calculated intracellular volume. As will be discussed later, there is reasonable evidence that between 25 and 30% of the cells in the slice are severely damaged, so that while they retain their dry weight, their membranes are leaky to the extracellular markers. Thus, about 25% of the dry weight in the slices may be associated with the calculated extracellular marker space or, put another way, a significant fraction of the extracellular marker space in the slices may actually be composed of very leaky cells. If this is so, then the tissue dry weight, which is actually associated with the calculated intracellular volume, is only about three-quarters of the total dry weight. Thus, treating the guinea pig hippocampal slice cells as an example, the actual value for intracellular volume/intracellular dry weight is 5.0 rather than 3.6. This compares with a value of about 3.2 for the tissue *in situ*. Thus, the cells that are impermeable to inulin are in fact probably quite swollen. Although there is a large increase in the extracellular marker space in the slices, much of this probably represents the space occupied by the leaky cells. The actual intercellular spaces are probably not dramatically enlarged. Other discussions of this increased extracellular marker space have been provided (i.e., Keesey *et al.*, 1965; Katzmann and Pappius, 1973) and Katzmann and Pappius consider the increased volume of brain slices to be largely due to an expanded intercellular space. Our analysis suggests the opposite point of view; that the slice volume increase is due to swelling of the intact cells of the preparation.

Table II summarizes data for ion concentrations of various slice and *in situ* preparations. (It is notable that allowing for the approximately 20% of the inulin, which is not distributed in extracellular fluid or in the fluid of damaged cells, significantly lowers calculated values for intracellular K^+ concentration. Although not evident from this table, it also elevates intracellular Na^+ concentrations by about 25%.) The table shows clearly that K^+ contents and intracellular concentrations in the noninulin spaces are significantly reduced in the slice compared to *in situ* tissue. Although more difficult to compare, intracellular sodium concentrations are significantly increased in the slices. Though these cation concentrations can be brought closer to *in situ* values by elevating extracellular K^+ to values close to 20 mM, no other manipulations have proved successful (Bachelard *et al.*, 1963). Thus, when the slice is bathed

Table II. Ion Contents of Brain Slices and of Brain Tissue in Situ

Tissue origin	K^+/Dry wt (mMole/kg)	K^+ (mM)	Na^+/Dry wt (mMole/kg)	Na^+ (mM)
Brain slices				
Guinea pig hippocampus (Lipton and Robacker, unpublished results)	369[a]	101[a]	510	50[a]
Rat hippocampus (Kass and Lipton, 1982)	396[a]	113[a]	—	73[a]
Guinea pig cortex (Keesey *et al.*, 1965)	350	120 (107)[a]	390	40
Cat cortex (Bourke and Tower, 1966)	468	136 (113)[a]	593	49
Rat cortex[c] (Franck *et al.*, 1968)	421	114	—	80
In situ brain				
Guinea pig hippocampus (Lipton and Robacker, unpublished results)	457	(145)[b]	211	(38)[b]
Guinea pig cortex (Keesey *et al.*, 1965)	480	—	220	—
Cat cortex (Bourke and Tower, 1966)	593	166	356	32
Rat cortex (Franck *et al.*, 1968)	600	(191)[b]	270	(45)[b]
Rat cortex (Baethmann and Sohler, 1975)	449	127	213	38

[a] Values calculated assuming that 18% of tissue inulin is *not* distributed in the extracellular space (see text).
[b] Values in parentheses are calculated by assumming that *in situ* water distribution is as determined by Amtorp (1979); see Figure 1.
[c] Calculated for second slice on basis of differences in inulin spaces and total ion contents between top (pial) slice and second slice that were determined by original authors.

in normal buffers, the intracellular K^+/Na^+ is significantly lower than in the *in situ* preparation. This and the cellular swelling are both indicative of a decrease in the ratio of pumping rate to the permeability for Na^+ ions. It is not possible, from these data, to determine which of these is, in fact, altered.

2.2. *Possible Bases for Compromised Function in the Brain Slice*

The previous section demonstrated that energy metabolism and ion homeostasis in the slice are both compromised with respect to *in situ* tissue. Certain parameters, in particular the contents of most of the compounds involved in high-energy phosphate metabolism, are reduced to between one-half and one-third their *in situ* values. The rate of tissue respiration is reduced to a similar degree.

Interestingly, the ratios of different metabolites, which are generally considered to be a measure of the integrity of oxidative phosphorylation, are, in certain cases, very similar in both *in situ* and *in vitro* preparations. However, other parameters that are indicative of cell integrity are clearly deficient in the slice. Foremost among these is ion and water homeostasis, where the data suggest that ion transport processes in slices are not optimal. AMP levels are also markedly elevated with respect to other metabolites. Other changes, including increased glycolysis and decreased protein synthesis (Lipton and Heimbach, 1978), probably result from deficiencies in energy metabolism and ion homeostasis.

In this section, we will discuss the alterations in the slice that could account for the observed metabolic deficits. We shall, firstly, consider the process of slice preparation with an emphasis on the step or steps that might be causing the damage. Then, we will consider the lesions within the tissue that might be responsible for the altered values of the parameters we measured. The damage will be described as consisting of two different components. One of these is, simply put, the "death" of a certain population of cells. This will have the effect of lowering the value of any parameter when expressed per dry weight of tissue, as such "dead cells" will present no barrier to metabolite diffusion and will not engage in oxidative phosphorylation. The other component of the damage is a malfunction of the remaining intact cells. The first component of the damage accounts for a substantial portion of the overall reduction in metabolite levels and respiration that is observed in the slice. The second component accounts for altered metabolite ratios as well as changes in ion homeostasis and ion-dependent parameters.

2.2.1. *Event Leading to Damage.* There are three phases of the process of preparing and incubating brain slices during which the slice might suffer damage. These are: (1) decapitation and subsequent hypoxia prior to incubating the slices, (2) the process of slicing the tissue, and finally, (3) the incubation of the slice in oxygenated buffer. The likelihood that each of these phases is responsible for the damage will now be considered.

2.2.1a. Decapitation and Hypoxia. During the preparation of brain slices, the tissue is generally ischemic and/or hypoxic for between 2 and 5 min. The question is whether this period is adequate to produce the metabolic profile that is observed in brain slices. There are several relevant considerations and the answer, unfortunately, is not a conclusive one.

(*i*) When *in situ* tissue is rendered ischemic (Ljunggren *et al.*, 1974) for 15 minutes and then returned to normal conditions, the energy charge and high-energy phosphates recover almost to preischemic levels within 90 min. Functionally, the tissue is deficient and there may be a significant (up to 40%) reduction is tissue oxygen consumption (Ljunggren *et al.*, 1974; Snyder *et al.*, 1975), which could well be a result of reduced neural activity. Although complete ischemia does not reproduce the metabolic alterations observed in the slice, *incomplete* ischemia does. Thus, while some controversy exists (Steen *et al.*, 1979), there is strong evidence that metabolic recovery 90 min after a 30-min episode of incomplete ischemia *in situ* is poor. Indeed, the metabolic profile of that tissue is quite similar to that observed in the slice (Nordstrom *et al.*, 1978). Conditions during most of the slice preparation procedure are probably analogous to incomplete ischemia in that there is generally some flow of oxygenated buffer present during the preparation period. Thus, these data are quite relevant to the problem of slice damage. However, in the same studies, when animals were exposed to only 15 min of incomplete ischemia, there was almost no change in metabolite levels following 90 min of recirculation. These studies were done under N_2O anesthesia, which could exert some protective effect. However, it is notable that slicing procedures are generally in the range of 2 to 6 min.

(*ii*) When the hippocampal slice is exposed to 5-min periods of anoxia, there is no permanent decrease in high-energy phosphate compounds after re-exposure to normoxic conditions (T. S. Whittingham and P. Lipton, unpublished observations) and even 10 min exposure of rat hippocampal slices to anoxia leads to only a 25% decline in high-energy phosphate compounds (Kass and Lipton, 1982). Thus, if ischemic damage occurs during slice preparation, it must either be due to factors that are present *in situ* but not in the slice, or it must be due to a process that has attained a maximal level as a result of the *in situ* insult and is, therefore, not accentuated by additional *in vitro* insults.

(*iii*) One of us (Lust *et al.*, 1982) has found that the final energy state of the slice is independent of the buffer in which the tissue is put immediately after its removal from the brain. In particular, substrate composition (absence or presence of glucose) and temperature (0° or 37°C) are not determinants of the final energy state. Thus, the rate of biochemical processes immediately after decapitation do not seem to influence the final composition of the tissue. This argues against the initial decapitation ischemia as being important.

An intriguing observation that also pertains to this issue is that the energy metabolite levels of synaptosomes, which are normally similar to those of the slice, are elevated to *in situ* values, if the synaptosomes are isolated from brains of animals that have been anesthetized with a barbiturate (Rafalowska *et al.*, 1980). This is important because barbiturates are thought to protect *in situ* brain tissue against effects of partial ischemia and anxoia (Nordstrom *et al.*, 1978; Nemoto *et al.*, 1982).

2.2.1b. Mechanical Damage as a Result of Slicing. During the slicing procedure, a zone on either edge of the slice will be severely disrupted as the cutting blade pulls through it. If this region retains most of its dry weight, but loses its ability to metabolize, then the process will lead to low-calculated values for all metabolites when expressed on a tissue protein or dry weight basis. Similarly, the calculated rate of respiration would be low. Histological studies support the supposition that there is a significant amount of this "edge damage." In different slice preparations, there is a 40- to 50-μm zone of destroyed tissue at the site of the cut (Garthwaite *et al.*, 1979; Misgeld and Frotscher, 1982) and disrupted tissue may extend up to 100 μm into the slice (Bak *et al.*, 1980). The sensitivity of the tissue to the slicing procedure has been stressed by electrophysiologists working with the hippocampal slice (Dingledine *et al.*, 1980; Schwartzkroin, 1981) and, in our experience, it is very difficult to record evoked responses in the first 50 microns of a slice. Thus, "edge damage" certainly exists. The degree to which it can account for the observed metabolic parameters of the slice depends upon how extensive it is. This will be discussed in the next section of this chapter.

2.2.1c. Slice Incubation Conditions. It is possible that the altered slice metabolism is due to the lack of some constituents normally present in CSF, to the loss of the spatial organization that results from the pressure of the cranium or, finally, to local anoxia. Adding high (1-mM) concentrations of creatine or adenosine to the buffer does elevate slice levels of phosphocreatine and ATP (Thomas, 1956; Thomas, 1957). However, these levels are still much below normal and there is no evidence that *in situ* extracellular concentrations of the added compounds approach the values required to elevate cell ATP and PCr. So, if constituents that are present in the extracellular fluid but lacking in the incubating buffer are responsible for the defective slice metabolism, they have yet to be identified. The question of the effect of the altered spatial organization has not been approached. However, it is possible that the physical disruption of the tissue due to its removal from the skull, or the reduction in hydrostatic pressure, could affect the function of the cells.

There may be a region in the center of the slice which is anoxic, an "anoxic core." However, this seems unlikely to be significant in our preparation, in which slice thicknesses range between 450 and 550 μm.

Thus, about 60 years ago, Warburg (1923) calculated that anoxia would not occur in liver slices up to a thickness of 470 μm. This should be an underestimate of the maximal thickness for brain slices because the rate of oxygen consumption that Warburg considered (0.5 ml/g per min) is about twice the value observed in the slices. Thus, according to Warburg's calculation the maximal thickness of a nonanoxic brain slice should be $470 \times \sqrt{2}$, or about 700 μm. Recent studies demonstrate empirically that olfactory cortex slices do not develop an anoxic core until they exceed 430 μm in thickness (Fujii *et al.*, 1982). In these studies, the slices were superfused very slowly (0.6 ml/min) so that the pO_2 at the surface was only half that in the original buffer. Our slice perfusion rates are far higher (20 to 30 ml/min) so that surface pO_2 should be greater than in those studies. This should increase the maximum thickness allowable before an anoxic core develops (Fujii *et al.*, 1982). Finally, it is relevant that the profile of slice high-energy metabolites that we observe (Table I) is not characteristic of hypoxic or anoxic tissue. Thus, those conditions lead to sharp decreases in the values of PCr/ATP and energy charge after 10 min of anoxia in slices (Kass and Lipton, 1982) and after 30 min of anoxia *in situ* (Siesjo and Nilsson, 1971). However, values of these parameters in the slice are very similar to their values *in situ*. Thus, the evidence suggests that there is not an appreciable anoxic core in the slices. Rigorous tests of the possibility that there is a functional hypoxic or anoxic region, however, have not been carried out. Such studies done several years ago on liver slices (Farr and Fuhrman, 1965) demonstrated that such a core developed for slice thicknesses somewhere between 330 and 670 μm.

2.2.1d. Summary. Three general possible bases for the damage to the slice have been considered. Of these, the probability of tissue trauma due to slicing seems the most certain to be leading to a substantial portion of the damage. Postdecapitation ischemia may be involved, but the evidence for this is not very strong on at least two counts. Finally, the possibility that conditions during incubation could be causing the deficits has been mentioned. Although tissue anoxia cannot be rigorously excluded for normal slice thicknesses, it seems unlikely. The possibility also exists that a lack of other factors in the incubating solution contributes to the observed deficits.

2.2.2. *Cell Changes Responsible for Altered Energy Metabolism*.

2.2.2a. Edge Damage. The most straightforward possibility is that all the reductions in metabolite levels and respiration result from the damage incurred upon slicing the tissue. If this were the case, then about 50% of the tissue would have to be damaged in this way to account for the reduction in energy metabolites. However, as indicated earlier, the data are more compatible with an "edge damage" of about half to two-thirds this amount. Thus, morphological studies (Garthwaite *et al.*, 1979; Bak *et al.*, 1980) indicate a layer of about 40 to 50 μm on each edge that is severely damaged and, in a 400-μm slice, this would amount to about 25% of the tissue.

Additional evidence comes from studies using purported extracellular markers, such as inulin, that indicate that there is a marker space in slices that does not exist *in situ* (see previous discussion). This space may well correspond to the water associated with damaged cells. In guinea pig hippocampus, the size of this "intermediate" inulin space is equal to about one-third of the noninulin space (Lipton and Heimbach, 1978), though it only represents 10 to 15% of the noninulin space in rat cortex slices (Lund-Anderson, 1974; Amtorp, 1979). However, in these slices, there is also a space that is accessible to sucrose but not accessible to inulin. It is equal in size to between 25 and 30% of the noninulin space (Hertz *et al.*, 1970; Goodman *et al.*, 1973; Amtorp, 1979). Thus, a volume equal to about one-third of the noninulin space in slices is quite accessible to large molecules and does not exist *in situ*. It is reasonable to postulate (i.e., Amtorp, 1979) that this space corresponds to the damaged cells at the edge of the cut slice. The size of this space is similar to the volume of tissue in which gross damage is observed. Thus, the data are consistent with the supposition that about one-fourth to one-third of the cells in the slice have been rendered very leaky as a result of the cutting process. If these cells have lost their metabolites and their ability to engage in biochemical reactions, but have retained their protein and dry weight, then their presence would lead to an approximate 25% decrease in the concentrations of energy metabolites, respiration, and related parameters. Still unexplained are the remainder of the losses in energy metabolites and respiration, as well as the high ratio of glycolysis to oxidation of glucose. In addition, the apparent cell swelling and reduction in K^+/Na^+ remain to be explained. It therefore appears that the metabolism of the nonleaky, or "healthy," cells must also be compromised.

2.2.2b. Mitochondrial or Membrane Damage. Two loci for the lesion(s) which come to mind are the cell membrane, where increased leakiness of Na^+ would lead to an increase in Na^+/K^+ intracellularly, and the mitochondria, where an inability to produce ATP at a normal rate could lead to decreased cell ATP content, respiration rate, and ion pumping. Such lesions could result from the artificial environment bathing the slices or from regions of hypoxia. They are both, however, argued against by the profile of the metabolic alterations observed in the slice. Thus, a mitochondrial or glycolytic lesion should be accompanied by a large decrease in the ratio of PCr to ATP; this invariably accompanies periods of hypoxia, anoxia, or hypoglycemia (Kass and Lipton, 1982; Lipton and Robacker, 1982; Lipton and Whittingham, 1982) and reflects the sensitivity of the creatine kinase equilibrium reaction to both a net decrease in high-energy phosphates (McGilvery and Murray, 1974) and to a decrease in pH_i (MacMillan and Siesjo, 1972). In addition, there is a large decrease in energy charge associated with inhibition of mitochondrial function (Whittingham *et al.*, 1981; Kass and Lipton; 1982). None of these changes occur significantly in the slice preparation (see Table I). An increased membrane permeability and ensuing increase in tissue Na^+/K^+ should lead to both a decrease in PCr/ATP and to an increased rate of respiration. This is what occurs when brain slices are exposed to treatments that increase Na^+ permeability, such as veratridine or electrical stimulation (McIlwain and Bachelard, 1971). Neither of these changes are observed in slices. In fact, exactly the opposite occurs. Thus, neither of these simple explanations readily fits the observed data.

2.2.2c. Postdecapitation Hypoxia. The only other reasonable explanation for the deficits, based on the discussion in Section 2.2.1, is that the postdecapitation hypoxia has affected the tissue. Although the anoxic episode is shorter than that required to produce reliable metabolic changes *in situ*, it is possible that the absence of normal circulation and structure in the slice makes it more sensitive to this process than is *in situ* tissue. Indeed, the profile of metabolites following recovery from 30 min of incomplete ischemia (Nordstrom *et al.*, 1978) quite closely approximates the profile in the slice. In particular, unlike during hypoxia, both PCr and ATP are reduced in such a way that PCr/ATP is actually increased in the incompletely ischemic tissue, as is also seen in the slice. Both ADP and AMP are elevated in the *in vivo* paradigm, while, in the slice, AMP is elevated and the percentage fall in ADP is far less than the percentage fall in ATP. The energy charge is lowered in both tissues, but this effect is far smaller in the slice. Although creatine levels in the *in vivo* paradigm were not measured, inspection of the changes in PCr, ATP, and ADP (Nordstrom *et al.*, 1978) indicates that pH_i, as calculated by the creatine kinase equilibrium, must be more alkaline in the *in situ* tissue that has recovered from the partial ischemia than in normal tissue. This apparent alkaline shift is also noted in the slice. The effect of partial ischemia on tissue respiration was not measured by Nordstrom *et al.* (1978), but it is notable that recovery from complete ischemia *in situ* is associated with a 40% decrement in respiration (Nordstrom and Rehncrona, 1977). In slices, respiration is decreased. Thus, it seems quite possible that the "healthy," or intact, cells are suffering from damage that is analogous to that caused by incomplete ischemia *in situ*. At least qualitatively, this would explain the altered metabolite levels and alkaline pH_i seen in brain slices. Glycolysis could also be activated by the elevated pH_i. This type of damage might also inhibit respiration, though the decreased respiration in brain slices may also be due to a decrease in spontaneous activity. The observed decrease in K^+/Na^+ and cellular swelling could well result from inhibition of the Na^+ pump, as a result of the decreased ATP content of the cells. Unfortunately, the actual cell mechanism(s) of damage due to partial ischemia are not known.

2.2.3. Summary. Mechanical damage to the slice that occurs during cutting probably accounts for between 50 and 60% of the reduction in the rate of energy metabolism and levels of metabolites that occurs when brain slices are prepared from *in situ* tissue. This damage is best thought of as a permeabilization of the membranes of the cells at or near the surfaces of the slice; it was described several years ago for liver slices (Farr and Fuhrman, 1965). In the brain slice, it includes about 25 to 30% of the cells when the slice is 400-μm thick. There may also be some reduction in metabolites due to a small anoxic core that could, if sustained, lead to cell death. However, this is probably not significant in a slice whose thickness is under about 550 μm. The most probable explanation for the remaining 40 to 50% of the reduction in energy metabolites is irreversible damage to the nonleaky cells that results from the short period of hypoxia during preparation of the tissue. The high-slice pH could also be a result of this damage. The defects in ion and water balance that exist in these cells could well be a consequence of decreased pump activity resulting from the decreased concentration of ATP in the cells. Although most of the decrease in respiration probably results from the lesions described above, it is possible that some of it is due to the absence of spontaneous neural activity in the preparation. As has been suggested at several points in the text, a great deal of experimental work on this question is necessary to transform the contents of this paragraph from informed speculation to fact.

Just how badly off is the slice? Between 25 and 30% of the cells are effectively dead. They need not be considered during analysis of phenomena except to the extent that their presence depresses the numerical values of all parameters that are expressed on a per mass basis. The remaining 75% of the tissue is composed of functioning cells. The analysis in this part of the chapter shows that they are associated with metabolites whose overall levels are reduced by 25%. (The other 25% of the reduction is due to the dead cells, as described above.) Thus, the intact cells in the slice contain about two-thirds of their normal contents of high-energy phosphates. Because the cells are swollen, the actual concentrations of these compounds are about 50% of their *in situ* values. The cells are swollen by about 40 to 50%, their K^+ concentration is reduced by about 25%, and their Na^+ is increased by about the same. Notwithstanding this apparently grim picture, the cells in brain slices maintain an almost steady metabolic state for many hours, show quite normal electrophysiological responses, and respond reversibly and predictably to a wide variety of metabolic perturbations. It is quite possible that they are "protected" by their reduced rate of respiration.

If the analysis of the cause of slice damage is correct, then certain precautions during slice preparation should be beneficial. Aside from using gentle cutting techniques, which has, at least, produced morphological improvements (Garthwaite *et al.*, 1979), prevention of irreversible anoxic damage would be worthwhile. Barbiturate anesthetics protect against the damage resulting from partial ischemia *in vivo* (Nordstrom *et al.*, 1978), so they might be administered prior to decapitation. Net entry of calcium seems to be a major source of irreversible anoxic damage (Siesjo, 1981; Kass and Lipton, 1982). Thus, inclusion of mM concentrations of cobalt, and/or removal of calcium from the preparatory buffer (I. S. Kass and P. Lipton, in preparation) might be quite beneficial.

3. MECHANISM OF ANOXIC DAMAGE

Reductions in oxidative energy metabolism have long been known to compromise brain function. A long-standing goal has been to understand the mechanisms involved in this process and, so, to understand the nature of the dependence of neural transmission on energy metabolism. The discussion in this part of the chapter will focus on one aspect of the process: the basis for the very rapid inhibition of neural transmission that occurs during cerebral anoxia. The hippocampal slice preparation has allowed several advances to be made in this area. Before describing these, we will provide the reader with a useful background by first describing the metabolic changes that occur during the early phases of anoxia and then discussing the way that these changes might affect neural transmission.

3.1. *Metabolic Changes during Anoxia*

Anoxia leads to several rapid metabolic changes in the *in situ* brain. There is a decline in PCr/Cr within 10 sec (King *et al.*, 1967), a somewhat delayed fall in ATP (Lowry *et al.*, 1964; Duffy *et al.*, 1972) and a very rapid fall in both extracellular pH (Urbanics *et al.*, 1978) and intracellular pH (Kaasik *et al.*, 1970). There is a somewhat delayed rise in cytosolic Ca^{2+} as a result of a net influx from the extracellular fluid (Siemkowicz and Hansen, 1981) and what may be an earlier cytosolic Ca^{2+} rise due to transport out of the mitochondria (Jundt *et al.*, 1975; Siemkowicz and and Hansen, 1981). There is a rapid increase in tissue adenosine concentration (Berne *et al.*, 1974) and in tissue cyclic AMP (cAMP) levels (Steiner *et al.*, 1972). The concentration of unsaturated free fatty acids also rises rapidly (Tang and Sun, 1982). All of these changes occur within 90 sec of the onset of anoxia or ischemia and so must be considered as possible causes of the electrophysiological deficits that begin within 10 to 60 sec (e.g., Siesjo, 1978). Even for those changes that occur slightly after the transmission deficits appear, there may be small, or local, changes which occur earlier and are not measurable (c.f., for ATP, Lipton and Whittingham (1982)).

3.2. *Effects of Metabolic Changes Occuring during Anoxia on Neurotransmission*

Neurotransmission depends on conduction of an impulse to the axon terminal, depolarization-dependent exocytosis of neurotransmitter, the interaction of neurotransmitter with the postsynaptic membrane and the generation of the postsynaptic potential. This may, in turn, lead to a postsynaptic action potential. Each of these steps is vulnerable to anoxia. In this section, we shall briefly consider the mechanisms by which the metabolic changes described in Section 3.1 might affect one or more of these steps in a way that would tend to block neurotransmission. The process should be sensitive to *membrane depolarization*, which would tend to inactivate sodium channels, decrease the rate of electrotonic conduction, decrease the amount of transmitter released during exocytosis and, postsynaptically, decrease the magnitude of the excitatory postsynaptic potential (EPSP). It should also be sensitive to *hyperpolarization* which would move the membrane potential away from threshold and, finally, the transmission process will be sensitive to any direct effect on the *exocytotic step* itself. We shall now consider effects of each of the metabolite changes discussed previously.

3.2.1. PCr/Cr. The rapid decrease in this ratio has generally been viewed as simply a reflection of the decrease in total high-energy phosphates and pH. Thus, the activity of creatine kinase is high in the brain (Booth and Clark, 1978) and the ability of the creatine kinase catalyzed reaction to buffer tissue ATP levels in this situation has been docu-

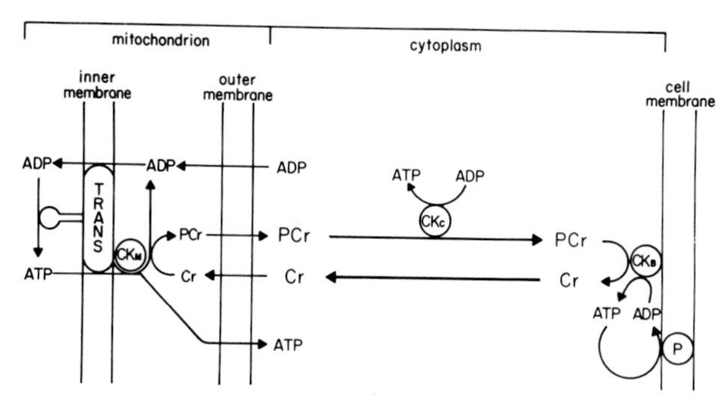

Figure 2. Model of the PCr–Cr energy shuttle. ATP is generated at the mitochondrial inner membrane and most of it reacts with the creatine kinase localized on the outer surface of this membrane (Saks, 1980). In this reaction, creatine is phosphorylated to phosphocreatine and ADP is regenerated for mitochondrial oxidative phosphorylation. PCr diffuses through the cell rapidly because of its high concentration and acts to rephosphorylate ADP at cytosolic or membrane sites of ATP utilization. This step is activated by the cytosolic and membrane-bound creatine kinases (Saks, 1980). The ATP is thus regenerated from local use. The creatine that is generated at these sites diffuses back to the mitochondria rapidly because of its high concentration and is then phosphorylated by the mitochondrial creatine kinase. As the arrows show, some ATP may well diffuse directly to utilization sites in the cytosol. The studies backing this model have been made in heart muscle cells (Jacobus, 1980).

mented on several occasions (Vincent and Blair, 1970; McGilvery and Murray, 1974). However, studies during the past several years have suggested that the creatine kinase system may function as more than simply a high-energy phosphate reserve pool during metabolic stress. Thus, several groups (e.g., Saks *et al.*, 1978; Jacobus, 1980; Seraydarian, 1980) have shown that PCr and Cr may function as an energy shuttle in heart muscle during the steady state. The key features of this model, which is diagrammed in Figure 2, are, firstly, that ATP synthesized in the mitochondria reacts very efficiently with a mitochondrially localized creatine kinase to form PCr and ADP. This PCr then acts as the carrier of high energy phosphate to many, or all, cell processes that utilize ATP. The PCr then donates its phosphate to ADP generated by the ATPases, via peripherally localized creatine kinase. This reaction, then, restores ATP that has been broken down via the energy-consuming processes and so acts to maintain ATP concentrations at cytosolic utilization sites. The model leads to a more sophisticated role for the creatine kinase catalyzed system than simply as a high-energy phosphate pool. Thus, while the total cell contents of the reactants may well be in equilibrium (e.g., MacMillan and Siesjo, 1972), there must be a rapid steady-state production of ADP and PCr at the mitochondria and an equally rapid steady-state production of ATP and Cr at ATP utilization sites. Saks *et al.* (1978) have emphasized the implication of this for heart muscle, where it is possible that a small pool of ATP at the myofibril is continually regenerated by creatine kinase located there. Steady-state levels of ATP in this pool will, then, be strongly determined by the concentration of PCr and the activity of the creatine kinase. Indeed, there is good evidence that levels of PCr in the muscle cell regulate the rate of ATP-dependent processes without measurable changes in ATP (Seraydarian and Artaza, 1976; Saks *et al.*, 1977). Little work has been done on this model in brain tissue but the distribution of creatine kinase suggests it may be applicable (Booth and Clark, 1978). If so, then the drop in PCr, which occurs early during anoxia, may lead to a steep drop in the rate of regeneration of peripheral ATP and, so, of the rate of ATP-dependent processes. The consequences of such a drop will be discussed in the next section, but the above analysis suggests that it may be necessary to look at PCr/Cr levels as true determinants of ATP concentrations at utilization sites.

3.2.2. *Decreased ATP.* A decrease in ATP should inhibit the sodium pump, eventually leading to a membrane depolarization because of ion shifts. It may also directly inhibit the exocytosis process. It is important, however, to consider the *extent* to which these processes will be inhibited by a certain fall in ATP in order to assess potential effects of the decreased ATP that occurs during anoxia.

The ATP dependence of the Na^+ pump appears to result from two interactions: enzyme phosphorylation at a very high affinity site, whose Michaelis constant (Km) is about 1 μM, and an allosteric binding that aids in the dissociation of K^+ from the enzyme. This second step has a much lower affinity; in the squid axon its Km is about 0.2 mM (Beaugé and diPolo, 1981). Estimates of the apparent overall Km for ATP in brain microsomal ATPases range from 0.8 to 1.2 mM (Robinson, 1967; Bertoni and Siegel, 1978; Kimelberg *et al.*, 1978). If these values represent actual *in situ* values, then it is apparent that slice ATP values in the intact cells are very close to the Km for the pump while values *in situ* are about $2\frac{1}{2}$ times the Km. In both cases, a simple analysis of Michaelis–Menten kinetics parabola shows that a 10% decrease in ATP will lead to a 5% decrease in the rate of pumping. Twenty percent decreases in ATP will inhibit pumping by about 15%. Using ouabain to inhibit the pump, we determined that an approximate 25% inhibition of the pump is required to block transmission at the same rate as it is blocked during anoxia (Lipton and Whittingham, 1979,1982).

The nature of the ATP dependence of presynaptic exocytosis has not been determined. However, the apparent relationship that exists between protein phosphorylation and exocytosis (Burke and DeLorenzo, 1982) suggests that exocytosis should be ATP dependent. Recent studies (Baker and Knight, 1981) have demonstrated this dependence for the release of adrenalin from permabilized chromaffin cells, and have determined a Km for ATP of about 1 mM. Conversely, however, high K^+-induced release of transmitter from isolated cerebellar slices demonstrates a marked insensitivity to tissue ATP concentrations, the rate of transmitter release being unaffected until ATP levels fall to about 10% of their normal levels (Bosley *et al.*, 1983). This massive high K^+-induced release may not, however, be a good model for normal exocytosis.

3.2.3. pH. The effects of altered extracellular and intracellular pH on synaptic transmission have not been well demonstrated. However, decreases in extracellular pH do appear to inhibit transmission. Thus, several *in vivo* studies have demonstrated that increasing CO_2 levels to 10% or 20% depress cortical excitability (Krnjevic *et al.*, 1965) and inhibit transmission across CNS synapses (Morris, 1971; Carpenter *et al.*, 1974). It is not clear whether the effects of increasing CO_2 are mediated by extracellular acidity, intracellular acidity or by CO_2 itself (Carpenter *et al.*, 1974). Decreasing extracellular pH by 0.4 units without altering CO_2 severely inhibits transmission at the neuromuscular junction (Landau and Nachsen, 1975) and decreasing extracellular pH by 0.45 units, either by decreasing bicarbonate or increasing CO_2, inhibits transmission in the hippocampal slice by 50% (Lipton and Korol, 1981). Thus, although it is clear that changes in extracellular pH do depress transmission, it is not clear whether the action is due to an intracellular or an extracellular change in pH. The notion that changes in buffer base concentration do not change intracellular pH in mammalian cells is no longer valid (i.e., Preissler and Williams, 1981).

In the hippocampal slice, the decreases in extracellular pH do not inhibit antidromic transmission, suggesting a synaptic site of action. Indeed, there is some direct evidence that decreased pH inhibits depolarization-induced Ca^{2+} entry in synaptosomes (Nachsen and Blaustein, 1979), and a decrease in pH also inhibits Na^+–Ca^{2+} exchange in cardiac sarcolemmal vesicles (Philipson *et al.*, 1982). It has been suggested that a large fraction of Ca^{2+} entry during synaptic depolarization occurs via the Na^+–Ca^{2+} exchange mechanism (Mullins and Requena, 1981). Other studies suggest that decreased intracellular pH inhibits exocytosis at a step that occurs after Ca^{2+} entry (Preissler and Williams, 1981). All of these effects are associated with pH changes of less than 1 unit, and so could occur when energy metabolism is inhibited.

3.2.4. Free Fatty Acids. The rapid increase in free fatty acids that occurs during anoxia (Tang and Sun, 1982) has only recently become recognized so that, to date, there have been no studies of the effects of these compounds on transmission. However, the addition of low concentrations of unsaturated fatty acids to brain slices does lead to a rapid swelling of the cells (Chan and Fishman, 1978). This may well result from an increase in sodium permeability and, if this is the case, one could expect accompanying depolarization and inhibition of transmission.

3.2.5. Calcium. There are, in principle, several ways in which increased cytosolic levels of calcium could inhibit transmission. Thus, increasing cytosolic Ca^{2+} increases membrane permeability to potassium in several cell types including, possibly, central neurons (Krnjevic, 1975). This increase will hyperpolarize neurons, rendering them less easily excited (Krnjevic, 1975). Increased intracellular calcium also blocks the Na^{2+} pump, even at concentrations as low as 10^{-5} M (Vincenzi, 1971). Finally, elevated intracellular Ca^{2+} levels apparently inhibit voltage-dependent calcium entry, at least in invertebrate preparations (Eckert and Tillotson, 1981). All of these effects occur at Ca^{2+} levels in the micromolar range, so a fairly small elevation in intracellular Ca^{2+} could, in principle, lead to inhibition of transmission by one of these mechanisms. Studies that directly address this possibility have not been reported.

3.2.6. Cyclic AMP. Although many studies have shown that exogenous cAMP or its analogues inhibit central neural transmission, there is little evidence that these effects are due to an increase in cell cAMP levels (Phillis, 1977). Thus, agents that produce large increases in hippocampal cAMP, such as norepinephrine and histamine, do not affect evoked responses in the hippocampal slice (Dunwidde and Hoffer, 1980). Although extracellular cAMP and its derivatives do inhibit transmission in the slice, the effect is almost certainly due to their conversion to adenosine via a nucleotidase (Dunwidde and Hoffer, 1980). Thus, evoked slice responses do not seem to be affected by intracellular cAMP, but recent studies in which cAMP has been injected into hippocampal pyramidal cells suggest that the nucleotide can hyperpolarize such cells (Benardo and Prince, 1982). The effect apparently takes several minutes to develop. These studies are consistent with a large body of work demonstrating that monoamine neurotransmitters may act to hyperpolarize neurons by slow-acting increases in cAMP (Seeman, 1980; Benardo and Prince, 1982). Thus, although cAMP may indeed cause inhibition in neural systems, endogenous levels do not appear to be affecting the normal evoked response in the hippocampal slice. cAMP's role in anoxic damage in other preparations is questionable because of the slow time course for onset of its action.

3.2.7. Adenosine. Low concentrations of adenosine profoundly inhibit evoked responses in many cerebral preparations (Fredholm and Hedqvist, 1980), including the hippocampal slice. Both excitatory and inhibitory (Lee and Schubert, 1982) responses are affected. When 50-μM adenosine is added to the buffer perfusing hippocampal slices synaptic transmission via the perforant path and the Schaeffer collateral path is completely blocked (Dunwidde and Hoffer, 1980). Although the adenosine receptor involved in this process appears to be the same one that increases the activity of adenylcyclase (Fredholm *et al.*, 1982), the inhibition of transmission does not appear to be mediated by an increase in cAMP (Okada and Saito, 1979; Reddington and Shubert, 1979; Dunwidde and Hoffer, 1980). Adenosine seems to inhibit transmission by acting on a receptor, which leads to inhibition of inward Ca^{2+} movement during synaptic depolarization (Wu *et al.*, 1982). Thus, if extracellular adenosine concentrations rise to micromolar levels during anoxia, then the nucleoside should noticeably inhibit transmission in many, and possibly all, cerebral pathways.

3.2.8. Summary. It is apparent that there are several consequences of inhibited oxidative metabolism that have the potential of affecting neural transmission. Thus, each of the metabolic parameters described above, with the possible exception of cAMP, must be considered as potential effectors of transmission. Clearly this list is not complete; there are, for example, profound alterations in the concentrations of glycolytic intermediates early during ischemia (Lowry *et al.*, 1964) and anoxia (Norberg and Siesjo, 1975) and there may be many other rapidly changing parameters. The above discussion serves two purposes. It provides possible mechanisms for the extreme sensitivity of synaptic transmission to energy metabolism. It also provides a basis for analyzing experiments aimed at determining the interaction between energy metabolism and electrophysiological phenomena. Although our studies have largely been directed at acute effects of anoxia, the above analysis clearly applies equally well to steady-state hypoxic conditions. However, in this case, the effect of reduced energy metabolism or oxygen on neurotransmitter synthesis (Gibson *et al.*, 1981) also needs to be considered. Neurotransmitter pools are unlikely to be affected during short periods of anoxia.

3.3. *Experimental Evidence Concerning the Mechanism of Synaptic Transmission Failure during Compromised Oxidative Phosphorylation*

We have been using the hippocampal slice preparation to study the mechanisms involved in the rapid block of transmission during anoxia. In particular, the evoked response between the perforant path and dentate granule cells is abolished following about 4 min of anoxia in the guinea pig hippocampal slice (Lipton and Whitthingham, 1979), and after about 3 min in the rat hippocampal slice (Kass and Lipton, 1982). This is approximately 25% longer than the time required for decay of hippocampal- (Andersen, 1960) and cortical-evoked (Grossman and Williams, 1971) responses in anesthetized animals *in situ*. This time relationship is interesting, given the large reduction in basal levels of high-energy nucleotides in slice tissue.

A particular advantage of our studies is that the molecular layer of the dentate gyrus can be frozen rapidly and is easily dissected from other hippocampal tissue, to be assayed for metabolites. The changes in metabolite concentration in this region, composed mainly of perforant-path–dentate-granule cell dendrite synapses (Matthews *et al.*, 1976), can then be correlated with alterations in synaptic transmission in this pathway.

3.3.1. *Decreased ATP as a Cause of Inhibition of Transmission.* We have tried to determine whether a decrease in ATP levels is a specific cause of anoxic transmission block. Perhaps surprisingly, the answer to this simple question has not been determined. Thus, several studies have demonstrated depressed neural function in brain during anoxia or hypoxia in the absence of a fall in ATP (Schmahl *et al.*, 1966; Siesjo and Nilsson, 1971; Duffy *et al.*, 1972). We examined the relationship between ATP changes and synaptic transmission in the hippocampal slice, reasoning that in this preparation we could ascertain metabolic changes in the same regions that we were measuring electrophysiological changes. Thus, we could make more accurate temporal and spatial correlations than were possible in the *in vivo* experiments previously mentioned.

Figure 3 shows the changes that occur in ATP and PCr during 4 min of anoxia in the hippocampal slice exposed to elevated potassium buffer. Also shown in the bars at the top of the graphs, are different stages in the decay of the evoked electrophysiological response. It is apparent that transmission in these conditions is compromised prior to a decrease in measured ATP values, a finding consistent with the above-mentioned *in vivo* studies. The finding leads to the conclusion that decreases in ATP are not responsible for the rapid attenuation of neural activity during anoxia. However, when we assayed ATP concentrations in the molecular layer of the tissue at the time at which electrical transmission is attenuated, then a different pattern emerged. As seen in Table III, the rate of fall of ATP is faster in the molecular layer than in the slice as a whole and, most importantly, ATP concentration is lowered by the time inhibition of synaptic transmission begins. Because the molecular layer is the region in which the synapses of the pathway occur,

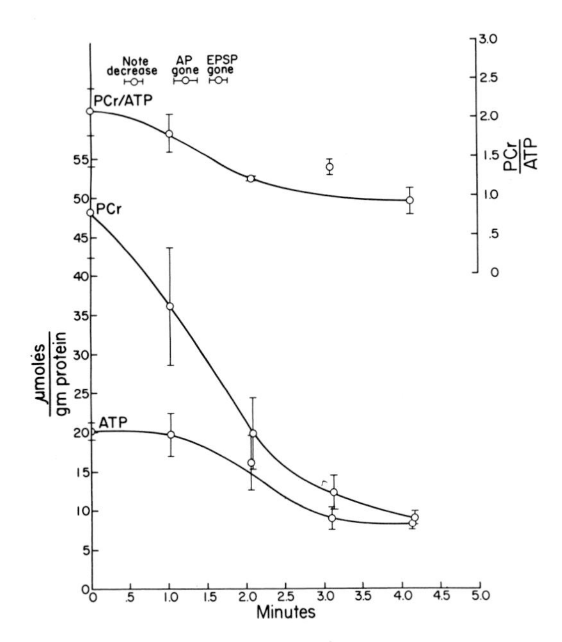

Figure 3. Changes in ATP, PCr, and evoked response during anoxia in the hippocampal slices. Hippocampal slices were exposed to anoxic buffer at 0 min. Recordings were made from the dentate granule cell layer following stimulation of the perforant path. Tissue was removed and analyzed at the times indicated. The horizontal bars at the top of the graphs denote, from left to right, the time at which the height of the population spike just begins to fall, the time at which the population spike vanishes and, finally, the time at which the field EPSP vanishes. The high-K^+ buffer is used to accelerate the rate of fall of the electrophysiological parameters (Lipton and Whittingham, 1982, p. 59). Reprinted with permission from *J. Physiol.* (*London*).

it is reasonable to assume that transmission should be most sensitive to metabolite levels in structures within that region. Thus, the data are now consistent with the conclusion that a fall in tissue ATP is, in fact, one of the factors leading to inhibition of transmission. Parenthetically, it is interesting that the concentrations of PCr in the molecular layer, as well as the rate of its utilization during anoxia, are higher than in the bulk of the tissue. These results suggest, as has also been suggested by other workers (Maker *et al.*, 1973; Sokoloff, 1981), that the rate of energy metabolism is higher in the neuropil region of cerebral tissue than it is elsewhere.

Although the preceding studies demonstrate that a fall in ATP content could be causing anoxic transmission block, they do not establish any causal relationship. However, we were able to use the slice prep-

Table III. ATP and Phosphocreatine (PCr) in Trimmed Slice and Molecular Layer during Normoxia and Hypoxia[a]

	Normoxic	Hypoxic
Trimmed slice values (nmole/mg protein)		
ATP	18.9 ± 0.8	17.0 ± 1.2
PCr	37.6 ± 1.5[b]	31.1 ± 2.7[b]
Molecular layer values (nmole/mg protein)		
ATP	23.5 ± 1.1[c]	19.6 ± 0.8[c]
PCr	55.6 ± 3.1[c]	36.0 ± 1.7[c]

[a] Tissue was frozen either immediately before exposure to hypoxia (normoxia) or during hypoxia at the time the evoked potential showed its first significant decrease (mean = 26 sec of hypoxic perfusion). Twelve experiments for each value; values are means ± SEM.
[b] Differences between normoxic and hypoxic means are significant at $p < 0.025$.
[c] Differences between normoxic and hypoxic means are significant at $p < 0.005$.

aration to establish a causal relationship between the fall in ATP and the inhibition of transmission with some confidence. The data in Figure 3 is typical of studies by many groups, which show that the major reduction in high-energy nucleotides early during anoxia is absorbed by PCr. Thus, the PCr–Cr couple buffers cell ATP content (McGilvery and Murray, 1974). We reasoned that if we could increase the cell contents of PCr and Cr, and so increase the ATP buffering capacity of this couple, then ATP should be maintained for a longer period during anoxia. If the fall in ATP was leading to the anoxic blockage of transmission, then transmission should be sustained for a longer period of time than previously observed. Cr and PCr can be considerably elevated in heart cells by incubating the tissue for several hours in a buffer containing a high concentration of creatine (Seraydarian *et al.*, 1974).

Figure 4 shows that prolonged incubation in 25-mM creatine-containing buffer substantially increases hippocampal cell PCr concentration without affecting ATP concentration. Importantly, the increase is associated with a prolonged (by 350%) survival time of the evoked response during anoxia (Whittingham and Lipton, 1981). The data in Table IV supports the idea that this effect is due to a prolonged maintenance of ATP during anoxia. In the table, ATP and PCr concentrations in the molecular layer of tissue incubated for 3 hr in creatine-containing buffer

Table IV. Molecular Layer ATP and PCr in Normoxia and Hypoxia in the Presence of Creatine[a]

	A		B	
	Normoxia	Hypoxia	Normoxia	Hypoxia
ATP (nmole/mg protein)	26.3 ± 2.7	25.6 ± 2.1	24.7 ± 1.4[b]	21.1 ± 0.8[b]
PCr (nmole/mg protein)	209.2 ± 14.0	180.9 ± 14.0	189.0 ± 14.3[b]	138.4 ± 11.7[b]

[a] Tissue was exposed to 25 mM creatine for 3 hr. In A, tissue was frozen either immediately before exposure to hypoxia (Normoxia) or during hypoxia at the time the evoked potential had first decreased prior to the creatine incubation (mean = 30 sec of hypoxic perfusion). In B, tissue was frozen either immediately before exposure to hypoxia (Normoxia) or during hypoxia at the time the evoked potential first decreased (mean = 74 sec of hypoxic perfusion). Seven experiments for each value; values are means ± SEM.

[b] Differences between hypoxic and normoxic means are significant at $p < 0.01$.

are shown for two different situations. In the first (A), the tissue was frozen for analysis at the time when transmission begins to fail in the absence of extracellular creatine. There is no fall in tissue ATP. (There is a fall in the absence of creatine, see Table III.) In the second situation (B), tissue was frozen at the time that transmission actually begins to fail in the presence of creatine. At this point, there is a 15% fall in tissue ATP, comparable to that measured at the time transmission began to fail in the absence of creatine. Thus, there is a good correlation between the fall in ATP and in synaptic transmission in two different conditions. Maintaining ATP maintains transmission, and a fall in transmission in two different situations is associated with the same fall in ATP.

Although these correlations lend strong support to an important role for a decline in ATP-producing transmission block, they do not establish the connection rigorously. Thus, it is possible that the increased PCr and Cr levels in the cells attenuate some other metabolic change that may be involved in transmission failure. In particular, because the creatine kinase-catalyzed reaction is one of the more important intracellular pH buffers (Siesjo *et al.*, 1972), it is quite possible that acidification during anoxia is attenuated by the high levels of Cr and PCr. Although it does not seem probable that the PCr and Cr elevation would affect processes other than ATP or H^+ ion equilibration, it is possible that the observed effects on transmission result from binding of intracellular ions, such as Ca^{2+} or Mg^{2+}, by the high levels of the creatine and its phosphate (Veech *et al.*, 1979). In the absence of further experiments, we take the simplest interpretation of our results to date and conclude that a fall in ATP is a factor leading to inhibition of transmission.

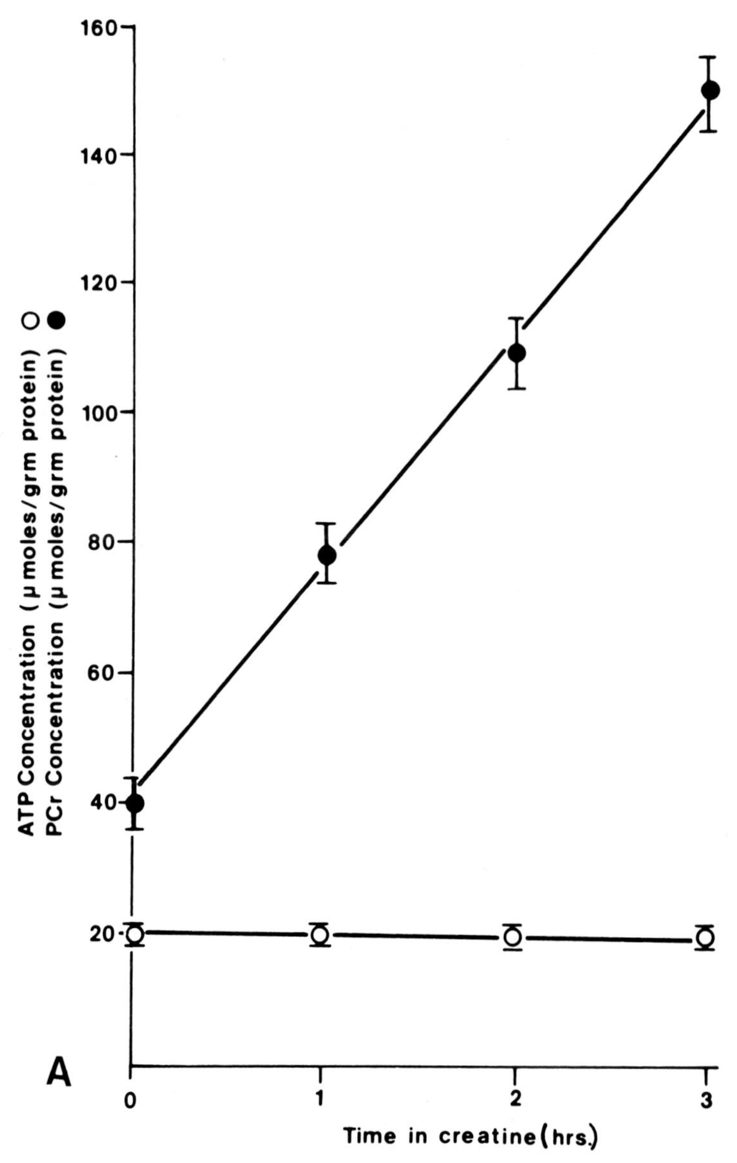

Figure 4. (A) Molecular layer PCr and ATP during a 3-hr incubation with 25 mM creatine. At time 0, slices were exposed to 25 mM-creatine in standard perfusate containing 13.4 mM-K^+. Slices were removed into liquid nitrogen at the times shown and then microdissected to yield the molecular layer tissue. This was analyzed for ATP and PCr. ○ = PCr, ● = ATP. Value are averages ±SEM determined from eight experiments. (Lipton and Whittingham, 1982, p. 59.) Printed with permission from *J. Physiol.* (*London*). (B) Effect of creatine incubation on the decay of the evoked response during anoxia. The abscissa denotes the duration during which the slice is incubated in buffer containing 25 mM creatine. The closed circles show the time required to observe the first noticeable decrease in the population spike during anoxia. The open circles show the time required to observe the complete block of the field EPSP.

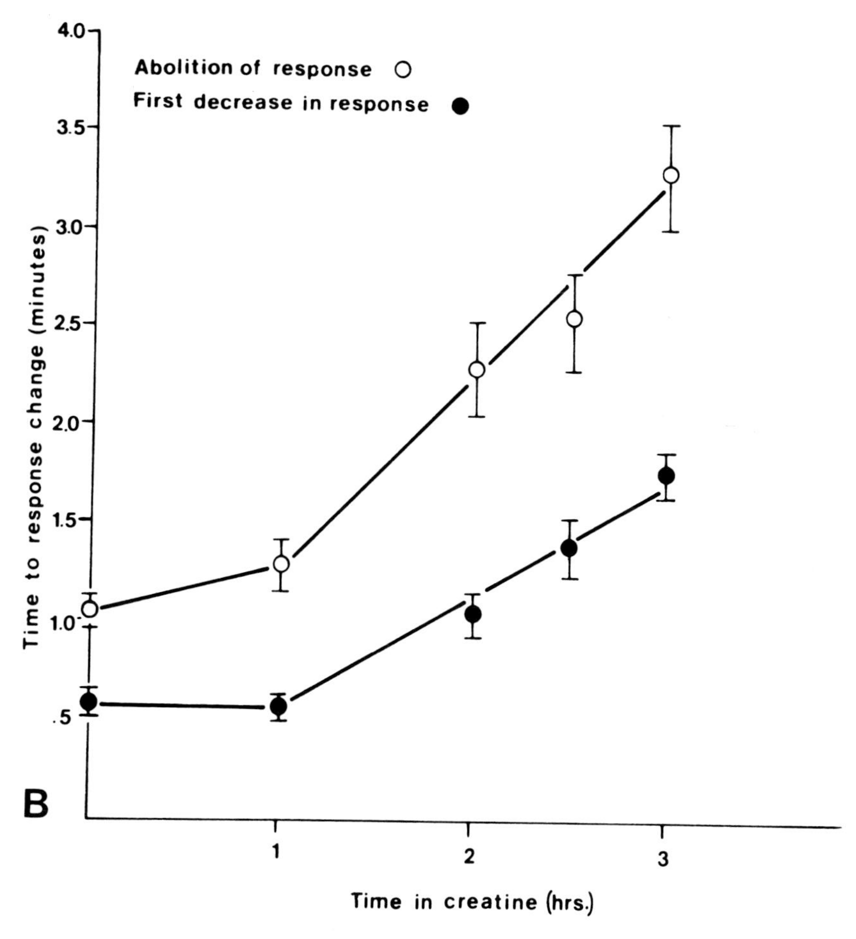

Figure 4. *(Continued).*

3.3.2. *Effects of the Reduced ATP on Transmission during Anoxia: Membrane Depolarization.* Our studies have suggested that there are two consequences of decreased ATP that are acting to inhibit transmission (Lipton and Whittingham, 1979; Lipton and Robacker, 1982). One of these is the depolarization of the membranes of the neural processes, presumably as the result of pump inhibition. Thus, we carried out a series of experiments in which we demonstrated that maintaining the tissue in depolarizing buffers (buffers in which K^+ was elevated up to 15 mM, or in which Cl^- was replaced by a less permeant anion) led to a much faster inhibition of transmission during anoxia (Lipton and Whit-

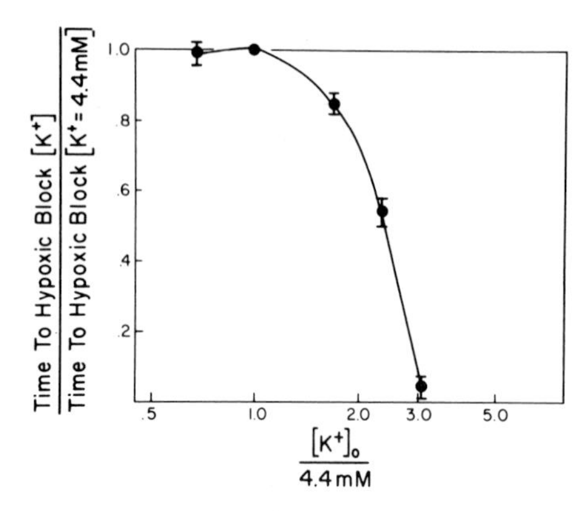

Figure 5. The effect of increasing the concentration of extracellular K^+ on the time required to abolish the evoked response during anoxia. The abscissa shows the ratio of extracellular $[K^+]$ to the normal value of extracellular $[K^+]$ which is 4.4 mM. Note the marked decrease in time required for the response to block during anoxia as extracellular K^+ in the incubation medium increases (Lipton and Whittingham, 1979, p. 59). Reprinted with permission from *J. Physiol. (London)*.

tingham, 1979). This is shown in Figure 5. We demonstrated that this was not due to a change in the rate of ATP loss during anoxia (Lipton and Whittingham, 1982) and concluded that depolarization of one or both of the pre- or postsynaptic neural elements to a critical point was responsible for transmission failure. Thus, prior depolarization of the tissue accelerates transmission failure because the extent of depolarization required to reach the critical blocking potential is reduced. This is illustrated in Figure 6. A similar explanation for transmission block in the spinal cord (Carregal, 1975) and at the neuromuscular junction (Krnjevic and Miledi, 1959) has been provided, using different approaches. Using ouabain, we demonstrated that an approximate 25% inhibition of the Na^+ pump would block transmission at the same rate as it is blocked during anoxia (Lipton and Whittingham, 1979).

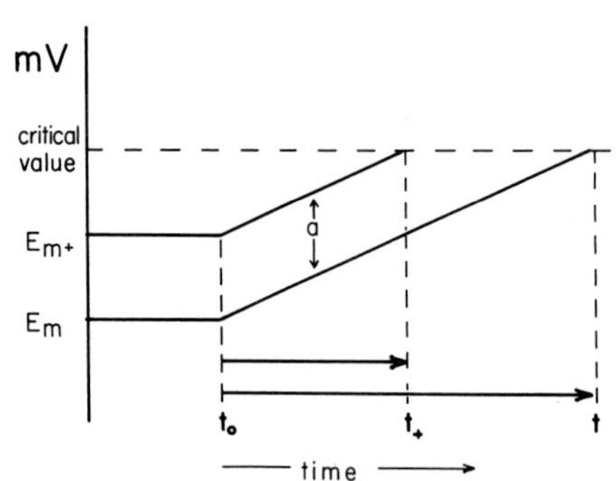

Figure 6. Model for dependence of time taken for anoxic transmission block on level of membrane depolarization prior to anoxia. There is (postulated to be) a "critical value" of the membrane potential at which block occurs in the neuronal processes (axon endings or dendrites). The sloped line denotes the postulated membrane potential is a function of time during anoxia, which begins at t_0. $E_m{}^+$ represents tissue which is in a depolarizing (e.g., high K^+) buffer; a represents the magnitude of this prior depolarization. The *rate* of depolarization is assumed to be about the same in each case so that the time taken to block is reduced in the tissue that is previously depolarized.

Although the analysis of the preceding experiments seems correct, measurements of the actual change in membrane potential of the neuronal elements cannot yet be made. In fact, measurements of the membrane potential of the cell bodies of cortical neurons *in vivo* (Krnjevic, 1975) and of hippocampal pyramidal cells in the slice (Hansen *et al.*, 1982; Misgeld and Frotscher, 1982) during anoxia show that the cells become transiently hyperpolarized. Krnjevic (1975) suggested such a hyperpolarization would be a result of increased tissue calcium and a resultant increase in potassium permeability. These measurements can be reconciled with our studies by assuming that while the soma of the pyramidal cells are hyperpolarizing due to an increase in P_K, the neuronal processes are simultaneously depolarized as a result of pump inhibition. (That this is not far-fetched is indicated by Hansen's studies, in which the maximum hyperpolarization is seen to be only 4 mV and of a transient nature. Thus, within 4 min of the onset of tissue anoxia, the membranes begin to depolarize. After 6 min, they are depolarized by 10 mV.) The explanation suggests that either or both the membranes and metabolism are quite different in the two cell regions. We found that ATP does, indeed, decline more rapidly in the neuropil. Thus, it is quite possible that pump inhibition dominates in this region, while increased cell Ca^{2+} and resultant increased P_K dominates in the cell bodies for a short time. It is very difficult to reconcile our results with the supposition that membrane hyperpolarization causes anoxic transmission block.

3.3.3. *Effects of the Reduced ATP on Transmission during Anoxia: Adenosine Release.* Further studies that one of us has carried out (Lipton and Robacker, 1982) implicate liberation of tissue adenosine as an important component of response inhibition during anoxia. Increases in brain adenosine have been measured as early as 1 min after the onset of anoxia (Berne *et al.*, 1974) and the purine nucleoside appears in the cerebral spinal fluid after a further 5 min. Although more rapid measurements have not been made, it is notable that adenosine levels are increased twofold after only 5 sec of ischemia in the heart (Berne *et al.*, 1971). The precise pathway for adenosine synthesis and release during anoxia has not been established, but the most likely possibility is that it results from the rapid formation of AMP (Lowry *et al.*, 1964) and the ensuing dephosphorylation of this nucleotide by 5′-nucleotidase, to the nucleoside. Adenosine probably leaves the cell by leakage across the cell membrane (Pull and McIlwain, 1972). Although adenosine has been implicated as a vasodilator in response to anoxia (Rubio *et al.*, 1975), its role as an inhibitor of transmission in this situation has not been studied.

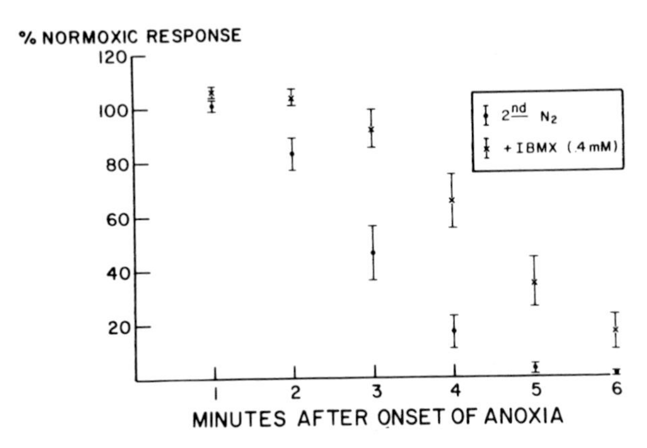

Figure 7. Effect of an adenosine receptor blocker on the rate at which the population spike decays during anoxia. The circles denote the rate of response decay without the blocker, in the presence of normal buffer. The crosses denote the decay of the response in tissue exposed to buffer to which 0.4 mM isobutylmethylxanthine (IBMX) had been added 20 min prior to anoxia. Note the protection against decay offered by the drug (n = 8 experiments; the experiments were paired).

We have used blockers of the extracellular adenosine receptors and adenosine deaminase to study the role of the nucleoside during anoxia. Thus, such agents as isobutylmethylxanthine (IBMX) and 8-phenyltheophylline (8-PT), which are adenosine receptor antagonists (Fredholm *et al.*, 1982), block the inhibition of hippocampal transmission that adenosine produces (Dunwidde and Hoffer, 1980). We reasoned that if adenosine contributes to the decay of the response during anoxia, then inclusion of these blockers, or adenosine deaminase, in the superfusate should slow the decay of the response. Figure 7 shows that 0.4 mM IBMX does, indeed, slow the rate at which the evoked response is blocked during anoxia. In particular, the onset of the response decay is delayed by 60 sec and the half-time for the decay is increased by 90 sec. Significant but smaller effects were seen with 0.1 mM 8-PT and with 2 U/ml of adenosine deaminase (Lipton and Robacker, 1982).

Adenosine is known to increase cell cAMP levels, and the adenosine receptor blockers also block cell phosphodiesterases. Thus, it seemed possible that agents we used were exerting their action by maintaining an elevated energy state inside the cell during anoxia. The results of metabolite measurements, shown in Table V, show that this is not the case. Thus, in the presence of the adenosine blockers, the fall in ATP during anoxia is actually greater than the normal decline. This suggests that adenosine is acting directly to inhibit transmission during anoxia,

Table V. Effect of Adenosine Blockers on Energy Metabolites

Conditions	ATP (nmoles/mg protein)	PCr (nmoles/mg protein)	cAMP (pmoles/mg protein)
Normoxia	15.0 ± 0.5	27.3 ± 1.1	9.7 ± 2.3
4′ Anoxia	13.9 ± 0.5*[,a]	13.9 ± 0.7**[,a]	13.1 ± 1.8
4′ Anoxia with 0.4 mM IBMX	11.4 ± 0.4*[,a]	9.2 ± 0.3**[,a]	10.9 ± 1.5
4′ Anoxia with 2 μm/ml Adase	12.1 ± 0.7+[,a]	15.1 ± 1.2	7.1 ± 0.9

[a] Differences significant at $p < 0.05$ using student's *t*-test. The protocol was as described in the legend to Figure 7. Tissue was analyzed as described by Lowry *et al.*, 1964. (*, **, +) Used to designate specific pairs.

and is also acting to maintain cell ATP levels during this time. This could well be the result of an activation of glycogenolysis due to the adenosine-induced increase in cAMP (Wilkening and Makman, 1977).

Adenosine, then, appears to play a dual role during anoxia. The first is to inhibit neural transmission and the second is to help maintain cell ATP levels, presumably by activating glycogenolysis. It acts, perhaps, as a "survival switch," to allow basal cell functions to continue in a condition of metabolic stress.

3.3.4. *Summary.* These studies demonstrate that both membrane depolarization and the release of adenosine, two consequences of decreased ATP concentration, contribute to the rapid decay of the hippocampal-evoked response during anoxia. There may well be other, still unidentified, processes that are also contributing to the decay. However, if there are not, then we can use Figure 7 to estimate the contributions of the two effects, membrane depolarization and adenosine binding, to the overall inhibition of the response. This is shown in Figure 8. The total decay of the response is the product of the two individual effects shown. It is apparent that adenosine accounts for the early portion of the decay and appears to act, throughout, more rapidly than does membrane depolarization. This is easily understood because, while adenosine will be effective as soon as it is liberated, the early depolarization of the membrane may be somewhat excitatory. Only when the level of depolarization approaches the level required to produce conduction failure will it begin to actually inhibit transmission. Thus, at high initial extracellular K^+ levels, inhibition of the response is greatly accelerated, as the membrane potential is much closer to this level (see Figure 5).

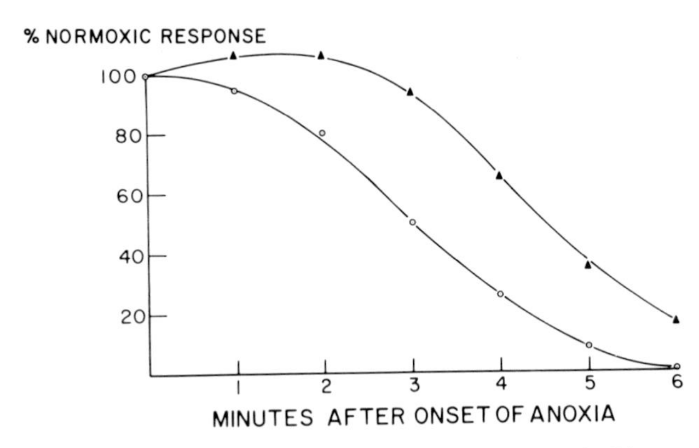

Figure 8. Components of the population spike decay during anoxia. The two curves are derived from the data in Figure 7. The effect of membrane depolarization (triangles) is identical to the crosses in Figure 7—that is, the decay of the response when the action of adenosine is blocked. The effect of adenosine (open circles) is calculated by assuming that the total fractional decay of the normoxic response (circles in Figure 7) is the product of the effect of membrane depolarization (crosses in Figure 7) and the effect of adenosine.

Using the hippocampal slice, then, we have been able to elucidate some of the mechanisms involved in a process that plays a very important role in the *in situ* brain. The problem is a hoary one and has defied solution for many years; the ability to control conditions carefully has enabled some progress to be made using the slice preparation. It now appears that a fall in ATP is, indeed, important in causing rapid anoxic effects on neural function, and that both membrane depolarization and adenosine liberation subsequent to this fall, are effectors of the actual response inhibition.

Do the results we obtained apply to the brain *in vivo*? We do not know. However, it is very encouraging that in spite of the fact that the metabolite levels are low in the slice, and that the ion balance is already somewhat defective, the process of anoxic decay is not any faster than it is *in situ*. This, at least, suggests that we are not seeing effects that are only meaningful because the slice is starting from an inhibited state. If that were the case, then the block of transmission in the slice would be expected to be more rapid than *in situ*. This is certainly not the case. As is the case for the problem posed in the first part of the chapter, this problem is also far from solved. Decreased energy metabolism clearly leads to changes in a large number of important parameters. The elucidation of how all these changes interact to finally affect transmission will require a great deal of study.

4. REFERENCES

Amtorp, O., 1979, Distribution of inulin, sucrose, and mannitol in rat brain cortex slices following *in vivo* or *in vitro* equilibration, *J. Physiol.* (*London*) **294:**81–89.

Andersen, P., 1960, Interhippocampal impulses. II. Apical dendritic activation of CA1 neurons, *Acta Physiol. Scand.* **48:**178–189.

Arieff, A. I., Kerian, A., Massry, S. G., and DeLima, J., 1976, Intracellular pH of brain: Alterations in acute respiratory acidosis and alkalosis, *Am. J. Physiol.* (*London*) **230:**804–812.

Astrup, J., Sorensen, P. M., and Sorensen, H. K., 1981, Oxygen and glucose consumption related to Na-K transport in canine brain, *Stroke* **12:**726–730.

Bachelard, H. S., Campbell, W. J., and McIlwain, H., 1963, The sodium and other ions of mammalian cerebral tissues maintained and electrically stimulated *in vitro, Biochem. J.* **84:**225–237.

Baethmann, A. and Sohler, K., 1975, Electrolyte and fluid spaces of rat brain *in situ* after infusion with dinitrophenol, *J. Neurobiol.* **6:**73–84.

Bak, I. J., Misgeld, U., Weiler, M., and Morgan, E., 1980, The preservation of nerve cells in rat neostriatal slices maintained *in vitro*: A morphological study, *Brain Res.* **197:**341–353.

Baker, P. F. and Knight, D. E., 1981, Calcium control of exocytosis and endocytosis in bovine adrenal medullary cells, *Philos. Trans. R. Soc. London, Ser. B* **296:**83–103.

Beaugé, L. and di Polo, R., 1981, The effects of ATP on interactions between monovalent cations and the sodium pump in dialysed squid axons, *J. Physiol.* (*London*) **314:**457–480.

Benardo, L. S. and Prince, D. A., 1982, Dopamine action on hippocampal pyramidal cells, *J. Neurosci.* **2:**415–423.

Benjamin, A. M. and Verjee, Z. H., 1980, Control of aerobic glycolysis in the brain *in vitro, Neurochem. Res.* **5:**921–934.

Berne, R. M., Rubio, R., and Duling, B. R., 1971, Vasoactive substances affecting the coronary circulation in *Myocardial Ischemia*, in: *Excerpta Medica*, (R. S. Ross and F. Hoffman, eds.) Elsevier North Holland, Amsterdam pp. 28–43.

Berne, R. M., Rubio, R., and Curnish, R. R., 1974, Release of adenosine from ischemic brain, *Circ. Res.* **35:**262–271.

Bertman, L., Dahlgren, N., and Siesjo, B. K., 1979, Cerebral oxygen consumption and blood flow in hypoxia: Influence of sympathoadrenal activation, *Stroke* **10:**20–30.

Bertoni, J. M. and Siegel, G. J., 1978, Development of Na–K ATPase in rat cerebrum: correlation with Na-dependent phosphorylation and K-paranitrophenylphosphatase, *J. Neurochem.* **31:**1501–1511.

Booth, R. F. G. and Clark, J. B., 1978, Studies on the mitochondrially bound form of rat brain creatine kinase, *Biochem. J.* **170:**145–152.

Bosley, T. M., Woodhams, P. L., Gordon, R. D., and Balazs, R., 1983, Effects of anoxia on the stimulated release of amino acid neurotransmitters in the cerebellum *in vitro, J. Neurochem.* **40:**189–201.

Bourke, R. S. and Tower, D. B., 1966, Fluid compartmentation and electrolytes of cat cerebral cortex *in vitro*. II Sodium, potassium and chloride of mature cerebral cortex, *J. Neurochem.* **13:**1099–1117.

Burke, B. E. and DeLorenzo, R. J., 1982, Ca and calmodulin dependent phosphorylation of endogenous synaptic vesicle tubulin by a vesicle-bound calmodulin-kinase system, *J. Neurochem.* **38:**1205–1218.

Carpenter, D. O., Hubbard, J. H., Humphrey, D. R., Thompson, H. K., and Marshall, W., 1974, Carbon dioxide effects on nerve cell function, in: *Carbon Dioxide and Metabolic Regulation* (G. Nahas and K. E. Shaefer, eds.) Springer-Verlag, New York.

Carregal, E. J. A., 1975, The site of anoxic block in the spinal monosynaptic pathway, *J. Neurobiol.* **6**:103–113.

Chan, P. H. and Fishman, R. A., 1978, Brain edema:Induction in cortical slices by polyunsaturated fatty acids, *Science* **201**:358–360.

Cohen, S. R., 1974, The dependence of water content and "extracellular" marker spaces of incubated mouse brain slices on thickness, *Exp. Br. Res.* **20**:435–457.

Dingledine, R., Dodd, J., and Kelly, J. S., 1980, The *in vitro* brain slice as a useful neurophysiological preparation for intracellular recording, *J. Neurosci. Methods* **2**:323–362.

Duffy, T. E., Nelson, S. R., and Lowry, O. H., 1972, Cerebral carbohydrate metabolism during acute hypoxia and recovery, *J. Neurochem.* **19**:959–977.

Dunwidde, T. V. and Hoffer, B. J., 1980, Adenine nucleotides and synaptic transmission in the *in vitro* rat hippocampus, *Br. J. Pharmacol.* **69**:59–68.

Eckert, R. and Tillotson, D. L., 1981, Calcium-mediated inactivation of the calcium conductance in caesium-loaded giant neurons of *aplysia californica*, *J. Physiol.* **314**:265–280.

Elliot, K. A. C. and Wolfe, L., 1962, Brain tissue respiration and glycolysis, in: *Neurochemistry* (K. A. C. Elliot, L. Wolfe, and J. H. Quastel, eds.), Thomas, Springfield, Ill.

Farr, D. A. and Fuhrman, F. A., 1965, Role of diffusion of oxygen in the respiration of tissues at different temperatures, *J. Appl. Physiol.* **20**:637–646.

Folbergova, J., Ingvar, M., and Siesjo, B. K., 1981, Metabolic changes in cerebral cortex, hippocampus and cerebellum during sustained bicuculline-induced seizures, *J. Neurochem.* **37**:1228–1238.

Franck, G., Cornette, M., and Schoeffeniels, E., 1968, The cationic composition of incubated cerebral cortex slices, *J. Neurochem.* **15**:843–857.

Fredholm, B. B. and Hedqvist, P., 1980, Modulation of neurotransmission by purine nucleotides and nucleosides, *Biochem. Pharmacol.* **29**:1635–1643.

Fredholm, B. B., Janzen, B., Lindgren, E., and Lindstrom, K., 1982, Adenosine receptors mediating cyclic AMP production in the rat hippocampus, *J. Neurochem.* **39**:165–175.

Fujii, T., Baumgartl, H., and Lubbers, D. W., 1982, Limiting section thickness of guinea pig olfactory cortical slices studied from tissue pO2 values and electrical activities, *Pflüger Arch.* **393**:83–87.

Garthwaite, J., Woodhams, P. L., Collins, M. J., and Balazs, R., 1979, On the preparation of brain slices: Morphology and cyclic nucleotides, *Brain Res.* **173**:373–377.

Ghajar, J. B. G., Plum, F., and Duffy, T. E., 1982, Cerebral oxidative metabolism and blood flow during acute hypoglycemia and recovery in anesthetized rats, *J. Neurochem.* **38**:397–409.

Gibson, G. E., Peterson, C., and Sansone, J., 1981, Decreases in amino acid and acetylcholine metabolism during hypoxia, *J. Neurochem.* **37**:192–201.

Goodman, F. R., Weiss, G. B., and Alderice, M. T., 1973, On the measurement of extracellular space in slices prepared from different rat brain areas, *Neuropharmacology* **12**:867–873.

Grossman, R. G. and Williams, V. E., 1971, Electrical activity and ultrastructure of cortical neurons and synapses in ischemia, in: *Brain Hypoxia* (G. Brierly and B. Meldrum, eds.), Lippincott, Philadelphia.

Hansen, A. J., Hounsgaard, J., and Jahnsen, H., 1982, Anoxia increases potassium conductance in hippocampal nerve cells, *Acta Physiol. Scand.* **115**:301–310.

Hawkins, R. A., Williamson, D. H., and Krebs, H. A., 1971, Ketone body utilization by adult and suckling rat brain *in vivo*, *Biochem. J.* **122**:13–18.

Hertz, L., Schousboe, A., and Weiss, G. B., 1970, Estimation of ionic concentrations and intracellular pH in slices from different areas of rat brain, *Acta Physiol. Scand.* **79:**506–515.

Jacobus, W. E., 1980, Myocardial energy transport: Current concepts of the problem, in: *Heart Creatine Kinase* (W. E. Jacobus and J. S. Ingwall, eds.), Williams and Wilkins, Baltimore.

Jundt, H., Parzig, H., Reuter, H., and Stucki, J. W., 1975, The effect of substances releasing intracellular calcium ions on sodium-dependent calcium efflux from guinea-pig auricles, *J. Physiol.* (*London*) **246:**229–241.

Kaasik, A. E., Nilsson, L., and Siesjo, B. K., 1970, The effect of asphyxia upon the lactate, pyruvate and bicarbonate concentrations of brain tissue and cisternal CSF, and upon the tissue concentrations of phosphocreatine and adenine nucleotides in anesthetized rats, *Acta. Physiol. Scand.* **78:**433–447.

Kass, I. S. and Lipton, P., 1982, Mechanisms involved in irreversible anoxic damage to the *in vitro* hippocampal slice, *J. Physiol.* (*London*) **332:**459–472.

Katzman, R. and Pappius, H. M., 1973, *Brain Electrolytes and Fluid Metabolism* Williams and Wilkins, Baltimore.

Keesey, J. C., Wallgren, H., and McIlwain, H., 1965, The sodium, potassium and chloride of cerebral tissues: Maintenance, change on stimulation and subsequent recovery, *Biochem. J.* **95:**289–300.

Kimelberg, H. K., Biddlecome, R., Narumi, S., and Bourke, R. S., 1978, ATPase and carbonic anhydrase actuities of bulk-isolated neurons, glia and synaptosome fractions from rat brain, *Brain Res.* **141:**305–323.

King, L. J., Schoepfle, G. M., Lowry, O. H., Passonneau, J. V., and Wilson, S., 1967, Effects of electrical stimulation on metabolites in brain of decapitated mice, *J. Neurochem.* **14:**613–618.

Kobayashi, M., Lust, W. D., and Passonneau, J. V., 1977, Concentrations of energy metabolites and cyclic nucleotides during and after bilateral ischemia in the gerbil cerbral cortex, *J. Neurochem.* **29:**53–59.

Krnjevic, K., 1975, Coupling of neuronal metabolism and electrical activity, Alfred Benzon Symposium (D. H. Ingrar and N. A. Hassen, eds.), Marksgaard, Copenhagen, pp. 65–78.

Krnjevic, K., and Miledi, R., 1959, Presynaptic failure of neuromuscular propagation in rats, *J. Physiol.* (*London*) **149:**1–22.

Krnjevic, K., Randic, M., and Siesjo, B. K., 1965, Cortical CO_2 tension and neuronal excitability, *J. Physiol.* (*London*) **176:**105–122.

Lai, Y. L., Atteberg, B. A., and Brown, E. B. Jr., 1973, Intracellular adjustments of skeletal muscle, heart and brain to prolonged hypercapnia, *Respir. Physiol.* **19:**115–122.

Landau, E. M. and Nachson, D. A., 1975, The interaction of pH and divalent cations at the neuromuscular junction, *J. Physiol.* **251:**775–790.

Lee, K. and Schubert, P., 1982, Modulation of an inhibitory circuit by adenosine and AMP in the hippocampus, *Brain Res.* **246:**311–314.

Lipton, P. and Heimbach, C. J., 1977, The effect of extracellular potassium concentration on protein synthesis in guinea pig hippocampal slices, *J. Neurochem.* **28:**1347–1354.

Lipton, P. and Heimbach, C. J., 1978, Mechanism of extracellular potassium stimulation of protein synthesis in the *in vitro* hippocampus, *J. Neurochem.* **31:**1299–1307.

Lipton, P. and Korol, D., 1981, Evidence that decreases in intracellular pH rapidly inhibit transmission in the guinea pig hippocampal slice, *Abstr. Soc. Neurosci.* **7:**440.

Lipton, P. and Robacker, K. M., 1982, Adenosine may cause early inhibition of synaptic transmission during anoxia, *Abstr. Soc. Neurosci.* **8:**982.

Lipton, P. and Whittingham, T. S., 1979, The effect of hypoxia on evoked responses in the *in vitro* hippocampus, *J. Physiol. (London)* **287**:427–438.

Lipton, P. and Whittinghahm, T. S., 1982, Reduced ATP concentration as a basis for synaptic transmission failure during hypoxia in the *in vitro* guinea pig hippocampus, *J. Physiol. (London)* **325**:51–65.

Ljunggren, B., Ratcheson, R. A., and Siesjo, B. K., 1974, Cerebral metabolic state following complete compression ischemia, *Brain Res.* **73**:291–307.

Lowry, O. H., Passonneau, J. V., Hasselberger, F. X., and Schulz, D. W., 1964, Effect of ischemia on known substrates and cofactors of the glycolytic pathway in brain, *J. Biol. Chem.* **239**:18–30.

Lund-Andersen, H., 1974, Extracellular and intracellular distribution of inulin in rat brain cortex slices, *Brain Res.* **65**:239–254.

Lust, W. D., Whittingham, T. S., and Passonneau, J. V., 1982, Effects of slice thickness and method of preparation on energy metabolism in the *in vitro* hippocampus, *Abstr. Soc. Neurosci.* **8**:1000.

MacMillan, V., 1975, The effects of acute carbon monoxide intoxication on the cerebral energy metabolism of the rat, *Can. J. Physiol. Pharmacol.* **53**:354–362.

MacMillan, V. and Siesjo, B. K., 1972, Intracellular pH of the brain in arterial hypoxemia, evaluated with the CO_2 method and from the creatine phosphokinase equilibrium, *Scand. J. Clin. Invest.* **30**:117–125.

Maker, H. S., Lehrer, G. M., Silides, D. J., and Weiss, C., 1973, Regional changes in cerebellar creatine phosphate metabolism during late maturation, *Exp. Neurol.* **38**:295–300.

Matthews, D. A., Cotman, C. A., and Lynch, G., 1976, An electron microscope study of lesion-induced synaptogenesis in the dentate gyrus of the adult rat, *Brain Res.* **115**:1–15.

McGilvery, R. W. and Murray, T. W., 1974, Calculated equilibria of phosphocreatine and adenosine diphosphates during utilization of high energy phosphate by muscle, *J. Biol. Chem.* **249**:5845–5850.

McIlwain, H., 1953, The effects of depressants on the metabolism of stimulated cerebral tissues, *Biochem. J.* **53**:403–412.

McIlwain, H. and Bachelard, H. S., 1971, *Biochemistry and the Central Nervous System*, Churchill Livingston, Edinburgh.

Meldrum, B. S. and Nilsson, B., 1976, Cerebral blood flow and metabolic rate early and late in prolonged seizures induced in rats by bicuculline, *Brain* **99**:407–418.

Misgeld, U. and Frotscher, M., 1982, Dependence of the viability of neurons in hippocampal slices on oxygen supply, *Brain Res. Bull.* **8**:95–100.

Morris, M. E., 1971, The action of carbon dioxide on synaptic transmission in the cuneate nucleus, *J. Physiol. (London)* **218**:671–688.

Mullins, L. J. and Requena, J., 1981, The "late" Ca channel in squid axons, *J. Gen. Physiol.* **78**:683–700.

Nachsen, D. A. and Blaustein, M. P., 1979, Regulation of nerve terminal calcium selectivity by a weak acid site, *Biophys. J.* **26**:329–334.

Nemoto, E. M., Shiu, G. K., Nemmer, J., and Bleyaert, A. L., 1982, Attenuation of free fatty acid liberation during global ischemia: A model for screening potential therapies for efficacy, *J. Cereb. Blood Flow Metab.* **2**:475–480.

Norberg, K. and Siesjo, B. K., 1975, Cerebral metabolism in hypoxic hypoxia. I. pattern of activation of glycolysis, A re-evaluation, *Brain Res.* **86**:31–44.

Norberg, K., Quistorff, B., and Siesjo, B. K., 1975, Effects of hypoxia of 10 to 45 seconds duration on energy metabolism in the cerebral cortex of unanesthetized and anesthetized rats, *Acta Physiol. Scand.* **95**:301–310.

Nordstrom, C. H. and Rehncrona, S., 1977, Postischemic cerebral blood flow and oxygen utilization rate in rats anesthetized with nitrous oxide or phenobarbital, *Acta Physiol. Scand.* **101**:230–240.

Nordstrom, C. H., Rehncrona, S., and Siesjo, B. K., 1978, Effects of phenobarbital in cerebral ischemia. II. Restitution of cerebral energy state, glycolytic metabolites, citric acid cycle intermediates and associated amino acids after incomplete ischemia, *Stroke* **9**:335–343.

Okada, Y. and Saito, M., 1979, Inhibitory action of adenosine, 5-HT and GABA on the post synaptic potential of slices from olfactory cortex and superior colliculus in correlation to the level of cyclic AMP, *Brain Res.* **160**:368–371.

Philipson, K. D., Bersohn, M. M., and Nishimoto, A. Y., 1982, Effects of pH on Na-Ca exchange in canine cardiac sarcolemmal vesicles, *Circ. Res.* **50**:224–229.

Phillis, J. W., 1977, The role of cyclic nucleotides in the CNS, *Can. J. Neurol. Sci.* **4**:153–182.

Preissler, M. and Williams, J. A., 1981, Pancreatic acinar cell function: measurement of intracellular ions and pH and their relation to secretion, *J. Physiol. (London)* **321**:437–448.

Pull, I. and McIlwain, H., 1972, Adenine derivatives as neurohumoral agents in the brain. The quantities liberated on excitation on superfused cerebral tissues, *Biochem. J.* **130**:975–981.

Rafalowska, U., Erecinska, M., and Wilson, D. F., 1980, Energy metabolism in rat brain synaptosomes from nembutal-anesthetized and non-anesthetized animals, *J. Neurochem.* **34**:1380–1386.

Reddington, M. and Shubert, P., 1979, Parallel investigations of the effects of adenosine on evoked potentials and cyclic AMP accumulation in hippocampus slices of the rat, *Neurosci. Lett.* **14**:37–42.

Rees, S., Cragg, B. G., and Everitt, A. V., 1982, Comparison of extracellular space in the mature and agine rat brain using a new technique, *J. Neurol. Sci.* **53**:347–357.

Robinson, J. D., 1967, Kinetic studies on a brain microsomal adenosine triphosphatase. Evidence suggesting a conformational change, *Biochemistry* **10**:3250–3258.

Rolleston, F. S. and Newsholme, E. A., 1967, Control of glycolysis in cerebral cortex slices, *Biochem. J.* **104**:524–533.

Roos, A. and Boron, W. F., 1981, Intracellular pH, *Physiol. Rev.* **61**:296–434.

Rubio, R. Berne, R. M., and Bockman, E. L., 1975, Relationship between adenosine concentration and oxygen supply in rat brain, *Am. J. Physiol.* **228**:1896–1902.

Saks, V. A., 1980, Creatine kinase isozymes and the control of cardiac contraction, in: *Heart Creatine Kinase: The Integration of Isozymes for Energy Distribution* (W. E. Jacobus and J. S. Ingwall, eds.), Williams and Wilkins, Baltimore, pp. 109–124.

Saks, V. A., Lipina, N. V., Sharov, V. G., and Chazov, E. I., 1977, The localization of the MM isozyme of creatine phosphokinase on the surface membrane of myocardial cells and its functional coupling to ouabain-inhibited (Na-K) ATPase, *Biochem. Biophys. Acta* **465**:550–558.

Saks, V. A., Rosenshtrauhk, L., Smirnov, V. and Chazov, E., 1978, Role of creatine kinase in cellular function and metabolism, *Can. J. Physiol. Pharmacol.* **56**:691–706.

Salford, L. G., Plum, F., and Siesjo, B. K., 1973, Graded hypoxia-oligemia in rat brain. I. Biochemical alterations and their implications, *Arch. Neurol.* **29**:227–233.

Schmahl, F. W., Betz, E. Dettinger, E., and Hohorst, H., 1966, Energiestoffwechs der grosshirnrinde und elektroencephalogram bei sauerstoffmangel, *Pflug. Arch. Gesamte Physiol.* **292**:46–59.

Schwartzkroin, P. A., 1981, To slice or not to slice, in: *Electrophysiology of Isolated Mammalian CNS Preparations* (G. A. Kerkut and H. V. Wheal, eds.), Academic Press, New York, pp. 15–50.

Seeman, P., 1980, Brain dopamine receptors, *Pharmacol. Rev.* **32**:229–313.

Seraydarian, M. W., 1980, The correlation of creatine phosphate with muscle function, in *Heart Creatine Kinase* (W. E. Jacobus and J. S. Ingwall, eds.) Williams and Wilkins, Baltimore, pp. 82–91.

Seraydarian, M. W., Artaza, L., and Abbot, B. C., 1974, Creatine and the control of energy metabolism in cardiac and skeletal muscle cells in culture, *J. Mol. Cell. Cardiol.* **6**:405–413.

Seraydarian, M. W. and Artaza, L., 1976, Regulation of energy metabolism by creatine in cardiac and skeletal muscle cells in culture, *J. Mol. Cell. Cardiol.* **8**:669–678.

Siemkowicz, E. and Hansen, A. J., 1981, Brain extracellular ion composition and EEG activity following 10 minutes ischemia in normo and hyperglycemic rats, *Stroke* **12**:236–240.

Siesjo, B. K., 1978, *Brain Energy Metabolism*, Wiley, New York.

Siesjo, B. K., 1981, Cell damage in the brain: A speculative synthesis, *J. Cereb. Blood Flow Metab.* **1**:155–185.

Siesjo, B. K. and Nilsson, L., 1971, The influence of arterial hypoxemia upon labile phosphates and upon extracellular and intracellular lactate and pyruvate concentrations in the rat brain, *Scand. J. Clin. Invest.* **27**:83–96.

Siesjo, B. K., Folbergova, J., and MacMillan, V., 1972, The effect of hypercapnia upon intracellular pH in the brain evaluated by the bicarbonate-carbonic acid method and from the creatine phosphokinase equilibrium, *J. Neurochem.* **19**:2483–2495.

Skrede, K. K. and Westgaard, R. H., 1971, The transverse hippocampal slice: A well defined cortical structure maintained *in vitro*, *Brain Res.* **35**:589–593.

Snyder, J. V., Nemoto, E. M., Carroll, R. G., and Safar, P., 1975, Global ischemia in dogs: Intracranial pressures, brain blood flow and metabolism, *Stroke* **6**:21–27.

Sokoloff, L., 1971, Neurophysiology and neurochemistry of coma, *Exp. Biol. Med.* **4**:15–23.

Sokoloff, L., 1981, Localization of functional activity in the central nervous system by measurement of glucose utilization with radioactive deoxyglucose, *J. Cereb. Blood Flow Metab.* **1**:7–36.

Steen, P. A., Michenfelder, J. D., and Milde, J. H., 1979, Incomplete versus complete ischemia: Improved outcome with a minimal blood flow, *Ann. Neurol.* **6**:389–398.

Steiner, A. L., Ferrendelli, J. A., and Kipnis, D. M., 1972, Radioimmunoassay for cyclic nucleotides. Effect of ischemia, changes during development and regional distribution of adenosine 3′-5′monophosphate and guanosine 3′-5′monophosphate in mouse brain, *J. Biol. Chem.* **247**:1121–1124.

Tang, W. and Sun, G. Y., 1982, Factors affecting the free fatty acids in rat brain cortex, *Neurochem. Int.* **4**:269–273.

Thomas, J., 1956, The composition of isolated cerebral tissues: Creatine, *Biochem. J.* **64**:335–339.

Thomas, J., 1957, The composition of isolated cerebral tissues: Purines, *Biochem. J.* **66**:655–658.

Urbanics, R., Leniger-Follert, E., and Lubbers, D. W., 1978, Extracellular K and H activities in the brain cortex during and after a short period of ischemia and arterial hypoxemia, *Adv. Exp. Biol. Med.* **94**:611–618.

Veech, R. L., Lawson, J. W. R., Cornell, N. W., and Krebs, H. A., 1979, Cytosolic phosphorylation potential, *J. Biol. Chem.* **254**:6538–6547.

Vincent, A. and Blair, J. McD., 1970, The coupling of the adenylate kinase and creatine kinase equilibria. Calculation of substrate and feedback signal levels in muscle, *FEBS Lett.* **7**:239–244.

Vincenzi, F. F., 1971, A calcium pump in red cell membranes, in: *Cellular Mechanisms for Calcium Transfer and Homeostasis* (G. N. Nicholls and R. H. Wasserman, eds.), Academic Press, New York, pp. 135–146.

Warburg, O., 1923, Versuche anüberebendem Carcinomgewebe (Methoden), *Biochem. Z.* **142**:317–350.

Whittam, R., 1962, The dependence of the respiration of brain cortex on active cation transport, *Biochem. J.* **82**:205–212.

Whittingham, T. S. and Lipton, P., 1981, Cerebral synaptic transmission during anoxia is protected by creatine, *J. Neurochem.* **37**:1618–1621.

Whittingham, T. S., Lust, W. D., Arai, H., Wheaton, A. O., and Passonneau, J. V., 1981, Changes in the energy profile and electrical response of hippocampal slices during decapitation ischemia and recovery *in vitro, Abstr. Soc. Neurosci.* **7**:458.

Wilkening, D. and Makman, M. H., 1977, Activation of glycogen phosphorylase in rat caudate nucleus slices by 1-isopropylnorepinephrine and dibutyrylcyclic AMP, *J. Neurochem.* **28**:1001–1007.

Wu, P. H., Phillis, J. W., and Thierry, D. L., 1982, Adenosine receptor agonists inhibit K-evoked Ca uptake by rat brain cortical synaptosomes, *J. Neurochem.* **39**:700–708.

Wu, T. F. L. and Davis, E. J., 1981, Regulation of glycolytic flux in energetically controlled cell free system, *Arch. Biochem. Biophys.* **209**:85–99.

Yamamoto, C. and Kurokawa, M., 1970, Synaptic potentials recorded in brain slices and their modification by changes in the level of tissue ATP, *Exp. Brain Res.* **10**:159–170.

6

Hippocampus

Electrophysiological Studies of Epileptiform Activity *in Vitro*

BRADLEY E. ALGER

1. INTRODUCTION

Studies on epileptic phenomena have multiplied rapidly with the growing realization that many of the persistent problems in the field can be attacked directly using the *in vitro* slice preparation. Recently, the hippocampal slice has been most widely used in studies of epilepsy for reasons that are detailed in the Appendix. While attempting to present an overview of electrophysiological studies of epilepsy in the hippocampal slice, this review will focus on data that has appeared since 1978. The fundamental hypothesis concerning the intrinsic nature of the epileptiform "burst" potential has been widely supported (see Prince, 1978; Schwartzkroin and Wyler, 1980). The chief areas of departure of the present discussion from previous reviews will be in consideration of data pertaining to: (1) the existence of "giant excitatory postsynaptic potentials (EPSPs)," (2) modes of inhibitory synaptic control over pyramidal cell dendrites, (3) the nature of various afterhyperpolarizations, (4) ways in which synchronization of firing within a neuronal population can occur, and (5) mechanisms that can enable epileptic discharges to spread.

BRADLEY E. ALGER • Department of Physiology, University of Maryland School of Medicine, Baltimore, Maryland 21201.

1.1. *Experimental Questions*

In the past, attention focused on the most prominent feature of the focal epileptic discharge, the interictal spike potential. Although the interictal spike has been a fruitful model and is an important phenomenon in its own right, there are many questions concerning epilepsy that may not be answered even once the interictal spike is fully understood. It may be useful to consider the events of epilepsy from the point of view of a number of reasonably distinct questions. Some of these questions have already been extensively studied and detailed answers to them are available. Answers to other questions are less complete.

1.1.1. *What Enables Some Cells to Fire Bursts Readily?* Some neuronal regions are prone to seizures while others are relatively resistant. The hippocampus is thought to be the structure in the brain with the lowest seizure threshold. Since the first electrophysiological anomaly that appears in models of focal epilepsy is the interictal discharge, involving burst firing by individual cells, one can ask why certain cells have a high susceptibility to bursting and therefore, presumably, to seizures. The answer to this question involves a detailed analysis of the burst response.

1.1.2. *Why Do Cells That Can Fire Burst Potentials Not Do So All the Time?* Even cells that can burst under certain circumstances do not always burst under normal conditions. What cellular mechanisms control this firing pattern? (See Section 3.)

1.1.3. *How Does Synchronization of Firing within a Population of Cells Occur?* Epilepsy involves the massive synchronous firing of a large group of neurons. Bursts might be fired by individual members of a cellular population, and yet epilepsy would not result.

1.1.4. *What Triggers the Switch from Interictal Spiking to Seizures?* From *in vivo* recordings, it appears that a seizure results when interictal spikes and afterdischarges occur so closely together in time that the individual spikes appear to merge. Therefore, the question of what triggers a seizure may be functionally equivalent to "What terminates an interictal spike?" or, since an interictal spike is the summated response of a group of neurons each individually firing burst potentials, to "What terminates a burst?" The obvious hypothesis is that whatever prevents the termination of an interictal spike leads to a seizure.

1.1.5. *How Can a Seizure Spread from Epileptic Tissue across Normal Tissue?* Primary epileptic anomalies are thought to occur in restricted regions of neural tissue called "foci." Outside of a focus, tissue appears to be normal. The existence of an epileptic focus, no matter how pathological, does not immediately imply that the hyperactivity should be able to spread from the focus. What changes occur that enable normal tissue to behave pathologically?

This review will consider what has been learned from studies on the hippocampal slice concerning the cellular events of epilepsy in the context of these questions.

1.2. *Terminology*

1.2.1. Interictal Spike. An EEG abnormality recorded in an intact animal in an epileptic focus. It involves discharge of a population of cells. The *in vitro* model for the interictal spike is the "synchronous" burst potential, i.e., a population event in which many cells fire burst discharges nearly simultaneously. Synchronous bursts may be evoked or may occur spontaneously. In certain regions, a single cell may burst spontaneously unaccompanied by simultaneous bursting of nearby cells. This is referred to as an "asynchronous" burst.

1.2.2. Burst. The complex epileptiform response of a single cell. It has two major components: a large, slow depolarizing wave and a series of regenerative action potentials riding on this wave. Small attenuated action potentials (see Section 1.2.4) may or may not be present.

1.2.3. Paroxysmal Depolarization Shift (PDS). According to the traditional definition, this term was applied to the large slow depolarizing wave that underlies and presumably triggers the repetitive firing of action potentials during the intracellularly recorded burst. In the original studies (e.g., Ayala *et al*, 1973; Prince, 1978) the bursts were probably always "synchronous", as defined in Section 1.2.1. However the "asychronous" burst also has a large slow depolarizing component. Although these bursts are probably not identical, evidence concerning the nature of the PDS can be gained by studying the events associated with the asynchronous burst.

1.2.4. Small Attenuated Action Potentials (Spikelets). Small all-or-none action potentials, also called *fast prepotentials* and *d-spikes*. The term prepotential seems inappropriate for an event that does not necessarily precede anything. They were called d-spikes because of their presumed site of origin in dendrites; however, it has been argued that they may originate in axons or may reflect action potentials in electrotonically coupled cells. Here they are simply referred to without prejudice as small attenuated action spikes or spikelets.

1.2.5. CA2/CA3 Region. Rigorous distinctions have not always been made between hippocampal subfields. Although such distinctions are increasingly seen as having physiological significance, the same regions have apparently not always been called by the same names. There is probably reasonable agreement about what constitutes "CA1," since there is agreement concerning the physiological properties of the cells. It is also agreed that "CA2" and "CA3" cells are quite different from CA1 cells. However, precise distinctions between CA2 and CA3 cells have not always been made. Therefore, unless the experimenters have clearly defined the regions from which they record, I will refer to all cells from "CA2" and "CA3" as from "CA2/CA3." This is not an ideal solution and more careful distinctions are required as the differences between the various fields become sharpened.

1.2.6. Afterdischrage. A state of excitability intermediate between an interictal spike and a seizure. An afterdischarge consists of a series of slow spikelike potentials following either an interictal spike or a burst. They can last from hundreds of milliseconds to a few seconds.

1.2.7. Seizure. The prolonged synchronous firing of afterdischarges and tonic depolarizations. A seizure lasts many seconds and is usually followed by a period of electrical inexcitability that then recovers over several minutes.

1.2.8. Epilepsy Models. Unless otherwise noted, all of the studies cited here on focal epileptogenic mechanisms were done on rat or guinea pig hippocampal slices made epileptogenic by a gamma aminobutyric acid (GABA) antagonist.

2. WHAT ENABLES SOME CELLS TO FIRE BURSTS READILY?

Among the hippocampal regions, there are differences in seizure and burst susceptibility. The dentate gyrus is generally acknowledged to have the highest seizure threshold while the CA2/CA3 region has the lowest. Intracellular recording has confirmed that dentate granule cells do not fire bursts while CA2/CA3 cells can burst spontaneously. CA1 cells can burst, but do not do so under ordinary circumstances. In order to understand such differences among cell types, it is necessary to consider what is known about the burst itself.

Many hypotheses have been offered to explain the burst, and, although the controversy has stimulated many important experiments, it seems that resolution of the problem involves parts of several proposals.

2.1. *Paroxysmal Depolarization Shift (PDS)*

2.1.1. Intrinsic Mechanisms. Convincing evidence that the PDS is largely an "intrinsic" potential has been obtained by demonstrations that very short depolarizing current pulses applied to certain hippocampal cells produces fully developed burst potentials that, once triggered, are independent of the duration and amplitude of the current pulse (Kandel and Spencer, 1961; Wong and Prince, 1979; cf. Fujita and Iwasa, 1977). These asynchronous bursts have both slow depolarizing components, like PDSs, and repetitive action potentials. This simple experiment demonstrates the important fact that when certain hippocampal membranes are adequately depolarized, they will express their innate capacity to burst. The ability to burst in this way is not universally distributed, however. CA1 cell dendrites can be induced to burst directly, but CA1 somata cannot, while both somata and dendrites of CA2/CA3 cells burst readily (Wong *et al.*, 1979).

With the information that the asynchronous burst is an intrinsic potential, its constituent parts can be analyzed. Hippocampal pyramidal cells have depolarizing after-potentials (DAPs) and the slow depolarization appears to represent the summation of these DAPs. Kandel and Spencer (1961) first observed that, following an action potential, a hippocampal pyramidal cell does not always hyperpolarize as a squid axon does. Rather, the active phase of repolarization stops at a level that is depolarized by several millivolts with respect to the resting membrane potential. DAPs have two components (Fujita, 1975; Wong and Prince, 1981). Wong and Prince (1981) showed that, from the point at which active repolarization stops, the membrane potential actually redepolarizes briefly before returning to baseline. This brief period of redepolarization is an active process and is sufficient to lead to the production of a second action potential that has a DAP that sums with the first, etc. The process usually shuts off after 4 to 8 action potentials for reasons discussed below.

The active phase of the DAP is due, at least in part, to a calcium (Ca^{2+}) influx since it is blocked by manganese (Mn^{2+}) or zero Ca^{2+} conditions, while it is facilitated by local application of barium (Ba^{2+}) or Ca^{2+} and is insensitive to tetrodotoxin (TTX). However, a contribution of other ions has not been ruled out and it is unclear whether this Ca^{2+} conductance mechanism is the same as the mechanism that leads to the production of Ca^{2+} spikes (see below). The active DAP is initiated by depolarization and can be prevented by strongly hyperpolarizing the cell. Such a conductance could be involved in "anomalous rectification" in hippocampal cells (Hotson *et al.*, 1979).

Although the active phase of the DAP can be turned on by the action potential, this may not always be the case. Subthreshold orthodromic stimulation of a CA1 cell in penicillin (Andersen *et al.*, 1978), picrotoxin, or bicuculline methiodide (Alger, 1983) may cause an EPSP followed by a late depolarizing bump. "Biphasic" EPSPs have been seen by others and may represent the same phenomenon (Schwartzkroin, 1975; Kandel and Spencer, 1961). The occurrence of this bump is facilitated by depolarization and prevented by hyperpolarization. It is, therefore, an intrinsic potential that appears very much like the active phase of the DAP. However, neither a large nor a small action potential necessarily precedes it. Since depolarizing current pulses into the cell soma do not produce the late bump, it may represent a dendritic event. Current injection into a dendrite can produce these events (R. K. S. Wong, personal communication).

The entire DAP, however, it not blocked either by Mn^{2+} or by hyperpolarization (Wong and Prince, 1981). Once the active DAP is blocked, there is a slow repolarization to baseline membrane potential following active spike repolarization. This residual DAP decays exponentially with a time constant equal to the membrane time constant, i.e., the remaining DAP appears to be a passive potential due to the decay of charge from the membrane following its charging by the action potential.

2.1.2. Giant EPSPs. By analogy, the studies of the asynchronous burst (discussed above) would suggest that the PDS potential is a purely intrinsic event. Alternatively, it has been proposed that the PDS *is* a "giant EPSP," due perhaps to the activation of recurrent excitatory circuitry. If the "giant EPSP hypothesis" of the PDS is stated as an identity in this way, then the work described above argues against the hypothesis; i.e., the PDS is probably not a giant EPSP. However, one can still ask if giant EPSPs occur and, if so, what their nature and significance might be. Strong evidence that giant EPSPs do occur has recently been presented (Johnston and Brown, 1981; Wong and Traub, 1983). Using cesium-sulfate ($CsSO_4$) filled electrodes that permit extreme depolarization of cells, Johnston and Brown (1981) demonstrated that it is possible to reverse a component of the PDS in CA1 cells. This would be impossible if the PDS were entirely a voltage-activated response. Moreover, spontaneous reversed EPSPs also occurred at the same frequency as the spontaneous bursts recorded in unclamped cells, indicating that the frequency of the EPSP and, hence, of the spontaneous bursts, was imposed by presynaptic elements and was not an intrinsic property of the cells. Spontaneously occurring EPSPs in the presence of penicillin were much larger than spontaneous EPSPs in normal media. Therefore, it can be concluded that a giant EPSP does occur coincident with a PDS. Hablitz and Andersen (1982) showed the slow depolarizing potential to be grossly unaffected by reductions in extracellular Na^+ which attenuated somatic action potentials. They considered this to be evidence that the slow depolarization is a synaptic potential. In view of the possibility that the slow depolarization might represent a voltage-dependent wave possibly dendritic in orgin, these data are not conclusive on this point. They do suggest that postsynaptic Na^+-dependent action potentials are not a prerequisite for the slow depolarization.

However, it is not possible at present to determine the degree to which a giant EPSP contributes to the PDS. Once an excitatory postsynaptic current (EPSC) has ceased, then the potential caused by the current can only decay passively at a rate determined by the membrane time constant(s). During a burst, the membrane is not quiescent. There are active conductances associated with Na^+ and Ca^{2+} action potentials, the active phase of the DAP, as well as with K^+ potentials. The occurrence of these active conductances and associated potential changes will distort the effects of the EPSC *per se*. The possible contribution of an EPSP to the PDS will depend to a large extent on the actual duration of the EPSC, and detailed EPSC measurements are not yet available.

It is useful to note a distinction between evoked and spontaneous bursts occurring in CA1. When the burst is evoked in penicillin, the underlying EPSP is not especially large (e.g., Wong and Prince, 1979; Schwartzkroin and Prince, 1980a). These evoked EPSPs have a relatively rapid time to peak (10–15 msec) and overall duration (half time of decay of 5 to 10 msec). Thus, it is unlikely that the time course of the underlying EPSC is extremely prolonged. For example, when the burst is prevented by hyperpolarizing the cell (and inhibitory postsynaptic potentials (IPSPs) are also blocked), the evoked EPSP decays passively in accordance with the membrane time constant (Wong and Prince, 1979). The close agreement between the decay of a purely electrotonic potential and of the EPSP is very good evidence that there is no substantial "residuum" of transmitter action after the EPSP peak (see Eccles, 1964).

However, for the giant EPSCs underlying spontaneous PDSs, one can estimate a different time course from the records of Johnston and Brown (1981). Here the EPSCs may have 20 to 40 msec times to peak and apparent overall durations of 40 to 80 msec. Evidently, the giant EPSC may contribute more to the build-up of the PDS than the evoked EPSC. One prediction of this hypothesis would be that giant EPSPs underlying spontaneous bursts would have longer times to peak than EPSPs evoked by direct stimulation of afferent fibers. Support for this hypothesis can be found in published data showing that, in the same cell, the directly stimulated ("short-latency") EPSP may have a time to peak approximately twice as fast as the spontaneous ("delayed") EPSP (see Figure 4D). The delayed EPSP of Wong and Traub (1983) appears to be the same as the "giant EPSP" of Johnston and Brown (1981). Because of the rather different modes of activation, it may be that the evoked EPSP is an inappropriate model for the spontaneous giant EPSP. Study of the evoked EPSP was important in providing evidence against a possible explanation for giant EPSPs (e.g., Schwartzkroin and Prince, 1980a). Since penicillin did not cause the size of the evoked EPSP to become larger, it is likely that this drug does not increase either the excitability of presynaptic fibers or transmitter release from a single bouton. In view of the work of Wong and Traub (1983) (see Section 4 below), it is clear why this is so. Giant EPSPs occur because the cells that produce them begin to release excitatory transmitter nearly synchronously.

The significance of the giant EPSP is not that it obviates the need for IPSP depression to produce spontaneous bursting; all of the evidence to date indicates that IPSP depression is an absolute requirement. Nor are giant EPSPs sufficient to explain the PDS. Nevertheless, the giant EPSP does occur.

2.1.3. What Is the Function of the Giant EPSP? Giant EPSPs are not necessarily giant with respect to the size of evoked EPSPs. They are "giant" with respect to ordinarily occurring spontaneous EPSPs. They are important because ordinary spontaneous EPSPs are not adequate to trigger a burst in a CA1 cell. Because of the intrinsic nature of the burst potential, the DAP following a single action potential will be sufficient to cross threshold for the next action potential, etc, provided an IPSP does not abort the process. Thus, the question of the role of the giant EPSP can be reduced to the question of whether giant EPSPs are necessary to trigger spontaneous action potentials in CA1 dendrites. Inasmuch as spontaneous action potentials in healthy hippocampal cells *in vitro* are exremely rare, it can be concluded that ordinary spontaneous EPSPs are not sufficient to trigger action potentials, and thus, cannot trigger bursts. It is reasonable to assume that a synchronized or giant EPSP is necessary to trigger spontaneous bursts.

In conclusion, the PDS is probably largely an intrinsic event. The large slow depolarizing waves that result from direct current injection into CA2/CA3 cells are exclusively intrinsic events due to summation of DAPs. Slow depolarizations evoked synaptically by direct activation of afferent fibers are also primarily intrinsic potentials, since ordinary EPSCs are too brief to be a major factor. The PDS that is triggered spontaneously by giant EPSPs may have a component that is due to EPSP as well as to intrinsic potentials since the giant EPSC appears to be prolonged. The reason for this difference between events underlying synchronous as against directly evoked or asynchronous bursts will become clear from the discussion of synchronization and the origin of spontaneous bursts (see Section 4).

2.2. *Regenerative Components of the Burst*

A series of fast and slow action potentials is associated with a burst. Schwartzkroin and Slawsky (1977) found that TTX applied near the neuronal soma blocked fast, presumably Na^+-dependent action potentials, while the high threshold, broad action potentials remained. When TTX was applied to the dendrites, fast action potentials remained, but the slow ones were not present. TTX-resistant action potentials were enhanced by application of Ca^{2+} and Ba^{2+} and blocked by Mn^{2+}. Therefore, it was suggested that CA1 cell dendrites were capable of generating Ca^{2+}-dependent action potentials. Wong and Prince (1979) confirmed and extended these results. Using histological techniques, Wong *et al.* (1979) showed that electrodes could be placed into pyramidal cell dendrites as well as somata and that TTX-sensitive and TTX-resistant action potentials could be recorded in both. Calcium spikes were consistently larger when recorded in the dendrites while Na^+ spikes were always larger when recorded in somata, suggesting the primary action potential in the dendrite is produced by Ca^{2+} influx while Na^+ spikes predominate in the soma.

A slow voltage-activated inward current has been recorded in voltage-clamped CA2/CA3 cells (Johnston *et al.*, 1980). This current is enhanced by Ba^{2+} and decreased by Mn^{2+} and is thus probably carried by Ca^{2+}. The current produces a negative slope resistance in the range of -50 to -10 mV in CsC1-loaded cells in which most potassium (K^+) potentials have been blocked. It presumably underlies the repetitive Ca^{2+} potentials that are produced readily in CA2/CA3 cells.

In summary, a variety of data strongly suggest that Ca^{2+}-dependent action potentials contribute in a major way to the burst potential in the hippocampus. Conversely, cells less prone to fire burst potentials (for instance, the granule cells of the dentate gyrus (e.g., Prince, 1982)) have a much smaller capacity for firing Ca^{2+} potentials. Intradendritic recording from granule dendrites has shown that it is also not possible to trigger bursts simply with short depolarizing current pulses, as it is in CA1 and CA3. It appears that a complex of intrinsic specializations predisposes some cells to fire burst responses. The chief specializations include a heavy investiture of calcium channels especially in the dendritic regions and mechanisms that produce depolarizing afterpotentials. Spontaneous epileptiform bursting of a population of cells is usually triggered by giant EPSPs.

2.2.1. Small Attenuated Action Potentials (Spikelets). The occurrence of small attenuated action potentials has been suggested to be a prominent feature of epileptogenic tissue (Schwartzkroin and Prince, 1978,

1980a). First studied by Spencer and Kandel (1961), these events (see Section 1.2.4.) are small all-or-none depolarizations that are blocked by TTX. When they occur, they are capable of triggering full-sized action potentials and can participate in bursts. Their origin is still unknown. Although they may be initiated in pyramidal cell dendrites, it is apparently possible to produce spikelets by stimulation of the axon, so an axonal site of initiation is conceivable (Schwartzkroin and Prince, 1980a). Wong *et al.* (1979) placed two separate recording electrodes into two electrically coupled elements that they presumed were actually part of the same cell. A burst evoked directly by current passage through one electrode could trigger a burst that was recorded at the second (passive) electrode. When the electrode in the presumed dendrite activated the burst, spikelets were recorded in the soma one-for-one with dendritic action potentials. Others would argue that these results could be due to electrotonic coupling, however (see Section 4.3).

Small attenuated action potentials do not always occur with stimulation of hippocampal cells (e.g., Schwartzkroin, 1975; 1977) and they are most frequently seen in damaged cells. When they do occur, they do not always trigger full-sized action potentials (Spencer and Kandel, 1961) and, conversely, simple EPSPs are adequate to trigger action potentials in many cells, so the requirement for the small action potentials may be questioned. Schwartzkroin and Prince (1978, 1981) report that spikelets are more frequently seen in penicillin models of epilepsy. MacVicar and Dudek (1981) suggest that spikelets, representing electrotonic coupling, could help synchronize the neuronal firing seen in epilepsy. However, Schneiderman and Schwartzkroin (1981) found no evidence of greater electrotonic coupling in penicillin. The nature and function of these potentials remain obscure.

3. WHY DO CELLS THAT CAN FIRE BURST POTENTIALS NOT DO SO ALL THE TIME?

3.1. *Inhibitory Postsynaptic Potentials (IPSPs) Prevent Bursts*

Using intradendritic recording from CA1 cells, Wong and Prince (1979) showed that bursts in these cells are initiated primarily in the dendrites. Direct intradendritic current injection produced a burst potential that greatly exceeded the duration of the current pulse. In normal recording conditions, however, synaptic activation of the cell produced an EPSP–action potential–IPSP sequence, but no burst. In the presence of penicillin, synaptic stimulation also produced bursts. Wong and

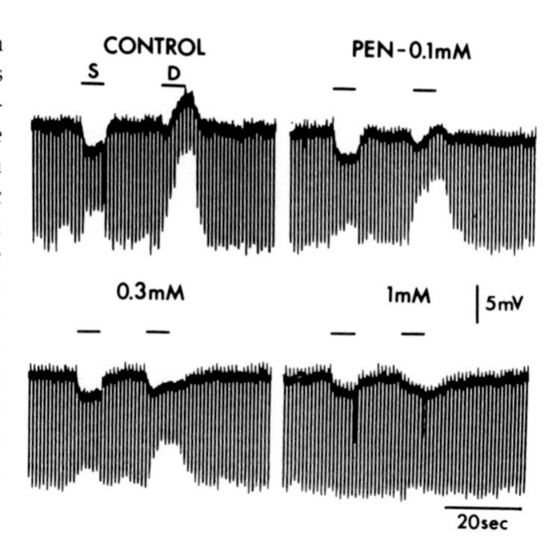

Figure 1. The effects of penicillin on iontophoretic GABA responses. In this cell, two GABA-containing iontophoretic pipettes were positioned near the cell. One pipette was in the soma region (S) and the other in the apical dendritic field (D). All records were taken from the same cell either in control saline or in saline containing the indicated concentrations of sodium penicillin G. Dendritic GABA application (20 nA) produces a large depolarization in control solution, but this is reduced rapidly by doses of penicillin that have relatively less effect on the hyperpolarizations produced by somatic GABA application. Modified from Alger and Nicoll, 1982a.

Prince inferred that the action of penicillin was to block IPSPs. EPSP amplitudes were not changed, while EPSP duration was prolonged, and the decay phase of the EPSP in penicillin closely paralleled that of a passive electrotonic potential, indicating that the IPSP conductance increase, which normally followed the EPSP, had been abolished. Moreover, either brief hyperpolarizing current pulses or IPSPs could abort a directly evoked burst. Since GABA is the major hippocampal inhibitory transmitter, and since penicillin is known to block GABA-mediated IPSPs in other systems, it appeared that penicillin could antagonize the action of GABA in the hippocampus. This hypothesis has since been confirmed (Dingledine and Gjerstad, 1980; Alger and Nicoll, 1982a; see Figure 1). Further support for the idea that the major epileptogenic action of penicillin is on IPSPs (rather than on EPSPs or recurrent circuitry *per se*) comes from observations that other treatments such as perfusion with the GABA antagonists bicuculline (Schwartzkroin and Prince, 1980a; Alger and Nicoll, 1982a), picrotoxin, and pentylenetetrazole (Alger and Nicoll, 1982a) will all affect IPSPs and all produce burst responses to synaptic activation of cells. Low extracellular Cl^- (Yamamoto, 1972; Alger and Nicoll, 1982a) or opiates (Nicoll *et al.*, 1980) also block IPSPs and produce bursting. At the concentrations used, these treatments do not affect EPSPs. Interestingly, in the initial stages of perfusion with Mn^{2+}, intense orthodromic stimulation can produce bursts in the absence of penicillin (Wong, 1982). Presumably at this stage disynaptic inhibitory pathways are more depressed than monosynaptic excitatory pathways.

Bath application of Ba^{2+} results in the appearance of spontaneous asynchronous bursts in hippocampal pyramidal cells (Hotson and Prince, 1981). The effects of Ba^{2+} are presumably due to its dual effects of passing readily through calcium channels and blocking potassium channels. Barium does not, however, promote the occurrence of synchronous bursting in the population, whether evoked or spontaneous. Barium also does not block IPSPs.

3.2. *Feedforward Dendritic Inhibition*

Although the issue was not directly raised by Wong and Prince (1979), it seemed possible that the small IPSPs they recorded intradendritically in fact were generated locally in the pyramidal cell dendrites. One difficulty in accepting this conclusion was that the only synaptic inhibitory system in the hippocampus that had been documented in detail was the recurrent somatic system first described by Kandel *et al.* (1961), and by Andersen *et al.* (1964a,b). This system is activated via recurrent collaterals from pyramidal cell axons and terminates on, or very close to, the pyramidal cell somata (Andersen, 1976; Ribak *et al.*, 1978). Indirect evidence using extracellular recording techniques suggested the existence of either a feedforward system or a dendritic inhibitory system (Andersen *et al.*, 1969; Leung, 1978; Lynch *et al.*, 1981). Alger and Nicoll (1982b) have recently presented evidence based on intracellular recording that argues strongly in favor of a distinct feedforward inhibitory system making synaptic contact with the pyramidal cells in their dendritic regions.

The demonstration of this system involved: (1) showing that orthodromic (i.e., synaptic) activation of the CA1 pyramidal cells was much more effective in producing IPSPs than antidromic stimulation. A given orthodromic population spike was associated with a larger IPSP in a given cell than the same-sized antidromic population spike when field potentials and IPSPs were measured concurrently. The hypothesis of recurrent somatic inhibition would have predicted that the IPSPs evoked in these conditions would have been the same size. Stimulation of fibers in stratum (s.) radiatum could produce pyramidal cell IPSPs in the absence of recordable field potential population spikes. (2) Orthodromic IPSPs differed in shape and duration from antidromic IPSPs, the former being complex multiphasic events while the latter are usually simple monophasic events (Figure 2). If there were only a single IPSP pathway, then IPSPs, no matter how activated, would have the same morphology. (3) Some of the GABA synapses activated by orthodromic stimulation were made in s. radiatum at sites clearly distinct from the

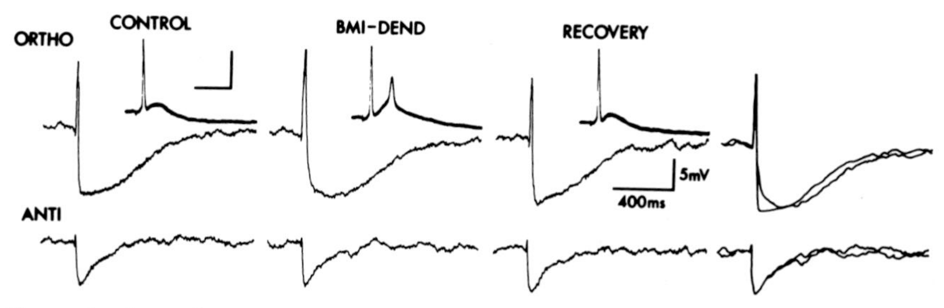

Figure 2. Iontophoretic application of a GABA antagonist (bicuculline methiodide) can block dendritic IPSPs independently of recurrent somatic IPSPs. Both antidromic (recurrent) and orthodromic IPSPs were recorded in the same cell in control saline and then bicuculline methiodide (10 mM) was injected at 100 nA from a pipette placed in the apical dendritic field. After 30 sec, the orthodromic response had changed and the cell had begun firing repetitively (see oscilloscope traces) before the antidromic IPSP was altered. Compare the overlapped control and experimental traces for ortho- and antidromic responses in the right-hand column. Calibrations for oscilloscope traces is 40 mV, 20 msec. From Alger and Nicoll, 1982b, used with permission.

GABA synapses activated by antidromic stimulation. This was demonstrated by applying bicuculline methiodide (either by pressure or iontophoresis) to a small region of the pyramidal cell dendrites (see Figure 2). Bicuculline methiodide reduced the early fast phase of the orthodromic IPSP before it had any effect on the antidromic IPSP. Most importantly, the cell began to fire repetitively (compare oscilloscope traces) when the injection had reduced the orthodromic, but not the antidromic, response. This confirms that dendritic bursting is to a substantial extent controlled by a dendritic IPSP, and that dendritic bursting can occur even when the recurrent somatic IPSP is intact (cf., Silfvenius *et al.*, 1980). Application of bicuculline methiodide to the bath or to the soma of the cells could block the antidromic IPSP. (The late slow hyperpolarizing phase of the orthodromic IPSP was not affected by GABA antagonists at any time and a variety of evidence indicates it is neither Cl^--dependent nor GABA-mediated (see Section 5.2).)

Pharmacological evidence further supports a distinction between ortho- and antidromically evoked IPSPs. Alger and Nicoll (1979) observed that, when the barbiturate pentobarbital was added to the bathing medium in anesthetic concentrations (10^{-4}M), orthodromic stimulation of the cells resulted in the appearance of a late slow depolarization ("depolarizing IPSP") (Figure 3). The hyperpolarizing peak of the early phase of the IPSP was not inverted but a depolarizing hump followed it and peaked in about 2 sec. The depolarizing response was enhanced

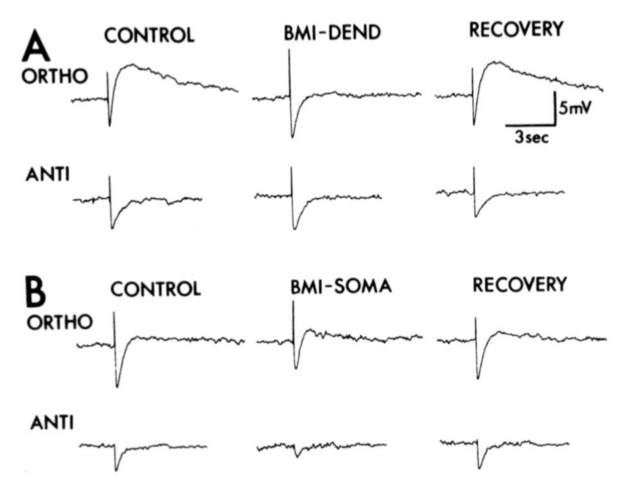

Figure 3. Iontophoretic application of bicuculline methiodide to two cells bathed in 10^{-4}M pentobarbital. Note the depolarizing IPSPs prior to bicuculline injection (control). When bicuculline was applied to the dendrites, the depolarization was affected prior to the antidromic IPSP (cell A). Conversely, when bicuculline was applied to the cell soma, the antidromic IPSP was blocked first (Cell B). From Alger and Nicoll, 1982b, used with permission.

by the low (approximately 30°C) temperatures used in most experiments (Alger and Nicoll, 1982b). However, pentobarbital, which is known to affect GABA receptors (e.g., Nicoll *et al.*, 1975), did not simply act in a generalized way to produce this depolarizing IPSP. This conclusion arose from observations that antidromic activation of the cells produced simple GABA-mediated hyperpolarizing IPSPs that, while prolonged by pentobarbital, were not succeeded by depolarizing IPSPs. Iontophoretic application of GABA to pyramidal cell dendrites results predominantly in depolarizing responses, whereas GABA application to the soma region results in hyperpolarizations (Langmoen *et al.*, 1978; Alger and Nicoll, 1979, 1982a; Andersen *et al.*, 1980). Thus, it appeared that the depolarizing IPSP in pentobarbital is due to the action of a GABA system that contacts the pyramidal cell dendrites synaptically. This hypothesis was tested by injecting bicuculline methiodide into s. radiatum. When a small quantity was injected, the depolarization was blocked before the antidromic IPSP was reduced (see Figure 3). The converse was true when bicuculline methiodide was injected into s. pyramidale. The depolarizing IPSP could also be selectively blocked by injections of TTX into s. radiatum (Alger and Nicoll, 1979; 1982b). Since TTX does not block

GABA receptors these results suggest that some parts of the neurons that release GABA are present in s. radiatum.

Further evidence has indicated that the two types of GABA responses, hyperpolarizing and depolarizing, are probably mediated by two different types of GABA receptors that have distinct pharmacological properties (Alger and Nicoll, 1982a). It is not clear if the depolarizing receptors are activated under ordinary physiological conditions. As discussed below, however, this is a possibility (see Section 6.3).

In summary, the evidence indicates that hippocampal CA1 cells do not ordinarily burst because the burst generating mechanism is inhibited by a GABA-mediated dendritic IPSP. This IPSP is activated by a feedforward inhibitory interneuronal system that is distinct from the recurrent somatic system. GABA-mediated IPSPs also prevent CA2/CA3 cells from bursting but in these cells somatic IPSPs are probably as important as dendritic ones, since the burst threshold is low in both soma and dendrites (Wong *et al.*, 1979). As stated previously (see Section 2.1.2), the evidence to date supports the conclusion that blockade of GABA-mediated IPSPs is an absolute requirement for the occurrence of synchronized bursting, whether evoked or spontaneous. Presumably however, this is true only because GABA-mediated IPSPs are ordinarily coactivated during excitation of the pyramidal cells. If excitatory synapses could be selectively activated then bursting would be expected, even if the GABA-mediated IPSPs were not specifically blocked.

4. HOW DOES SYNCHRONIZATION OF FIRING WITHIN A POPULATION OF CELLS OCCUR?

4.1. *Imposed Synchrony*

It is possible to determine if a population of cells is firing synchronously simply by recording the extracellular field potentials. In the hippocampus, the field potentials have been studied in detail and the population action potential has been shown to be due to the summated currents from a number of synchronously activated cells (Andersen *et al.*, 1971). Under normal recording conditions, one never records spontaneous field potentials (even though individual cells may fire spontaneously) from any region in the hippocampus. When IPSPs are blocked, spontaneous multipeaked field potentials can be recorded in CA1, CA2, and CA3 (e.g., Schwartzkroin, 1975). Simultaneous intracellular and field potential recordings suggest that these potentials represent the nearly synchronous firing of burst potentials by a group of neurons. Schwartzkroin and Prince (1978) first argued that the spontaneous activity in CA1 was actually driven by activity in CA2/CA3, since a cut through s. radiatum, which would sever fibers projecting from CA2/CA3 to CA1, abolished the spontaneous activity in CA1, while spontaneous activity continued in CA2/CA3 (see also Gjerstad *et al.*, 1981).

However, even synchronous bursting in CA2/CA3 cannot induce bursting in CA1 unless IPSPs in CA1 have first been blocked. To demonstrate this, Mesher and Schwartzkroin (1980) applied penicillin dissolved in an agar droplet focally to CA2/CA3. This method slows the diffusion of penicillin to distant regions of the slice. In this case, CA2/CA3 fired spontaneous bursts but CA1 did not. Individual CA1 cells recorded intracellularly did not respond with burst potentials, but rather with spontaneous "EPSP–IPSP" sequences (whether the hyperpolarization was entirely an IPSP is open to question, see Section 5.2). Mesher and Schwartzkroin did not report the occurrence of spontaneous EPSP field potentials in CA1 in these experiments. One would expect these to have occurred.

Interestingly, Silfvenius *et al.* (1980) had previously done the experiment that is complimentary to that of Mesher and Schwartzkroin. Silfvenius *et al.* placed penicillin, also fixed in an agar droplet, onto the CA1 dendritic layer, and then stimulated afferents in s. radiatum. The CA2/CA3 region was not affected and was not bursting spontaneously. Nevertheless, in this case, afferent stimulation did produce bursting of the CA1 cells. Silfvenius *et al.*, did not discuss their results in terms of IPSPs, but simply noted that penicillin "affected the dendritic properties of the cells." Their observations can now be explained by the hypothesis of GABA-mediated dendritic inhibition discussed earlier. The important point at the moment is that, CA1 cells do not burst synchronously, unless IPSPs in CA1 are blocked.

Wong and Traub (1983) focused on several aspects of the synchronous bursting pattern in CA1. In particular, they asked: (1) What accounts for the very long and variable latency of the burst discharge frequently recorded in CA1 when weak stimuli are delivered to CA2/CA3 cells? Stereotyped bursts with long and variable latencies have been noted as a puzzling aspect of depolarization shifts in penicillin-treated neocortex (e.g., Ayala *et al.*, 1973); (2) What is the role of the CA2/CA3 region in initiating spontaneous and delayed evoked synchronized bursting?; and (3) Are the same mechanisms involved in generating spontaneous and delayed synchronous bursts? Often, during pharmacological blockade of GABA-mediated IPSPs, a single stimulus to afferent fibers leading to CA1 will produce a double burst (see Figure 4) in a CA1 cell (Gjerstad,

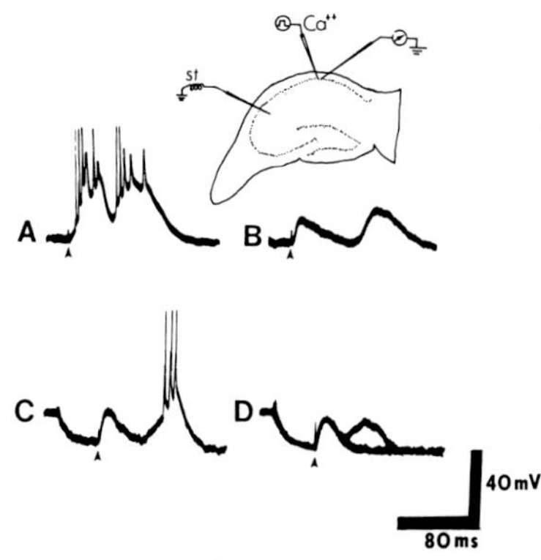

Figure 4. Double responses in a CA1 cell caused by a single stimulus applied to CA2/CA3 region. (A) At resting membrane potential, two bursts are produced. (B) Also at resting potential, action potentials have been blocked following the pressure injection of a bolus of $CaCl_2$ (10 mM) near the cell soma (see diagram). Presumably the action potential threshold was increased due to the screening action of Ca^{2+}. In (C) and (D), bursts were again blocked by hyperpolarizing the cell directly. EPSPs underlying both bursts are revealed when action potentials are prevented. (B and D). Note the second ("long-latency") EPSP has a slower time to peak and longer duration than the first EPSP. In one of the superimposed traces in (D), the long-latency EPSP failed to occur. From Wong and Traub, 1983a, used with permission.

et al., 1981; Wong and Traub, 1983). The initial burst follows directly from stimulation of the afferent fibers in CA1. Wong and Traub (1983) found the second, delayed, burst was an all-or-none event. It was not a purely intrinsic phenomenon since, when bursts were blocked with either hyperpolarizing current pulses or raised levels of extracellular Ca^{2+}, EPSPs underlying both first and second bursts were uncovered (although see Schwartzkroin and Prince, 1980a). The second EPSP could be activated via a variety of pathways but was only seen when the CA2/CA3 projection to CA1 was intact. For example, stimulation of the granule cell mossy fiber system to CA2/CA3 led to a burst in CA2/CA3 and an EPSP in CA1 which, unless prevented from doing so, triggered a burst there also. The mossy fibers do not project to CA1 and thus could not be directly responsible for CA1 EPSPs. Most importantly, when the synchronous all-or-none burst was recorded in CA2/CA3 cells the delayed EPSP recorded in CA1 cells was also all-or-none. That is, at a sufficient stimulus intensity a stereotyped burst occurred in CA2/CA3 and a stereotyped EPSP in CA1. With higher stimulus intensities there were no further changes in the morphologies of these potentials, but the latencies were shortened. Simultaneous recordings showed that each CA1 EPSP was preceded by a burst in CA2/CA3. Severing the connection between CA2 and CA1 abolished the EPSP in CA1 (cf, Gjerstad *et al.*, 1981). These results suggested that the delayed EPSP (and burst) in CA1 were due to prior activity in CA2/CA3 which was then projected to CA1 in the form of an EPSP. Since IPSPs had been blocked the EPSP resulted in a CA1 burst. The EPSP appeared to be all-or-none because, as discussed below, the projection cells in CA2 became activated nearly simultaneously. The double bursts in CA1 which result from stimulation of afferents in CA1 can be explained as follows: (1) transmitter liberation from excitatory afferent terminals onto CA1 cells results in the first EPSP (and burst) and (2) CA2/CA3 cells are also activated (perhaps via axon collaterals or excitatory interneurons) and spread of excitation throughout this region leads, after a delay, to reexcitation of fibers to CA1 and thus to the delayed EPSP and burst. Thus it appears that synchronous activity of CA1 cells is imposed on these cells by projection cells. The latencies of CA1 bursts are inversely dependent on the intensity of activation of CA2. Presumably the latencies of bursts in CA1 are determined by the rate at which a substantial proportion of CA2 cells can become active. Although not explicitly tested by Johnston and Brown (1981) it is likely that the giant EPSPs they recorded in CA1 were due to spontaneous synchronous activity in CA2/CA3.

4.2. *"Spontaneous" Synchrony*

Wong and Traub (1983) showed that the pacemaker cells for CA1 are actually in the CA2 region. They first noted that the synchronized field potentials, whether spontaneous or evoked, were largest, and had the shortest latencies in the small CA2 region, as compared to CA1 or CA3 subregions. Furthermore, they surgically isolated CA2 with microlesions at the CA1/CA2 and CA2/CA3 borders. This area, distinguished initially by Lorente de Nó (1934) on the basis of cellular morphology and connectivity, cannot be unambiguously identified simply by looking at a hippocampal slice under a dissecting microscope. Wong and Traub provisionally identified CA2 with gross measurements of the s. pyramidale region and then used histological and physiological techniques to confirm the identity of the region. They then found that, even in the isolated islands of CA2, spontaneous bursting could arise during perfusion with GABA antagonists, whereas spontaneous bursting did not occur in the nearby CA1 or CA3 areas. When a small bolus of K^+ was injected into the CA2 a complete synchronized burst involving large numbers of CA2 cells, occurred with a long and variable latency. The burst was very pronounced and yet the bolus of K^+ was so small that Wong and Traub (1983) could calculate that all of the cells in CA2 could not have been directly depolarized by it. This confirms that stimulation of a few cells can lead to activation of many cells. Thus a great deal of evidence indicates that, under conditions in which IPSPs are blocked, spontaneous firing does occur in the hippocampus. This synchrony appears to originate in the CA2 region and, since it can be recorded in isolated islands of CA2, it is clear that synchrony *can* arise spontaneously in a population of cells. It is still unclear how or why this occurs.

The observations of MacVicar and Dudek (1980a) suggest a means by which excitation can spread quickly throughout the CA2/CA3 population. Using paired intracellular recordings, they found that in a number of cases (5/88) excitatory cell-to-cell synaptic coupling was present in CA2/CA3. Action potentials produced by direct depolarization of one cell led to EPSPs produced in a simultaneously recorded cell nearby. Wong and Traub (1983) interpreted their own results as follows: activation of a few cells in CA2/CA3 spreads quickly via excitatory chemical synaptic transmission throughout this cell group. If a large number of CA2/CA3 cells are active simultaneously, then the target cells in CA1 will experience a "long-latency EPSP." Traub and Wong (1982, 1983) made a mathematical model of the CA2 region. They endowed 100 model cells with known properties, i.e., propensity to burst spontaneously, to fire both Ca^{2+} and Na^{+}-dependent action potentials, to respond with Ca^{2+}-dependent K^{+} potentials, etc. They also allowed a small degree of excitatory synaptic interconnectivity (each cell was assumed to be connected randomly with five other cells in the array). In the absence of synaptic inhibitory interconnections (analogous to epilepsy models in which IPSPs are pharmacologically blocked) simply causing the activation of four cells was sufficient to cause nearly all of the cells in the population to burst within about 80 msec. At the peak of the activity, 80% of the cells were bursting at once, thus providing a substantial "pulse" of activity to be projected to any target cells, i.e., CA1. This model produces realistic field potentials and cellular events. A remaining question is how the pacemaker activity in CA2/CA3 is initiated physiologically. In the computer model, it was necessary to provoke the pacemaker activity by activating a few cells simultaneously. Individual cells in this region do fire spontaneously and it is possible that many would be active close enough in time to trigger the rest of the population. It is not known whether the slice conditions themselves affect the tendency to fire spontaneous population bursts. For example, the level of potassium in the bathing medium can alter the readiness with which cells will burst (Ogata, 1975, 1976; Ogata *et al.*, 1976; Schwartzkroin and Prince, 1978; Oliver *et al.*, 1980a,b; Hablitz and Lundervold, 1981). It is conceivable that, *in vivo*, some other conditions might have to be present to enable the pacemaker region to begin firing.

The modeling is significant in that it demonstrates that only a few assumptions are necessary in order to produce a realistic representation of the physiologically recorded events. All of the phenomena that went into the model have been well supported with physiological measurements. This suggests that many of the major factors underlying synchronization within a small neuronal population may be in hand.

4.3. *Electrotonic Interactions*

Two types of electrotonic interaction among hippocampal cells have attracted interest recently. The first, called here *electrotonic coupling*, refers to cases in which activity in one cell is presumed to be transmitted directly to a limited number of other cells via specialized, anatomically identifiable junctions. The other type of electrotonic interaction is called *ephaptic transmission* and refers to the influence on a cell caused by extracellular current flow through the extracellular resistance, i.e., a field potential effect.

It has been suggested that electrotonic coupling between cells in the hippocampus contributes to the synchronization of firing between these cells as it does, for example, in the inferior olive (Llinás *et al.*, 1974b). Two kinds of evidence support this proposition: (1) experiments showing dye coupling between cells and (2) simultaneous recordings from pairs of pyramidal cells.

4.3.1. *Dye Coupling.* The intracellular dye, Lucifer Yellow, can traverse electrotonic junctions. It is not taken up in appreciable amounts from the extracellular space and does not cross chemical synaptic junctions. Therefore, if injection of the dye into one cell results in staining of two or more cells, this is assumed to be evidence that the cells are electrotonically coupled. Such evidence has now been found in the case of cells in CA2/CA3 region of the hippocampal slice and in the neocortex (Gutnick and Prince, 1981). Anatomical gap junctions have also recently been found between "CA3" pyramidal cells, although they were quite rare (Schmalbruch and Jahnsen, 1981). However, Lucifer Yellow diffuses very rapidly in the intracellular environment and there is a question as to whether the transient impalements inevitably experienced in searching for a stable cell may be enough to cause staining in more than one cell in a slice (Andrew *et al.*, 1982). A related and perhaps more serious problem is that of "dual impalement," noted in the Appendix. It occasionally happens that a single electrode appears to be recording from a neuron and a glial cell simultaneously (Alger *et al.*, 1983). Presumably a mechanical coupling is established by the electrode itself partially impaling both cells. These cases can be easily distinguished because of the very different properties of neurons and glia. Given the tight packing of pyramidal cells in the hippocampus, it seems possible that two neurons could be recorded in the same way and yet this would be very difficult to detect electrically. Some evidence that this may happen can be found when the intracellular stain, HRP, is used. HRP does not cross gap junctions and yet it is possible to find more than one cell stained when apparently only one was injected (B. E. Alger, unpublished observations). It has been reported in the CA1 region that multiple staining of cells with Lucifer Yellow can easily be found while evidence for electrotonic coupling is much harder to obtain (Knowles *et al.*, 1982; Taylor and Dudek, 1981).

4.3.2. Electrotonic Coupling. More convincing evidence for electrotonic coupling has been put forward by MacVicar and Dudek (1981) using simultaneous recordings from pairs of pyramidal cells. In these experiments, it was shown that action potentials at one electrode were recorded as small attenuated action potentials at the other and that current injected through one microelectrode could produce voltage deflections recorded by the other electrode. In some cases, the coupling ratio (the ratio of the magnitudes of the voltage deflections recorded by the passive and active electrodes) was high (approximately one) and this was assumed to be evidence that the two electrodes were in the same cell (although presumably they could have been in two very tightly coupled cells). In other cases, the coupling ratio was low (approximately 0.3) and it was assumed that, in these cases, the two electrodes were in two different cells. In several pairs with low-coupling ratios, HRP was present in both electrodes and, after the dye had been injected, two cells were found to have been stained. This is good evidence that CA2/CA3 cells may be electrotonically coupled. Nevertheless, some questions regarding electrotonic coupling that have not been fully answered include: (1) electrotonic "tightness" of coupling. Some data appears to indicate that cells are only weakly coupled, e.g., spikelets are small (Schwartzkroin, 1975; MacVicar and Dudek, 1981). Other data suggest a large degree of tight coupling (Taylor and Dudek, 1982). It is also puzzling why there is so limited a range of spikelet amplitudes if indeed there is variability in the degree of coupling. Small action potentials usually appear to be approximately 10 mV and apparently are not greater than 15 mV. (2) What accounts for the variety in morphology of d-spikes? Some decay quickly and others slowly. (3) How can studies on the electrotonic structure of hippocampal cells, which have pointed to a relatively simple electrotonic structure (Andersen *et al.*, 1980; Turner and Schwartzkroin, 1980; Brown *et al.*, 1981; Johnston, 1981), be reconciled with the evidence for electrotonic coupling, since coupling would produce a more complex electrotonic structure? (4) Can the discrepancy between the large numbers of dye-coupled cells in CA1 (Taylor and Dudek, 1981; Knowles *et al.*, 1982) and the apparent difficulty in demonstrating specific electrotonic coupling be reconciled? It is not likely that the controversy will be resolved until it can be shown unambiguously that two different cells have been recorded and that artifactual coupling can be eliminated as a problem. One possibility for doing this is to use two different impenetrable dyes in the two recording electrodes. Following a physiological demonstration of electrotonic coupling, there should be traces of only one dye in each cell.

4.3.3. *Function of Electrotonic Coupling.* MacVicar and Dudek (1981) propose that electrotonic coupling represents a means by which synchronization of firing can occur. In the model developed by Traub and Wong (1982, 1983) electrotonic coupling of the magnitude demonstrated by MacVicar and Dudek was neither necessary nor sufficient to produce synchronous firing of the CA2 cells. When assumed present, electrotonic coupling did enhance the synchrony of firing within the population to a small extent. As noted earlier, GABA antagonists enhance spontaneous synchronous firing. There is no clear evidence that these drugs affect electrotonic coupling in the hippocampus.

4.3.4. *Ephaptic Interactions.* Two groups have independently reported an unusual phenomenon seen when hippocampal slices are bathed for prolonged periods in a low calcium-containing solution (Jefferys and Haas, 1982; Taylor and Dudek, 1982). In these instances, despite blockade of all apparent chemical synaptic transmission, large synchronized field potentials can occur in the CA1 region. These rhythmic bursts of activity can last for many seconds, and may be either spontaneous or evoked by electrical stimulation. Interestingly, while they are present in an 0.2 mM Ca^{2+}, 4 mM Mg^{2+} containing saline, the spontaneous bursts are blocked when the Mg^{2+} concentration is raised to 6 mM. When simultaneous intracellular recordings are made, action potentials are found to occur in close synchrony with the extracellularly recorded population action potentials (Taylor and Dudek, 1982). Taylor and Dudek also made differential recordings of extracellular and intracellular potentials simultaneously and were able to show that extracellular field potentials do exert effects on the membrane potentials. These effects may not be apparent in the usual recording arrangement in which the intracellular potential is recorded with respect to the bath potential. Thus, it appears that this type of electrotonic interaction may be important in synchronizing pyramidal cell activity, along with electrotonic coupling and changes in ionic concentrations, in the absence of chemical synaptic transmission. Interesting questions concerning these observations include: (1) Why do the effects take so long to develop, since it can be shown that chemical synaptic transmission is blocked well before they occur and (2) How does spontaneous synchronized population activity arise? The relevance of these observations to epilepsy remains to be demonstrated.

5. WHAT TRIGGERS THE SWITCH FROM INTERICTAL SPIKING TO SEIZURES?

This question cannot be satisfactorily answered at present. However, recent evidence does allow some tentative conclusions to be drawn.

5.1. *Burst Afterhyperpolarizations (AHP)*

It has long been known that a burst is followed by a long-lasting hyperpolarization, or afterhyperpolarization (AHP). This AHP has been designated the AHP_s (for *synaptic* activation) to distinguish it from the AHP_d following *direct* activation of the cell (see below). Because this potential is hyperpolarizing and inhibitory, it was thought to be an IPSP. Although recent evidence suggests that this is not entirely true, the AHP_s is still an obvious candidate for the termination of the burst response. As noted in the Introduction, it is reasonable to assume that whatever prevents burst termination leads to a seizure (e.g., Spencer and Kandel, 1965; Ayala *et al.*, 1973).

It is first necessary to identify the ionic mechanism of the AHP_s. Remaining doubt that the AHP_s is a GABA-mediated IPSP was removed by showing the large AHP_s could be evoked in CA1 pyramidal cells following orthodromic activation of the cells in the presence of such high doses of GABA antagonists (e.g., bicuculline methiodide at 5×10^{-4}M) that responses to iontophoretically applied GABA were abolished as well as all GABA IPSPs (Alger and Nicoll, 1980a). As originally suggested by Yamamoto (1972), Alger and Nicoll found that the AHP_s was not dependent on the electrochemical gradient for chloride. Reversing the Cl^- gradient using either Cl^- injection into the cells or lowering the bath Cl^- concentration had no effect on the AHP_s, but in fact could initiate burst potentials and AHP_s in the absence of GABA antagonists. When potassium concentrations in the bath were altered, however, the AHP_s was markedly affected. Its reversal potential followed the Nernst potential for a potassium electrode closely over the range of 1 to 15 mM extracellular K^+. The AHP_s is also depressed by the potassium channel blockers, Ba^{2+} (Alger and Nicoll, 1980) and Cs^+ (B. E. Alger, unpublished observations). Thus, the AHP_s appears to be a K^+-dependent potential.

In CA1 cells, which are depolarized to fire a brief train of action potentials by direct current injection into the cells, a large AHP_d follows the train (Hotson and Prince, 1980). The AHP_d shows K^+ dependence and $C1^-$ independence similar to the synaptically evoked AHP_s in the presence of GABA antagonists. In addition, the AHP_d appears to be activated by Ca^{2+} since: (1) a cell that is depolarized in the presence of the Ca^{2+} antagonists, Mg^{2+}, Mn^{2+}, Co^{2+}, or Cd^{2+}, responds with a train of fast action potentials, but no slow AHP_d; (2) in cells injected with the calcium chelator, EGTA, the AHP_d is abolished (Alger and Nicoll, 1980a; Schwartzkroin and Stafstrom, 1980); and (3) increasing the bath concentration of Ca^{2+} increases the AHP_d amplitude (Gustafsson and Wigstrom, 1981). Thus, the AHP_d is a Ca^{2+}-activated K^+ potential. Although the AHP_s is also clearly K^+ dependent, its Ca^{2+} dependence has been harder to establish. For one thing, it is not possible to test for Ca^{2+} dependence by perfusing the slice with Ca^{2+} antagonists. This would abolish chemical synaptic transmission and field potentials and so block of the AHP_s would be trivial. When EGTA is injected into the cells, the AHP_s in CA1 is substantially affected, typically being shortened greatly in duration, although, in some cells, the amplitude of the AHP_s is depressed as well (Alger and Nicoll, 1980a). The early EGTA-resistant portion of the AHP_s is actually a synaptic potential, the late hyperpolarizing potential (LHP) discussed below. Similar results have been found by Hablitz (1981). However, Schwartzkroin and Stafstrom (1980) found a somewhat different effect when they injected EGTA into CA2/CA3 cells. To understand the problem, one must distinguish between different modes of activation of CA2/CA3 cells. These cells can burst in any of three different modes: (1) direct activation by a depolarizing current injection into the cell; (2) "asynchronous" spontaneous firing, i.e., the individual cell that is recorded fires spontaneously but is not accompanied by simultaneous firing of other cells in the CA2/CA3 population and simultaneous spontaneous field potentials are not recorded, and (3) "synchronous" spontaneous firing in which a large number of neighboring cells fire together with the recorded cell. The AHP that followed repetitive firing in modes (1) and (2) were blocked by EGTA. These bursts (so-called "EGTA-bursts") could be prolonged, lasting up to 1 sec, and yet were not followed by an AHP. However, when spontaneous firing of the EGTA-injected cell occurred in conjunction with a synchronous burst in the CA3 population [as in (3)] (so-called "penicillin-induced" bursts), an AHP_s followed the burst. Similar observations have been made in picrotoxin (B. E. Alger, unpublished observations). There is thus some difference between afterhyperpolarizations produced when a population of cells bursts at once and when a single cell fires alone. These observations led Schwartzkroin and Stafstrom (1980) to suggest that the penicillin-induced burst AHP_s was not a Ca^{2+}-activated K^+ potential. From the point of view of seizure development, the important issue is burst termination. Alger and Nicoll (1980a) reported that even in the CA1 cells, in which intracellular EGTA injection appeared to reduce the AHP_s, the duration of the burst was not significantly prolonged. However, when TEA (0.5 M) was co-injected into the cells with EGTA, then bursts were significantly prolonged in many cells. The same results have been obtained with co-injection of Cs^+ and EGTA. Intracellular TEA or Cs^+ by themselves prolong action potentials and cause repetitive firing but do not cause the extremely protracted bursts seen when EGTA is also present in the recording electrode.

To unify the superficially different results of Alger and Nicoll (1980a) (and Hablitz, 1981) and those of Schwartzkroin and Strafstrom (1980), I suggest that EGTA injection leads to a prolongation of the directly activated and single cell bursts in CA2/CA3 because these potentials are generated near the EGTA-containing electrode in the soma. These bursts are terminated by the Ca^+-dependent K^+ potentials that are turned on by Ca^{2+} action potentials in the soma. When the Ca^{2+}-dependent K^+ potentials are blocked, the bursts are prolonged. There are two factors that distinguish "synchronous" population-associated bursts from single-cell bursts. In the former case, there is (1) concurrent activation of a synaptically generated K^+ potential (the LHP, to be discussed in Section 5.2) and (2) an AHP_s generated in the dendritic region where Ca^{2+} action potentials are largest (e.g., Wong *et al.*, 1979) and that is distant from the site of EGTA injection in the soma. The ability of EGTA to diffuse to distant sites is unknown, but it is reasonable to suppose that the EGTA concentration in the dendrites is lower than that in the soma. Since the free EGTA concentration will determine the Ca^{2+} concentration after a sudden Ca^{2+} influx, this would result in a less effective antagonism of the dendritic Ca^{2+}-activated K^+ potentials than of the locally generated AHP.

5.2. *The Late Hyperpolarizing Potential (LHP) Is a Slow IPSP*

Orthodromic IPSPs are longer than antidromic IPSPs and have a different morphology (see Figure 5B) with a second late negative peak about 100 to 200 msec after the peak of the early IPSP. The late hyperpolarizing phase of the orthodromic response is associated with a conductance increase, but is immune to blockade by GABA antagonists and cannot be reversed by changes in the electrochemical gradient for $C1^-$, unlike the usual GABA-mediated IPSP (Alger and Nicoll, 1980a; Thalmann and Ayala, 1980; Nicoll and Alger, 1981; cf. Fugita, 1979). The LHP has a very negative null potential, in the region of −85 mV, which is identical to the potassium equilibrium potential (E_K) in these cells. With extracellular K^+ concentrations higher than 5.4 mM, the LHP can be reversed. These properties suggest the LHP is not a GABA-mediated, $C1^-$-dependent IPSP. Hori and Katsuda (1978) reported the LHP could be abolished by ouabain. A reexamination of the effects of ouabain (5 $\times 10^{-7}$ M to 10^{-6}) indicated that while ouabain does block the LHP, it also blocks $C1^-$-dependent GABA-mediated IPSPs (Alger, 1983). The effects of ouabain appeared to be largely "nonspecific" and may be partially due to changes in extracellular K^+, since ouabain also produces shifts in the IPSP reversal potential (E_{IPSP}). Increases in K^+ will affect intracellular $C1^-$ (Martin, Appendix to Matthews, and Wickelgren, 1979), Alger and Nicoll, 1983). The LHP appears to be due to a K^+ conductance increase.

Since the LHP could not be produced by directly depolarizing the cells, the possibility existed that the LHP might be triggered by synaptically mediated Ca^{2+} influx into the cells. Ascher *et al.* (1978) reported a similar effect following ACh application to *Aplysia* cells. The relevant afferent systems in the hippocampus are more likely to use an excitatory amino acid, probably glutamate or aspartate, as their neurotransmitter (Storm-Mathisen, 1977). In the hippocampus as in other systems (e.g., Constanti *et al.*, 1980), iontophoretically applied glutamate produced depolarizations that were succeeded by hyperpolarizations having reversal potentials above −80 mV (Nicoll and Alger, 1981). Glutamate hyperpolarizations, but not depolarizations, could be blocked with a variety of Ca^{2+} antagonists. This suggested that glutamate permitted Ca^{2+} entry into the cells and that Ca^{2+} activated a K^+ conductance. The glutamate-induced hyperpolarizations were independent of voltage-gated Ca^{2+} conductances. Two kinds of evidence supported this: first, hyperpolarizations could follow small membrane depolarizations that were below the threshold for activation of Ca^{2+} action potentials. Second, the hyperpolarizations continued to be produced even when the glutamate-induced depolarizations were entirely prevented by voltage clamping the cell. Thus, under these conditions, glutamate probably directly opened Ca^{2+} channels by a voltage-independent mechanism. The hypothesis suggested by these experiments was that synaptic excitation of the pyramidal cells by an excitatory amino acid caused a hyperpolarization and conductance increase via a Ca^{2+}-activated K^+ potential. However, recent experiments have not supported this proposal.

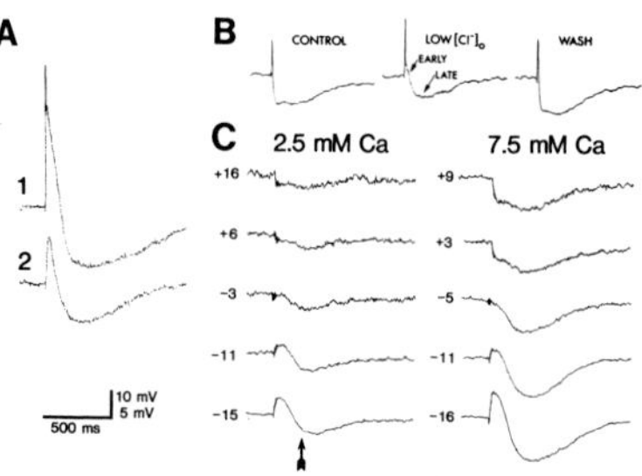

Figure 5. Ionic mechanisms of the LHP. (A) A CA1 cell recorded in the presence of picrotoxin (6×10^{-5}M). In A1, the stimulus was suprathreshold and the burst was followed by a long AHP. A2 is a subthreshold response recorded from the same cell. Note the EPSP–LHP sequence. (B) The LHP in normal medium (late phase) is not blocked when extracellular Cl^- is lowered by 90%, while the early hyperpolarizing IPSP is reversed by this treatment. (C) The LHP recorded in picrotoxin at depolarized membrane potentials is enhanced by increases in extracellular Ca^{2+}. The LHP was first recorded at several membrane potentials in normal Ca^{2+} (2.0 mM) and then measured again in the same cell after the bath Ca^{2+} concentration had been raised to 6.0 mM. Note increases in LHP at all potentials. Voltage calibrations are 5 mV for parts A and B; 10 mV for C. (Alger, 1983.)

In these experiments GABA-mediated IPSPs were routinely blocked by bath application of a GABA antagonist. The idea that the LHP is synaptically generated has been supported by experiments showing that the LHP continues to be evoked in cells in which regenerative voltage responses have been blocked (Alger, 1983; Figure 5). Electrodes filled with 2 M CsC1 permit cells to be depolarized to high levels (0 to +20 mV) at which all signs of regenerative activity are blocked. The EPSP itself can be reversed at these levels (Johnston and Brown, 1981; Hablitz and Langmoen, 1982). Nevertheless, the LHP continues to be produced upon synaptic stimulation of the cells (Figure 5C). When measurements were made in conditions in which EPSPs did not occur the LHP was found to have an onset latency of about 100 msec, a time to peak after onset of 200 msec, and an overall duration of 500–1000 msec. It is therefore a very slow potential when compared to EPSPs and GABA-mediated IPSPs. A late outward current, which may be responsible for the LHP, has been detected in orthodromically activated hippocampal cells voltage-clamped at depolarized levels (Johnston and Brown, 1981). If low stimulus strengths are used to stimulate cells in the presence of picrotoxin, then it is possible to produce subthreshold responses in CA1 cells, i.e., the cell that was recorded may respond with an EPSP but no burst. An LHP still follows the EPSP (Figure 5A). Indeed, when the CA2/CA3 pacemaker region is bursting spontaneously in the presence of GABA antagonists, regularly occurring EPSP–LHP sequences can occasionally be recorded in CA1 pyramidal cells (Figure 6A). It appears that these spontaneous responses are set up in the CA1 cells following activity in the fibers projecting from CA2/CA3 to CA1 and are probably due to the activation of interneurons by these fibers. The LHP is sensitive to ex-

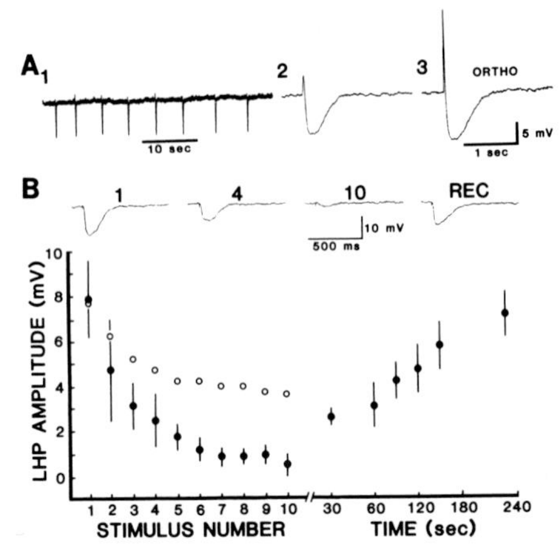

Figure 6. Properties of the LHP. (A) Spontaneous EPSP–LHP sequences can occasionally occur in cells bathed in GABA antagonists. The cell shown in A was bathed in 2 mM penicillin. The responses shown in A1 and A2 are spontaneous and are illustrated at different sweep speeds. The burst in A3 was evoked with orthodromic stimulation. (B) The LHP is depressed rapidly during a train of stimuli given at low frequencies. In (B), representative responses are shown for a cell depolarized to approximately E_{EPSP} and stimulated at 0.2 Hz. The responses were to the first, fourth, and tenth stimuli and to a pulse given 4 min after the end of the stimulus train. In the graph, the filled circles represent means and standard deviations of data from four experiments in which stimuli were given at 0.2 Hz. Data from one experiment in which the stimuli were given at 0.1 Hz are shown in the open circles. (Alger, 1983.)

tracellular Ca^{2+}, increasing as the Ca^{2+} concentration is raised over the range 2.0 to 6.0 mM (Figure 5C). These results are consistent with the hypothesis that the LHP is linked to the synaptic activation of the CA1 cells rather than to action potentials in the cells. LHP amplitudes are rapidly depressed when repetitive stimulation is given (Figure 6B). However, the LHP appears to be independent of the preceding EPSP since LHP and EPSP amplitudes are not found to be correlated (Alger, 1983). Taken together these data suggest the LHP is a slow inhibitory postsynaptic potential which results from the action of some as yet unidentified neurotransmitter released from an interneuron onto the pyramidal cells.

Although the actual mechanism of initiation of the LHP has not been established it is doubtful that it is activated by intracellular calcium. Experiments discussed above argue that the LHP is not due to voltage dependent Ca^{2+} influx into the cells. Furthermore, while in a few cells recorded with EGTA in the electrode the LHP was found to be quite small (<2 mV), in the majority of cells it has not been possible to demonstrate an unambiguous effect of EGTA on this potential (Thalmann, 1982; Schwartzkroin and Knowles, 1982; Alger, 1983). Given the limitations of the technique of intracellular EGTA injection (irreversibility of apparent effects; inability to determine EGTA concentration throughout a spatially complex cell; difficulty in obtaining detailed control data prior to EGTA leak from an electrode, etc), the relative paucity of positive effects do not constitute strong evidence in support of the hypothesis that the LHP is activated by intracellular calcium.

Moreover, it appears that bath application of cyclic AMP (cAMP) has a differential effect on the AHP_d and on the LHP (Newberry and Nicoll, 1982). cAMP reversibly blocks the former while having little effect on the latter. With a bath-applied drug, problems of control recordings and of accessibility to different parts of the cell are eliminated. Newberry and Nicoll were led to conclude that the LHP cannot be a Ca^{2+}-activated K^+ potential. However, until the actions of cAMP in blocking the AHP_d are explained, their conclusion cannot be fully accepted.

Nevertheless, differences between somatic and dendritic K^+-potentials could explain the different effects of intracellular EGTA injection on the burst duration of CA2/CA3 vs CA1 cells. A CA2/CA3 cell bursting alone (whether spontaneously or as a result of depolarizing current injection) is like a directly activated CA1 cell. The cell is dominated by voltage-dependent Ca^{2+} potentials and Ca^{2+}-dependent K^+ AHPs. These AHPs are ordinarily adequate to terminate the burst and, when they are prevented from occurring by EGTA, the burst is prolonged. When cells in the CA2/CA3 population are bursting synchronously, however, they are each receiving synaptic inputs (presumably from each other and from putative LHP-producing interneurons, see Section 4.2). These excitatory and inhibitory synaptic inputs, in addition to causing bursts to occur, initiate AHPs and LHPs in the dendrites of the target cells. The dendritic potentials are less susceptible to the effects of EGTA, and thus, EGTA does not maximally prolong such bursts.

The prolongation of bursts when TEA or Cs^+ is also present in the recording electrode indicates other K^+ conductances are involved in burst termination. However, since bursts are not maximally prolonged unless EGTA is present in the electrode, a Ca^{2+}-dependent K^+ potential probably always contributes to burst termination (see Schwartzkroin and Prince, 1980b).

In conclusion, it can be said that cells whose GABA-mediated IPSPs have been blocked do not seize because potassium potentials prevent it. The predominant potentials are activated in part by Ca^{2+} influx into the cell, but the LHP constitutes the initial portion of the AHP, and is probably independent of calcium influx. Nevertheless, the question for the genesis of seizures is, "What leads to the disappearance of K^+ potentials?"

5.3. *Seizure Development*

Suggestions that simple increases in extracellular potassium concentrations or decreases in extracellular Ca^{2+} concentrations might lead to a seizure by causing a depression in Ca^{2+}-dependent K^+ potentials now seem unlikely to be true. Repetitive activation of the systems in the presence of GABA antagonists leads to marked increases in $[K^+]_o$ and decreases in $[Ca^{2+}]_o$; however, seizures are not necessarily produced (Benninger *et al.*, 1980). Although sustained afterdischarges can be induced in the hippocampal slice, the firing rarely lasts longer than a second unless extreme measures are taken to prevent the activation of K^+ potentials (however, see Section 4.3.4).

It is not yet clear how this affects the hypothesis that the onset of seizures is due to a deficit in burst-termination processes. It may be that K^+ channels can be modulated by unknown endogenous mechanisms. The observations that cAMP (Newberry and Nicoll, 1982) and norepinephrine (Madison and Nicoll, 1982) can enhance cellular excitability by blocking Ca^{2+}-dependent K^+ potentials may be relevant.

6. HOW CAN A SEIZURE SPREAD FROM EPILEPTIC TISSUE ACROSS NORMAL TISSUE?

As discussed above, reduction of IPSPs permits susceptible tissue to begin firing interictal spikes. This tissue then serves as a focus for the spread of the seizure across otherwise normal tissue. Even though the events leading finally to a seizure are not yet understood, it is nevertheless possible to begin inquiring into the sorts of conditions that would allow spread to occur. Of course, one possibility is that whatever caused the pathology in the focus simply spreads and affects inhibitory systems nearby. However, in the alumina gel model studied by Ribak *et al.*, (1978), it was found that, while inhibitory interneuronal terminals had disappeared in the cortical region immediately under the focus, the GABA systems in nearby regions appeared unaffected. Seizures can develop in animals treated with alumina gel and can spread from the focus across neighboring regions. Thus, it is reasonable to seek some other enabling mechanism.

The evidence to date supports the idea that depression of IPSPs is necessary and sufficient to permit the development of interictal spikes. If these conclusions, developed using *in vitro* preparations, can be carried to the intact brain, then the question is how can IPSPs in normal CNS tissue be depressed. There are several possibilities.

6.1. *Endogenous Opiates*

Endorphins cause increases in hippocampal unit firing (Nicoll *et al.*, 1977). Injection of opiates into certain brain regions causes seizures (Urca *et al.*, 1977). A number of studies based on extracellular recording provided evidence that the mechanism of this excitation might involve depression of firing of inhibitory interneurons; in effect, the excitation would be a form of "disinhibition" (e.g., Zieglgansberger *et al.*, 1979; Corrigall and Linseman, 1980; Lee *et al.*, 1980). However, with extracellular studies, it is very difficult to distinguish depression of inhibition from increases in excitation. Nicoll *et al.*, (1980) used intracellular recordings in an attempt to resolve the issue. As reported by others, enkephalin applied by bath application resulted in a naloxone reversible increase in the amplitude of the orthodromic population spike and repetitive firing with no effect on the antidromic response (Dunwiddie *et al.*, 1980; Lynch *et al.*, 1981). IPSPs, both orthodromic and antidromic and their associated conductance increases, were reduced, also in a naloxone-reversible way (cf., Siggins and Zieglgansberger, 1981). Interestingly, the depolarizing phase of the orthodromic response produced in pentobarbital (see Section 3.2) appeared to be even more sensitive to the depression by opiates (see also Robinson and Deadwyler, 1981). Opiate effects are not postsynaptic since opiates do not affect pyramidal cell membrane properties or responses to iontophoretically applied GABA, even in cells in which GABA-mediated IPSPs are very much reduced. This suggestion of a presynaptic action of opiates was confirmed by the finding that spontaneous IPSPs were also blocked. Unlike spontaneous EPSPs recorded in CA3 cells (Brown *et al.*, 1979), these spontaneous IPSPs are not true "quanta," and are probably dependent on the firing of the inhibitory interneurons (Alger and Nicoll, 1980b). Therefore, the fact that spontaneous IPSPs are blocked strongly suggests that the effects of enkephalin are exerted directly on the presynaptic elements and not on the pyramidal cells.

6.2. *Acetylcholine*

ACh application can also produce interictal spikes. Hippocampal responses to iontophoretic application of ACh have received a great deal of attention recently.

Yamamoto and Kawai (1968) originally reported that ACh caused a response depression of field potentials in the dentate gyrus through a presynaptic inhibition of excitatory transmitter release (cf., Hounsgaard, 1978). Others have emphasized the slow muscarinic depolarization that increases hippocampal cell responsiveness (Benardo and Prince, 1981; Dodd *et al.*, 1981). Valentino and Dingledine (1981) showed that ACh does cause presynaptic inhibition of EPSPs when applied to CA1 s. radiatum and a slow depolarization when applied to the s. pyramidale. The slow muscarinic depolarization associated with a conductance decrease in hippocampal pyramidal cells has recently been studied in voltage clamp experiments (Halliwell and Adams, 1982). It has been found to be due to a suppression of the M current (Brown and Adams, 1980). The M current is a voltage-dependent K^+ current that is partially activated at rest. As in sympathetic ganglion cells, the blocking of this K^+ current results in a slow depolarization and repetitive firing of the cells in response to other inputs.

None of the previous mechanisms of ACh action would seem to be in agreement with the hypothesis discussed earlier that the depression of inhibition is necessary for the synaptic activation of bursting activity. However, several reports have suggested that the excitatory responses to iontophoretic ACh application (Krnjevic *et al.*, 1981) or to stimulation of medial septum are due to decreases in inhibition (Krnjevic and Ropert, 1981). Somatically applied ACh can depress IPSPs (Valentino and Dingledine, 1981; Haas, 1982). Therefore, ACh may not be anomalous in its ability to produce bursts.

The evidence for the mechanism of ACh action does not appear to be finally established and its role in seizures has been questioned. It nevertheless remains a possible candidate for contributing to seizure spread.

6.3. *Use-Dependent IPSP Depression*

Hippocampal IPSPs become depressed when hippocampal afferents are activated repetitively. Two hypotheses have been put forward to explain this.

6.3.1. Desensitization. Ben-Ari *et al.* (1979, 1981) have emphasized that repetitive stimulation may lead to disinhibition. IPSPs and brief GABA applications are ordinarily accompanied by a marked increase in membrane conductance. During repetitive activation of IPSPs or prolonged GABA application, these conductance increases decline. It is reported that there is no change in the reversal potential for IPSPs or GABA; however, these experiments were done with IPSPs that had been reversed by increases in intracellular $C1^-$ by diffusion from 3 M KC1 pipettes, so this is less certain. These observations would be consistent with a failure in the GABA receptor–ionophore system, i.e., GABA receptors may desensitize or, possibly, GABA conductance channels may otherwise inactivate. In any event, the results suggested that GABA no longer produces a response because it is unable to.

A possible complication of these experiments with iontophoretically applied GABA is posed by the observations that iontophoretically ap-

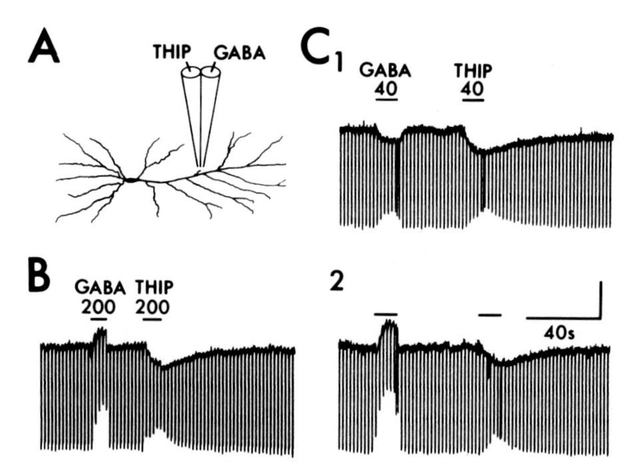

Figure 7. (A) Comparison of dendritic responses to iontophoretic application of GABA and the GABA receptor agonist, 4,5,6,7-tetrahydroisoxazolo[5,4-*c*]pyridin-3-ol (THIP). (B) When injected sequentially from adjacent barrels of a double-barreled iontophoretic pipette GABA typically depolarizes the cell, and THIP hyperpolarizes it. Both responses are Cl^- dependent and blocked by GABA antagonists (not shown). (C) Small movements of the pipette caused the GABA response to invert from a hyperpolarization to a depolarization (compare C1 with C2), while the THIP response was hyperpolarizing in both positions. Modified from Alger and Nicoll, 1982a.

plied GABA can have two effects on hippocampal pyramidal cells (Langmoen *et al.*, 1978; Alger and Nicoll, 1979, 1982a,b; Andersen *et al.*, 1980; Jahnsen and Mosfeldt-Laursen, 1981; Thalmann *et al.*, 1981, Figure 1). When applied to the dendritic region, GABA ordinarily depolarizes it. When iontophoresed to the soma, hyperpolarizations predominate, although with large GABA doses, hyperpolarizations are followed by depolarizations. The hyperpolarizing GABA responses at the soma (H responses) have the same reversal potential as IPSPs (Ben-Ari *et al.*, 1979; Andersen *et al.*, 1980; Alger and Nicoll, 1982a), supporting the hypothesis that GABA is the inhibitory transmitter for somatically generated IPSPs. There is general agreement that, while both types of response can be blocked by GABA antagonists, the depolarizing responses (D responses) are more sensitive (Thalmann *et al.*, 1980; Alger and Nicoll, 1982a; Wong and Watkins, 1982; Figure 1). This suggests that these two types of response are due to the activation of pharmacologically distinct GABA receptor types, a conclusion that has been strongly supported in a study of several other GABA agonist and antagonist drugs (Alger and Nicoll, 1982a; Figure 7). For example, the GABA agonist, 4,5,6,7-tetrahydroisoxazolo[5,4-*c*]-pyridin-3-ol (THIP) produces responses that are blocked by GABA antagonists and are Cl^- dependent and yet dendritic THIP responses are typically hyperpolarizing, even when GABA responses are depolarizing (Figure 7B). Pentobarbital has a dramatically greater effect in potentiating D responses than H responses and this undoubtedly accounts for the appearance of the late "depolarizing IPSP" that is seen with orthodromic stimulation in the presence of pentobarbital (Figure 3; see Section 3.2).

It is also possible to find conditions in which the same (small) GABA ejection current produces responses that can shift between hyperpolarizations and depolarizations with movements of the ejection pipette in the dendritic region (at right angles to the axis of the dendritic tree) of as little as 50 μm (Figure 7C). This and other evidence suggested that while the primary dendritic GABA response would be a depolarization, hyperpolarizations would also be produced in small "hot spots" for GABA. It was hypothesized that the H receptors are relatively more prevalent at the sites of actual GABA terminals ("hot spots") while D receptors are more heavily distributed at extrasynaptic sites. Some experimenters emphasize that D responses may require higher doses of GABA than H responses (Thalmann *et al.*, 1981; Wong and Watkins, 1982) and suggest that a single receptor type, whose response may be dose dependent, might be involved. The hypothesis that concentration differences of GABA account for differences between D and H responses would suggest that depolarizing responses would inevitably be preceded by hyperpolarizations. However, in many cases, purely depolarizing responses are seen (Andersen *et al.*, 1980; Alger and Nicoll, 1982a). (Note that a "purely" depolarizing response does not imply that a unitary conductance mechanism has been activated. In fact, hyperpolarizing iontophoretic GABA responses can probably always be unmasked by perfusing a cell with a low dose of GABA antagonist [Figure 1]. A "purely" depolarizing response simply implies that a preceding hyperpolarization need not occur and that the *net* current flowing during the combined activation of H and D receptors is inward. The net current depolarizes the cell; see Djorup *et al.*, 1981). The ideas of different spatial distributions of different GABA responses on the one hand, and of different affinities for GABA by different receptors on the other, are not mutually exclusive. Chestnut and Schwartzkroin (1982) have recently reported evidence from neonatal guinea pig hippocampus that supports the hypothesis of synaptic and extrasynaptic GABA receptors. Application of GABA to somata of pyramidal cells of immature animals produces depolarizing responses. IPSPs in these cells are also depolarizing. In mature animals, both IPSPs and somatic GABA responses are hyperpolarizing. The occurrence of hyperpolarizing responses appears to depend on the development of a subset of synaptic GABA receptors that can mediate these responses.

The possibility thus exists that in experiments using repetitive activation or long GABA pulses that D receptors were activated along with H receptors. This is a potentially important problem because it is not yet known if the D and H receptors respond identically to various experimental treatments. Following orthodromic tetanic stimulation, hyperpolarizing GABA responses appear to invert to depolarizations and gradually recover (Wong and Watkins, 1982).

Wong and Watkins (1982) found that prolonged periods (over 10 sec) of GABA application resulted in a decrease of the GABA-induced conductance increase, which they attributed to desensitization. They suggest that both H and D receptors desensitize. However, in view of the likely overlap of activation of these receptors when GABA is iontophoresed, the relative extent to which they may desensitize is unclear. These considerations make the interpretation of IPSP depression with repetitive activation very difficult. In addition, increases in extracellular K^+ concentration were found to increase the size of the D response, and presumably decrease the size of H responses (Wong and Watkins, 1983). Alger and Nicoll (1982a, 1983) found that increasing extracellular K^+ shifts E_{IPSP} in the depolarizing direction. Thus, it is likely that the increases in extracellular K^+ that occur as a result of stimulation also affect IPSPs (cf., Krnjevic, *et al.*, 1982).

6.3.2. Frequency-Dependent Depression of IPSPs. Andersen *et al.* (1969) suggested on the basis of extracellular recordings that repetitive activation of afferent systems to the hippocampus might result in the block of the inhibitory pathway, perhaps by a depolarization block of inhibitory interneurons. Knowles and Schwartzkroin (1981) discovered a possible mechanism for failure to follow high frequencies of stimulation by recording from synaptically coupled pairs of cells, one pyramidal cell and one interneuron. When the pyramidal cell was stimulated directly with a depolarizing current injection through the microelectrode, EPSPs were evoked in the interneurons. With repetitive stimulation, the EPSPs quickly diminished. It was suggested that IPSPs would drop out because of a decreased input to the interneurons.

In summary, repetitive activation can cause a decrease in the amplitude and presumably in the inhibitory effectiveness of IPSPs. Several mechanisms appear able to account for this depression; however, details remain to be clarified.

6.4. *Ammonia*

Clinical findings of elevated levels of ammonia in cases of hepatic encephalopathies suggested ammonia might be a causative agent in seizures (e.g., Lockwood *et al.*, 1979). Ammonia can cause a depression of IPSPs in mammalian CNS. Intravenous ammonia perfusion results in shifts in the reversal potentials for Cl^--dependent IPSPs of spinal (Lux, 1971) and trochlear motoneurons (Llinás *et al.*, 1974a) and in neocortical neurons (Raabe and Gumnit, 1975). The evidence indicated that under normal conditions an inward Cl^- gradient is maintained by the action of a Cl^- extrusion mechanism: an outward Cl^- pump. After experimentally increasing intracellular Cl^- concentration, Cl^- extrusion from the cell can be measured indirectly. Ammonia markedly slowed this extrusion process. Furthermore, shortly after beginning ammonia treatment, E_{IPSP} shifted to the resting potential. The IPSP conductance increases still occurred, i.e., inhibitory receptor–channel complexes still functioned, but the IPSP gradient appeared to have dissipated. There were no changes in resting membrane potential properties or in potassium-dependent potentials. Therefore, it was suggested that ammonia blocked an outward Cl^- pump that lead to a passive redistribution of Cl^- across the membrane. The resulting shift in E_{IPSP} caused a decrease in the efficacy of the IPSP (due to a reduction in the driving force on Cl^-). Ammonia would thus seem to be an attractive candidate for promoting epileptiform activity.

However, Allen *et al.* (1977) failed to replicate these findings for IPSPs in the intact hippocampus. Ammonia did not cause a reduction in IPSP amplitude. Indeed, no evidence was found for the existence of a Cl^- transport mechanism dependent on Cl^- levels in these cells. The ammonia-sensitive mechanism in other systems is dependent on the intracellular Cl^- concentration. Because of the possibility that technical difficulties limited the investigation *in vivo*, Alger and Nicoll (1983) reexamined the action of ammonia (in the form of $NH_4^+Cl^-$ or NH_4^+-acetate) on hippocampal IPSPs *in vitro*. The major finding of the *in vitro* study was that ammonia at concentrations fully effective in blocking IPSPs in other systems (<2 mM) has only relatively weak effects on hippocampal IPSPs or on the action of iontophoretically activated GABA responses (Figure 8). At higher concentrations (4 to 8 mM), ammonia did produce measureable effects on the IPSP reversal potential. However, at these higher concentrations, ammonia also depressed synaptic transmission nonspecifically; i.e., in addition to depressing IPSPs, ammonia also depressed EPSPs and presynaptic fiber potentials. It was further possible to demonstrate a disruption between pre- and post-

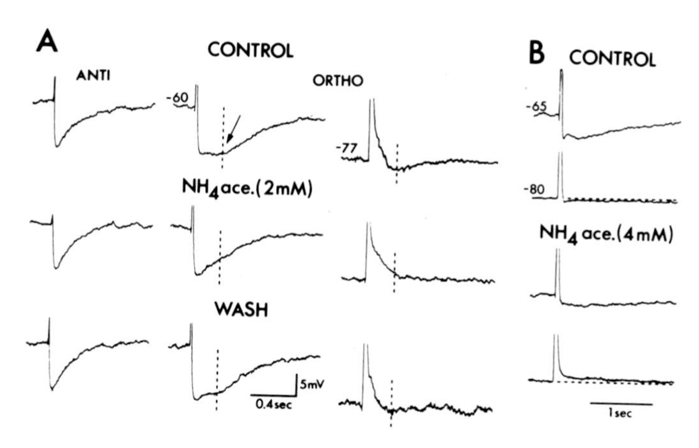

Figure 8. Effects of ammonium acetate on hippocampal potentials. In (A), after recording anti- and orthodromic IPSPs at the resting potential and the LHP near its null potential [right-hand column in (A)], 2 mM NH_4^+ acetate was applied to the cell for 15 min. In this cell, the NH_4^+ had no effect on the resting potential, and caused only a slight decrease in the IPSP and shift in E_{IPSP}. There was also a shift in the null potential for the LHP (compare amplitudes of LHP measured at the dotted lines), which was reversible (lowest row). (B) In another cell, 4 mM NH_4^+-acetate reduced the AHP_d measured at the resting potential (first and third traces) and shifted E_{AHP} (compare second and fourth traces). Voltage calibration in (A) applies to all traces. Time calibrations are indicated. (Alger and Nicoll, 1983.)

synaptic responses, i.e., a given presynaptic fiber spike was associated with a smaller postsynaptic response in the presence of ammonia than in control.

The depression in IPSP conductance resulting from higher doses of ammonia appeared to be due to a presynaptic effect, probably to a reduction in inhibitory transmitter release. This conclusion was suggested by the fact that these doses of ammonia did not cause a decrease in the conductance increase that accompanied the iontophoretic application of GABA. Some of the generalized effects of high doses of ammonia could be attributed to the effects of ammonia on potassium levels in the slice and probably also to K^+-like effects of ammonia itself; i.e., ammonia can probably pass through hippocampal pyramidal cell K^+ channels as it does other K^+ channels (Binstock and Lecar, 1969). Ammonia does depolarize pyramidal cells. Direct measurements of K^+ concentration with K^+-sensitive microelectrodes revealed that ammonia releases K^+ into the extracellular space. These effects of ammonia on K^+ and K^+-dependent processes probably explain much of the effect of ammonia on IPSPs in the hippocampus (Figure 8B). Increased extracellular K^+

will result in increased intracellular Cl^- through Donnan-equilibrium-like effects.

In spinal cord and neocortex maximal disinhibition is found at ammonia concentrations of about 0.7 mM (Iles and Jack, 1980). Indeed, intravenous ammonium acetate infusion into rats has an LD_{50} of about 1.2 mM (Erlich *et al.*, 1980). This fact together with the relative insensitivity of hippocampal IPSPs to blockade by ammonia levels below 2 mM make it unlikely that effects on hippocampal IPSPs are a major factor in the convulsions caused by ammonia.

7. CONCLUSIONS

Use of the hippocampal slice preparation has contributed a great deal to the understanding of the events of epilepsy. From the studies discussed in this review, the following conclusions seem justified:

1. The firing of a burst potential is largely an intrinsic capacity of a cell. Once activated in the appropriate conditions, some cells will inevitably burst and they do so by virtue of specialized biophysical properties, the most important being a capacity for producing voltage-activated calcium potentials.

2. Control over cellular bursting is exerted by influences projected on a cell by other cells. Inhibitory postsynaptic potentials, usually mediated by GABA, are the most important in preventing the expression of synchronous bursting in a neuronal population, even in susceptible cells. Indeed, the present evidence supports the hypothesis that reduction of IPSPs is the necessary and sufficient condition for permitting the synaptic activation of the burst potential. However, except in highly specialized regions, even when IPSPs are blocked, spontaneous bursting does not develop unless cells are driven by spontaneous large ("giant") EPSPs.

3. Synchronization of neuronal firing, which is the most prominent feature of epilepsy, originates within regions of the brain characterized by excitatory interconnections among the principal cells. Spontaneous activity may develop in this population if the individual cells have a relatively low threshold for burst firing. Synchronization of firing is enhanced by the occurrence of prolonged intrinsic Ca^{2+}-dependent K^+ potentials. The giant EPSPs that are produced in other regions are the result of the synchronous excitatory output of these "pacemaker" cells.

4. Bursts are ordinarily terminated by K^+ potentials. There appear to be several different types of K^+ potentials that are important. Besides intrinsic voltage- and Ca^{2+}-activated potentials, there are synaptically

activated K^+ potentials. Depending on the circumstances, all of these probably contribute to burst termination. Ca^{2+}-dependent K^+ potentials are especially important in the timing of the interburst interval. Afterdischarges can be seen under conditions in which most K^+ potentials have been blocked. Nevertheless, it is very difficult to produce seizures in a hippocampal slice. The factors underlying the transition from interictal spike to seizure are not understood.

5. Spread of a seizure from a focus across normal brain tissue first requires the depression of IPSPs in the normal tissue in order to permit the expression of burst discharges. Several endogenous mechanisms are known by which IPSPs can be reduced including enkephalin, ACh, blockade of inhibitory interneurons, and desensitization of GABA receptors. Ammonia does not seem to be a likely candidate for this effect in the hippocampus, although it could contribute to the process in other regions. Nevertheless, there is, at present, no persuasive evidence that any of these processes is actually responsible for the propagation of the seizure.

ACKNOWLEDGMENTS. I thank Drs. R. K. S. Wong and R. D. Traub for helpful discussions. Supported by NIH Grant NS17539 and the McKnight Foundation.

8. REFERENCES

Alger, B. E., 1983, Characteristics of a slow hyperpolarizing synaptic potential in rat hippocampal pyramidal cells, *J. Neurophysiol.*, in press.

Alger, B. E. and Nicoll, R. A., 1979, GABA-mediated biphasic inhibitory responses in hippocampus, *Nature (London)* **281:**315–317.

Alger, B. E. and Nicoll, R. A., 1980a, The epileptiform burst afterhyperpolarization: A calcium-dependent potassium potential in hippocampal pyramidal cells, *Science*, **210:**1122–1124.

Alger, B. E. and Nicoll, R. A., 1980b, Spontaneous inhibitory post-synaptic potentials in hippocampus, *Brain Res.* **200:**195–200.

Alger, B. E. and Nicoll, R. A., 1981, Epileptiform burst termination and the AHP in hippocampal CA1 pyramidal cells, *Neurosci Abstr.* **7:**629.

Alger, B. E. and Nicoll, R. A., 1982a, Pharmacological evidence for two kinds of GABA receptor on rat hippocampal pyramidal cells studied *in vitro*, *J. Physiol. (London)* **328:**123–141.

Alger, B. E. and Nicoll, R. A., 1982b, Feedforward dendritic inhibition in rat hippocampal pyramidal cells studied *in vitro*, *J. Physiol. (London)* **328:**105–123.

Alger, B. E. and Nicoll, R. A., 1983, Ammonia does not selectively block IPSPs in rat hippocampal pyramidal cells, *J. Neurophysiol.* **49:**1381–1390.

Alger, B. E., McCarren, M., and Fisher, R. S., 1983, On the possibility of simultaneously recording from two cells with a single microelectrode in the hippocampal slice, *Brain Res.*, **270:**137–141.

Allen, G. I., Eccles, J. C., Nicoll, R. A., Oshima, T., and Rubia, F. J., 1977, The ionic mechanisms concerned in generating the IPSPs of hippocampal pyramidal cells, *Proc. R. Soc. London, ser. B* **198:**363–384.

Andersen, P., 1976, Some properties of synapses near to and far from the soma of hippocampal pyramids, *Exp. Brain Res.* **1**(Suppl.):202–206.

Andersen, P., Eccles, J. C., and Loyning, Y., 1964a, Location of postsynaptic inhibitory synapses on hippocampal pyramids, *J. Neurophysiol.* **27:**592–607.

Andersen, P., Eccles, J. C., and Loyning, Y. 1964b, Pathway of postsynaptic inhibition in hippocampal pyramids, *J. Neurophysiol.* **27:**608–619.

Andersen, P., Gross, G. N., Lomo, T., and Sveen, O., 1969, Participation of inhibitory and excitatory interneurones in the control of hippocampal cortical output, in: *The Interneuron*, UCLA Forum for Medical Science, No. 11 (M. A. B. Brazier, ed.), University of California Press, Los Angeles, pp. 415–467.

Andersen, P., Bliss, T. V. P., and Skrede, K. K., 1971, Unit analysis of hippocampal population spikes, *Exp. Brain Res.* **13:**208–221.

Andersen, P., Gjerstad, L., and Langmoen, I. A., 1978, A cortical epilepsy model *in vitro*, in: *Abnormal Neuronal Discharges* (N. Chalazonitis and M. Buisson, eds.), Raven Press, New York, pp. 29–36.

Andersen, P., Dingledine, R., Gjerstad, L., Langmoen, I. A., and Mosfeldt-Laursen, A., 1980, Two different responses of hippocampal pyramidal cells to application of gamma-aminobutyric acid, *J. Physiol. (London)* **305:**279–296.

Andrew, R. D., Taylor, C. P., Snow, R. W., and Dudek, F. E., 1981, Coupling in rat hippocampal slices: Dye transfer between CA1 pyramidal cells, *Brain Res. Bull.* **8:**211–222.

Ascher, P., Marty, A., and Neild, T. O., 1978, Lifetime and elementary conductance of the channels mediating the excitatory effects of acetylcholine in *Aplysia* neurones, *J. Physiol. (London)* **278:**177–206.

Ayala, G. F., Dichter, M., Gumnit, R. J., Matsumoto, H., and Spencer, W. A., 1973, Genesis of epileptic interictal spikes: New knowledge of cortical feedback systems suggests a neurophysiological explanation of brief paroxysms, *Brain Res.* **52:**1–17.

Benardo, L. S., and Prince, D. A., 1981, Acetylcholine induced modulation of hippocampal pyramidal neurons, *Brain Res.* **211:**227–234.

Ben-Ari, Y., Krnjevic, K., and Reinhardt, W., 1979, Hippocampal seizures and failure of inhibition, *Can. J. Physiol. Pharmacol.* **57:**1462–1466.

Ben-Ari, Y., Krnjevic K., Reinhardt, W., and Ropert, N., 1981, Intracellular observations on the disinhibitory action of acetylcholine in the hippocampus, *Neuroscience* **6:**2475–2484.

Benninger, C., Kadis, J., and Prince, D. A., 1980, Extracellular calcium and potassium changes in hippocampal slices, *Brain Res.* **187:**105–182.

Binstock, L. and Lecar, H., 1969, Ammonium ion currents in the squid giant axon, *J. Gen. Physiol.* **53:**342–361.

Brown, D. A. and Adams, P. R., 1980, Muscarinic suppression of a novel voltage-sensitive K^+ current in a vertebrate neurone, *Nature (London)*, **283:**673–676.

Brown, T. H., Wong, R, K. S., and Prince, D. A., 1979, Spontaneous miniature synaptic potentials in hippocampal neurons, *Brain Res.* **177:**194–199.

Brown, T. H., Fricke, R. A., and Perkel, D. H., 1981, Passive electrical constants in three classes of hippocampal neurons, *J. Neurophysiol.* **46:**812–827.

Chestnut, R. M. and Schwartzkroin, P. A., 1982, Responses to GABA in developing rabbit hippocampus, *Neurosci. Abstr.* **8:**326.

Constanti, A., Connor, J. D., Galvan, M., and Nistri, A., 1980, Intracellularly-recorded effects of glutamate and aspartate on neurones in the guinea-pig olfactory cortex, *Brain Res.* **195:**403–420.

Corrigall, W. A. and Linseman, M. A., 1980, A specific effect of morphine on evoked activity in the rat hippocampal slice, *Brain Res.* **192:**227–238.

Dingledine, R. and Gjerstad, L., 1980, Reduced inhibition during epileptiform activity in the *in vitro* hippocampal slice, *J. Physiol.* (*London*) **305:**297–313.

Djorup, A., Jahnsen, H., and Mosfeldt-Laursen, A., 1981, The dendritic response to GABA in CA1 of the hippocampal slice, *Brain Res.* **219:**196–201.

Dodd, J., Dingledine, R., and Kelly, J. S., 1981, The excitatory action of acetylcholine on hippocampal neurones of the guinea pig and rat maintained *in vitro*, *Brain Res.* **207:**109–127.

Dunwiddie, T., Mueller, A., Palmer, M., Stewart, J., and Hoffer, B., 1980, Electrophysiological interactions of enkephalins with neuronal circuitry in the rat hippocampus. I. Effects on pyramidal cell activity, *Brain Res.* **184:**311–330.

Eccles, J. C., 1964, *The Physiology of Synapses*, Springer-Verlag, New York.

Erlich, M., Plum, F., and Duffy, T. E., 1980, Blood and brain ammonia concentrations after portacaval anastomosis. Effects of acute ammonia loading, *J. Neurochem.* **34:**1538–1542.

Fujita, Y., 1975, Two types of depolarizing after-potentials in hippocampal pyramidal cells of rabbits, *Brain Res.* **94:**435–446.

Fujita, Y., 1979, Evidence for the existence of inhibitory postsynaptic potentials in dendrites and their functional significance in hippocampal pyramidal cells of adult rabbits, *Brain Res.* **175:**59–69.

Fujita, Y. and Iwasa, H., 1977, Electrophysiological properties of so-called inactivation response and their relationship to dendritic activity in hippocampal pyramidal cells of rabbits, *Brain Res.* **130:**89–100.

Gjerstad, L., Andersen, P., Langmoen, I. A., Lundervold, A., and Hablitz, J. J., 1981, Synaptic triggering of epileptiform discharges in CA1 pyramidal cells *in vitro*, *Acta Physiol. Scand.* **113:**245–252.

Gustafsson, B. and Wigstrom, H., 1981, Evidence for two types of afterhyperpolarization in CA1 pyramidal cells in the hippocampus, *Brain Res.* **206:**462–468.

Gutnick, M. J. and Prince, D. A., 1981, Dye coupling and possible electrotonic coupling in the guinea pig neocortical slice, *Science* **211:**67–70.

Haas, H. L., 1982, Cholinergic disinhibition in hippocampal slices of the rat, *Brain Res.* **233:**200–204.

Hablitz, J. J., 1981, Effects of intracellular injections of chloride and EGTA on postepileptiform-burst hyperpolarizations in hippocampal neurons, *Neurosci. Lett.* **22:**159–163.

Hablitz, J. J. and Andersen, P., 1982, Effect of sodium ions on penicillin-induced epileptiform activity *in vitro*, *Exp. Brain Res.* **47:**154–157.

Hablitz, J. J. and Langmoen, I. A., 1982, Excitation of hippocampal pyramidal cells by glutamate in the guinea pig and rat, *J. Physiol.* (*London*) **325:**317–331.

Hablitz, J. J. and Lundervold, A., 1981, Hippocampal excitability and changes in extracellular potassium, *Exp. Neurol* **71:**410–420.

Halliwell, J. V. and Adams, P. R., 1982, Voltage clamp analysis of muscarinic excitation in hippocampal neurons, *Brain Res.* **250:**71–92.

Hori, N. and Katsuda, N., 1978, Electrophysiological studies on the depolarization shift of hippocampal pyramidal cells *in vitro*: The nature of the long duration hyperpolarization, in: *Integrative Functions of the Nervous System*, Volume 1 (M. Ito, N. Tsukahara, K. Kubota, and K. Yagi, eds.), Kodansha, Tokyo, pp. 345–347.

Hotson, J. R. and Prince, D. A., 1980, A calcium-activated hyperpolarization follows repetitive firing in hippocampal neurons, *J. Neurophysiol.* **43**:409–419.

Hotson, J. R. and Prince, D. A., 1981, Penicillin- and barium-induced epileptiform bursting in hippocampal neurons: Actions on Ca^{++} and K^{+} potentials, *Ann. Neurol.* **10**:11–17.

Hotson, J. R., Prince, D. A., and Schwartzkroin, P. A., 1979, Anomalous inward rectification in hippocampal neurons, *J. Neurophysiol.* **42**:889–895.

Hounsgaard, J., 1978, Presynaptic inhibitory action of acetylcholine in area CA1 of the hippocampus, *Exp. Neurol.* **62**:787–797.

Iles, J. F. and Jack, J. J. B., 1980, Ammonia: Assessment of its action on postsynaptic inhibition as a cause of convulsions, *Brain* **103**:555–578.

Jahnsen, H. and Mosfeldt-Laursen, A., 1981, The effects of a benzodiazepine on the hyperpolarizing and the depolarizing responses of hippocampal cells to GABA, *Brain Res.* **207**:214–217.

Jefferys, J. G. R. and Haas, H. L., 1982, Synchronized bursting of CA1 hippocampal pyramidal sells in the absence of synaptic transmission, *Nature* **300**:448–450.

Johnston, D. and Brown, T. H., 1981, Giant synaptic potential hypothesis for epileptiform activity, *Science* **211**:294–297.

Johnston, D. J., 1981, Passive cable properties of hippocampal CA3 pyramidal neurons, *Cell. Mol. Neurobiol.* **1**:41–55.

Johnston, D., Hablitz, J. J., and Wilson, W., 1980, Voltage clamp discloses slow inward current in hippocampal burst-firing neurones, *Nature* (*London*) **286**:391–393.

Kandel, E. R. and Spencer, W. A., 1961, Electrophysiology of hippocampal neurons. II. Afterpotentials and repetitive firing, *J. Neurophysiol.* **24**:243–259.

Kandel, E. R., Spencer, W. A., and Brinley, F. J., 1961, Electrophysiology of hippocampal neurons. I. Sequential invasion and synaptic organization, *J. Neurophysiol.* **24**:225–242.

Knowles, W. D. and Schwartzkroin, P. A., 1981, Local circuit synaptic interactions in hippocampal brain slices, *J. Neurosci.* **1**:318–322.

Knowles, W. D., Funch, P. G., and Schwartzkroin, P. A., 1982, Electrotonic and dye coupling in hippocampal CA1 pyramidal cells *in vitro, Neuroscience* **7**:1713–1722.

Krnjevic, K. and Ropert, N., 1981, Septo-hippocampal pathway modulates hippocampal activity by a cholinergic mechanism, *Can. J. Physiol. Pharmacol.* **59**:911–914.

Krnjevic, K. K., Reiffenstein, R. J., and Ropert, N., 1981, Disinhibitory action of acetylcholine in the rat hippocampus: Extracellular observations, *Neuroscience* **6**:2465–2474.

Langmoen, I. A., Andersen, P., Gjerstad, L., Mosfeldt-Laursen, A., and Ganes, T., 1978, Two separate effects of GABA on hippocampal pyramidal cells *in vitro, Acta Physiol. Scand.* **102**:C27.

Lee, H. K., Dunwiddie, T., and Hoffer, B., 1980, Electrophysiological interactions of enkephalins with neuronal circuitry in the rat hippocampus. II. Effects of interneuron excitability, *Brain Res.* **184**:331–342.

Leung, L. S., 1978, Hippocampal CA1 region demonstration of antidromic dendritic spike and dendritic inhibition, *Brain Res.* **158**:219–222.

Llinás, R., Baker, R., and Precht, W., 1974a, Blockage of inhibition by ammonium acetate action on C1-pump in cat trochlear motoneurons, *J. Neurophysiol.* **37**:522–533.

Llinás, R., Baker, and Sotelo, C., 1974b, Electrotonic coupling between neurons in cat inferior olive, *J. Neurophysiol.* **38**:541–560.

Lockwood, A. H., McDonald, J. M., Reiman, R. E., Gelbard, A. S., Laughlin, J. S., Duffy, T. E., and Plum, F., 1979, The dynamics of ammonia metabolism in man, *J. Clin. Invest.* **63**:449–460.

Lorente de Nó, R., 1934, Studies on the structure of the cerebral cortex. II. Continuation of the study of the ammonic system, *J. Psychol. Neurol.* **46**:113–177.

Lux, H. D., 1971, Ammonium and chloride extrusion: Hyperpolarizing synaptic inhibition in spinal motoneurons, *Science* **173**:555–557.
Lynch, G. S., Jensen, R. A., McGaugh, J. L., Davila, K., and Oliver, M. W., 1981, Effects of enkephalin, morphine and naloxone on the electrical activity of the *in vitro* hippocampal slice preparation, *Exp. Neurol.* **71**:527–540.
MacVicar, B. A. and Dudek, F. E., 1980a, Local synaptic circuits in rat hippocampus: Interactions between pyramidal cells, *Brain Res.* **184**:220–223.
MacVicar, B. A. and Dudek, F. E., 1980b, Dye coupling between CA3 pyramidal cells in slices of rat hippocampus, *Brain Res.* **196**:494–497.
MacVicar, B. A. and Dudek, F. E., 1981, Electrotonic coupling between pyramidal cells: A direct demonstration in rat hippocampal slices, *Science* **213**:782–784.
Madison, D. V. and Nicoll, R. A., 1982, Noradrenaline blocks accommodation of pyramidal cell discharge in the hippocampus, *Nature (London)* **299**:636–638.
Matthews, G. and Wickelgren, W. O., 1979, Glycine, GABA and synaptic inhibition of reticulospinal neurones of lamprey, *J. Physiol.* **293**:393–415.
Mesher, R. A. and Schwartzkroin, P. A., 1980, Can CA3 epileptiform discharge induce bursting in normal CA1 hippocampal neurons? *Brain Res.* **183**:472–476.
Newberry, N. R. and Nicoll, R. A., 1982, Properties of the late hyperpolarizing potential in hippocampal pyramidal cells *in vitro*, *Neurosci. Abstr.* **8**:412.
Nicoll, R. A. and Alger, B. E., 1981, Synaptic excitation may activate a calcium dependent potassium conductance in hippocampal pyramidal cells, *Science* **212**:957–959.
Nicoll, R. A., Eccles, J. C., Oshima, T., and Rubia, F., 1975, Prolongation of hippocampal inhibitory postsynaptic potentials by barbiturates, *Nature (London)* **258**:625–627.
Nicoll, R. A., Siggins, G. R., Ling, N., Bloom, F. E., and Guillemin, R., 1977, Neuronal actions of endorphins and enkephalins among brain regions: A comparative microiontophoretic study, *Proc. Natl. Acad. Sci. USA* **74**:2584–2588.
Nicoll, R. A., Alger, B. E., and Jahr, C. E., 1980, Enkephalin blocks inhibitory pathways in the vertebrate central neurons system, *Nature (London)* **287**:22–25.
Ogata, N., 1975, Ionic mechanisms of the depolarization shift in thin hippocampal slices, *Exp. Neurol.* **46**:147–155.
Ogata, N., 1976, Mechanisms of the stereotyped high-frequency burst in hippocampal neurons *in vitro*, *Brain Res.* **103**:386–388.
Ogata, N., Hori, N., and Katsuda, N., 1976, The correlation between extracellular potassium concentration and hippocampal epileptic activity *in vitro*, *Brain Res.* **110**:371–375.
Oliver, A. P., Carman, J. S., Hoffer, B. J., and Wyatt, R. J., 1980a, Effect of altered calcium ion concentration on interictal spike generation in the hippocampal slice, *Exp. Neurol.* **68**:489–499.
Oliver, A. P., Hoffer, B. J., and Wyatt, R. J., 1980b, Kindling induces long-lasting alterations in the responses of hippocampal neurons to elevated potassium levels *in vitro*, *Science* **208**:1264–1265.
Prince, D. A., 1978, Neurophysiology of epilepsy, *Annu. Rev. Neurosci.* **1**:395–415.
Prince, D. A., 1982, Epileptogenesis in hippocampal and neocortical neurons, in: *Physiology and Pharmacology of Epileptogenic Phenomena* (M. R. Klee, H. D. Lux, and E-J. Speckman, eds.), Raven Press, New York, pp. 151–161.
Raabe, W. and Gumnit, R. J., 1975, Disinhibition in cat motor cortex by ammonia, *J. Neurophysiol.* **38**:347–356.
Ribak, C. E., Vaughn, J. E., and Saito, K., 1978, Immunocytochemical localization of glutamic acid decarboxylase in neuronal somata following colchicine inhibition of axonal transport, *Brain Res.* **140**:315–332.
Robinson, J. H. and Deadwyler, S. A., 1981, Intracellular correlates of morphine excitation in the hippocampal slice preparation, *Brain Res.* **224**:375–387.

Schmalbruch, H. and Jahnsen, H., 1981, Gap junctions on CA3 pyramidal cells of guinea pig hippocampus shown by freeze-fracture, *Brain Res.* **217:**175–178.

Schneiderman, J. H. and Schwartzkroin, P. A., 1981, Evidence that penicillin-induced synchrony is not accompanied by increased electrotonic coupling, *Neurosci. Abstr.* **7:**590.

Schwartzkroin, P. A., 1975, Characteristics of CA1 neurons recorded intracellulary in the hippocampal slice, *Brain Res.* **85:**423–435.

Schwartzkroin, P. A., 1977, Further characteristics of CA1 cells *in vitro, Brain Res.* **128:**53–68.

Schwartzkroin, P. A. and Prince, D. A., 1978, Cellular and field potential properties of epileptogenic hippocampal slices, *Brain Res.* **147:**117–130.

Schwartzkroin, P. A. and Prince, D. A., 1980a, Changes in excitatory and inhibitory synaptic potentials leading to epileptogenic activity, *Brain Res.* **183:**61–76.

Schwartzkroin, P. A. and Prince, D. A., 1980b, Effects of TEA on hippocampal neurons, *Brain Res.* **185:**169–181.

Schwartzkroin, P. A., and Slawksy, M. 1977, Probable calcium spikes in hippocampal neurons, *Brain Res.* **135:**157–161.

Schwartzkroin, P. A. and Stafstrom, C. E., 1980, Effects of EGTA on the calcium activated afterhyperpolarization in hippocampal CA3 pyramidal cells, *Science* **210:**1125–1126.

Schwartzkroin, P. A. and Wyler, M. D., 1980, Mechanisms underlying epileptiform burst discharge, *Annu. Rev. Neurol.* **7:**95–107.

Siggins, G. R. and Zieglgansberger, W., 1981, Morphine and opioid peptides reduce inhibitory synaptic potentials in hippocampal pyramidal cells *in vitro* without alteration of membrane potential, *Proc. Nat. Acad. Sci. USA* **78:**5230–5235.

Silfvenius, H., Olofsson, S., and Ridderheim, P-A., 1980, Induced epileptiform activity evoked from dendrites of hippocampal neurons, *Acta Physiol. Scand.* **108:**109–111.

Spencer, W. A. and Kandel, E. R., 1961, Electrophysiology of hippocampal neurons. IV. Fast prepotentials, *J. Neurophysiol.* **24:**272–284.

Spencer, W. A. and Kandel, E. R., 1965, Synaptic inhibition in seizures, in: *Basic Mechanisms of the Epilepsies* (H. H. Jasper, A. A. Ward, Jr., and W. Pope, eds.), Little, Brown and Co., Boston, pp. 575–604.

Storm-Mathisen, J., 1977, Localization of transmitter candidates in the brain: The hippocampal formation as a model, *Prog. Neurobiol.* **8:**119–181.

Taylor, C. P. and Dudek, F. E., 1981, Physiological evidence for electrotonic coupling between CA1 pyramidal cells in rat hippocampal slices, *Neurosci. Abstr.* **7:**519.

Taylor, C. P., Dudek, F. E., 1982, Synchronous neural afterdischarges in rat hippocampal slices without active chemical synapses, *Science* **218:**810–812.

Thalmann, R. H., 1982, Is the late hyperpolarization which follows synaptic stimulation of hippocampal pyramidal neurons calcium-dependent?, *Neurosci. Abstr.* **8:**797.

Thalmann, R. H. and Ayala, G. F., 1980, A picrotoxin-resistant hyperpolarizing response is elicited by orthodromic stimulation of hippocampal neurons, *Neurosci. Abstr.* **6:**300.

Thalmann, R. H., Peck, E. J., and Ayala, G. F., 1981, Biphasic response of hippocampal pyramidal neurons to GABA, *Neurosci. Lett.* **21:**319–324.

Traub, R. D. and Wong, R. K. S., 1982, Cellular mechanism of neuronal synchronization in epilepsy, *Science* **216:**745–747.

Traub, R. D. and Wong, R. K. S., 1983, Synchronized burst discharge in the disinhibited hippocampal slice. II. Model of the cellular mechanism, *J. Neurophysiol.* **49:**459–471.

Turner, D. A. and Schwartzkroin, P. A., 1980, Steady state analysis of intracellularly stained hippocampal neurons, *J. Neurophysiol.* **44:**184–199.

Urca, G., Frenk, H., Liebeskind, J. C., and Taylor, A. N., 1977, Morphine and enkephalin: Analgesic and epileptic properties, *Science* **197:**83–86.

Valentino, R. J. and Dingledine, R., 1981, Presynaptic inhibitory effect of acetylcholine in the hippocampus, *J. Neurosci.* **1**:784–792.

Wong, R. K. S., 1982, Postsynaptic potentiation mechanism in the hippocampal pyramidal cells, in: *Physiology and Pharmacology of Epileptogenic Phenomena* (M. R. Klee, H. D. Lux and E.-J. Speckman, eds.), Raven Press, New York, pp. 163–173.

Wong, R. K. S. and Prince, D. A., 1978, Participation of calcium spikes during intrinsic burst firing in hippocampal neurons, *Brain Res.* **159**:385–390.

Wong, R. K. S. and Prince, D. A., 1979, Dendritic mechanisms underlying penicillin-induced epileptiform activity, *Science* **204**:1228–1231.

Wong, R. K. S. and Prince, D. A., 1981, Afterpotential generation in hippocampal pyramidal cells, *J. Neurophysiol.* **45**:86–97.

Wong, R. K. S. and Traub, R. D., 1983, Synchronized burst discharge in the disinhibited hippocampal slice. I. Initiation in the CA2-CA3 region, *J. Neurophysiol.* **49**:442–458.

Wong, R. K. S. and Watkins, D. J., 1982, Cellular factors influencing the GABA response in hippocampal pyramidal cells, *J. Neurophysiol.* **48**:938–951.

Wong, R. K. S., Prince, D. A., and Basbaum, A. I., 1979, Intradendritic recordings from hippocampal neurons, *Proc. Nat. Acad. Sci. USA* **76**:986–990.

Yamamoto, C., 1972, Intracellular study of seizure-like afterdischarges elicited in thin hippocampal sections *in vitro, Exp. Neurol.* **35**:154–164.

Yamamoto, C. and Kawai, N., 1968, Generation of the seizure discharge in thin sections from the guinea pig brain in chloride free medium *in vitro, Jpn. J. Physiol.* **18**:620–631.

Zieglgansberger, W., French, E. D., Siggins, G. R., and Bloom, F. E., 1979, Opioid peptides may excite hippocampal pyramidal neurons by inhibiting adjacent inhibitory interneurons, *Science* **205**:415–417.

7

Correlated Electrophysiological and Biochemical Studies of Hippocampal Slices

GARY LYNCH, MARKUS KESSLER, and MICHEL BAUDRY

1. INTRODUCTION

The nervous system operates with impulses and transmissions that have time scales in the millisecond range, and yet is called upon to store information for periods of years. Evidently, the patterns of electrical activity that speed through brain circuitries, on some occasions, must modify the properties of the elements that transmit them. Understanding the nature of these modifications, the physiological forms they take, and the cellular chemistries that bring them into existence constitutes one of the major problems of neurobiology. Studies of the relatively simple nervous systems of invertebrates by Kandel and others have located synapses that are modified by experience (see Kandel, 1981, for a review). Comparable efforts on mammalian central nervous system (CNS), hampered as they are by the extraordinary complexity of the brain, are still at the stage of conclusively pinning down sites that show lasting traces of experience.

Beyond the problem of finding and defining the physiological aftereffects of experience there lies the difficulty of unraveling the requisite

GARY LYNCH, MARKUS KESSLER, and MICHEL BAUDRY • Department of Psychobiology, University of California, Irvine, California 92717.

biochemical processes. Here a formidable array of problems, both conceptual and technical, will have to be overcome. In fact, biochemical studies of neurophysiological events in well-defined components of mammalian CNS are rare (release of putative transmitters is the chief exception). There is much literature describing experiments in which drugs or high ion concentrations are applied to brain slices or "synaptosomes" in an effort to mimic synaptic transmission and the effects of these manipulations on various biochemical indices measured. Although much has been learned from this, the stimulus conditions used in this type of work are only very crude approximations of "real-world" events. But these types of experiments have been so extensively used simply because electrical stimulation of the type that produces typical physiological potentials is not well suited for biochemical research. Biochemical assays require substantial quantities of tissue, which means that stimulation must induce measurable changes in a vastly greater population of neurons than physiologists work with. Moreover, the events to be studied are induced in living systems, but subsequently analyzed under conditions that are antithetical to physiological activity; thus, the hoped-for biochemical changes must be quite hardy and there is no reason to assume that nature will be so accommodating. These problems, and there are others, suggest that simplifying assumptions and preparations are needed if we are to attempt to measure the biochemical consequences of physiological events.

Physiologically active brain slices may be of use in this regard. They retain many of the functional properties found in the brain *in situ*, but can be transferred to test-tubes in a matter of seconds. It is also possible to rapidly stimulate large numbers of fibers by using multiple stimulating electrodes, particularly in laminated structures such as the hippocampus or cerebellum. In the following sections, we will review our efforts to exploit these advantages of slices in experiments intended to identify biochemical correlates of an unusual physiological phenomenon.

2. MODIFICATION OF STIMULATION PROCEDURES AND SLICE TECHNIQUES FOR BIOCHEMICAL EXPERIMENTS

As mentioned, a major problem in studying the biochemical effects of physiological events is causing those events to occur in a population of elements large enough to allow the measurement of biochemical parameters by whatever assays are to be used. There are two aspects to this: first, the absolute amount of tissue needed for most biochemical tests is relatively large and second, the percentage of affected elements within the tissue sample needs to be sufficient so that any induced perturbations are not masked by elements that did not experience the physiological manipulation. These considerations necessarily influence the choice of stimulation paradigms as well as the tissue to be selected for analysis. The simplest stimulation strategy is to use conventional electrodes and very high currents. This approach was followed in two of the experiments to be discussed below. Unfortunately, high-stimulation currents can cause undesirable side effects that may not be evident to a distant recording electrode and that may produce biochemical consequences unrelated to normal neuronal operation. An alternative is to employ multiple stimulating electrodes in well-laminated structures such as hippocampus. A brief description of the anatomy of field CA1 is necessary to explain the rationale behind the experimental design we have used as well as the problems it encounters (Figure 1). The pyramidal cells form the great majority of neurons in CA1 and nearly all of these have dendritic fields aligned in the manner shown in the figure. Three regions provide most of the afferents that innervate these cells: (1) the entorhinal cortex (a bilateral projection), (2) field CA3 (again, a bilateral projection), and (3) the subiculum. Each of these projections runs at right angles to the axis of the dendrites and forms an extremely dense synaptic field in a well-defined dendritic layer (Figure 1B). Theoretically then, it should be possible to place stimulating electrodes in the positions indicated in the drawing and stimulate the afferents of successive proximodistal segments of the CA1 dendrites. However, it is most unlikely that the majority of slices will be cut exactly along the trajectory of the afferent fibers; therefore, the full length of most fibers will not be contained in a single slice. This results in the situation illustrated in Figure 2. Note that axons transverse the slice for variable distances and leave (or enter) it at various points along the mediolateral extent of field CA1. This point is well illustrated by the fact that over the course of dozens of extracellular recording experiments using routine stimulation currents we have never succeeded in eliciting paired-pulse facilitation between two electrodes located at opposite ends of the CA1 zone (Dunwiddie and Lynch, 1978). Therefore, it is necessary to use multiple arrays of electrodes or multiple "drops" across the path of the fibers of interest if one is to stimulate the majority of the axons.

The optimal electrode is a matter for further research. We have used electrodes with four small tips in our more recent experiments and, by moving these quickly about the slice, stimulate the points illustrated in Figure 2. It may well prove to be the case that large bipolar electrodes with widely spaced tips aligned along the trajectory of the axons are more effective in activating a larger axonal field (J. B. Ranck, personal

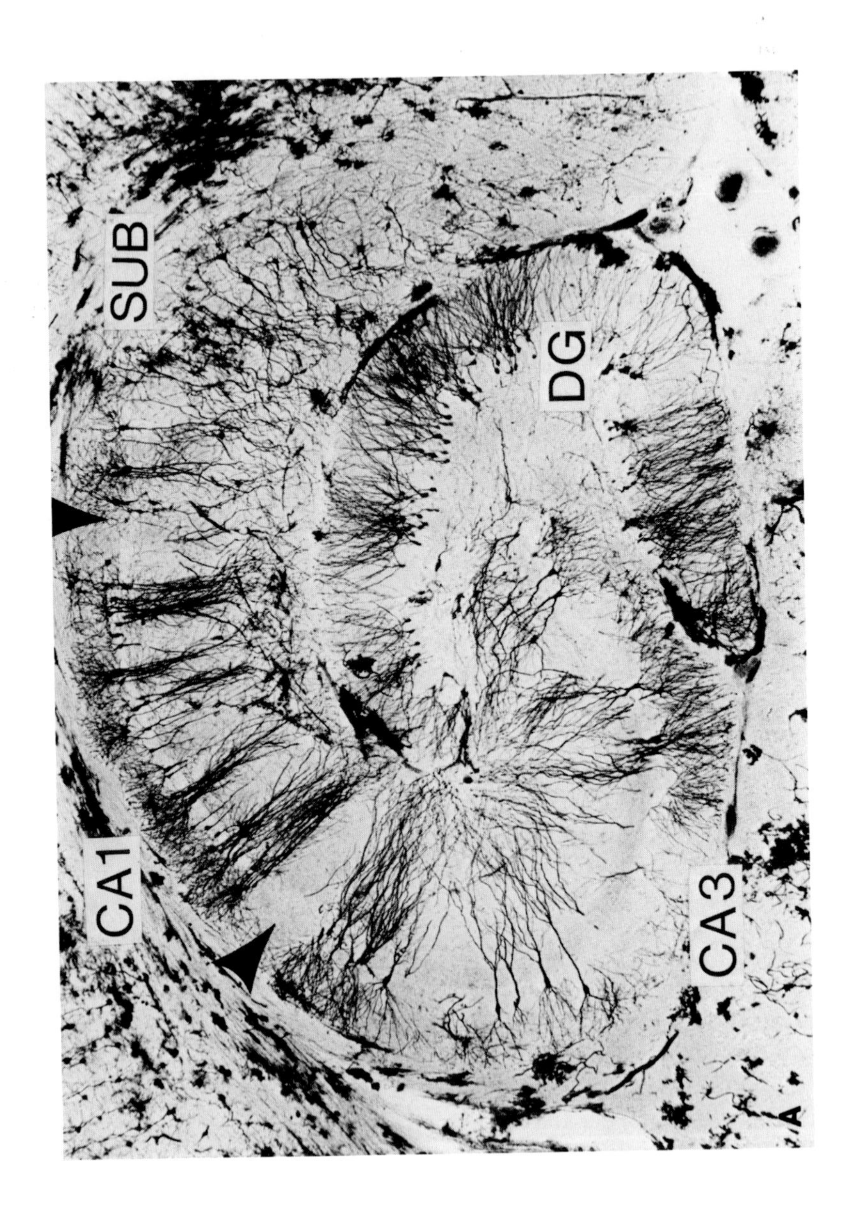
SUB
DG
CA1
CA3
A

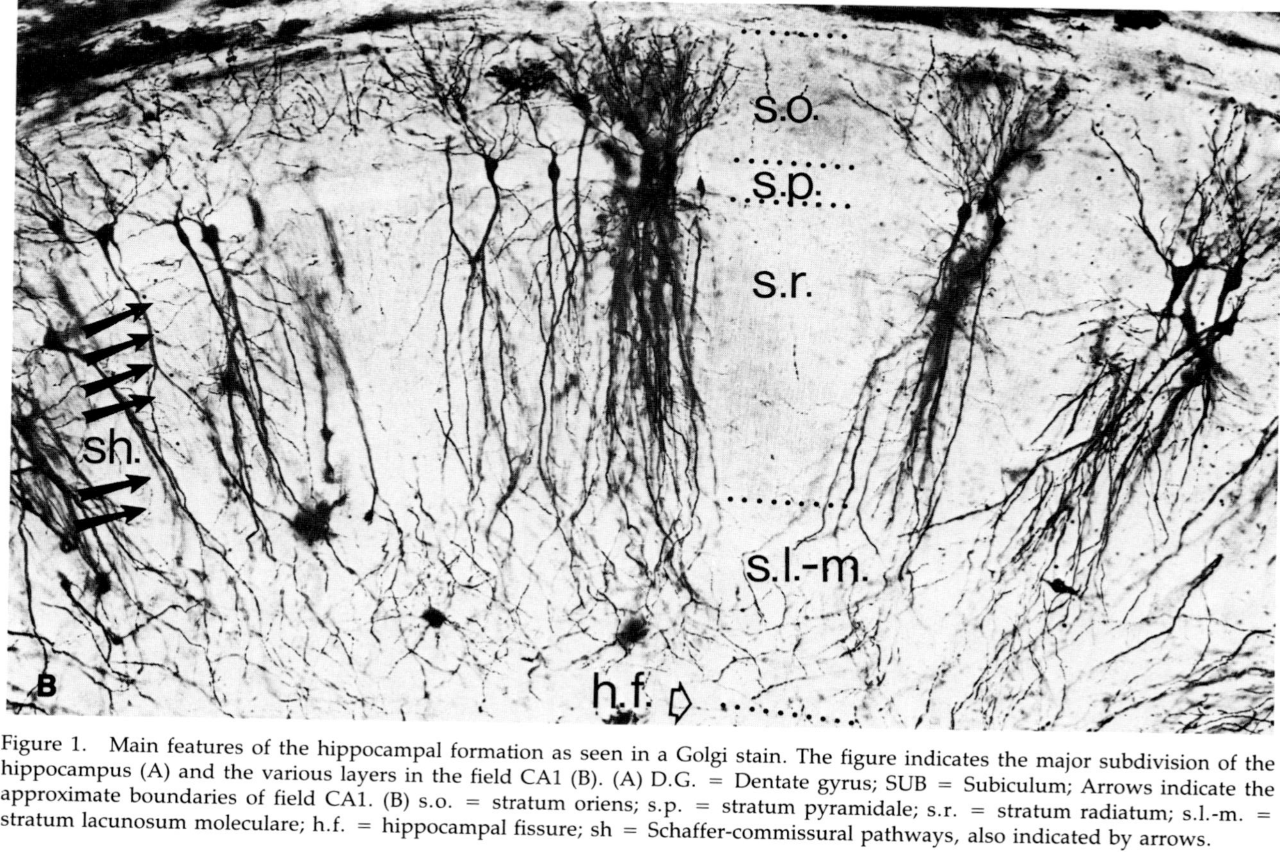

Figure 1. Main features of the hippocampal formation as seen in a Golgi stain. The figure indicates the major subdivision of the hippocampus (A) and the various layers in the field CA1 (B). (A) D.G. = Dentate gyrus; SUB = Subiculum; Arrows indicate the approximate boundaries of field CA1. (B) s.o. = stratum oriens; s.p. = stratum pyramidale; s.r. = stratum radiatum; s.l.-m. = stratum lacunosum moleculare; h.f. = hippocampal fissure; sh = Schaffer-commissural pathways, also indicated by arrows.

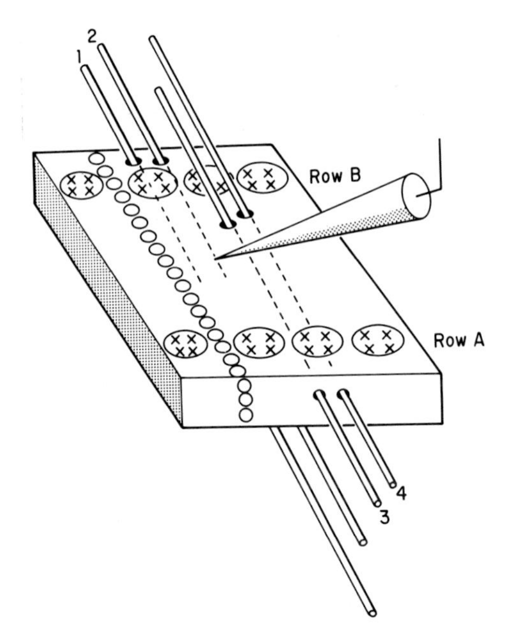

Figure 2. Relationships between fiber bundles in the minislice of field CA1 and stimulating and recording electrodes. Bundles 1 and 2 enter the slice at the surface of its distant edge, fall into the field stimulated by one group of electrodes (Row B) and leave the slice before reaching the second group of electrodes (Row A). An opposite situation occurs for bundles 3 and 4. From Lynch *et al.*, 1982.

communication). By using paired-pulse facilitation between two electrodes, it should be possible to arrive at an estimate of the size of the axonal field stimulated by a given electrode configuration. Such techniques should also be useful in establishing that stimulation arrangements do not produce seizures, depressions, and other pathophysiological states.

Another approach to increasing the percentage of stimulated axons and synapses in the slice is to dissect the preparation so that only those regions that contain electrode sites are taken for assay. Thus, one could stimulate, remove the slice, and then cut away the zones outside the stimulation field. In doing this, we have encountered problems in defining the points at which the electrodes were lowered and in making the appropriate cuts. This is not to say that the technique is unworkable, only that it has problems. As an alternative, we began to dissect the slices before they were placed in the recording chamber. This allowed us to select only the slices that were appropriately sectioned and to place our stimulation electrodes in accordance with the shape of the "minislice." Certain of the anatomical features of this preparation merit comment. First, the minislice contains an extremely homogeneous population of synapses. The apical dendritic zone is innervated almost exclusively by the CA3 pyramidal cells; the basal dendrites receive most of their input from the same source with an additional contribution from the subiculum (see above). The target cells include a number of interneuron types but the dominant cell type is the CA1 pyramidal neuron. Thus, the minislice is, for the most part, a multiplied version of a pyramidal-neuron-to-pyramidal-neuron connection. This homogeneity may be of importance in biochemical studies.

The minislices display most of the physiological properties of the whole slice and hence of the intact hippocampus. The results of experiments on this preparation have been described in the literature (Lynch *et al.*, 1982) and are discussed in a later section of the present review.

Although multiple electrodes undoubtedly stimulate a substantial percentage of the synapses in the minislice, we still have no useful estimate of what that percentage might be. The problem is compounded by the presence of degenerating elements in the slice. Electron microscopic studies have shown that the two surfaces of the slice are severely necrotic for 50 μm into the slice and that the core contains numerous pathological dendrites, axons, and synapses (Lee *et al.*, 1981). How much of this material is carried over into biochemical assays is unknown. Thus, the possibility exists that a substantial portion of a given sample consists of "dead" tissue. This would not only serve to dilute any changes elicited by stimulation but, depending on the index being studied, could also contribute pathological neuronal biochemistries.

3. HIPPOCAMPAL LONG-TERM POTENTIATION

Long-term potentiation (LTP) is an extremely stable facilitation of synaptic transmission that follows short (30-msec) periods of high-frequency stimulation (Bliss and Gardner-Medwin, 1973; Bliss and Lomo, 1973). LTP provides evidence that neuronal circuitries do modify their operating characteristics following particular patterns of activity and it thus may be pertinent to the issues raised in the introduction. The biochemical experiments to be described in the later sections of this review were concerned with finding the cellular processes that trigger and maintain LTP and it is appropriate that we briefly discuss it here. Fuller consideration of the subject can be found in Lynch and Baudry (1983).

LTP, as mentioned, is extremely stable and has been followed for days and weeks in rats with chronic electrodes (Douglas and Goddard, 1975; Barnes, 1979). In slices, the effect is essentially nondecremented for hours (Dunwiddie and Lynch, 1978). Experiments using multiple inputs to the same dendritic zones have established that LTP is not the result of generalized changes in the target cell (Dunwiddie and Lynch,

1978), although careful recording of the fiber volley suggests that it is not due to the recruitment of additional axons (Andersen *et al.*, 1977). The idea that LTP is not an axonal effect also finds support in the observations that repetitive stimulation performed in slices maintained in incubation media in which synaptic but not axonal responses are blocked does not yield LTP when normal conditions are restored (Dunwiddie *et al.*, 1978; Dunwiddie and Lynch, 1979).

LTP has been reported to occur in the mossy fiber–CA3 synapse (Yamamoto and Chujo, 1978). This is an important result since the mossy axons do not extensively branch in their target zones and the recipient spines emerge from primary dendritic branches; this combination indicates that LTP is not likely to reflect either decrease in branch point failures or a biophysical alteration in fine oblique dendritic branches. Taken together, the available evidence points to a synaptic locus for the effect.

There is conflicting evidence as to whether the potentiation is pre- or postsynaptic. Dolphin *et al.* (1982) have recently reported that ^{3}H-glutamate release from the perforant path (the primary afferent of the dentate gyrus) is increased following LTP inducted by high-voltage stimulation in intact animals. The LTP illustrated in their paper declines slowly over an hour to a new baseline that is elevated above control values; the release data were collected during the period when the responses were still declining and thus it is not clear if the increased release is related to the stable LTP (or variant of LTP). Work from this laboratory has shown that high-frequency stimulation in the presence of aminophosphonobutyric acid (APB), a drug that blocks glutamate binding sites (Baudry and Lynch, 1981a) and synaptic transmission in field CA1, does not produce potentiation of responses tested after the drug has been removed from the slice (Dunwiddie *et al.*, 1978). Although this result certainly suggests that LTP involves postsynaptic receptors, APB may have effects other than receptor blockade. However, it should be noted that APB does *not* affect the depolarizing responses of slices to a variety of excitatory amino acids and thus must be reasonably selective in its effects (Fagni *et al.*, 1983b). Indirect evidence for a postsynaptic locus of LTP is found in neuroanatomical studies of the consequences of high-frequency stimulation on dendritic ultrastructure. van Harreveld and Fifkova (1975) reported that potentiation of the perforant path caused a marked swelling of the spines contacted by that projection. Since these authors did not record responses in their experiments and employed a rather unusual stimulation procedure, we repeated this experiment with neurophysiological controls in slices and anesthesized rats. We found that spine morphology was indeed altered but in a manner somewhat different from that reported by Fifkova and van Harreveld (Lee *et al.*, 1980, 1981). Although these results do not establish a causal relationship between spine changes and LTP, the very fact that the stimulation does produce postsynaptic changes lends credence to the idea that the potentiation is due to a modification of target neurons.

4. INFLUENCES OF HIGH-FREQUENCY STIMULATION ON ^{3}H-GLUTAMATE BINDING TO SYNAPTIC MEMBRANES

The properties of LTP suggest that it is caused by a rapidly appearing yet enduring change at the synapse. A change in the number or affinity of postsynaptic receptors would match this description. Given that high frequency stimulation produces structural changes in spines, an effect that might well be expected to modify cell surface properties, we decided to investigate possible alterations in glutamate binding sites following LTP. Glutamate was chosen because first, there is good reason to believe that it, or an allied compound, is the transmitter in the Schaffer-commissural system (Storm-Mathisen, 1977; Wieraszko and Lynch, 1979) and second, as discussed immediately below, biochemical experiments had identified a binding site for this amino acid that appeared to be a good candidate for a synaptic receptor.

4.1. *Characteristics of ^{3}H-Glutamate Binding to Hippocampal Synaptic Membranes*

Following the procedures developed in other systems for the study of neurotransmitter receptors (Snyder and Bennett, 1976), several laboratories have used ^{3}H-glutamate as a ligand to label possible glutamate receptors (Michaelis *et al.*, 1974; Roberts, 1974; Foster and Roberts, 1978; Baudry and Lynch, 1979a, 1981a; De Barry *et al.*, 1980). These studies revealed that one class of ^{3}H-glutamate binding sites, with an affinity of about 500 nM, could be detected in synaptic membranes from various brain regions, and possessed several properties of a postsynaptic glutamate receptor (Baudry and Lynch, 1981b; Table I). Developmental studies indicate that the maximal density of sites is reached before the appearance of the majority of glial cells, suggesting that they are located on neurons (Baudry *et al.*, 1981a). This site is enriched in purified synaptic junctions (Foster *et al.*, 1981) and is not modified by lesions of the main afferent systems to the hippocampus, suggesting that it is located postsynaptically. Moreover, the pharmacological properties of this site match the pharmacological profile of the synaptic receptor, at least in

Table I. Characteristics of Na^+-Independent ^{3}H-Glutamate Binding

Properties	Characteristics
Saturability	$K_D = 0.5\ \mu M$ $B_{max} = 6.5$ pmol/mg protein $n_h = 1.05$
Reversibility	Association $K_{on} = 0.60 \times 10^6\ M^{-1}\ min^{-1}$ Dissociation $K_{off} = 0.30\ min^{-1}$ $K_D = K_{off}/K_{on} = 0.5\ \mu M$
Localization	Enriched in synaptic junctions Not decreased after lesions of major hippocampal afferents Developmental pattern parallels synapse formation
Pharmacology	Antagonized by most excitatory amino acids, except kainic acid Antagonized by most of the "classical" antagonists of excitatory amino acids except Glutamate Diethylester
Other properties	Selectively inhibited by low concentrations of sodium Stimulated by micromolar concentrations of calcium

the hippocampus; thus, two blockers of synaptic transmission, α-aminoadipate and aminophosphonobutyrate, inhibit ^{3}H-glutamate binding, whereas a variety of excitatory amino acids, such as quisqualate and homocysteate are also effective displacers of glutamate (Baudry and Lynch, 1981a). The apparent discrepancies between the pharmacological profile of the binding site and of the physiological site studied by iontophoretic applications of glutamate appears to be resolved by our recent demonstration that exogenous glutamate probably stimulates an extrajunctional rather than a synaptic receptor (Fagni *et al.*, 1983a,b).

4.2. *Effect of High-Frequency Stimulation on ^{3}H-Glutamate Binding*

In a first series of experiments, we induced long-term potentiation in whole hippocampal slices, by using high-voltage stimulation in order to activate a maximum number of synapses. Under these conditions, we pooled several slices, prepared crude synaptic membranes and determined the characteristics of ^{3}H-glutamate binding (Baudry *et al.*, 1980). We found that membranes prepared from stimulated slices exhibited an increase in the maximal number of binding sites without significant changes in the apparent affinity of ^{3}H-glutamate for the binding site. However, two criticisms could be addressed to this study. First, as noted above, high-voltage stimulation can be expected to produce a variety of side effects, making it difficult to attribute the change in binding to LTP alone. Second, the need to pool several slices to generate enough material to determine ^{3}H-glutamate binding properties prevented us from correlating physiological and biochemical alterations in the same preparation. The second series of experiments, therefore, was intended to answer these two criticisms by using the minislice preparation described in Section 2. This enabled us to use stimulation currents that produced typical evoked responses. It was necessary to miniaturize the procedure for preparing crude synaptic membranes by combining microhomogenization and microcentrifugation as shown in Figure 3. Under these conditions, one minislice provided enough material for both ^{3}H-glutamate binding and protein assay.

We first confirmed that most of the physiological properties of the minislices (paired-pulse facilitation, long-term potentiation, etc) and most of the biochemical properties of ^{3}H-glutamate binding (affinity, sensitivity to ions, and pharmacological profile) in this new preparation were comparable to those previously described. We then replicated the study of the effects of high-frequency stimulation on ^{3}H-glutamate binding (Lynch *et al.*, 1982). Three separate experiments, each with 10 to 20 slices, confirmed that four 30- to 300-msec bursts delivered at a frequency of 300 Hz produced an increase in ^{3}H-glutamate binding compared to matched nonstimulated controls. This effect was not obtained in slices receiving random or low-frequency stimulation nor in slices receiving high-frequency stimulation in low calcium, high magnesium medium, conditions that prevent stimulation-induced LTP (Dunwiddie *et al.*, 1978). The increase in ^{3}H-glutamate binding was detected within 5 min following the stimulation (the shortest interval that could be tested under our experimental procedure) and was still present and of the same magnitude 50 to 60 min after a high-frequency stimulation train. Finally, in a certain number of cases, the high-frequency stimulation did not induce long-term potentiation, a fact that has been described by several laboratories (Alger and Teyler, 1976). In synaptic membranes prepared from these slices, ^{3}H-glutamate binding was not increased as compared to control unstimulated slices (Figure 4). This set of data, therefore, makes a very strong case that the induction of long-term potentiation is accompanied by an increase in the number of ^{3}H-glutamate binding sites.

4.3. *Calcium Regulation of ^{3}H-Glutamate Binding to Hippocampal Membranes*

The results described immediately above raise questions about the types of biochemical events that high-frequency stimulation might trig-

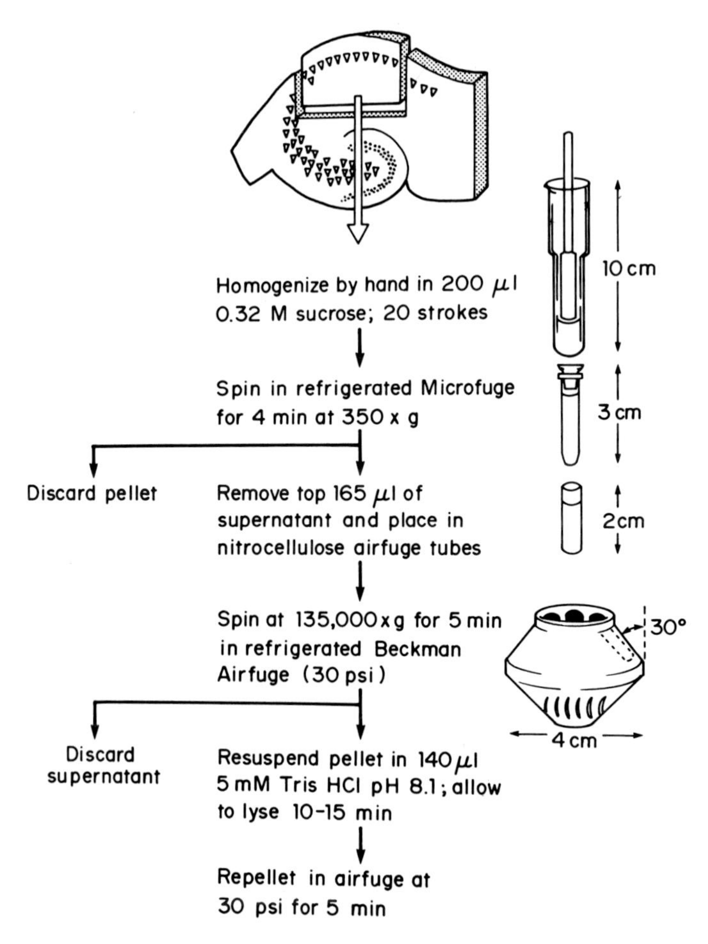

Figure 3. The membrane preparation for minislices. After stimulation, each minislice is removed from the *in vitro* chamber and homogenized individually in 200 μl of ice-cold 0.32 M sucrose. The final pellet is resuspended in 165 μl of tris-HCl pH 7.4 for the subsequent ^{3}H-glutamate binding assay. The entire procedure takes about 45 min. From Lynch *et al.*, 1982.

ger so as to produce an increase in glutamate binding. Biochemical experiments using purified membranes have suggested one possibility and indicate one direction that might be followed in future studies using electrical stimulation of hippocampal slices.

Whereas monovalent cations inhibit ^{3}H-glutamate binding, some divalent cations stimulate the binding to hippocampal membranes

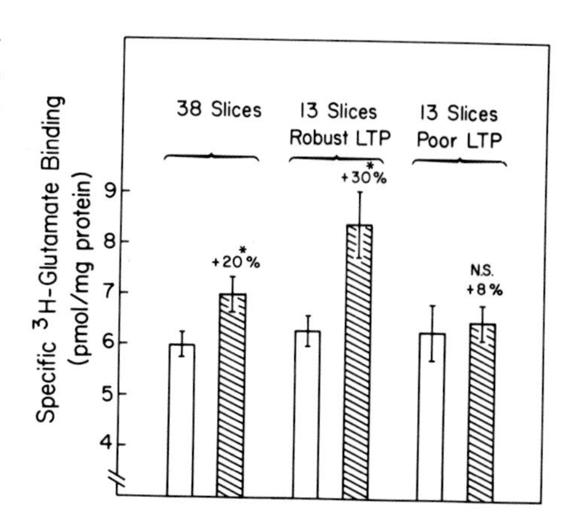

Figure 4. Effects of high-frequency electrical stimulation on ^{3}H-glutamate receptor binding. Results of 13 experiments with 3 stimulated (hatched bars) and 3 control (open bars) (total of 38 pairs) slices are shown on the left. The data were then subdivided between slices exhibiting robust LTP (corresponding to the first stimulated slice; 13 pairs) and poor LTP (corresponding to the third stimulated slice; 13 pairs). Results expressed in pmol/mg protein are means ± SEM of the number of experiments. $^*p < 0.02$ (Student's *t*-test 2 tails).

(Baudry and Lynch, 1979b). In particular, calcium ions induce a two- to threefold increase in the number of binding sites without significant changes in the affinity of glutamate for the binding sites; the half-maximal effect occurs at a calcium concentration of 30 μM and the effect of calcium ions is strongly cooperative with a Hill coefficient of about 2 (Baudry *et al.*, 1981b). Moreover, the effect of calcium is partly irreversible; thus, it is possible to preincubate hippocampal membranes with calcium, eliminate calcium by dilution and centrifugation, and still find an increased number of binding sites compared to control membranes preincubated with buffer. In trying to understand the mechanism(s) by which calcium ions unmask "cryptic receptors," we were fortunate to discover that compounds that inhibit calcium-activated neutral thiol-proteinases (CANP or calpain) prevent the stimulation of ^{3}H-glutamate binding by calcium (Baudry and Lynch, 1980). Leupeptin, which, at a concentration of 80 μM, is a very selective inhibitor of calpain (Toyooka *et al.*, 1978), totally blocks the calcium-induced stimulation of ^{3}H-glutamate binding without modifying basal binding (Baudry *et al.*, 1981c; see also Vargas *et al.*, 1980). Calpain activity has been found in many tissues including brain and peripheral nerves, and two forms of this enzyme have been purified: calpain I is half-maximally stimulated by about 30 μM calcium while calpain II requires about 0.5 to 1.0 mM for half-maximal stimulation (Murachi *et al.*, 1981a,b). Although the nature of the substrates for these two enzymes is not yet known, we found that low concentrations of calcium induce the proteolysis of a high molecular weight doublet protein in hippocampal and cortical membranes (Baudry *et al.*, 1981c; Figure 5). The migration properties of this doublet protein in gel electrophoresis suggest that it is similar to one of the

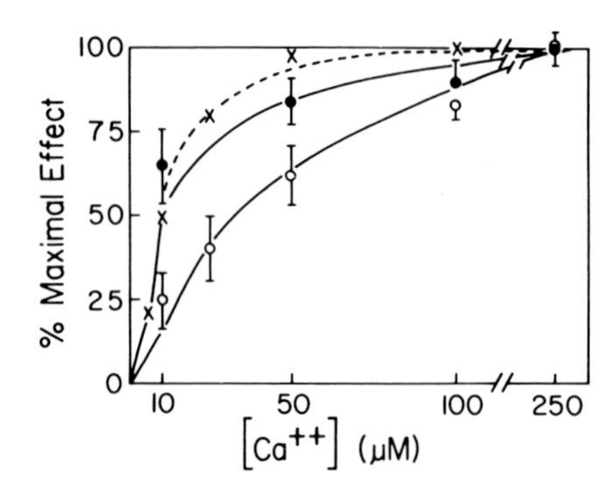

Figure 5. Comparison of the effects of calcium on ^{3}H-glutamate binding stimulation, fodrin degradation, and calpain I activity. The stimulation by calcium of ^{3}H-glutamate binding to cortical membranes (open circles) is compared to its stimulation of a high molecular weight peptide degradation in cortical membranes (closed circles). Also shown is the calcium dependency of calpain I activity from rat brain soluble fraction (crosses; data from Kishimoto *et al.*, 1981).

components of the cytoskeletal proteins called "fodrin" by Levine and Willard (1981), a polypeptide that may be involved in the regulation of receptors in lymphocyte membranes. More recently, we have obtained direct evidence that hippocampal and cortical membranes possess a proteolytic activity, the properties of which are identical to calpain I (Siman *et al.*, 1983). Combining these data, we proposed that the stimulation by calcium of ^{3}H-glutamate binding, mediated by an activation of calpain I, could be the mechanism by which repetitive electrical stimulation of hippocampal synapses induces a long-lasting increase in synaptic efficacy (Baudry and Lynch, 1980).

5. HIGH-FREQUENCY STIMULATION AND PROTEIN PHOSPHORYLATION

Brain tissue contains a large number of proteins that are rapidly phosphorylated when homogenates are incubated with ^{32}P-labeled ATP. The function of most of these proteins is still elusive, but there is indirect evidence linking protein phosphorylation to synaptic transmission (Greengard, 1978). Protein phosphorylation differs from most of the other known posttranslational protein modifications by being readily reversible due to the presence of protein phosphatases. Since long-term potentiation is extremely stable, it is unlikely *a priori* that it would simply reflect an altered phosphorylation state of some synaptic component. However, phosphorylation might activate processes transiently, which in turn lead to the final irreversible alterations. For this reason, we have searched for changes in the pattern of protein phosphorylation in slices that had been homogenized immediately after application of the high-frequency stimulation. Using a *post-hoc* phosphorylation assay, we found that high-frequency stimulation of hippocampal slices resulted in a decreased incorporation of ^{32}P into a protein with an apparent molecular weight of 40,000 daltons (40 K protein) (Browning *et al.*, 1979). This protein was subsequently identified as the α subunit of pyruvate dehydrogenase (PDH), a mitochondrial enzyme that converts pyruvate to acetyl-CoA (Morgan and Routtenberg, 1980; Browning *et al.*, 1981a; Magilen *et al.*, 1981). The activity of PDH is controlled by the phosphorylation of its α subunit with the enzyme complex being inactive in the phosphorylated state.

The first stimulation studies used above-normal stimulation voltages in order to activate the greatest possible numbers of fibers. Beyond that, the procedures used to prepare subcellular fractions were not intended to purify mitochondria (identification of the 40 K protein as α-PDH followed the stimulation experiments). Accordingly, we have reexamined the effects of high-frequency stimulation on protein phosphorylation using multiple stimulation electrodes and modified separation-assay conditions. Since these results have not been previously reported, we will describe the experiments in some detail.

Slices were tested for viability and if stimulation of the Schaffer pathway did not elicit a maximum population spike of at least 8 mV, the slice was discarded. We then lowered a double bipolar stimulating electrode (four intertwined Nichrome wires, two connected to each polarity) into the Schaffer pathway, and selected a voltage that elicited a 1 to 2 mV population spike. A train of a hundred pulses at a frequency of 100 Hz was delivered, followed after 3 sec by a second train with inverted polarity (so that each wire acted once as a cathode).

The population spike was monitored for 3 to 10 min; only those slices that showed stable potentiation over this period were used. The stimulating electrode was placed in 4 to 5 positions successively across the s. radiatum and s. oriens, at the CA1/CA2 border, and in the same number of positions at the subicular border of the CA1, and in each position the same stimulation scheme was applied as in the test phase except that the stimulation voltage was doubled. This phase of stimulation lasted 120 to 150 sec. Immediately after the last train, the slice was removed from the chamber and transferred into a dish with ice-cold medium. The CA1 field of the slice was dissected free with a scalpel and transferred into 3 ml of ice-cold 0.32 M sucrose. Three to five samples were pooled. After each stimulated slice, a nonstimulated control slice was removed from the chamber and processed the same way. Control slices were also checked for viability, but were not subjected to any high-frequency stimulation.

The samples were homogenized with 10 strokes in a motor-driven glass/teflon homogenizer. A P_2 fraction was prepared (first spin: 1000 g

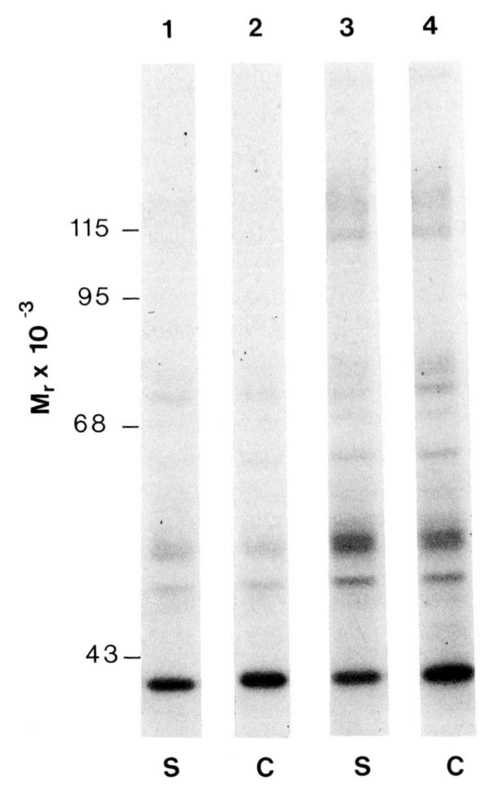

Figure 6. Effects of high-frequency electrical stimulation on protein phosphorylation. 20 μl aliquots of the mitochondrial fraction, containing 14 μg protein, were preincubated for 5 min at 30°C in 50 mM Hepes/tris (pH 7.4), 10 mM $MgCl_2$, with (Lane 3,4) or without (Lane 1,2) 0.5% Triton X-100. Phosphorylation was started by adding 5 μCi ^{32}P-ATP (20 μM). The reaction was stopped after 20 sec with 15 μl of the solubilization buffer used for the subsequent electrophoresis (final concentrations: 2.3% sodium dodecylsulfate, 5% β-mercaptoethanol, 10% glycerol, 60 mM tris/HCl (pH 6.8)). ^{32}P incorporation into the 40 K band was reduced by 25% (−Triton) or 35% (+Triton). Total ^{32}P incorporation into all the remaining bands with mol. wt. $<$100,000 was changed by −2% (−Triton) and −7% (+Triton). S = stimulated slices; C = control slices.

for 10 min; supernatant spun at 18,000 g for 20 min) and lysed in 2 ml of 1 mM tris/HCl, 50 μM $CaCl_2$ (pH 8.1) for 30 min. After addition of 0.5 ml of 1.2 M sucrose, the solution was layered on top of a sucrose density gradient (3 ml each of 0.8 M, 1.0 M, and 1.2 M sucrose, all sucrose solutions containing 50 μM $CaCl_2$) and centrifuged at 130,000 g (average) for 80 min. The mitochondrial pellets were resuspended by mild sonication in 150 μl 50 mM Hepes/tris (pH 7.4), 10 mM $MgCl_2$; the suspensions were then equated according to protein concentration and immediately used for the phosphorylation assay. The assay was conducted according to procedures described in the chapter by Browning *et al.* (1979). Samples were then run on polyacrylamide gels and autoradiograms prepared and scanned with a densitometer. In agreement with our earlier findings, less ^{32}P was incorporated into the α subunit of PDH of high-frequency stimulated slices (Figure 6 and Table II). In 20 out of 28 experiments, the reduction was substantial (on the average, −16%), and in only one experiment was an increase larger than 5% found. Figure

Table II. Effects of High-Frequency Stimulation on α-PDH Phosphorylation[a]

Changes in the phosphorylation of the 40 K band after high-frequency stimulation (%)	Number of experiments
−40 to −5	20
−5 to +5	7
+11	1
Averaged change in all 28 experiments: −11%	

[a] Duplicate phosphorylation assays were done on each sample-pair from stimulated and control slices. Care was taken to equalize the protein concentration prior to the phosphorylation assay. The gels were stained with Coomassie Blue and scanned on a densitometer to monitor any differences in the protein concentration in the gel. Phosphorylation of the 40 K band was determined by measuring the peak-height in the densitometric scans of the X-ray films; we made sure that peak-widths at the base and at half the peak-height were the same within each sample pair. For this table, absolute peak-heights of the duplicates were averaged and the percent change in the stimulated sample was computed. In addition, the peak heights of the 40 K band were normalized to (1) the protein content of each gel-lane as determined from summing up all major peaks in the Coomassie-Blue scanning and (2) to the sum of all other clearly discernible peaks on the autoradiogram. Both methods of normalization introduced some minor changes within single experiments, but they did not affect significantly the values shown in the table.

6 shows an experiment in which the reduction of PDH phosphorylation was larger than average, and demonstrates that the phosphorylation of other bands was much less affected by the high-frequency stimulation. Indeed, none of the other bands in either the mitochondrial or the synaptosomal fraction have so far shown any consistent change in ^{32}P incorporation. The decrease in PDH phosphorylation was also observed when the phosphorylation assay was conducted in the presence of Triton X-100 (Figure 6, lanes 3 and 4); this argues against the possibility that the difference arose from a differential ATP entry into mitochondria or from different concentrations of calcium or cofactors inside the mitochondria. Identical results were also obtained in the presence of Triton X-100 and EGTA and at shorter incubation times (t = 10 sec; data not shown), suggesting that the reduced phosphorylation is not due to differential activities of PDH phosphatase.

These experiments employed a "backward" assay procedure, which assumes that the amount of ^{32}P incorporation in the assay reflects the number of sites that were dephosphorylated in the slice (Figure 7). Thus, a decreased phosphorylation in the assay is interpreted as increased net phosphorylation in the slice. We have tested the validity of this assumption in various ways: (1) Hippocampal slices were incubated with dichloroacetate (DCA), an inhibitor of PDH–kinase (Leiter *et al.*, 1978), then washed free of the inhibitor and homogenized. Both PDH activity and ^{32}P incorporation into the 40 K band were similarly increased when

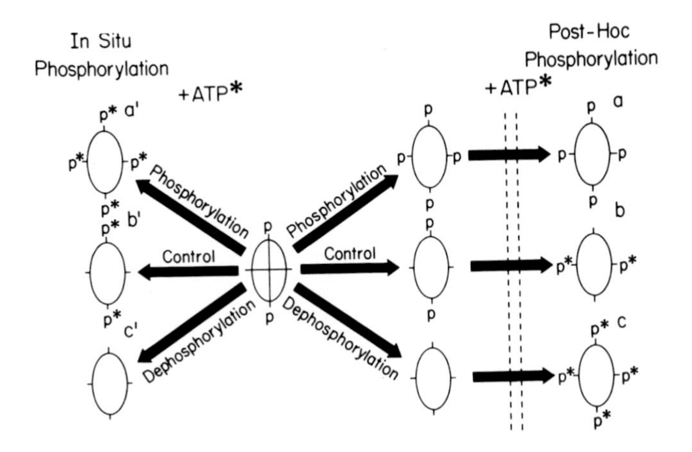

Figure 7. Schematic representation of the difference between the "*post-hoc*" and *in situ* phosphorylation. The effects of various conditions (phosphorylation, dephosphorylation) on the incorporation of labeled phosphate (*) provided by ^{32}P-ATP are analyzed under two conditions: (1) *in situ* phosphorylation refers to the presence of the labeled ATP at the place and time when the proteins are phosphorylated or dephosphorylated; (2) *post-hoc* phosphorylation refers to the addition of the labeled ATP after the changes in the phosphorylation took place and under conditions preventing the dephosphorylation of the proteins. The pattern of incorporation of labeled phosphate under the two conditions are clearly images of each other.

compared to homogenates from control slices, whereas ^{32}P incorporation into the 40 K band was markedly decreased when purified mitochondria were incubated with ^{32}P-ATP in the presence of DCA (Baudry *et al.*, 1982a; Figure 8). (2) Mitochondria from partially deafferented hippocampi were found to have a lower fraction of the total PDH in the active

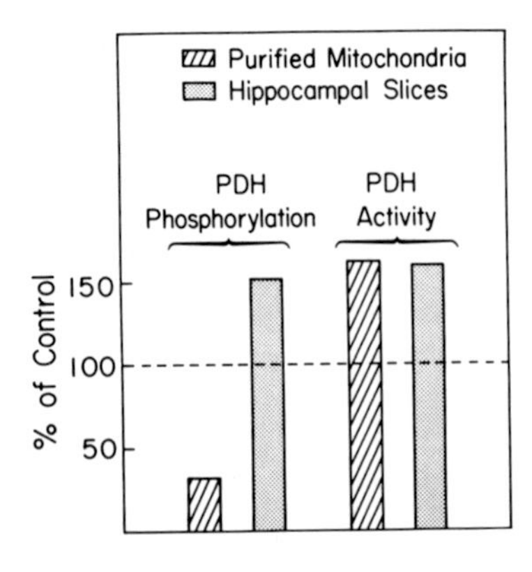

Figure 8. Effect of DCA on PDH phosphorylation and activity in slices and in purified mitochondria. Purified mitochondria were incubated in the presence of 2.5 mM dichloroacetate (DCA) and either ^{32}P-ATP to measure PDH phosphorylation or ^{14}C-pyruvate to measure PDH activity as described in Browning *et al.* (1981b). Hippocampal slices were incubated for 5 min with 10 mM DCA, washed and homogenized. PDH phosphorylation and activity were then determined as described in Baudry *et al.* (1982b). Results are expressed as per cent of the respective control values.

Table III. Effects of Succinate or Dinitrophenol on PDH Activity and Phosphorylation in Mitochondrial Fraction[a]

Preincubation with:	Change in PDH activity (%)	Change in PDH phosphorylation (%)
Succinate	−30	−24
Dinitrophenol	+41	+25

[a] A P_2 fraction was prepared and resuspended in 210 mM mannitol, 70 mM sucrose, 0.1 mM EGTA, and 10 mM PIPES/KOH (pH 7.0). Aliquots were incubated for 8 min at 20°C in the presence of (1) 10 mM succinate, (2) 0.1 mM dinitrophenol, or (3) no further addition (control). The samples were solubilized with 0.5% Triton X-100 and immediately used for the phosphorylation assay (see legend to Figure 6) and for the PDH assay (Baudry *et al.*, 1982b). The values listed in the table are the changes compared to the control incubation. Additional aliquots withdrawn from the three incubations at the beginning of the incubation period yielded values identical to those of the control incubation (±5%), indicating that succinate and dinitrophenol did not interfere with the subsequent phosphorylation and PDH assays.

state, and a reduced incorporation of ^{32}P in the 40 K band in the "backwards" assay (Baudry *et al.*, 1982b, 1983). (3) If mitochondria were preincubated with either the rapidly metabolized substrate succinate, thereby decreasing the ADP/ATP ratio, or with dinitrophenol, which depletes mitochondria of ATP, and then solubilized with Triton X-100, both PDH activity, and ^{32}P incorporation into the 40 K band were similarly decreased or increased respectively (Table III).

These experiments provide evidence that the "backwards" assay gives valid information about the endogenous state of phosphorylation of PDH. However, a few more considerations need to be added. The observed reduction in PDH phosphorylation in stimulated slices might also be explained by a stable decrease of PDH-kinase activity. This, however, seems an unlikely explanation, since any difference in the concentration of a modulator of PDH-kinase, as it might have existed within the mitochondria after high-frequency stimulation, should be abolished in the presence of Triton X-100. Only the existence of a modulator with extremely slow dissociation rate might account for the observed effect. However, if high-frequency stimulation had produced such a slowly dissociating inhibitor, it would also have led to an increased number of dephosphorylated sites, which, in our "backwards" assay, would have at least partially compensated the reduced activity of the kinase.

In a further study, we have looked for changes in PDH phosphorylation and PDH activity in homogenates and P_2 fractions rather than in purified mitochondria. In these experiments, we have so far not found significant changes of any sort after long-term potentiation. This inconsistency could be explained in two ways: (1) subfractionation on the sucrose density gradient separates out a subpopulation of mitochondria that is involved in the development of LTP from another pool of mitochondria that remains unaffected. (2) LTP does not influence the ratio active PDH/total PDH; instead, it affects the distribution of mitochondria on the sucrose density gradient, e.g., through the degree of swelling, thereby yielding a mitochondrial fraction that contains less PDH. Measurements of PDH activity, particularly after complete activation, could determine which of these explanations is correct. Accordingly, we have recently begun to measure PDH activity in slices. Preliminary experiments have shown a correlation between reduction in PDH phosphorylation and reduction in PDH activity after long-term stimulation: from seven experiments in which phosphorylation was reduced by more than 10% (on the average 14%), six experiments also showed a reduced PDH activity (average: −10%) while only one experiment had increased PDH activity (+4%).

Having obtained these results, the question arises as to what role they might play in producing the elevated glutamate binding and long-term potentiation. Pyruvate dehydrogenase is an obligatory and possibly rate-limiting step for the tricarboxylic acid cycle and the activity of the enzyme should be tightly correlated with mitochondrial function. In addition to ATP synthesis, it is known that mitochondria have a high-capacity, low-threshold uptake process for calcium. Therefore, it might be expected that phosphorylation of PDH (which, it will be recalled, renders the enzyme complex inactive) would produce a depression of mitochondrial sequestration of calcium. We have confirmed this prediction using purified brain mitochondria (Browning *et al.*, 1981b). This raises the possibility that changes in PDH phosphorylation induced by high-frequency stimulation leads to a transient alteration in the buffering of free cytosolic calcium levels provided by the mitochondria; if a substantial calcium influx were to occur under these conditions, then local concentrations of the cation might reach the low micromolar value needed to activate calpain I.

6. SUMMARY

The previously described experiments show that slices can be used to assess at least some biochemical after-effects of repetitive synaptic activity. Further developments of the techniques we have used are badly needed. First, intracellular recording studies will have to be performed to provide further assurances that the multiple stimulation procedure is not producing subtle physiological changes that are not found when using modest stimulation to a single electrode. Second, the stimulating electrodes we have used could certainly be improved. Our arrangement required us to move electrodes to several sites in the slice, and was cumbersome in terms of wire leads and providing for sequential application of current. It was not possible to stimulate the entire slice in less than 2 to 3 min and this is certainly long enough for effects initiated at the beginning of the stimulation period to begin to dissipate. It may prove feasible to simplify the situation by using much larger electrodes that cover more of the slice and thus greatly reduce the stimulation period. Experiments directed at this are in progress. Third, and as discussed, studies are needed to assess the size of the field of axons stimulated by a particular procedure. This, in turn, should allow us to make an estimate, albeit crude, of the percentage of the total axonal population activated by a given set of electrodes. Finally, considerable variability was evident in the extent to which stimulation influenced the biochemical indices so far tested. The fluctuations in the effects of stimulation on phosphorylation (see Table II) could not be related to electrophysiological variables or the extent or stability of long-term potentiation. Put simply, unknown and evidently significant factors are at work in the types of experiments described above.

With regard to the specific question of long-term potentiation, our experiments produced one result that offers a reasonable explanation of the effect; namely, an increase in postsynaptic receptors. The stimulation-induced increase in binding requires further characterization and definition. Beyond this, we will attempt to develop treatments that manipulate the changes in binding sites and measure the effects of these on long-term potentiation.

The mechanism proposed to be responsible for stimulation-induced changes may be accessible to testing with slices. To the extent that the calpain hypothesis is correct, intracellular injections of proteinase inhibitors should prevent the development of LTP. Moreover, the slice offers the possibility of directly testing for the activation of calpain by electrical stimulation. The present assays for the enzyme are almost certainly too insensitive to detect small changes in activity but changes in the intensity of staining of specific proteins in one- or two-dimension polyacrylamide gels after high-frequency stimulation might be measureable.

The idea that changes in mitochondrial metabolism serve as triggering events for LTP by producing changes in cytosolic free calcium has found some support in our experiments. Using multiple stimulation sites and reasonable stimulation currents, we replicated our earlier result

(Browning *et al.*, 1979) that high-frequency stimulation modifies the endogenous phosphorylation of pyruvate dehydrogenase. Furthermore, the assumption underlying the *post-hoc* assay has been tested and confirmed.

In all then, we feel that the experimental results described in this review provide grounds for a cautious optimism about the use of hippocampal slices to correlate physiology with biochemistry. Given the importance of this problem, not only to questions of plasticity but to the study of synaptic chemistry in general, further study and development of slice techniques is clearly warranted.

7. REFERENCES

Alger, B. E. and Teyler, T. J., 1976, Long-term and short-term plasticity in CA3 and dentate regions of the rat hippocampal slice, *Brain Res.* **110**:463–480.

Anderson, P., Sundberg, S. H., Sveen, O., and Wigstrom, H., 1977, Specific long-lasting potentiation of synaptic transmission in hippocampal slices, *Nature* (*London*) **266**:736–737.

Barnes, C. A., 1979, Memory deficits associated with senescence: A neurophysiological and behavioral study in the rat, *J. Comp. Phys. Psychol.* **93**:74–104.

Baudry, M. and Lynch, G., 1979a, Two glutamate binding sites in rat hippocampal membranes, *Eur. J. Pharmacol.* **58**:519–521.

Baudry, M. and Lynch, G., 1979b, Regulation of glutamate receptors by cations, *Nature* (*London*) **282**:748–750.

Baudry, M. and Lynch, G., 1980, Regulation of hippocampal glutamate receptors: Evidence for the involvement of a calcium-activated protease, *Proc. Natl. Acad. Sci. USA* **77**:2298–2302.

Baudry, M. and Lynch, G., 1981a, Hippocampal glutamate receptors, *Mol. Cell. Biochem.* **38**:5–18.

Baudry, M. and Lynch, G., 1981b, Characterization of two ^{3}H-glutamate binding sites in rat hippocampal membranes, *J. Neurochem.* **36**(3):811–820.

Baudry, M., Oliver, M., Creager, R., Wieraszko, A., and Lynch, G., 1980, Increase in glutamate receptors following repetitive electrical stimulation in hippocampal slices, *Life Sci.* **27**:325–330.

Baudry, M., Arst, D., Oliver, M., and Lynch, G., 1981a, Development of glutamate binding sites and their regulation by calcium in rat hippocampus, *Dev. Brain Res.* **1**:37–48.

Baudry, M., Smith, E., and Lynch, G., 1981b, Influences of temperature, detergents, and enzymes on glutamate receptor binding and its regulation by calcium in rat hippocampal membranes, *Mol. Pharmacol.* **20**:280–286.

Baudry, M., Bundman, M., Smith, E., and Lynch, G., 1981c, Micromolar levels of calcium stimulate proteolytic activity and glutamate receptor binding in rat brain synaptic membranes, *Science* **212**:937–938.

Baudry, M., Fuchs, J., Kessler, M., Arst, D., and Lynch, G., 1982a, Entorhinal cortex lesions induced a decreased calcium transport in hippocampal mitochondria, *Science* **216**:411–413.

Baudry, M., Kessler, M., Smith, E. K., and Lynch, G., 1982b, The regulation of pyruvate dehydrogenase activity in rat hippocampal slices: Effect of dichloroacetate, *Neurosci. Lett.* **31**:41–46.

Baudry, M., Gall, C., Kessler, M., Alapour, H., and Lynch, G., 1983, Denervation-induced decrease in mitochondrial calcium transport in rat hippocampus, *J. Neurosci.* **3**:252–259.

Bliss, T. V. P. and Gardner-Medwin, A. T., 1973, Long-lasting potentiation of synaptic transmission in the dentate area of the unanaesthetized rabbit following stimulation of the perforant path, *J. Physiol.* (*London*) **232**:357–374.

Bliss, T. V. P. and Lomo, T., 1973, Long-lasting potentiation of synaptic transmission in the dentate area of the anaesthetized rabbit following stimulation of the perforant path, *J. Physiol.* (*London*) **232**:331–356.

Browning, M., Dunwiddie, T., Bennett, W., Gispen, W., and Lynch, G., 1979, Synaptic phosphoproteins: Specific changes after repetitive stimulation of the hippocampal slice, *Science* **903**:60–62.

Browning, M., Baudry, M., Bennett, W., and Lynch, G., 1981a, Phosphorylation-mediated changes in pyruvate dehydrogenase activity influence pyruvate-supported calcium accumulation by brain mitochondria, *J. Neurochem.* **36**:1932–1940.

Browning, M., Bennett, W., Kelly, P., and Lynch, G., 1981b, The 40,000 Mr brain phosphoprotein influenced by high frequency synaptic stimulation is the alpha subunit of pyruvate dehydrogenase, *Brain Res.* **218**:255–266.

De Barry, J., Vincendon, G., and Gombos, G., 1980, High affinity glutamate binding during postnatal development of rat cerebellum, *FEBS Lett.* **109**:175–179.

Dolphin, A. C., Errington, M. L., and Bliss, T. V. P., 1982, Long-term potentiation of the perforant path *in vivo* is associated with increased glutamate release, *Nature* (*London*) **297**:496–498.

Douglas, R. M. and Goddard, G. V., 1975, Long-term potentiation of the perforant path – granule cell synapses in the rat hippocampus, *Brain Res.* **86**:205–215.

Dunwiddie, T. V. and Lynch, G. S., 1978, Long-term potentiation and depression of synaptic responses in the rat hippocampus: Localization and frequency dependency, *J. Physiol.* (*London*) **276**:353–367.

Dunwiddie, T. V. and Lynch, G., 1979, The relationship between extracellular calcium concentration and the induction of hippocampal long-term potentiation, *Brain Res.* **169**:103–110.

Dunwiddie, T. V., Madison, D., and Lynch, G. S., 1978, Synaptic transmission is required for initiation of long-term potentiation, *Brain Res.* **150**:413–417.

Fagni, L., Baudry, M. and Lynch, G., 1983a, Desensitization to glutamate does not affect synaptic transmission in rat hippocampal slices, *Brain Res.* **261**:167–171.

Fagni, L., Baudry, M. and Lynch, G., 1983b, Classification and properties of acidic amino acid receptors in hippocampus. I. Electrophysiological studies of an apparent desensitization and interactions with drugs which block transmission, *J. Neuroscience*, **3**:1538–1546.

Foster, A. C. and Roberts, P. J., 1978, High-affinity L-^{3}H-glutamate binding to postsynaptic receptor sites on rat cerebellar membranes, *J. Neurochem.* **31**:1467–1477.

Foster, A. C., Mena, E. E., Fagg, G. E., and Cotman, C. W., 1981, Glutamate and aspartate binding sites are enriched in synaptic junctions isolated from rat brain, *J. Neurosci.* **1**:620–626.

Greengard, P., 1978, Phosphorylated proteins as physiological effectors, *Science* **199**:146–152.

Kandel, E., 1981, Neuronal plasticity and the modification of behavior, in: *Handbook of Physiology*, Section I: The Nervous System (J. M. Brookhart, V. B. Mountcastle, E. R.

Kandel, and S. R. Geiger, eds.), American Physiological Society, Baltimore, pp. 1137–1182.

Kishimoto, A., Kajikawa, N., Tabuchi, H., Shiota, M., and Nishizuka, Y., 1981, Calcium-dependent neutral proteases, widespread occurrence of a species of protease active at lower concentrations of calcium, *J. Biochem.* **90:**889–892.

Lee, K., Schottler, F., Oliver, M., and Lynch, G., 1980, Brief bursts of high-frequency stimulation produce two types of structural change in rat hippocampus, *J. Neurophysiol.* **44:**247–258.

Lee, K., Oliver, M., Schottler, F., and Lynch, G., 1981, Electron microscopic studies of brain slices: The effects of high frequency stimulation on dendritic ultrastructure, in: *Electrical Activity in Isolated Mammalian CNS Preparations* (G. Kerkut, ed.), Academic Press, New York, pp. 189–212.

Leiter, A. B., Weinberg, M., Isohashi, F., Utter, M. F., and Linn, T., 1978, Relationship between phosphorylation and activity of pyruvate dehydrogenase in rat liver mitochondria and the absence of such a relationship for pyruvate carboxylase, *J. Biol. Chem.* **253:**2716–2723.

Levine, J. and Willard, M., 1981, Fodrin: Axonally transported polypeptides associated with the internal periphery of many cells, *J. Cell. Biol.* **90:**631–643.

Lynch, G. and Baudry, M., Origins and Manifestations of Neuronal Plasticity in the hippocampus, in: *Clinical Neurosciences* (W. Willis, ed.), Churchill-Livingstone Publishers, New York, in press.

Lynch, G., Halpain, S., and Baudry, M., 1982, Effects of high-frequency synaptic stimulation in glutamate receptor binding studied with a modified *in vitro* hippocampal slice preparation, *Brain Res.* **244:**101–111.

Magilen, G., Gordon, A., Au, A., and Diamond, I., 1981, Identification of a mitochondrial phosphoprotein in brain synaptic membrane preparations, *J. Neurochem.* **36:**1861–1864.

Michaelis, E. U., Michaelis, M. L., and Boyarsky, L. L., 1974, High-affinity glutamic acid binding to brain synaptic membranes, *Biochem. Biophys. Acta* **367:**338–348.

Morgan, D. G. and Routtenberg, A., 1980, Evidence that a 41,000 dalton brain phosphoprotein is pyruvate dehydrogenase, *Biochem. Biophys. Res. Comm.* **95:**569–576.

Murachi, T., Hatanaka, M., Yasumoto, Y., Hakayata, N., and Tanaka, K., 1981a, A quantitative distribution study on calpain and calpastatin in rat tissues and cells, *Biochem. Int.* **2:**651–656.

Murachi, T., Tanaka, K., Hatanaka, M., and Murakami, T., 1981b, Intracellular Ca^{2+}-dependent protease (calpain) and its high-molecular weight endogenous inhibitor (calpastatin), *Adv. Enzyme Regul.* **19:**407–424.

Roberts, P. J., 1974, Glutamate receptors in rat central nervous system, *Nature* (*London*) **252:**399–401.

Siman, R., Baudry, M. and Lynch, G., Purification from synaptosomal plasma membranes of calpain I, a thiol-protease activated by micromolar calcium concentrations, *J. Neurochem.*, in press.

Snyder, S. H. and Bennett, J. P., 1976, Neurotransmitter receptors in the brain: Biochemical identification, *Annu. Rev. Physiol.* **38:**153–175.

Storm-Mathisen, J., 1977, Localization of transmitter candidates in the brain: The hippocampal formation as a model, *Prog. Neurobiol.* **8:**119–181.

Toyo-oka, T., Shimizu, T., and Masaki, T., 1978, Inhibition of proteolytic activity of calcium-activated neutral protease by leupeptin and antipain, *Biochem. Biophys. Res. Comm.* **82:**484–491.

van Harreveld, A. and Fifkova, E., 1975, Swelling of dendritic spines in the fascia dentata after stimulation of the perforant path fibers as a mechanism of post-tetanic potentiation, *Exp. Neurol.* **49:**736–749.

Vargas, F., Greenbaum, L., and Costa, E., 1980, Participation of cysteine proteinase in the high-affinity Ca^{++}-dependent binding of glutamate to hippocampal synaptic membranes, *Neuropharmacology* **19**:791–794.

Wieraszko, A. and Lynch, G., 1979, Stimulation-dependent release of possible transmitter substances from hippocampal slices studied with localized perfusion, *Brain Res.* **160**:372–376.

Yamamoto, C. and Chujo, T., 1978, Long-term potentiation in thin hippocampal sections studied by intracellular and extracellular recordings, *Exp. Neurol.* **58**:242–250.

8

Optical Monitoring of Electrical Activity

Detection of Spatiotemporal Patterns of Activity in Hippocampal Slices by Voltage-Sensitive Probes

A. GRINVALD and M. SEGAL

1. INTRODUCTION

1.1. *Preview*

This chapter describes a novel approach to investigate the spatiotemporal distribution of electrical activity in nervous systems. Using voltage-sensitive dyes and an electro-optical measuring system, it has recently become possible to monitor electrical activity simultaneously from multiple sites on the processes of single nerve cells, either in culture or in an intact central nervous system (CNS) *in vitro*, to detect the activity of many individual neurons controlling a behavioral response in invertebrate ganglia, or to follow the activity of populations of neurons at many neighboring loci in mammalian brain slices or in the intact brain. Employing optical recordings and a display processor, the images of nerve cells light up on a TV monitor when they are electrically active. Thus, the spread of electrical activity can literally be visualized in slow

A. GRINVALD and M. SEGAL • The Weizmann Institute of Science, Rehovot 76100, Israel.

motion. This chapter describes recent progress in the implementation of this new technique.

1.2. *The Limitations of Current Intracellular Electrical Recording Techniques*

Classical electrophysiological techniques have been very effective in studies of electrical properties of neuronal cell bodies and of large axons or dendrites as well as in studies of the mechanisms of synaptic transmission. Unfortunately, these techniques suffer from two main drawbacks: (1) recording from small neuronal objects is difficult and (2) simultaneous intracellular recording from many cells is impossible. Thus, determination of the electrical properties of fine dendrites and axons is very difficult, mostly because simultaneous recordings from many sites on the arborization of single cells is at present virtually impossible. At the other extreme of complexity, the study of the cellular basis of the function of a given network, the necessary simultaneous recording from many individual cells in that network is very tedious. Even mapping the synaptic connections in a simple invertebrate ganglion containing only 300 neurons would require more than 60,000 pairs of microelectrode impalements—practically an impossibility.

A powerful alternative approach may be offered by optical monitoring of neuronal activity. In this direction, the efforts of Cohen and his collaborators (Cohen *et al.*, 1974; Ross *et al.*, 1977; Gupta *et al.*, 1981), together with the chemist Waggoner (Waggoner, 1979), and Tasaki and his collaborators (Tasaki *et al.*, 1968), have laid the foundation for this promising technique (For review see Cohen and Salzberg, 1978; Cohen *et al.*, 1978; Grinvald *et al.*, 1980).

1.3. *The Principle of the Optical Recording Technique*

In principle, the optical technique is simple. Voltage-sensitive probe molecules bind to the excitable membrane and act as molecular transducers transforming changes in membrane potential into optical signals. The dye molecules are added to the physiological solution to vitally stain the preparation. The preparation is then illuminated, and small changes in color or in emitted fluorescence of the stained cells are monitored with light measuring devices positioned at the point where a real magnified image is formed by a microscope objective. The striking similarity between optical and intracellular electrical signals is illustrated in Figure 1. Evidently, if optical signals can be detected by a single photodetector, then by using an array of photodetectors positioned in the microscope

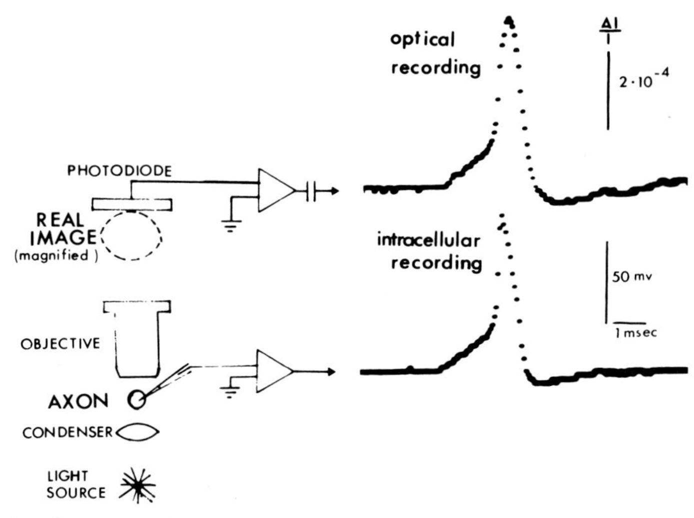

Figure 1. Comparison between optical and intracellular recordings from squid giant axon. Schematic diagram showing the experimental arrangement for both methods. A *single sweep* optical recording (top) of an action potential and an electrical recording (bottom) are shown. Note: The time course of the optical recording is nearly identical to that of the electrode recording. Even though the optical signal is exceedingly small ($\Delta I/I = 4 \times 10^{-4}$), it could be recorded with an excellent signal-to-noise ratio. Figure modified from Ross *et al.*, 1977.

image plane, the activity of many individual targets, (e.g., neuronal processes, individual neurons, or brain regions) can be detected simultaneously.

2. OPTICAL MONITORING OF CHANGES IN MEMBRANE POTENTIAL

2.1. *The Apparatus*

The optical monitor was described in detail elsewhere (Grinvald *et al.*, 1981a,b; Grinvald *et al.*, 1982b). The apparatus is depicted in Figure 2. A Zeiss Universal microscope was rigidly mounted on a vibration–isolation table. A 12 V/100 W tungsten/halogen lamp was the light source. Slices were viewed with a long working-distance 40× water immersion objective. Transmitted light was detected by a 10 × 10 square array of photodiodes, each 1.4 × 1.4 mm. Each photodiode received light from a 45 × 45 μm area of the microscope objective field, and was coupled to a current-to-voltage converter and amplifier. The amplifiers

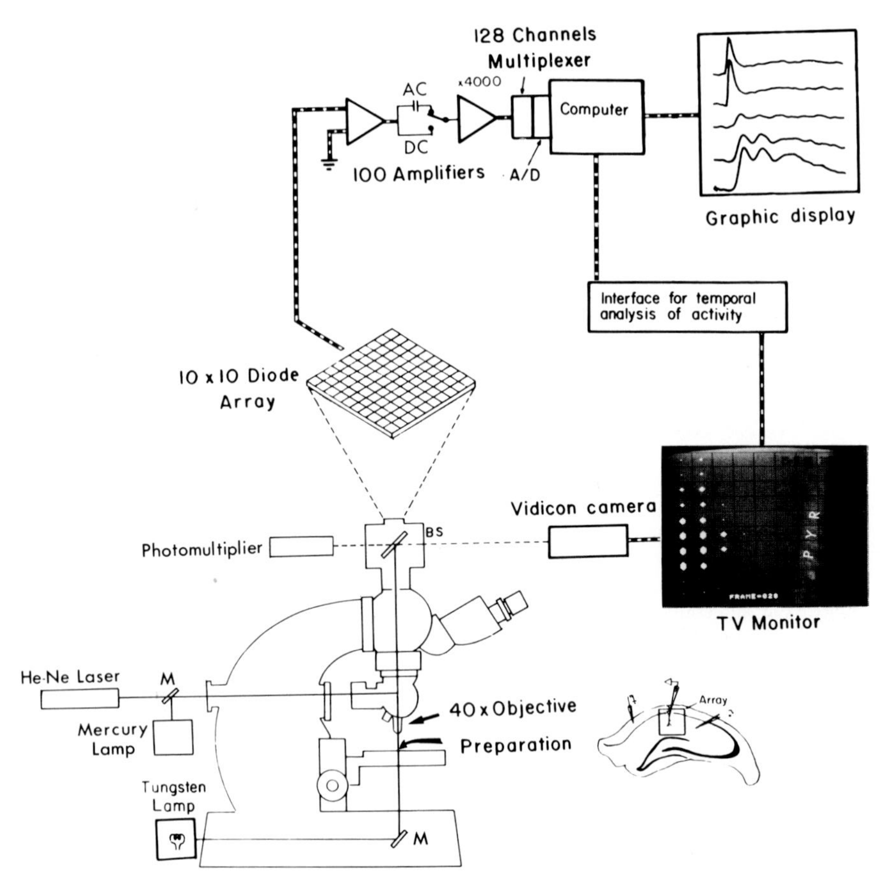

Figure 2. The computer-controlled optical apparatus used to record electrical activity from multiple locations. During a transmission experiment, the preparation is illuminated by a tungsten lamp and optically detected electrical activity is monitored with a 10 × 10 photodiode array. In fluorescence experiments, a He–Ne laser or a 100-W mercury lamp is employed as a light source, and a photomultiplier is used. The output of the detectors is displayed on the graphic display terminal or on a TV monitor. The scheme on the TV screen shows a slice preparation. The bright hexagon illustrates electrical activity 0.8 msec after the stimulation. Figure modified from Grinvald *et al.*, 1982.

filtered the optical data with resistance-capacity (RC) filters having a time constant of 0.7 msec. The signals were AC-coupled with a switch-selected time constant of 100 msec or 900 msec and amplified again (× 1000 to 4000). In most of the experiments, the AC time constant was 100 msec. With AC coupling of 100 msec, fast signals (<5 msec) are not distorted while slower signals (>10 msec) are. The output of the amplifiers was multiplexed and digitized by two multiplexer 8-bit A/D converter cards, and deposited in the PDP 11/34 computer memory. The recording from each channel was displayed on the screen of a graphic display terminal. The object field was viewed with a Vidicon camera; the video image was stored on videotape for further localization of the photodiode array with respect to the slice.

In order to combine electrophysiological recordings with the optical experiment, up to four hydraulic Narashige micromanipulators were mounted on the moveable stage of the Universal microscope. Recently, we have also used an inverted microscope for the slice experiments. The inverted microscope was mounted on an X–Y positioner and its stage replaced by a large flat stage rigidly attached to the vibration isolation table top. This arrangement offers two advantages: (1) Heavy manipulators can be mounted on the fixed stage. The insertion of microelectrodes into the slice was done with the aid of a swing-in stereomicroscope that replaced the swing-out condenser of the inverted microscope. With a 0.63 NA condenser, the microelectrode angle would be as large as 45° (instead of the 27° permitted by the water immersion objective used in the Universal microscope). (2) It is easy to change the objectives under the preparation to attain the optimal magnification.

Some years ago, Sherrington suggested metaphorically that if nerve cells emitted tiny flashes of light when activated, it would be possible to understand the workings of the brain (Sherrington, 1953). To make this metaphor real, a computer-to-video display processor was designed in order to visualize the spread of electrical activity. It projects the calibrated outlines of the array on the TV monitor, superimposed on the picture of the preparation. In addition, it displays a hexagon in each of the 100 elements of the picture. The size of each hexagon is proportional to the amplitude of the activity at a given time. The interface displays the simultaneous activity from all the detectors in slow motion.

2.2. *Design and Synthesis of Improved Optical Probes*

The key to the success of optical monitoring of neuronal activity has been the design of adequate voltage-sensitive molecular probes. Of more than 1400 dyes already tested (Cohen *et al.*, 1974; Ross *et al.*, 1977; Gupta *et al.*, 1981; Grinvald *et al.*, 1982a), less than two hundred have proven to be sensitive indicators of membrane potential while causing minimal pharmacological side effects or light-induced photochemical damage to the neurons (Ross *et al.*, 1977). A significant difficulty of the optical probes was first reported by Ross and Reichardt (1979), who observed that a dye's sensitivity can vary from one preparation to the next and even among different species of the same genus. Thus, for a new preparation, careful selection of the best probe may be still required. There has recently been considerable improvement in the quality of fluorescent probes. The best fluorescence dye that we have designed is designated RH 421. For neurons in tissue culture (neuroblastoma cells), the change in fluorescence with this dye is 25%/100 mV of membrane potential change (Grinvald *et al.*, 1983b). This value is 120 times greater than the sensitivity obtained with leech neurons in the pioneering experiments of Salzberg *et al.* (1973); the change can probably be directly monitored by the eye. In the design of transmission-voltage-sensitive probes, the progress has been slower, presumably because the sensitivity of the best probes is already close to the theoretical maximum (Waggoner and Grinvald, 1977). However, very large differences in sensitivity among various preparations were observed, and the present sensitivity for brain slices can probably be improved considerably, by evaluating the properties of more probes.

2.3. *Example of Application to the Study of Single Cells and the Study of Invertebrate CNS*

Examples of recent applications of optical methods to monitor neuronal activity on three different levels are now presented. The first example consists of studies of the processes of individual cells, either in tissue culture or in the intact invertebrate CNS. The investigation of electrical responses of individual neurons in "simple" central nervous systems will then be discussed, followed in the next section by studies of the electrical behavior of populations of cells in mammalian brain slices. The examples taken from tissue culture and invertebrate studies are presented here because similar types of experiments are probably feasible also in slices.

2.3.1. Fluorescence Recordings from Neuronal Processes. Nerve cells maintained in culture offer a simplified preparation that is more accessible to a variety of research tools. However, studies of the electrical properties of processes or growth cones are difficult, and can benefit from the use of the optical method. Theoretical considerations suggest that recording the changes in fluorescence, rather than transmission, should yield the largest optical signals from processes of single nerve cells (Waggoner and Grinvald, 1977). This prediction is especially true for the study of isolated cells maintained in a monolayer culture, since, under these conditions, the background fluorescence is minimal and only the target of interest is stained with the fluorescent dye.

For the experiments, the tissue culture medium is replaced with appropriate balanced salt solution containing 1 to 10 $\times$ 10^{-6} M of the fluorescent probe. After 1 to 5 min, the cells are sufficiently stained for the fluorescence experiments.

The part of the cell in which the measurement is to be made is positioned at the center of the field of view, directly under the laser or mercury monitoring microbeam. An electronic shutter affords illumination of the preparation for the period during which the cell is stimulated, and the computer records changes in fluorescence intensity. Figure 3 illustrates typical results of such an experiment using neuroblastoma cells.

When the fluorescence recording is made from the same site as the electrical recordings, the two signals have strikingly identical time courses (Figure 3A). Thus, measurements of the time course of the light intensity change are reliable, and can be used also to monitor electrical responses from neuronal parts that are *not* easily accessible to microelectrode recording, such as neurites (Figure 3B) or growth cones (Figure 3C).

Figure 3C illustrates the simultaneous recording of Ca^{2+} action potentials obtained from the cell body (by means of an electrode) and from the growth cone (using fluorescence). Note that even though the Ca^{2+} action potential was evoked at the cell body (by passing current through the electrode), the growth cone Ca^{2+} action potential precedes the cell body action potential. (Compare the rising phases of the corresponding action potentials.) These results imply that active Ca^{2+} channels, having a low threshold, are present at or near the growth cone (Grinvald and Farber, 1981). With recent improvements of the fluorescence apparatus, action potentials in 1 μm process can be detected in a single trial with a signal-to-noise ratio of 30 (Grinvald *et al.*, 1983b).

The optimal procedure for optical recording of electrical activity and synaptic responses from the processes of single nerve cells in intact CNS preparations is different from that in isolated cells in culture. Fluorescent voltage-sensitive dyes should be injected, to stain only the cell under investigation. Preliminary experiments in this direction indicate that such experiments are useful for recording electrical activity and synaptic responses directly from the sites of the synapses in the neuropile (Agmon *et al.*, 1982; Grinvald *et al.*, 1982c).

2.3.2. *Simultaneous Recordings from Multiple Sites on the Arborization of Single Cells.* Using a 10 $\times$ 10 array of photodetectors positioned in the microscope image plane, the activity of many individual targets can be detected simultaneously. For technical reasons, such recordings had been obtained formerly only with transmission measurements rather

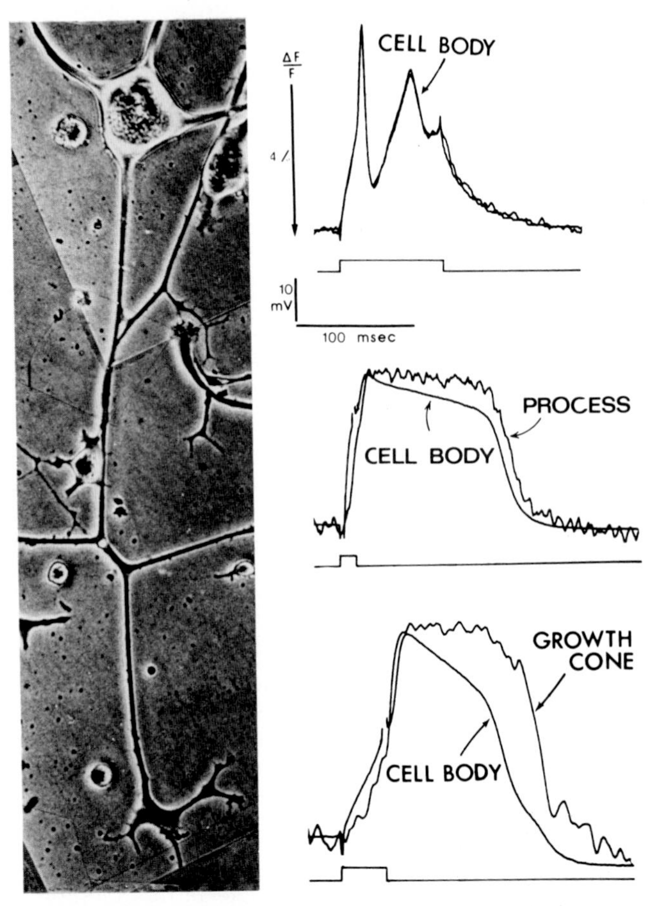

Figure 3. Electrical and fluorescence recordings from the cell body, a process, and a growth cone. Cells were impaled with microelectrodes to stimulate the cell and to record the voltage response. To record from a given point, the cell could be moved with respect to the fixed microbeam by means of the microscope stage, while maintaining stable electrode penetration. (A) Comparison between electrical recording (thin trace) and fluorescence recording (noisy trace) from the soma. A single trial was recorded. (B) Comparison between the time course of Ca^{2+} action potential in the soma and in a process. The optical recording was made from a segment of a 4 μm-wide process. Four sweeps were averaged. (C) Comparison of the time course of Ca^{2+} action potential in the cell body and in the growth cone, recorded from a third cell. (In both (B) and (C), the measurement was made in normal medium to which 10^{-6} M tetrodotoxin (TTX) and 15 mM tetraethylammonium (TEA) were added to block the sodium and potassium channels respectively). A single trial was recorded; process length: 600 μm; growth cone diameter: 50 to 60 μm; neurite diameter: 5 μm. The photograph shows a typical growth cone of an N1E-115 neuroblastoma cell. (Figure modified from Grinvald and Farber, 1981; Grinvald *et al.*, 1982a.)

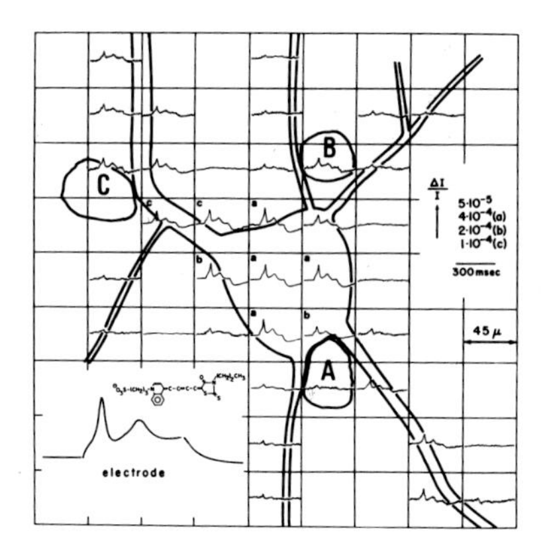

Figure 4. Simultaneous transmission changes from the arborization of a single cell recorded by a 10 × 10 photodetector array. The array is positioned over the real magnified image of a stained N1E-115 neuroblastoma cell. The image of the cell has been superimposed over the array outline. The soma was stimulated with a microelectrode and 50 sweeps were averaged. Traces in the individual boxes show the optically detected activity. Some records that are not from the cell or its processes are presented as a control at a high gain to show the noise level. Figure modified from Grinvald *et al.*, 1981b.

than with fluorescence. With the recently improved apparatus, the light intensity was sufficiently increased so that the photodiode array can be used also for fluorescence experiments (Grinvald *et al.*, 1983c). Figure 4 shows the projected image of a large neuroblastoma cell superimposed upon the array. The cell body was stimulated with a microelectrode and the optically-detected electrical responses from multiple sites are displayed at their appropriate locations.

The stimulation of the cell evoked a fast all-or-none response followed by a second smaller deflection known to be due largely to the entry of Ca^{2+} ions (see the electrode trace). Inspection of the optical recordings indicates that the second signal is proportionately smaller in the thin processes than in the cell body and thick processes. This observation gave us the first hint that functional Ca^{2+} channels are less abundant along the processes of neuroblastoma cells than at the cell body (Grinvald *et al.*, 1981b). Similar types of experiments in intact ganglia were recently reported by Krauthamer and Ross (1981) who studied the properties of the arborization of single cells in the supraesophageal ganglion of the giant barnacle and such experiments are also directly applicable to slices under conditions when single cells are stimulated.

2.3.3. *Localization of Neurons Controlling a Behavioral Response in the Giant Barnacle CNS.* Another application of the photodiode array is to characterize patterns of electrical activity in neuronal networks (Grinvald *et al.*, 1981a). When simple invertebrate central nervous systems are studied, the activity of the neurons that control behavioral responses of the animal can be detected optically. This application was pioneered by Cohen and his colleagues at Yale University (Salzberg *et al.*, 1977; Grinvald *et al.*, 1981a).

The giant barnacle, *Balanus nubilus*, has a stereotypic defensive withdrawal behavior known as the shadow response. The barnacle responds to passing shadows by terminating feeding behavior, rapidly withdrawing its cirri, and closing its opercular plates. This "off-response" is controlled in part by the median ocellus (a primitive eye with four photoreceptors) and the supraesophageal ganglion, which receives the synaptic input from the photoreceptors. Experiments were carried out to locate large motor neurons in the ganglion that spike in response to reduced illumination of the ocellus (i.e., an artificial shadow) and command the appropriate muscles to contract. To avoid interference between the light used for monitoring activity and the light used to stimulate the photoreceptors, the ganglion and ocellus were pinned out in separate pools of a chamber made of black Lucite (Figure 5A). For the experiment illustrated in Figure 5B, the ocellar illumination (bottom trace) was turned off just after the start of the measurement. The off-response was monitored in an isolated ocellus-ganglion preparation by measuring the effect of reduced ocellar illumination, via suction electrode recordings from both a connective to the subesophageal ganglion and a nerve to the periphery. Analysis of the optical recordings from one hemiganglion showed five neurons that responded to turning off the light (top traces, Figure 5B). Also shown are the suction electrode recordings from the connective and antennular nerve. In this experiment, the off-response recorded on these suction electrodes after staining was similar in number and pattern of spikes to that recorded before staining. Thus, there were no irreversible pharmacologic effects of the dye even on a sensitive behavioral response known to be mediated by a polysynaptic pathway.

This experiment demonstrates the potential of the optical recording technique to localize spiking neurons. The same type of experiment can also be used to search for "all" of the postsynaptic cells to a given neuron: small (>1 mV) synaptic responses evoked by the stimulation of that neuron can be detected with extensive averaging of many trials to increase the sensitivity (Grinvald *et al.*, 1981a).

3. OPTICAL RECORDING FROM BRAIN SLICES

The organized structure of the hippocampus is particularly amenable to an analysis of the physiological properties of single neurons and of the synaptic interactions among them. The advent of the slice preparation (Yamamoto, 1972) facilitated the study of neural interactions since it allows prolonged intracellular recording and control of the cel-

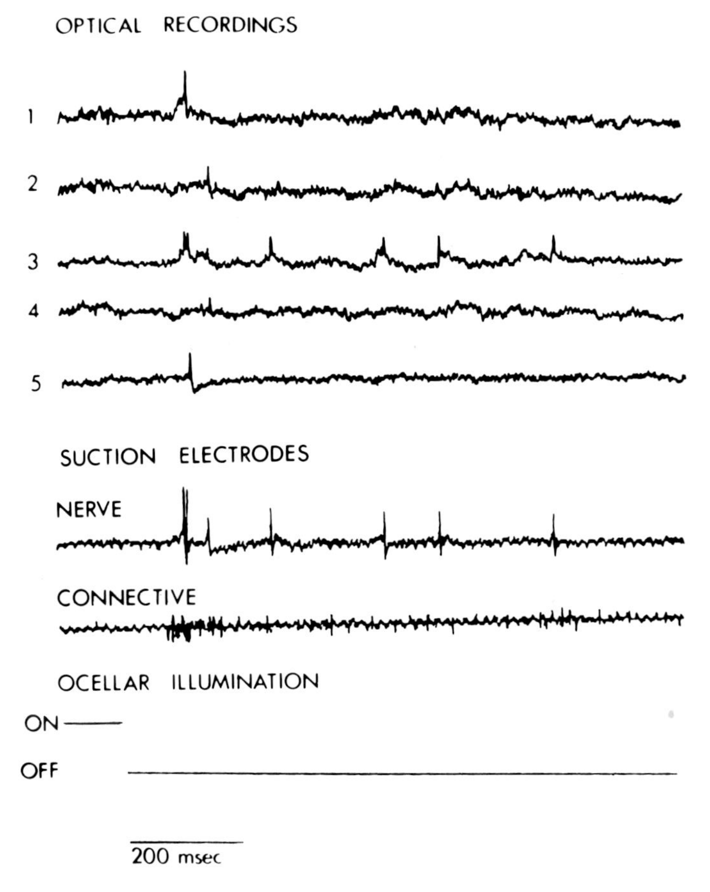

ocellus was illuminated for 20 sec prior to the beginning of the sweep. Suction electrode recordings from the ipsilateral connective and antennular nerve are shown below the optical traces. The ocellus was illuminated through a light guide embedded in the lid of the ocellar pool. Figure modified from Grinvald *et al.*, 1981a.

The stimulus duration was 0.3 to 0.5 msec in the rat slices experiments and 0.03 to 0.1 msec in the guinea pig experiments. The current was adjusted to yield a 1 to 15 mV extracellular population spike as measured in the s. pyramidale. To facilitate comparison with the optical recordings, which were filtered, the extracellular recording was also filtered with an RC filter having a time constant of 0.7 msec. (The filtration and the low time resolution, used for these computerized recordings from 100 channels, resulted in extracellular recordings that are somewhat distorted relative to their traditional presentation.) Intracellular recordings were also attempted, but stable recordings were not obtained, probably because of the low angle of the electrode under the water immersion objective (27°), and the type of manipulator used.

The Krebs solution contained 124 mM of NaCl, 5 mM of KCl, 1.25 mM of KH_2 PO_4, 26 mM of $NaHCO_3$, 2mM of $MgSO_4$, 2 mM of $CaCl_2$, and 10 mM of *d*-glucose. The solution was oxygenated with a gas mixture of 95% O_2, and 5% CO_2; the pH was 7.4. In some experiments a "low Ca^{2+}" medium was used, containing 0.2 mM Ca^{2+} and 4.0 mM Mg^{2+}. In a "low Cl^-" medium, the NaCl was replaced by Na^+-propionate. The rat experiments were performed at room temperature (24 to 28°C). The guinea pig experiments were carried out at a temperature of 35°C.

In most experiments, the stimulating and recording electrodes were positioned in the microscope field of view. The dye, WW401 (Grinvald *et al.*, 1981b; Gupta *et al.*, 1981) [5[1-γ-sodium sulfopropyl-4(1H)quinolylidene]-3-propylrhodanine], a merocyanine–rhodanine derivative, was then applied at a concentration of 0.2 mM in normal medium, for 20 min (the perfusion was stopped). The staining solution was washed away for additional 20 min during which the size of the evoked potentials, which were often reduced during the dye-staining, was restored. Using this dye, voltage-sensitive optical signals can be obtained at a wavelength range of 510 to 780 nm. The largest optical signals were obtained at a wavelength of 690 nm and a band width of 30 nm (Grinvald *et al.*, 1981b). In more recent experiments, we have found a new dye designated RH-155 (a dipyrozolone oxonol dye, designed in our laboratory) that gave 3 to 5 times larger signals (excitation wavelength 720). The dyes RH-27 and WW 433, both analogs of WW 401, were also more sensitive than WW 401 (J. Kuhnt and A. Grinvald, in preparation). To minimize bleaching of the dye molecules and photodynamic damage to the tissue (Ross *et al.*, 1977), a shutter in the light path was opened only for the duration needed to measure the optical responses (10 to 30 sec for averaging 10 to 30 trials using a 1-Hz stimulus).

3.2. *The Correction Procedure for Light-Scattering Signals*

Light-scattering optical signals were detected in unstained slices in response to electrical stimulation (see Section 3.3). Their time course was independent of the wavelength of the light passing through the tissue, whereas the maximal voltage-sensitive dye signals were recorded with a 690-nm interference filter (with WW 401). Virtually no dye-related response was seen when a 810-nm filter was used. Thus, in most of the recordings presented below, light-scattering responses measured with a 810-nm filter (multiplied by a correction factor) were subtracted from responses obtained with a 690-nm filter to yield a net dye-related signal. The proper correction factor for the light-scattering signals was calculated from the measurements of the transmitted light intensities at 690 and 810 nm. (The scattering signal measured at 690 nm was 1.3 times larger than the one observed at 810 nm, for equal excitation intensities.) In the more recent guinea pig experiments, the interference from the light-scattering signals was almost negligible because the dye-related signals were 3 to 5 times larger than those in the rat experiments.

3.3. *Optical Signals from a Stained Slice*

Stimulation of the s. radiatum produced a short latency (2 to 4 msec) fast signal (3.5 to 6 msec) in a strip 90 to 180 μm wide, corresponding to the Schaffer collateral-commissural system (two top traces of Figure 6). The signals could be traced in adjacent detectors for a long distance, but only along the collateral system (not shown). These fast signals probably reflect action potentials in the Schaffer collateral axonal fibers.

A second wave of excitation was detected in the same region with a latency of 4 to 15 msec, but this wave "traveled" along the dendritic tree towards the s. pyramidale and s. oriens. These slow signals were assumed to represent the dendritic excitatory postsynaptic potentials (EPSPs). [The possibilities that the dendrite can fire action potentials (MacVicar and Dudek, 1981; Spencer and Kandel, 1961) or that there exist other contributions to the slow signals, are discussed elsewhere (Grinvald *et al.*, 1982b).]

The dendritic depolarization triggered multiple optical spikes at the pyramidal cell bodies. These optical signals probably represent the action potential discharges there. The optical signals propagated into the s. oriens with an average conduction velocity of 0.1 m/sec. Their passive spread back into the apical dendrites was noticeable only over a short distance, under normal conditions (Figure 6). The optical signals detected at s. oriens probably reflect the activity of the basal dendrites and axons of the pyramidal cells, as well as the activity of interneurons there.

There was an excellent correlation between the electrically-recorded field potential and the optical signals from s. pyramidale; whenever there was a large population spike in the electrical recording, there were also large optical signals. Furthermore, a single stimulus would occasionally generate secondary and tertiary or even quaternary population spikes on both the electrical and optical recordings from the s. pyramidale. The peaks of the field potentials preceded the peaks of the

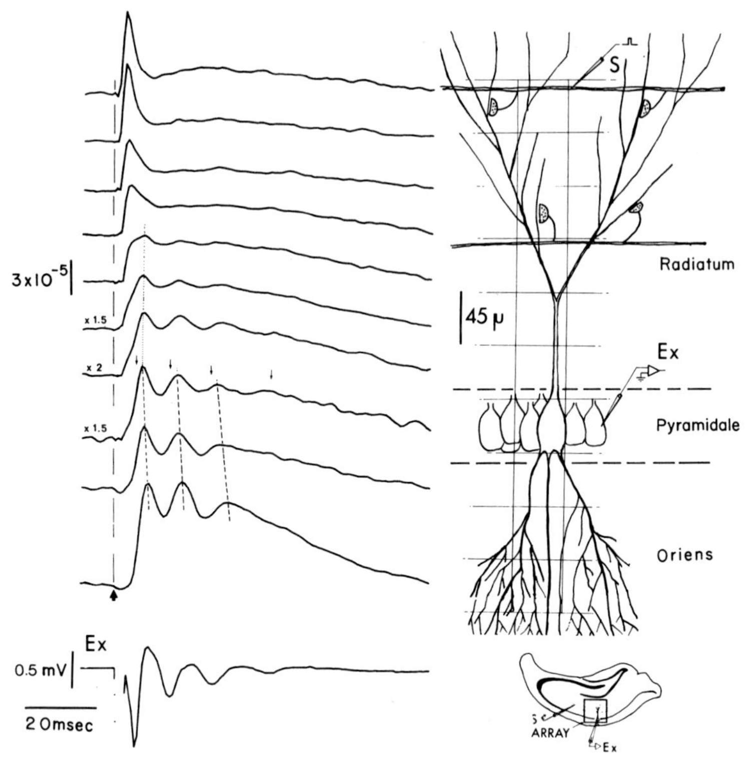

Figure 6. Optical recordings from 10 loci along the long axis of CA1 pyramidal cells. The scheme shows the various areas along the cell axis, the position of the stimulating (S) and recording (Ex) electrodes and the relative position of the 10 individual photodetectors (ten squares). The scale for the amplitude of the optical signals, in this and the following figures, shows the fractional change in transmission. In three traces, the amplitude was multiplied arbitrarily (see Discussion). The thin arrows show the time of the peak of the electrically recorded field potential presented at the bottom. The dotted line shows the latency of the optically detected potentials at s. radiatum. The stimulating electrode was placed 100 μm away from the recording area. The time of the stimulus is marked by a thick arrow. A scheme of the slice in the bottom right corner shows the optically monitored area (square) and the approximate position of stimulating and recording electrodes. Ten trials were averaged.

optical spike by 0.8 to 1.6 msec. Such small time differences were frequently observed, and are discussed in the following sections. Both the electrical and optical responses were markedly reduced by reversing the polarity of the stimulus. Finally, various drugs (see the following sections) affected the two types of recording in a similar manner.

The results obtained in a series of experiments designed to clarify the origin of the optical signals are given in the following sections.

3.4. *Light-Scattering Signals from Unstained Slices*

The light-scattering response to the stimulation had characteristics that were clearly different from the dye responses. The time course of the light-scattering signal was independent of wavelength. It was present also at 690 nm, and presumably added to the dye response measured at that wavelength. The light-scattering signals produced by the stimulation had a rather slow time course, outlasting any known changes in neuronal membrane potential. The time constant for the decay of the light-scattering signal was also measured using the slower AC coupling of 900 msec, rather than the fast AC coupling (100 msec) (see Section 2.1). A value of 250 msec was obtained for the decay of the light-scattering signals. Evidently, the slow light-scattering signals are not linearly related to membrane potential changes, although their spatial and temporal characteristics were correlated with electrical activity. The signals were generated first in s. radiatum, had a longer latency in the s. pyramidale, and the longest in s. oriens. Interestingly, the largest magnitude change was found in the s. pyramidale (Figure 7A2). Both the magnitude and latency-to-onset of the light-scattering response were dependent on the stimulation intensity. With low stimulation intensity, the light-scattering responses were small and had a slow rise-time, whereas high stimulation intensity resulted in light-scattering responses with shorter latencies, faster rise-times, and larger amplitudes (compare Figure 7A1 with Figure 7A2). The onset of the light-scattering responses in the s. pyramidale had the same latency as that of the population spike (Figure 7A2).

When twin pulse stimulation was applied, the light-scattering response to the second stimulus was often augmented in comparison with that of the first stimulus. This was more noticeable in the s. radiatum far away from the stimulation site than next to the electrode, and only at moderate stimulus strength (Figure 7B). Reversal of the stimulus polarity led to a marked decrease in the size of the light-scattering signals as well as in that of the electrically-recorded population spike. Application of tetrodotoxin (TTX) abolished the light-scattering signals (see Section 3.6).

In summary, the light-scattering signals were triggered by the physiological activity, and were not merely an artifact of stimulation. The signals were similar to those recorded previously in squid giant axons

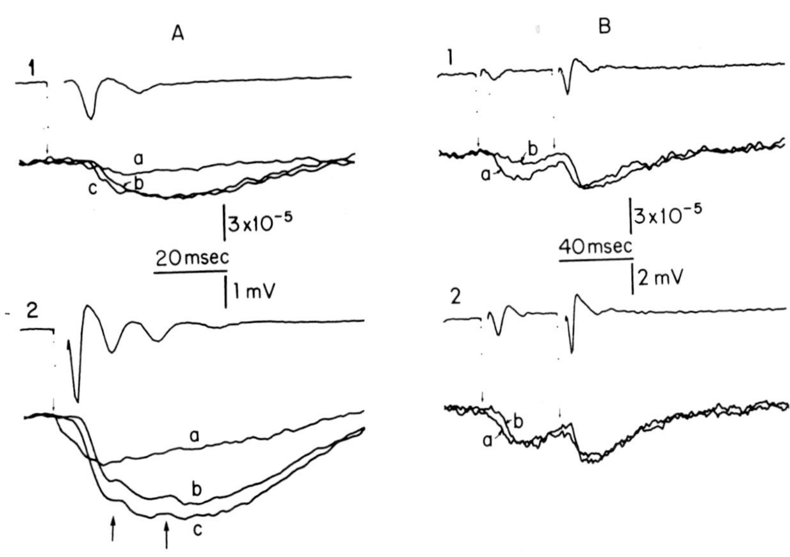

Figure 7. Light-scattering signals in the slice. Stimulation was applied at s. radiatum and responses from an unstained slice were recorded. (A) The effect of stimulus strength on light-scattering signals. (A1) Weak stimulus, top trace, shows the extracellular recordings from the pyramidal layer. Bottom traces are optical recordings from: (a) the s. radiatum; (b) the s. oriens; and (c) the s. pyramidale. (A2) Stronger stimulus, (a–c) same positions as (A1). The signals are larger than those in (A1) and the differences in the latencies of the light-scattering signals in the different areas are clearer. The thick arrows show the onset of second and third peaks of multiple discharges on the slow light-scattering signals. The thin arrows show time of stimulus onset. (B) Pulse-pair facilitation of the light-scattering signals. (B1) Responses to weak twin pulses; (a) is from the s. radiatum 50 μm away from the stimulating electrode. (b) is 300 μm away from the stimulating electrode; the electrical field potential recording is from the s. pyramidale. (B2) Light-scattering responses to stronger stimuli. Same as (B1), except that the facilitation is smaller. Twenty trials were averaged.

and crab nerves (Cohen *et al.*, 1968, 1972; Cohen and Keynes, 1971), and cerebral slices (Lipton, 1973).

3.5. Ca^{2+} *Dependency of the Fast and Slow Responses*

To test if the fast signals are action potentials, and if the long-latency slow signals indeed represent EPSPs generated in the apical dendrites of CA1 pyramidal neurons, a low Ca^{2+} (0.2 mM)–high Mg^{2+} (4.0 mM) solution was used to block postsynaptic activity. The field potential was monitored continuously and found to be completely suppressed after 20 min, at which time optical responses were measured (Figure 8). While

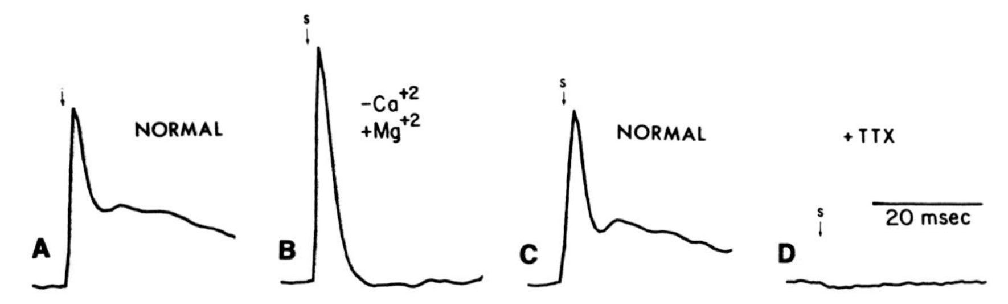

Figure 8. The effect of Ca^{2+} removal and tetrodotoxin (TTX) on the optically detected evoked activity. (A) Optical and electrical recordings in normal medium from the s. radiatum. (B) Optical recording in a medium containing 0.2 mM Ca^{2+} and 4.0 mM mg^{2+}. Note that the size of the fast signals was increased while the slow response disappeared. (C) Optical recording 20 min after the low Ca^{2+} medium was replaced by the normal medium. The stimulating electrode was less than 150 μm away from the optical recording site. The field potential was recorded from the s. pyramidale. (D) The effects of TTX on the optical responses. Recordings were made from the s. radiatum. Small arrows marked by s: time of stimulation.

the initial fast response was somewhat enhanced by the low Ca^{2+} medium, the long-latency slow response was completely blocked. The occurrence of fast signals was restricted to the region of the Schaffer collaterals (over an inspected distance larger than 900 μm). The increase in the spike size may reflect recruitment of additional presynaptic elements in the low Ca^{2+} medium, possibly due to increased excitability. These effects were reversible and the original response was restored by perfusion with the normal medium.

3.6. *Effects of Tetrodotoxin on the Fast Responses*

If indeed the fast signals represent voltage-activated Na^+ action potentials, they should be blocked by the Na^+ channel blocker, TTX. TTX (5×10^{-6} M) was applied to the slice and the disappearance of the field potential was monitored. The optical responses diminished simultaneously with the disappearance of the electrical responses (Figure 8). These experiments illustrate that the fast signals are indeed Na^+, and not locally evoked dendritic Ca^{2+}, action potentials.

3.7. *Optical "Artifacts" near the Stimulating Electrode*

A small local response could still be detected in the presence of TTX. It decayed rapidly in space and was not seen 150 μm away from the stimulating electrode. These small signals may partially represent local passive polarizations, the sign of which depends upon stimulus polarity. Such local and direct depolarizations are present near the stimulating electrode, and are probably responsible for the fact that, in a few cases, the latency for the fast spike was zero. The same type of local response was also observed using intracellular electrical recording if the stimulating and recording electrodes were close and the stimulus intensity employed was high. However, some local signals persisted also at a wavelength of 810 nm, where dye signals cannot be detected. Hence, these signals may be due to a mechanical effect of the stimulation on the slice (movement artifact) or to electrophoretic effects on bound dye molecules, which may unbind and change their color, giving rise to an optical signal. Thus, it seems preferable to record optically far away from the stimulating electrode. However, when a moderate stimulus is used and a correction for light scattering is made, optical recordings near the stimulating electrode are only slightly distorted.

3.8. *Properties of the Unmyelinated Axons in the Hippocampus Slice*

3.8.1. Effects of Tetraethylammonium and 4-Aminopyridine. Tetraethylammonium (TEA) blocks K^+ channels when applied intracellularly to hippocampal neurons (Schwartzkroin and Prince, 1980). TEA (2 mM) was applied in the incubation medium and its effect on the evoked optical responses was assessed (Figure 9A). A broadening of the fast responses was observed; there was no change in latency or magnitude of the evoked optical response, but it decayed more slowly than under control conditions. The control response could be restored after 20 to 30 min of washing with normal medium (Figure 9A). Incubation of the slices in a solution containing 10 mM of 4-aminopyridine (4-AP) also led to a pronounced broadening of the presynaptic signal as well as an augmentation of the postsynaptic response (U. Kuhnt and A. Grinvald, unpublished results).

3.8.2. Conduction Velocity. The conduction velocity along the Schaffer collateral system was estimated by measuring the delay between peaks of the fast responses measured at different locations along the collaterals. Due to the limited time resolution of the present apparatus (0.7 msec), an accurate estimate could be obtained only across fairly long distances. The microscope field of view was shifted, and an adjacent field was sampled. In one such case, three adjacent fields were sampled, yielding a total length of over 1 mm of the fibers (Figure 9B); a conduction velocity of 0.2 m/sec was determined. Similar conduction velocities were found in other slices over shorter distances. The fast optical response

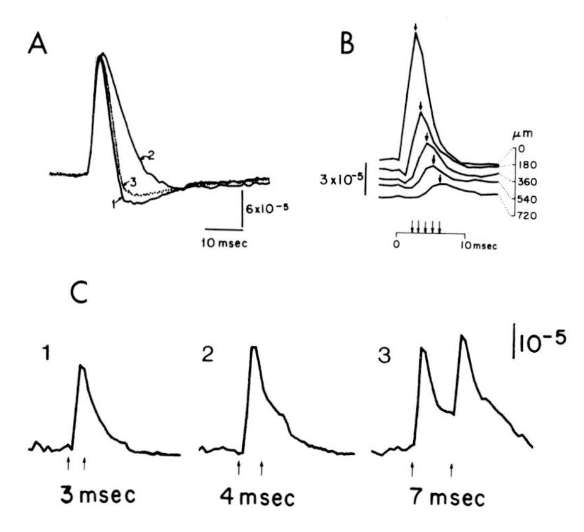

Figure 9. Properties of unmyelinated axons. (A) The effect of tetraethylammonium (TEA) on action potentials in the s. radiatum. The stimulating electrode was placed 150 μm away from the recording point. (A1) The average response in a medium containing low Ca^{2+}. (A2) Ten min after application of 2 mM TEA. (A3) Twenty min after the TEA solution was washed out. (B) Measurement of conduction velocity in the Schaffer colateral system. The Schaffer colaterals were stimulated 100 μm away from the first shortest latency recording point. The experiment was done in a low-Ca^{2+}, high-Mg^{2+} medium. The distance from one to the next monitored location is 180 μm. The calculated conduction velocity is 0.2 m/sec. (C) Measurements of the refractory period of action potentials in the Schaffer collaterals. The Schaffer collaterals were stimulated 120 μm away from the recording point. Recording was carried out in normal medium. Traces from locations with minimal postsynaptic responses were selected for this figure. (C1) Interpulse interval of 3 msec; (C2) 4 msec; (C3) 7 msec.

gradually broadened and diminished in amplitude the farther the recording site was from the stimulation site. These results are essentially identical to earlier observations made using extracellular recordings (Andersen *et al.*, 1978). This result can be attributed to a possible reduction in the number of intact fibers activated away from the stimulating electrodes (due to the slice dissection angle), and to nonuniformity in velocity. (Obviously, near the stimulating electrode, a small contribution of activity from other neuronal elements is added to the fibers signal. However, in a few experiments, the size of the presynaptic action potential was fairly constant over the 450-μm field of view.)

3.8.3. *Refractory Period.* The refractory period of the activated axons was estimated by applying two pulses at various interpulse intervals (Figure 9C). There was almost no response to the second stimulus when it was applied less than 4 msec after the first one, and a partial response when it was applied 4 msec after the first one; a full response was obtained with a 7-msec delay between the pulses. It appears that the refractory period is in the range of 3 to 4 msec. The lack of a second response in Figure 9C also indicates that the contribution of local direct polarization to the fast and short latency signals was minimal.

3.9. *Postsynaptic Responses*

3.9.1. The Excitatory Postsynaptic Responses. It was already suggested that the longer latency slow responses probably represent EPSPs, generated at the apical dendrites of CA1 pyramidal neurons, because such a response was absent in a low Ca^{2+} medium (see Figures 6 and 8). No delay could be observed between the peaks of the slow responses at s. radiatum and the peaks of the fast responses detected at the s. pyramidale. In fact, the gradual diminution of the "dip" between the two spikes (from the pyramidale toward apical dendrites) may suggest that the fast spikes were initiated at s. pyramidale and their passive deflection is superimposed on the slow EPSP. It is possible that presynaptic and postsynaptic activity of other interneurons also contributes to the slow signals detected at the s. radiatum. The EPSP observed at s. pyramidale was often much smaller than expected. This probably cannot be attributed solely to the attenuation caused by passive spread from the point of origin; the EPSP was probably also masked by hyperpolarizing synapses at this level.

3.9.2. Hyperpolarizing Inhibitory Potentials. Hyperpolarizing potentials could be detected when strong stimuli were applied. Under such conditions (Figure 10A), the EPSP was followed by a long-lasting hyperpolarization of short (1 to 3 msec) latency. This was pronounced in the s. pyramidale and also in the transition area between the s. radiatum and s. pyramidale. Inhibitory GABA synapses are indeed found on the somata of the pyramidal cells, and thus, these hyperpolarizing signals probably reflect IPSPs.

Antidromic stimulation of CA1 pyramidal cells via the alveus is known to activate a recurrent inhibitory pathway (Andersen *et al.*, 1964). Figure 10B demonstrates that upon antidromic stimulation, IPSPs were recorded at some loci, whereas in response to orthodromic stimulation only depolarizations were detected at these sites. Because of the nature of intracellular population recordings, it is clear that the lack of a net hyperpolarization does not imply the absence of hyperpolarizing inhibitory responses in the slice; they may be masked by a larger depolar-

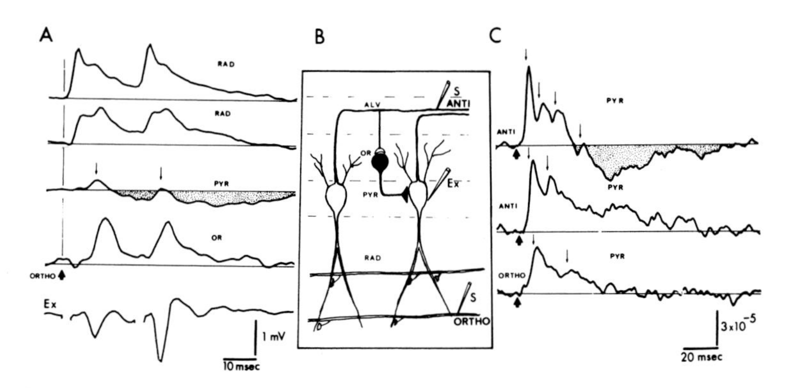

Figure 10. Inhibitory postsynaptic potentials in the s. pyramidale. (A) Activity was evoked by stimulation of the s. radiatum 100 μm away from the recording site. Four optical traces are shown (top to bottom): from the s. radiatum, 300 μm away from the layer of cell somata; from the s. radiatum, 150 μm away from the s. pyramidale; from the s. pyramidale; from the s. oriens. The extracellular recording electrode was placed 200 μm away from the site of the optical recording of pyramidale activity. (B) A simplified diagram of a recurrent inhibitory circuit in the CA1 area. The dark neuron is the inhibitory interneuron; it accepts excitatory input from axonal collaterals of CA1 pyramidal cells, and its inhibitory synapses are located at the s. pyramidale. ALV: alveus; OR: s. oriens; PYR: s. pyramidale; RAD: s. radiatum; S: stimulating electrode; Ortho: orthodromic stimulation at the s. radiatum; Anti: antidromic stimulation at the alveus. (C) Optical responses at the s. pyramidale to stimuli at different loci. Top trace: the response to a strong antidromic stimulus of the alveus. Thick arrows: time of stimulus. Thin arrows show the time of the s. pyramidale field potentials (500 μm away from the optical recording site). Middle trace: the response to a three-fold weaker stimulus, with inhibition no longer evident. Bottom trace: the response, at the same location, to orthodromic stimulation.

ization. Pharmacological manipulations are helpful in demonstrating the various types of synapses (see Section 3.11).

3.10. *The Cellular Discharges*

Large signals with fast rise-time were observed in both the s. pyramidale and the s. oriens. The optical signal at its rise-time appeared to coincide with the extracellular electrically-recorded population spike from the same area (e.g., Figure 6). These action-potential-like fast signals were not as evident in the s. radiatum. They arose in the s. pyramidale, and were yet more evident in the transition zone between s. pyramidale and s. oriens, where they had the fastest rise-time. They occurred only with high stimulus intensity, and were occasionally followed by secondary and tertiary spikes. These, too, were not as prominent in the s. radiatum. It appears that these fast signals represent action potentials generated in the axon hillocks, which spread into the s. oriens, s. pyramidale, and s. radiatum. Interneuronal activity probably also contributes to the optical signals detected at the s. oriens. [In the guinea pig experiment, the amplitude of the population activity at s. oriens was reduced relative to that of the rat. We did not yet clarify whether this observation reflects a difference in the viability of the slices (e.g., temperature effects) or some differences in functional organization.]

3.11. *Examples of Pharmacological Studies: The Effect of Picrotoxin*

Picrotoxin, a GABA antagonist, was applied in the superfusion medium at a concentration of 10 μM. In the presence of picrotoxin, the stimulation of the s. radiatum produced a greater depolarization than under control conditions. In addition, the stimulation produced an oscillatory response consisting of 3 to 4 successive peaks (Figure 11). This oscillation was most pronounced in the s. oriens and s. pyramidale, but was also seen in the s. radiatum. The timing of the optical signals at the s. pyramidale was similar to that of the field potentials. If indeed the oscillatory response represents repetitive synchronous discharges originating in the axon hillock region, then their appearance in the s. radiatum indicates their passive or active spread from the border zone between s. pyramidale and s. oriens. Their effective spread back into the dendrites is compatible with the electrotonic length of about 1 calculated for the apical dendrites (Turner and Schwartzkroin, 1980; Brown *et al.*, 1981). Similar results were obtained when the experiment was carried out in a low-chloride medium (data not shown).

3.12. *Visualization of the Spread of Activity in Slices*

Figure 12 demonstrates the main advantage of optical recording, i.e., the feasibility of simultaneous recording from many (hundreds of) neighboring loci. The figure illustrates the spread of a focal excitation along the Schaffer collateral-commissural axons, and subsequent postsynaptic spread down the apical dendrites of the CA1 neurons initiated the multiple discharge at the axonal hillock region of the pyramidal cells, which continued toward the s. oriens. Occasionally, the size of the dendritic depolarization was not the largest at the location of the synapses in the s. radiatum (e.g., top of columns 5 to 7). One possible explanation is that Ca^{2+} action potentials were evoked by the EPSPs. The slow afterhyperpolarization and the shape of the dendritic responses support this interpretation, in line with other recent observations (Schwartzkroin and Slawsky 1977; Wong *et al.*, 1979; Llinás and Sugimori, 1980).

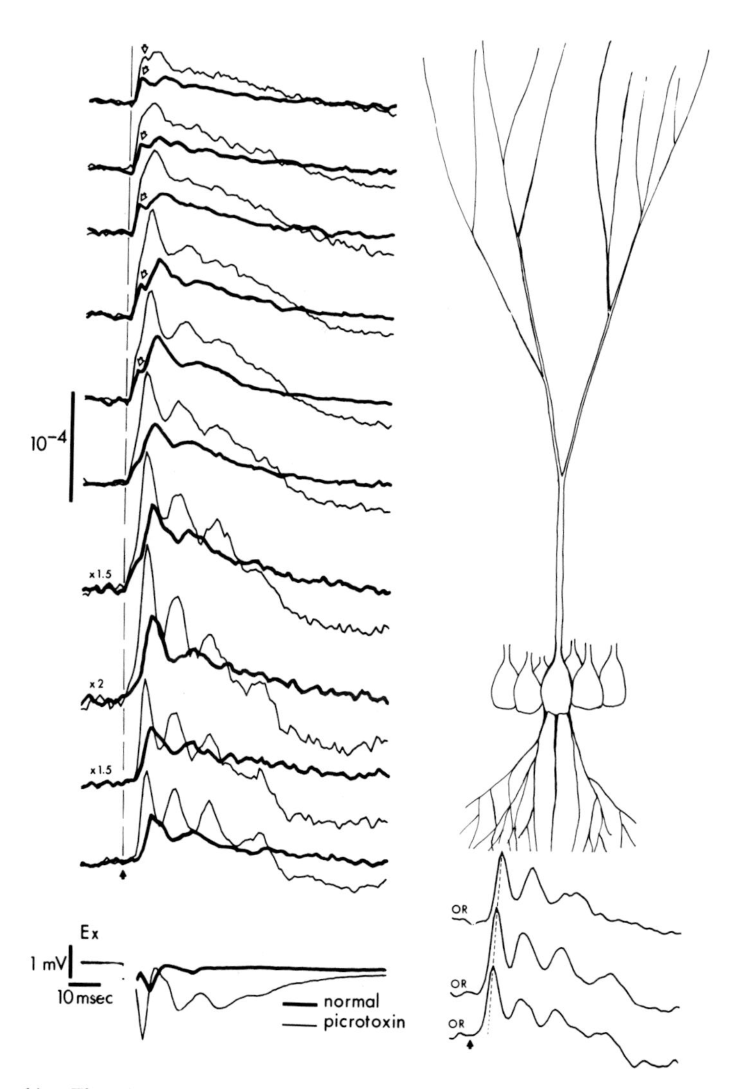

Figure 11. The effect of picrotoxin on the spread of excitation along pyramidal cells. The scheme at the right shows the locations from which recordings were made. Thick traces, responses in normal medium; thin traces, 15 min after application of 10^{-5} M picrotoxin; see Figure 6 for more details. Bottom right: The slow lateral spread at the s. oriens 45 μm away from s. pyramidale. Each recording is 90 μm away from its neighbor.

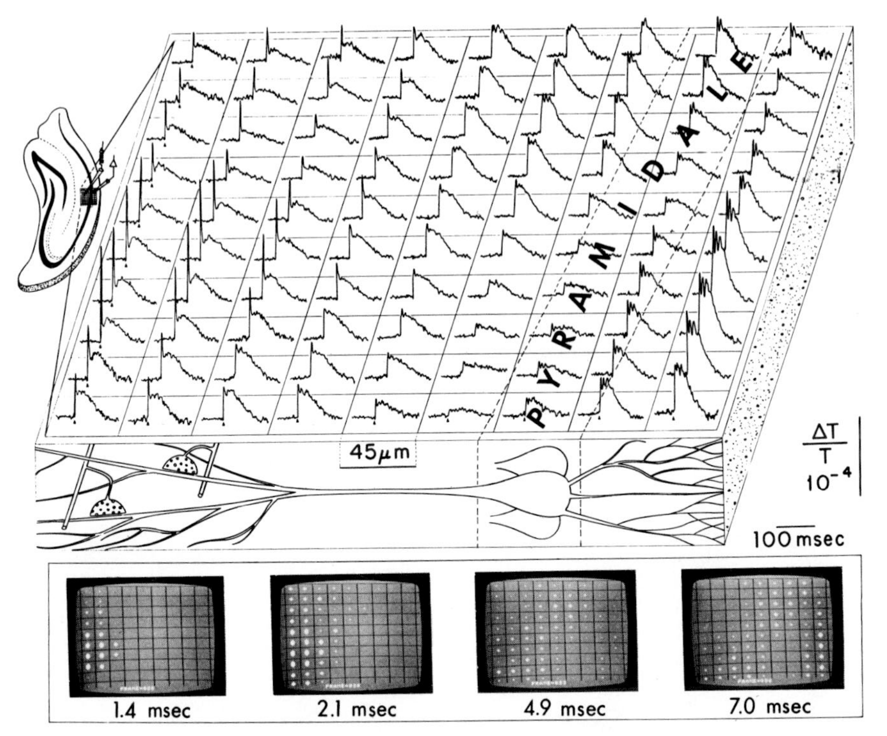

Figure 12. Simultaneous optical recordings of the electrical activity at multiple loci. Top left: a scheme of the slice showing the area monitored by the array. Top: The optical traces from 9 × 10 detectors are shown at their appropriate locations in the field of view. Activity was evoked by stimulation of the s. radiatum. The fast-rising short-latency signals (left side) are the optically detected presynaptic action potentials. The slow waves represent mostly the postsynaptic responses in the apical dendrites of the pyramidal cells. The EPSPs spread along the dendrites, resulting in action potential discharges of the pyramidal cells. The last two columns on the right show the activity of neuronal elements in the s. oriens, i.e., basal dendrites, axons of CA1 pyramids and other interneurons. The results of 20 trials were averaged. The ordinate bar represents the fractional change in light intensity. Each detector samples a square area of 45 × 45 μm. The optical signals were corrected for light-scattering. (Traces from four bad photodetectors were replaced by their neighbors, and a few traces with excessive noise were retouched). Bottom: Visualization of the activity presented in the top figure. The four TV frames show the pictures formed by the display processor. They show the spread of activity from the s. radiatum to the s. oriens. The appropriate time interval is marked below each frame. (The superimposed TV picture of the preparation was omitted for clarity; see Figure 2.)

The dynamic patterns of electrical activities within cell populations are heterogeneous even in a relatively small area of the highly ordered CA1 region. Further experiments are required to determine the synaptic networks and morphological correlates underlying the observed patterns of activity. Analysis of the data presented in Figure 12 is not trivial. There are 10^6 bits of information there, which can be collected in less than 1 sec. Detailed analysis is feasible by using conventional approaches and multiple data presentation similar to Figure 6. This approach is obviously time-consuming and requires memorizing 20 to 40 such figures for a single experiment. Therefore, we have designed a display processor that affords on-line slow motion visualization of the spread of electrical activity (see Section 2.1). The display of electrical activity at four different time intervals is demonstrated at the bottom of Figure 12.

3.13. *Comparison between Field Potentials and Optical Recordings*

The above experiments were designed to verify that the optical signals reflect changes in membrane potential. First, a correlation could be established between the optical recording and the extracellular recording from the s. pyramidale, in terms of size and pattern of responses. Second, the results deduced from the optical signals in response to various pharmacological treatments (TTX, TEA, 4-AP, picrotoxin, low Cl^-, low Ca^{2+}) were similar to those obtained with conventional methods. Third, the conduction velocity, refractory period, facilitation, and the particular loci of a given optical signal support our identification of the origin of the various optical signals. It should be noted that, in principle, some of the slow optical signals may come from glia, rather than neurons.

Although the optical and electrical signals share many characteristics, certain differences exist between the two types of signals. The presynaptic volleys recorded by Andersen *et al.* (1978) are shorter in duration than the present optical signals. This is due to the following: (1) Extracellular spikes are usually narrower than intracellular ones. (2) The spatial resolution of the extracellular recordings is smaller than that of the optical recordings, which picked up the activity from all the elements in a "box" of $45 \times 45 \times \sim 200$ μm. The activities in these widely spread elements are probably less synchronous than those of a smaller volume monitored by the extracellular electrode. (3) The temperature in many of the present experiments was lower than that used by Andersen *et al.* (1978). (4) The present time resolution of 0.7 msec may also artifactually increase the width of the signals. (5) Averaged responses may be wider than those detected in a single sweep.

The patterns of optical and extracellular field potential were identical. However, the optical signals usually precede the electrical signal by 0.7 to 2.1 msec. This is approximately the expected timing relationship for extracellular and intracellular recordings, as verified by simultaneous intracellular and extracellular recordings of action potentials from single cells in culture. Small differences in timing between the two types of records may also arise from the fact that the two recordings cannot be done from precisely the same area. (It is also difficult to visualize the exact position of the electrode tip, and the differences in timing between neighboring areas can be quite large, i.e., 1 to 2 msec per 100 μm separation.)

A major advantage of the intracellular population recording is that the optical signals are restricted to their site of origin whereas the extracellular field potentials may spread over large, often unpredicted, distances and their interpretation is therefore more difficult.

3.14. *Limitations and Advantages*

3.14.1. *Instrumental Difficulties.* The present instrument has a time resolution of 0.7 msec. The duration of each sweep was limited to 350 msec and, in averaging experiments, the light shutter had to be opened at least 1.5 sec prior to each of the recording periods to allow for the "settling" of the AC-coupled signals. These limitations are now being improved with modified hardware and software.

In the newer models of the amplifiers, each individual amplifier contains a sample-and-hold circuit to measure the light intensity at the onset of the measurements, after the shutter was opened. This value is then differentially subtracted from the light signal. In this way, a high-gain DC-offset recording can be obtained, and the shutter can be opened only for the duration of the measurements. The improved time resolution of the new system (0.3 msec) can be achieved by the construction of a digital multiplexer collecting sequential data from four 32-channel A/D cards and moving it to the memory using a direct memory access interface (DMA), or by the use of faster multiplexer A/D cards which are now available.

3.14.2. *Present Limitations of Optical Measurements.*

3.14.2a. Optical Probes for Brain Slices. The size of the optical signals was small. Therefore, spontaneous activity of single cells could not be unequivocally identified. The signal-to-noise ratio for optical detection of an action potential in single mammalian neurons maintained in culture, or in invertebrate neurons, was in the range of 10 to 20 under optimal conditions (Grinvald *et al.*, 1981a,b). The signal-to-noise ratio for action potential from a population of pyramidal cells was only 4 to 8 in the present work. It has been reported that the sensitivity of an optical probe may depend on the preparation (Ross and Reichardt, 1979). Because only 25 dyes were tested in these transmission experiments, it is likely that even more sensitive optical probes already exist for the present preparation. Fluorescence measurements form slices were not yet reported. However, recent studies of neuromal activity in intact brain structures *in vitro* indicated that fluorescence may give very large signals, thus eliminating the need for signal averaging (see Section 3.15).

3.14.2b. Light-Scattering Signals and Optical Signals. Because the signals are presently small, the activity-dependent light-scattering signals from the slice distort the voltage-sensitive optical signals. (These light-scattering signals were subtracted from the optical response; in fact, in the guinea pig experiment with the more sensitive dye RH-155, the light-scattering signals were small relative to the optical signals)

3.14.2c. Pharmacological Side Effects. These side effects can be expected when extrinsic probe molecules are bound to neuronal membranes. We did not observe significant pharmacological side effects in the present studies, using field potential recordings from s. pyramidale. However, more stringent tests using intracellular recording are required to assess pharmacological effects. A variety of voltage-sensitive probes are available, with very different chemical structures and net charges. It is, therefore, unlikely that all of them (more than 100) will cause similar pharmacological side effects. An example has been reported (Grinvald *et al.*, 1981a) in which a proper choice of probe eliminated pharmacological side effects on a behavioral reflex mediated by a polysynaptic pathway.

3.14.2d. Bleaching. During the course of an experiment, some of the dye bleaches, resulting in diminution of the optical signals. To minimize bleaching, the exposure time of the slice to light should be reduced to a minimum.

3.14.2e. Photodynamic Damage. The dye molecules, in the presence of intense illumination, sensitize the formation of reactive singlet oxygen. These reactive radicals attack membrane components and damage the cells (Cohen *et al.*, 1974; Ross *et al.*, 1977). Such photodynamic damage limits the duration of the experiments. Using the present probes, continuous measurements can be made for 1 to 5 min wihout marked damage. If the duration of each trial is 50 msec, then more than 1000 trials (with the new apparatus) can be carried out before significant damage occurs.

3.14.2f. Resting Potential. The resting potential cannot readily be calculated or manipulated. The inability to evaluate reversal potentials for various EPSPs and IPSPs limits the interpretation of some observed responses.

3.14.2g. Optical Signal Size. The size of optical signals is also related to the membrane area and the extent of binding. There are differences in "concentrations" of membrane elements across the slice. In distal dendrites, there are many processes and, therefore, a larger membrane area. There is much less membrane in the somata layer, and the size of the optical signals from the s. pyramidale is roughly estimated to be three to four times smaller than those from equal potential change within s. radiatum. Furthermore, a lower density of bound dye molecules in some parts of the tissue will result in a smaller signal size. Thus, a direct comparison of the amplitudes of optical signals in different regions may not be straightforward; the interpretation must rely principally on the time course of the signals.

3.14.2h. Analysis of Intracellular Population Activity. The recording is made from the entire depth of the slice (250 to 300 μm) which consists of different neural elements. The signal size as a function of the distance of the focal plan from the target follows a Gaussian distribution function. Signals from out-of-focus targets cause distortions of in-focus signals. In the highly organized hippocampal slice the present three dimensional resolution leads to apparent broadening of the region occupied by signals from a given neuronal strata, and signals are mixed along the borders. The depth-of-field resolution (Salzberg *et al.*, 1977) could be increased by using an objective with a higher numerical aperature. Although the hippocampus was selected because of its clear stratification, neurons of various types coexist within strata, and signals from these neurons reaching the same detectors can undoubtedly obscure signals from any particular population of neuronal elements, i.e., axons, dendrites, pyramidal somata, etc. Thus, the activity detected at the s. oriens reflects the activity of the CA1 pyramidal axons as well as that of the interneurons there. The activity detected at s. radiatum reflects the action potentials of presynaptic elements, the postsynaptic responses of apical dendrites, the passive spread from pyramidal cell somata, as well as the activity of other interneurons. However, proper manipulation of stimulus size, location, frequency, and pharmacological manipulation, together with intracellular stimulation of single cells, should permit a separation of the various components.

Development or discovery of more sensitive dyes, improvement of the resolution of the optics, and the possible iontophoretic injection of a suitable fluorescent dye into single cells (Grinvald *et al.*, 1982d), could overcome some of the above difficulties.

3.14.3. *Advantages.* Compared with electrical recording, optical methods have several inherent advantages. First, there is no lower limit to the size of the neuronal element recorded, provided a voltage-sensitive dye can attach to its membrane. (However, a large number of such elements may have to be synchronously active or, alternatively, a large number of responses may have to be averaged, in order to detect responses from small elements.) Intracellular–optical recordings can thus be obtained from unmyelinated axons as well as from very thin dendrites. These can rarely be recorded intracellularly in mammalian preparations. Second, this method is a noninvasive one: there is no need to impale the membrane and risk injuring it in order to record. For this reason, recording can be obtained from the same elements for a considerable length of time (e.g., 6 hr). Third, the detection and localization of synchronously active populations of neurons are feasible. Finally, the major advantage is that a large number of detectors can be placed side-by-side to monitor simultaneously the electrical activity from hundreds of loci. This advantage offers two clear benefits. First, it is relatively easier to search for a given responding cell(s) (a search for a needle in a haystack). Second, the patterns of spatiotemporal spread of electrical responses can be investigated, and often the optical data also yields "anatomical" information (i.e., the location of the presynaptic Schaffer collaterals).

3.15. *Future Prospects*

The spatial and temporal resolution of optical recording techniques will continue to improve and it is likely that spike activity of single cells will also be detectable in brain slices. In addition, the same technique can be used to study brain structures *in vitro* and *in vivo* as demonstrated by the studies of the goldfish optic tectum (Anglister and Grinvald, 1983) and the olfactory bulb (Orbach and Cohen, 1983). In preliminary studies, visually evoked responses from "superficial" layers in the rat visual and somatosensory cortex were also detected with fluorescence measurements (Orbach *et al.*, 1982, 1983). More recently, patterns of responses from the frog optic tectum were recorded, in response to well-defined patterns of visual stimuli (Grinvald *et al.*, 1983a). Figure 13 illustrates the fluorescence signals which were recorded from the frog's optic tectum in response to a 150 msec light flash presented to the contralateral eye (signal averaging was not used). Evidently fluorescence measure-

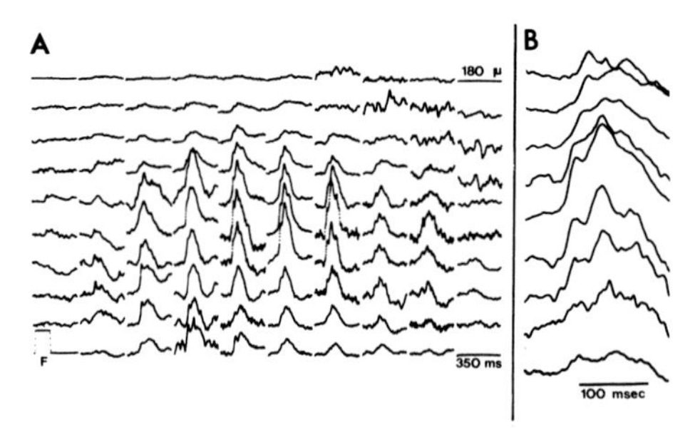

Figure 13. Fluorescence signals from the optic tectum *in vivo*. (A) The tectum was stained for 20 min with a styryl dye, designated RH-414, by topical application on the exposed tectum (the dura mater was removed). Signals were evoked by a 150 msec light flash presented to the contralateral eye. Signals from 96 photodetectors are shown in their appropriate location; each photodetector covered an area of 180 × 180 μ. (B) Signals from nine photodectors along one column are shown on an expanded scale. Usually the number of resolved peaks on the optical signal was identical to the number of peaks of the electrically-recorded evoked potential (not shown).

ment from brain slices may give larger signals than the absorption signals presently obtained. Moreover it is relatively easy to further improve the signal-to-noise ratio in fluorescence experiments by a factor of 10 to 50 by using the recommended technical improvements (Grinvald *et al.*, 1982a; Grinvald *et al.*, 1983). Thus fluorescence measurements from slices should also be attempted. In addition, the properties of dendrites of intracellularly stained single cells can be studied in slices as was recently reported for invertebrate preparation (Grinvald *et al.*, 1983c). The cost of most of the specific hardware is continuously coming down despite improved performance. Therefore, optical recording of neuronal activity should become a practical tool, which in combination with electrophysiological methods, will contribute to the study of the electrical and pharmacological properties of single cells and local circuits in brain slices, as well as the evaluation of the functional organization and information processing in the intact vertebrate brain.

ACKNOWLEDGMENT. Special thanks to U. Kuhnt who participated in many of our slice experiments. Supported in part by grants from the USPHS NS-14716, and a grant from the U.S.–Israel Binational Science Foundation to A.G.

4. REFERENCES

Agmon, A., Hildesheim, R., Anglister, L., and Grinvald, A., 1982, Optical recording from processes of individual leech CNS neurons iontophoretically injected with new fluorescent voltage-sensitive dyes, *Neurosci. Lett.* **10:**535.

Anglister, L., and Grinvald, A., 1983, Real-time visualization of the spatio-temporal spread of electrical responses in the optic tectum of vertabrates, *Israel J. Med. Sci.*, in press.
Andersen, P., Eccles, J. C., and Loyning, Y., 1964, Pathway of postsynaptic inhibition in the hippocampus, *J. Neurophysiol.* **27**:608–619.
Andersen, P., Silfvenius, H., Sundberg, S. H., Sveen, O., and Wigstrom, H., 1978, Functional characteristics of unmyelinated fibers in the hippocampal cortex, *Brain Res.* **144**:11–18.
Brown, T. H., Fricke, R. A., and Perkel, D. H., 1981, Passive electrical constants in three classes of hippocampal neurons, *J. Neurophysiol.* **46**:812–827.
Cohen, L. B. and Keynes, R. D., 1971, Changes in light-scattering associated with the action potential in crab nerve, *J. Physiol.* (*London*) **212**:259–275.
Cohen, L. B. and Salzberg, B. M., 1978, Optical measurement of membrane potential, *Rev. Physiol. Biochem. Pharmacol.* **83**:35–88.
Cohen, L. B., Keynes, R. D., and Hille, B., 1968, Light scattering and birefringence changes during nerve activity, *Nature* (*London*) **218**:438–441.
Cohen, L. B., Keynes, R. D., and Landowne, D., 1972, Changes in axon light-scattering that accompany the action potential: Current dependent components, *J. Physiol.* (*London*) **224**:727–752.
Cohen, L. B., Salzberg, B. M., Davila, H. V., Ross, W. N., Landowne, D., Waggoner, A. S., and Wang, C. H., 1974, Changes in axon fluorescence during activity; Molecular probes of membrane potential, *J. Membrane Biol.* **19**:1–36.
Cohen, L. B., Salzberg, B. M., and Grinvald, A., 1978, Optical methods for monitoring neuron activity, *Annu. Rev. Neurosci.* **1**:171–182.
Grinvald, A. and Farber, I., 1981. Optical recording of Ca^{+2} action potentials from growth cones of cultured neurons using a laser microbeam, *Science* **212**:1164–1169.
Grinvald, A.. Salzberg, B. M. and Cohen, L. B., 1977, Simultaneous recording from several neurons in an invertebrate central nervous system, *Nature* (*London*) **268**:140–142.
Grinvald, A., Ross, W. N., Farber, I., Saya, D., Zutra, A., Hildesheim, R., Kuhnt, U., Segal, M., and Kimhi, Y., 1980, Optical methods to elucidate electrophysiological parameters, In: *Neurotransmitters and Their Receptors* (U. Z. Littauer, ed.), John Wiley and Sons, New York, pp. 531–546.
Grinvald, A., Cohen, L. B., Lesher, S., and Boyle, M. B., 1981a, Simultaneous optical monitoring of activity of many neurons in invertebrate ganglia, using a 124 element "Photodiode" array, *J. Neurophysiol.* **45**:829–840.
Grinvald, A., Ross, W. N., and Farber, I., 1981b, Simultaneous optical measurements of electrical activity from multiple sites on processes of cultural neurons, *Proc. Natl. Acad. Sci. USA*, **78**:3245–3249.
Grinvald, A., Manker, A., and Segal, M., 1982b, Visualization of the spread of electrical activity in rat hippocampal slices by voltage sensitive optical probes, *J. Physiol.* (*London*), **333**:269–291.
Grinvald, A., Hildesheim, R., Farber, I. C., and Anglister, L., 1982a, Improved fluorescent probes for the measurement of rapid changes in membrane potential, *Biophys. J.*, **39**:301–308.
Grinvald, A., Anglister, L., Hildesheim, R., and Freeman J. A., 1983a, Optical monitoring of naturally evoked dynamic patterns of neuronal activity from the frog optic tectum, *Neurosci. Abstr.* **9**:540.
Grinvald, A., Fine, A., Farber, I. C., and Hildesheim, R., 1983b, Fluorescence monitoring of electrical responses from small neurons and their processes, *Biophys. J.* **42**:195–198.
Grinvald, A., Hildesheim, R., Agmon, A., and Fine A., 1982c, Optical recording from

neuronal processes and their visualization by iontophoretic injection of new fluorescent voltage sensitive dyes, *Neurosci. Abstr.* **8**:491.

Gupta, R., Salzberg, B. M., Grinvald, A., Cohen, L. B., Kamino, K. Boyle, M. B., Waggoner, A. S., and Wang, C. H., 1981, Improvements in optical methods for measuring rapid changes in membrane potential, *J. Membrane Biol.* **58**:123–138.

Krauthaimer, V. and Ross, W. N., 1981, Optical measurement of potential changes in axons and processes of neurons of a barnacle ganglion, *Neurosci. Abstr.* **7**:114.

Lipton, P., 1973, Effects of membrane depolarization on light scattering by cerebral slices, *J. Physiol.* (*London*) **231**:365–383.

Llinás, R. and Sugimori, M., 1980, Electrophysiological properties of *in vitro* Purkinje cell dendrites in mammalian cerebellar slices, *J. Physiol.* (*London*) **305**:197–213.

MacVicar, B. A. and Dudek, S. E., 1981, Electrotonic coupling between pyramidal cells: A direct demonstration in rat hippocampal slices, *Science* **213**:782–785.

Orbach, S. H. and Cohen, L. B., 1983, Simultaneous optical monitoring of activity from many areas of the salamander olfactory bulb. A new method for studying functional organization in the vertebrate CNS, *J. Neurosci*, in press.

Orbach, S. H., Cohen, L. B., and Grinvald, A., 1982, Optical monitoring of evoked activity in the visual cortex of the rat, *Biol. Bull.*, **163**:389.

Orback, S. H., Cohen, L. B., and Grinvald, A., 1983, Optical monitoring of electrical responses in the rat somatosensory and visual cortex, *Israel J. Med. Sci.*, in press.

Ross, W. N. and Reichardt, L. F., 1979, Species-specific effects on the optical signals of voltage sensitive dyes, *J. Membrane Biol.* **48**:343–356.

Ross, W. N., Salzberg, B. M., Cohen, L. B., Grinvald, A., Davila, H. V., Waggoner, A. S., and Wang, C. H., 1977, Changes in absorption, fluorescence, dichroism and birefringence in stained axons: Optical measurement of membrane potential, *J. Membrane Biol.* **33**:141–183.

Salzberg, B. M., Davila, H. V., and Cohen, L. B., 1973, Optical recording of impulses in individual neurons of an invertebrate central nervous system, *Nature* (*London*) **246**:508–509.

Salzberg, B. M., Grinvald, A., Cohen, L. B., Davila, H. V., and Ross, W. N., 1977, Optical recording of neuronal activity in an invertebrate central nervous system: Simultaneous monitoring of several neurons, *J. Neurophys.* **40**:1281–1291.

Schwartzkroin, P. A. and Prince, D. A., 1980, Effects of TEA on hippocampal neurons, *Brain Res.* **185**:169–181.

Schwartzkroin, P. A. and Slawsky, M., 1977, Probable calcium spikes in hippocampal neurons, *Brain Res.* **135**:157–161.

Sherrington, C. S., 1953, *Man on His Nature*, Doubleday and Company, Inc., Garden City, New York, p. 183.

Spencer, W. A. and Kandel, E. R., 1961, Electrophysiology of hippocampal neurons. IV. Fast prepotentials, *J. Neurophysiol.* **24**:272–285.

Tasaki, I., Watanabe, A., Sandlin, R., and Carnay, L., 1968, Changes in fluorescence turbidity and birefringence associated with nerve excitation, *Proc. Natl. Acad. Sci. USA*, **61**:883–888.

Turner, D. A. and Schwartzkroin, P. A., 1980, The steady-state electrotonic analysis of intracellularly stained hippocampal neurons, *J. Neurophysiol.* **44**:184–199.

Waggoner, A. S., 1979, Dye indicators of membrane potential, *Annu. Rev. Biophys. Bioeng.* **8**:47–63.

Waggoner, A. S. and Grinvald, A., 1977, Mechanisms of rapid optical changes of potential sensitive dyes, *Ann. N. Y. Acad. Sci.* **303**:217–242.
Wong, R. K. S., Prince, D. A., and Basbaum, A. I., 1979, Intradendritic recordings from hippocampal neurons, *Proc. Natl. Acad. Sci. USA* **76**:986–990.
Yamamoto, C., 1972, Intracellular study of seizure-like afterdischarges elicited in thin hippocampal sections *in vitro*, *Exp. Neurol.* **35**:154–164.

9

Probing the Extracellular Space of Brain Slices with Ion-Selective Microelectrodes

JØRN HOUNSGAARD and CHARLES NICHOLSON

1. INTRODUCTION

In this chapter, we illustrate the use of ion-selective microelectrodes (ISMs) in brain slices. We have confined ourselves to extracellular K^+ measurements since these have been at the focus of our interests, although intracellular measurements are becoming feasible (see Syková *et al.*, 1981).

2. THE BRAIN CELL MICROENVIRONMENT

The interpretation of the functional role of the extracellular space in the brain is changing. The extracellular space is unquestionably the main pathway for the transfer of metabolic substances between cells and between the brain and the blood vessels. In considering neuronal function, however, the extracellular space has been viewed as a homeostat that provided a stable homogenous ionic milieu for synaptic transmis-

JØRN HOUNSGAARD • Department of Neurophysiology, Panum Institute, University of Copenhagen, 2200 Copenhagen N, Denmark. CHARLES NICHOLSON • Department of Physiology and Biophysics, New York University Medical Center, New York, New York 10016.

sion and impulse generation. The first challenge to this concept was the demonstration by Frankenhaeuser and Hodgkin (1956) that the homeostat was less than perfect even for a simple inorganic ion like K^+. Baylor and Nicholls (1969) further showed that in the central nervous system (CNS) of the leech such perturbations of ion homeostasis might affect neuronal function. In the intervening years, it has been repeatedly demonstrated that synchronous activation of mammalian neurons also leads to considerable transient changes in the ionic composition of the extracellular space (see Nicholson 1979, 1980; Syková *et al.*, 1981 for references). These findings, together with the fact that the brain extracellular space geometrically is like a thin, highly convoluted film that separates cells by less than a few hundred Angstroms, justified the introduction of the term *cell microenvironment* to designate the microcompartment of the extracellular space that interacts with any given cell (Schmitt and Samson, 1969; Nicholson 1979, 1980). It is this working hypothesis: namely that the brain cell microenvironment is a communication channel between the members of neuronal aggregates, that has motivated our present research.

We have approached the extracellular space of brain slices at three levels of analysis: (1) the passive biophysical properties of the extracellular space (size and parameters for diffusion), (2) ion changes induced by the simultaneous activity of ensembles of neurons, and (3) ion changes produced by a single cell. The dual purpose of the experiments has been to evaluate the degree of normality of the extracellular space of brain tissue *in vitro* and to explore the brain slice as a vehicle for studying the functional significance of the dynamics of the brain cell microenvironment.

The Passive Properties of the Extracellular Space in Brain Slices

In the brain, ISMs have been used extensively to monitor the concentration of naturally occurring ions. Recent work from our laboratory has extended the use of ISMs to monitor the dispersal of membrane-impermeable probe ions released from the tip of a point source and show that the resulting ion dispersal is describable in terms of simple Fickian diffusion, provided that the volume fraction α and the tortuosity λ of the extracellular space are taken into account (Nicholson and Phillips, 1981). We have adapted this approach to the cerebellar slice (Hounsgaard and Nicholson, 1983).

The experimental arrangement that we have used to apply the iontophoretic technique in cerebellar slices is shown schematically in Figure 1. An iontophoretic electrode containing the probe ion and an ISM sensitive to the probe ion are glued together in an array with a known tip separation of between 80 and 120 μm. Fabrication and characteristics of the ISMs, as well as the underlying theory, have been described elsewhere (Nicholson and Phillips, 1981; Hounsgaard and Nicholson, 1983). From among the useful probe ions available, we chose tetramethylammonium (TMA^+) for the present experiments. Cerebellar slices, 400 μm thick, were equilibrated with physiological medium containing 0.2 mM TMA chloride. The experimental chamber also included a submerged microcup containing 0.5% agar in Ringer. For each evaluation of α and λ, two sets of records were obtained, one from the agar cup and one from the slice. In the slice, the electrode array was positioned 200 μm deep, midway between the cut surfaces. Throughout the experiment, a positive bias current was passed through the iontophoretic electrode to secure a constant transport number for the TMA^+. On top of this current, a positive step current of 100 nA was applied for 40 sec. A typical set of records is shown in Figure 1B. The agar measurements gave the diffusion characteristics for TMA^+ in free medium (i.e., $\alpha = \lambda = 1$). From such records, the values of D (the diffusion coefficient) and n (the transport number of the iontophoretic electrode) could be extracted. Using these parameters, together with the diffusion curves in the slice, the values of α and λ were determined from the slice record. The rising phase during the iontophoretic pulse was used for the analysis. The values for D, n, α, and λ then allowed the theoretical curve for the diffusion equation to be constructed. In Figure 1B, this was superimposed on the experimental record. The fit during the falling phase served as an independent and sensitive test of the validity of the diffusion paradigm. The details and the justification of these procedures have been given elsewhere (Nicholson and Phillips, 1981).

In 24 measurements, we found $\alpha = 0.28 \pm 0.02$ and $\lambda = 1.84 \pm 0.05$ (mean ± SEM) (Hounsgaard and Nicholson, 1983). These values suggest that the geometric dimensions and the diffusion characteristics of the extracellular space in slices are comparable to *in vivo* conditions. The slightly higher extracellular volume we found in cerebellar slices (28% against 21% in the rat cerebellum *in vivo*) may be a consequence of occasional local disruption during preparation of the tissue. Although the influence of the free medium surrounding the slice cannot be totally ignored, it seemed to have minimal influence on the measurements, on the timescale that we have used, since the values of α and λ were independent of the segment of the rising phase used for the evaluation. Occasionally, there was a considerable disagreement between the theoretical and the experimental curves and, in these cases, the values of α and λ obviously lose their meaning. Such data have been excluded.

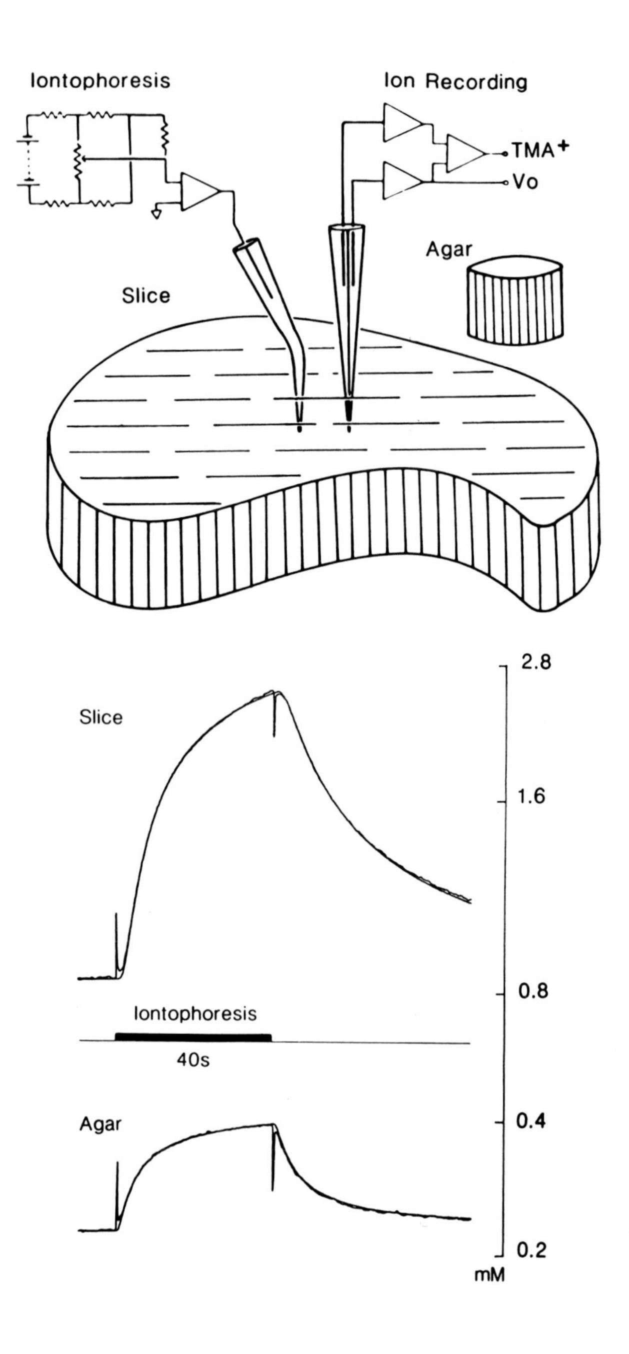
Iontophoresis
Ion Recording
TMA+
Vo
Agar
Slice
Slice
2.8
1.6
0.8
Iontophoresis
40s
Agar
0.4
0.2
mM

At present, we do not know whether these anomalies are due to technical errors or to real divergences from the model in some areas of the tissue. We emphasize that such anomalies are found both *in vivo* and *in vitro* and that no systematic trends have been obvious.

It is interesting that the values of α and λ appear to be similar in the granular layer, the Purkinje cell layer, and the molecular layer, despite the distinct cytological differences between these regions. This finding suggests that the macroscopic characteristics of diffusion studied here are predominantly determined by parameters that are invariant in the three regions. This suggests the local "granularity" of the brain exists on the scale of the order of a micrometer or so and these irregularities are smoothed out when diffusion is considered over dimensions of the order to tens of micrometers or more. Indeed, electron microscopy confirms that most cellular elements of brain (e.g., axons, dendrites, synapses) are only a micrometer or so in size.

Experimental approaches and techniques related to those described here can readily be used to investigate changes in the parameters of the extracellular space in response to changes under conditions such as ischemia (Hansen and Olsen, 1980), epilepsy (Dietzel *et al.*, 1980), and spreading depression (Phillips and Nicholson, 1979; Hansen and Olsen, 1980; Nicholson *et al.*, 1981). The main limitations are introduced by the spatial and temporal resolution of the measurements. Small changes in the electrode spacing when the array is positioned in agar compared to

---

Figure 1. Diffusion measurements in the slice. Upper panel shows the experimental setup. An ISM selective to TMA is glued with tip separation 70 to 100 μm from a bent iontophoresis micropipette. The two electrodes are lowered into a cerebellar slice and a long pulse of TMA ejected. The resulting signal is recorded on the ISM. The array of electrodes can also be inserted in a block of agar gel, adjacent to the slice, for similar control measurements to determine iontophoretic electrode transport number, n, and free diffusion coefficient, D. Using these control parameters, with the slice, measurements of the volume fraction, α, and tortuosity, λ, can be calculated (Nicholson and Phillips, 1981). Lower panel shows data obtained from such an experiment in agar (lower record) and slice (upper record) using an 80 nA pulse of TMA with a duration of 40 sec. Electrode separation was 95μm. The records plotted on a logarithmic scale together with the superimposed theoretical curves (see text) generated from the n, D, α, and λ parameters extracted from experimental curves. An excellent fit is seen between experimental and theoretical curves. Note that the falling phase theoretical curve is generated by adding an equal and opposite curve with a 40-sec delay. The difference in baseline between agar and slice is due to the presence of a constant bias current emitting TMA at all times to ensure iontophoretic electrode linearity (Nicholson and Phillips, 1981). Since the signal is enhanced by a factor λ^2/α in the slice relative to the agar, the baseline is similarly elevated in the slice. From Nicholson and Hounsgaard, 1983.

those in the slice introduce significant inaccuracies when tip separations below 50 μm are used. In addition, the rising phase of the diffusion curve, in our experience, has to last more than 10 sec to permit reliable estimates of the parameters.

The most important conclusions from this work are that the properties of the extracellular space in brain slices are well preserved and that the dispersal of membrane impermeable ions on the scale studied here occurs in a similar manner to that seen *in vivo*.

3. EXTRACELLULAR ION CHANGES PRODUCED BY SIMULTANEOUS ACTIVITY OF ENSEMBLES OF NEURONS

We now turn to the extracellular ion changes produced by ion fluxes across cell membranes. Although the extracellular ion concentration in the most superficial part of slices by necessity is "clamped" by the superfusing medium, transient changes, as we have seen, are unaffected deeper in the tissue. Several investigators have shown that stimulation of afferent fibers in a variety of noncerebellar slices evokes changes in $[K^+]_0$ and $[Ca^{2+}]_0$ closely similar to those see *in vivo* (Fritz and Gardner-Medwin, 1976; Alger and Teyler, 1978; Benninger *et al.*, 1980; Gutnik and Segal, 1981; King and Somjen, 1981). This is a further indication of a normal cell microenvironment in slices. Figure 2A shows how $[K^+]_0$ changed in the Purkinje cell layer in response to electric activation of afferents in the white matter in cerebellar slices. The amplitude of the $[K^+]_0$ increase was related to the stimulus frequency and there was a poststimulus undershoot lasting 1 to 2 sec. As with *in vivo* studies, this undershoot was interpreted as an indication of active cellular uptake of K^+ (Heinemann and Lux, 1975; Nicholson *et al.*, 1978). Similar $[K^+]_0$ changes were seen in response to glutamate applied iontophoretically in the molecular layer (Figure 2B). The increase in $[K^+]_0$ is quite substantial even with low iontophoretic current strength. Possible indirect effects produced by extracellular ion changes deserve careful consideration whenever the direct receptor-mediated effects of iontophoretically applied drugs are considered. Iontophoretically applied substances unavoidably attain a high concentration near the tip of the electrode and reach many surrounding cells by diffusion. The procedures traditionally used to test for indirect effects of applied substances may miss possible effects of induced extracellular ion changes (see Hosli *et al.*, 1978). One acknowledged advantage offered by the *in vitro* slice preparations is that indirect actions mediated by secondary synaptic activation can be tested for in control experiments by including blockers of synaptic transmission

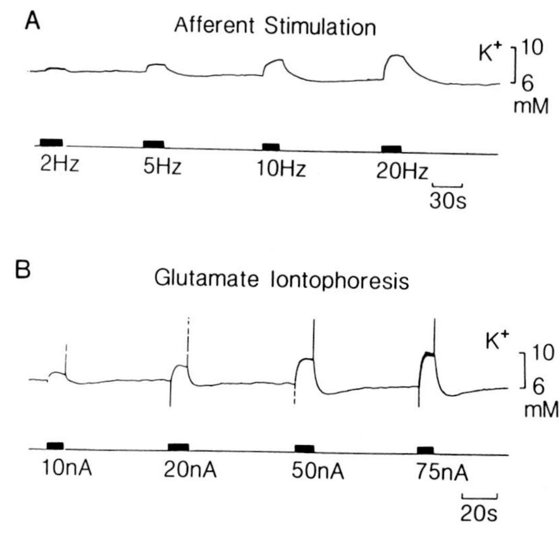

Figure 2. $[K^+]_0$ recorded in the Purkinje cell layer in response to electric stimulation of white matter (A) an ionophoretic application of glutamate in the molecular layer (B). In (A), electric stimuli were delivered to the white matter in trains of 2,5,10, and 20 Hz as indicated by the bars below the $[K^+]_0$ record. The $[K^+]_0$ response increased with stimulus frequency. In (B), increasing $[K^+]_0$ responses were recorded as the amplitude of the current pulses passed through a glutamate-containing electrode in the distal molecular layer was increased through 10, 20, 50, and 75 nA. Note the decrease in $[K^+]_0$ below baseline immediately after each activation. The magnitude of this undershoot was related to the preceding $[K^+]_0$ increase. From Hounsgaard and Nicholson, 1983.

in the bathing medium. The $[K^+]_0$ changes produced by iontophoretically applied glutamate were, however, unaffected by 2 mM Mn^{2+} added to the medium. In fact, the $[K^+]_0$ change was insensitive to Mn^+, tetrodotoxin (TTX), and 4-aminopyridine (4-AP) (Figure 3). The most likely explanation is that the flux of K^+ was mediated predominantly by the synaptic channels located in the membranes of neuronal elements in a wide region around the iontophoretic electrode.

4. EXTRACELLULAR ION CHANGES EVOKED BY INDIVIDUAL CELLS

With few exceptions, studies with ISMs in the intact mammalian brain and in slices have been confined to looking at ion changes induced by large populations of neurons. Although these studies have given valuable information on the lability of extracellular ion concentration

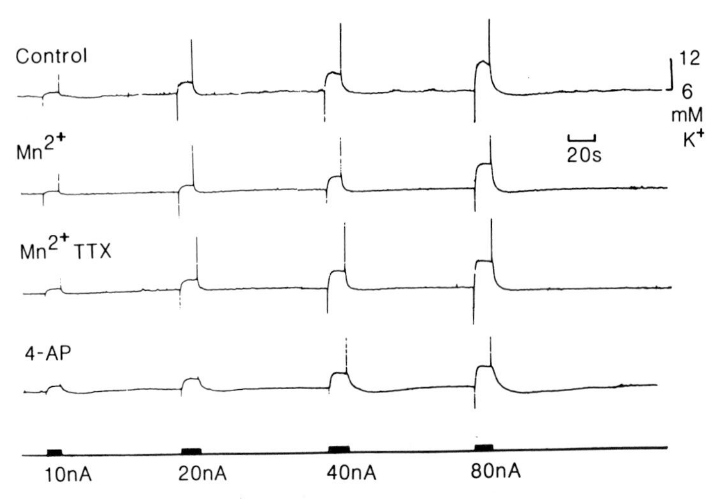

Figure 3. Lack of effect of blocking agents on glutamate-induced $[K^+]_0$ changes. Control shows potassium response of slice to pulses of glutamate ionophoresed with currents shown at bottom of figure. Subsequent application of 2 mM $MnCl_2$, 2 mM $MnCl_2$ + 10^{-7}M TTX or 0.1 mM 4-aminopyridine (4-AP) failed to alter the glutamate induced K^+-changes. (Note that the 4-AP began to affect the K^+—ISM characteristics.)

under various physiological and pathological conditions, detailed studies of the cellular sources and mechanisms have been impossible. It is not clear either how strongly the activity of a single neuron influences its immediate microenvironment. The results summarized in the previous sections suggested to us that the brain slice might be a favorable and legitimate preparation for such an analysis. As a first step, we have looked at K^+ changes.

The Purkinje cells in cerebellar slices offered exceptional advantages for our purpose. They can be visualized during experiments (Yamamoto and Chujo, 1978; Llinás and Sugimori, 1980a,b) and, because of the pacemaker properties of the cell membrane, they remain spontaneously active *in vitro* (Hounsgaard, 1979). The activity of a single cell can thus be monitored without simultaneous activation of neighboring neurons. The spontaneous activity of Purkinje cells is asynchronous in agreement with the fact that no excitatory synapses, electrical or chemical, have been identified between Purkinje cells (Palay and Chan-Palay, 1974). In addition, some Purkinje cells regularly alternate between silent and active periods (Yamamoto, 1974; Hounsgaard, 1979; Llinás and Sugimori, 1980a,b) that, in recordings with ISMs, allow measurements of ion changes during different phases of activity.

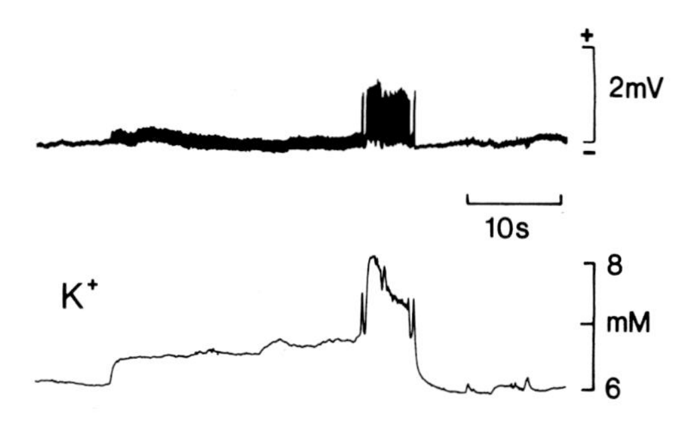

Figure 4. Na^+ and Ca^{2+} spikes recorded adjacent to Purkinje cell soma. The activity begins with a prolonged sequence of Na^+ spikes accompanied by a small but significant and augmenting elevation of $[K^+]_0$ (lower record). The Na^+ spikes are terminated by a burst of Ca^{2+} spikes that generate a much larger $[K^+]_0$ increase. Small groups of Ca^{2+} spikes preceded and follow the main burst and are reflected in transient $[K^+]_0$ increases. Note that the magnitudes of the Na^+ spike potentials (upper record) are artificially reduced below those of the Ca^{2+} spikes due to the frequency cut-off of the chart recorder.

4.1. *Spontaneous Purkinje Cell Activity*

An active period recorded with a K^+-ISM from the cell body of a Purkinje cell is shown in Figure 4. The upper record is the electrical activity recorded through the reference barrel and the lower record is the simultaneous $[K^+]_0$ response. There is a close correlation between the two records. At the onset of firing, $[K^+]_0$ rose to a slowly augmenting plateau. The electrical activity in this phase consisted of a long train of fast, TTX-sensitive, somatic action potentials. The activity was terminated by a burst of dendritic Ca^{2+} spikes during which there was an additional large increase in $[K^+]_0$. (The true relation between the amplitudes of the two spike types is reversed in this figure due to the low frequency response of the chart recorder). The close correlation between $[K^+]_0$ and firing pattern was a general finding in these studies. The level of $[K^+]_0$ during activity was affected by a number of factors: the proximity of the electrode to the cell from which the recording was made (as indicated by the extracellular spike amplitude), the firing frequency, and the type of action potential generated. The surprising finding, that the greatest level of $[K^+]_0$ recorded from the soma was associated with Ca^{2+} spikes generated in the dendrites, will be discussed in more detail below. The powerful effects of these spikes on $[K^+]_0$ was, however, clearly demonstrated by the dramatic transients associated with single, isolated dendritic spikes in Figure 4. A similar pattern was recorded from dendrites in the molecular layer, as shown in Figure 5.

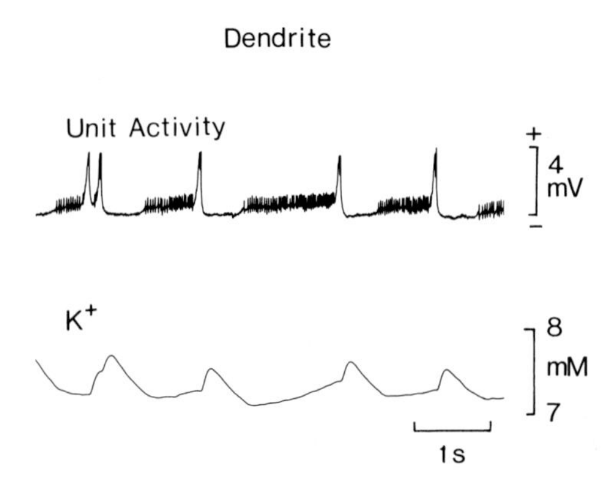

Figure 5. $[K^+]_0$ and extracellular potential records from the dendrite of spontaneously active Purkinje cell. $[K^+]_0$ records were closely related to the spike activity in the extracellular potential records (lower records). $[K^+]_0$ increased rapidly at the onset of Na^+ spike activity to a slowly increasing plateau level. A distinct additional increase was seen during the terminating burst of Ca^{2+} spikes. Note the unitary $[K^+]_0$ increases produced by individual Ca^{2+} spikes. Modified from Hounsgaard and Nicholson, 1983.

The $[K^+]_0$ increase recorded from single Purkinje cells was about 1 mM during fast somatic Na^+ spikes and 2 to 4 mM during Ca^{2+} spikes. These levels were recorded both in standard slice Ringer (124 mM NaCl, 5 mM KCl, 1 mM KH_2PO_4, 2 mM $MgCl_2$, 2 mM $CaCl_2$, 25 mM $NaHCO_3$, 10 mM glucose) and in medium more closely resembling the composition of cerebrospinal fluid (as shown above but with 3 mM K^+, 1.3 mM Ca^{2+} and 1.3 mM Mg^{2+}). The amplitude of extracellularly recorded action potentials under such conditions was 1 to 4 mV, suggesting that the electrode was in the immediate extracellular space of the Purkinje cell but not sealed to the membrane. High resistance seals were occasionally obtained, as indicated by spikes of 50 to 60 mV, with a typical "intracellular shape" and a "resting membrane potential" of 40 to 50 mV. As opposed to true intracellular recordings, no baseline increase in K^+ was recorded and a slight retraction of the electrode often restored the typical extracellular record. Slight advancement of the electrode resulted in penetration of the cell and an abrupt increase in K^+. Possibly the siliconization of the ISM tip during fabrication facilitated the seal formation.

It is often assumed that ISMs create a substantial cavity around their tips that forms a dead space in which the true extracellular ion changes are attenuated. We have found no evidence for this assumption either in the analysis of the diffusion parameters or in recordings from single cells. In particular, the fast time course of $[K^+]_0$ changes produced by single Ca^{2+} spikes (Figure 4) suggests that the extracellular space approaches normal dimensions around the electrode tip. When the electrode was moved within the slice, it appeared to glide through the extracellular space pushing the cells aside. Only rarely was there evidence of membrane rupture as indicated by injury discharge, DC-shifts, or sudden $[K^+]_0$ increases.

The main source of artifact in recordings from single cells originates from the subtraction of the reference signal from the signal recorded through the ion barrel, because of the high frequency components involved (recently discussed by Kleine and Kupersmith, 1982). In our experiments, we adjusted the subtraction by using a test pulse applied to the headstages of the two amplifiers. The response time of the two barrels were then matched by adjusting the capacitive feedback of the headstage of the reference amplifier. Occasionally, it was necessary to pass the two signals through a lowpass filter before subtraction to avoid artifacts associated with fast action potentials.

4.2. *Changes in* $[K^+]_0$ *during Simultaneous Intracellular Recording and Current Passage*

Extracellular recordings in cerebellar slices rely on the spontaneous firing pattern of the neurons and leave little room for experimental manipulation and analysis of the membrane properties that govern the ion fluxes responsible for changes in the extracellular ion concentration. It was, therefore, a significant improvement when we found it possible to make simultaneous intracellular records together with the extracellular ion recording. In this way, $[K^+]_0$ could be measured while monitoring the membrane potential and controlling the activity of the cell by intracellular current injection. This arrangement also made it possible to apply drugs that drastically changed or abolished spike generation.

In our experience, simultaneous intra- and extracellular recordings from Purkinje cells are difficult to obtain with independently manipulated electrodes. Most successful experiments were done with the electrodes in an array preadjusted so that the tip separation was 5 to 10 μm with the intracellular potential-recording electrode slightly protruding ahead of the ISM. The $[K^+]_0$ responses recorded during simultaneous intracellular recording of potential were usually of smaller amplitude than could be obtained with an independently manipulated K^+–ISM, since only minor readjustments of electrode position were possible after the intracellular electrode penetrated the cell membrane.

Figure 6 shows a set of records from the cell body of a Purkinje cell. $[K^+]_0$ was not measurably affected by hyperpolarizing or subthreshold depolarizing current pulses (Figures 6A and 6B). With increasing amplitude of suprathreshold, depolarizing current pulses, the spike frequency increased and generated a parallel increase in the $[K^+]_0$ response (Figures 6C–6G). In Figure 6G, the threshold for dendritic spikes was reached, producing additional $[K^+]_0$ responses for each low amplitude dendritic spike. Thus, it is clear that the $[K^+]_0$ level is determined by

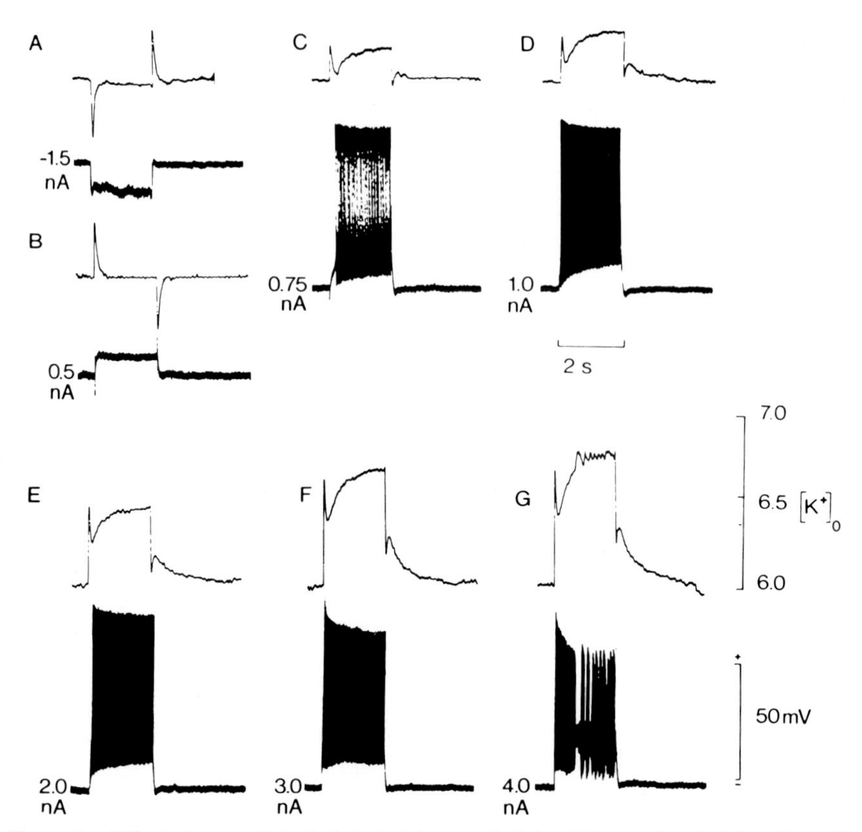

Figure 6. Effect of intracellularly injected hyperpolarizing (A) and depolarizing (B to G) current pulses on $[K^+]_0$ recorded extracellularly from the immediate vicinity of the soma. $[K^+]_0$ is shown in the upper trace and intracellular potential (Vi) in the lower trace in A–G. The amplitude of the applied current is indicated with each record. The lower record is the intracellular potential, Vi. Hyperpolarizing (A) and subthreshold (B) depolarizing currents had no effect on $[K^+]_0$. Suprathreshold depolarizing currents evoked a train of spikes and an increase in $[K^+]_0$ (C–G). The amplitude of the $[K^+]_0$ increase was related to the frequency of spikes (C–F). In (G), the threshold for dendritic Ca^{2+} spikes was exceeded and each Ca^{2+} spike was associated with a unitary $[K^+]_0$ response. Same concentration, voltage, and time calibrations apply to each record. From Hounsgaard and Nicholson, 1983.

the spike frequency of the immediately underlying membrane and also by the activity of more distant membrane regions, in this case the dendritic membrane.

We have analyzed the ionic conductances responsible for the $[K^+]_0$ responses. Figure 7 shows a recording from a neuron in which the fast

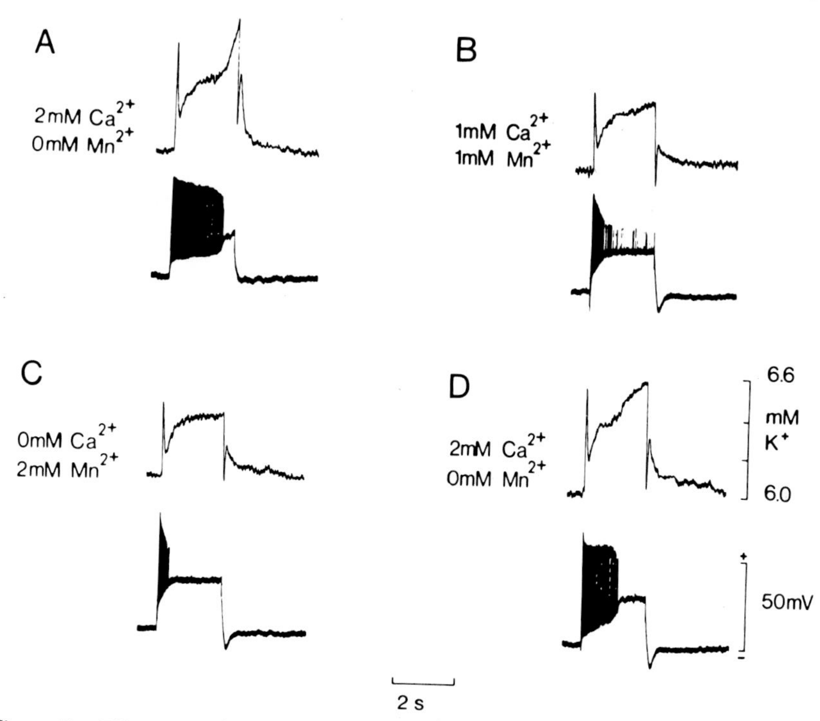

Figure 7. Effect of Mn^{2+} on the $[K^+]_o$ response during plateau potentials evoked by current injection in Purkinje cells. Upper records: $[K^+]_o$ response: lower records: intracellular potential. (A) In normal Ringer, an additional rapid increase in $[K^+]_o$ occurred during the plateau potential following inactivation of Na^+ spikes. (B and C) As Ca^{2+} replaced Mn^{2+}, the latter rapid $[K^+]_o$ increase was blocked even though the plateau potential persisted. (D) The late rapid $[K^+]_o$ response during the plateau potential recovered in normal Ringer. From Hounsgaard and Nicholson, 1983.

soma spikes inactivated during the terminating phase of a long-lasting depolarizing pulse leaving a plateau depolarization. In normal medium, $[K^+]_o$ increased rapidly during the plateau (Figure 7A). This additional $[K^+]_o$ increase was blocked by Mn^{2+} although the plateau still persisted (Figures 7B and 7C) and recovered in normal medium (Figure 7D). This experiment thus identified a Ca^{2+} dependent component of the somatic $[K^+]_o$ response characterized by a high threshold, slow activation, and the fact that it was not directly related to the production of action potentials. This conclusion raises a number of interesting interpretational questions. A somatic location of the involved conductances would provide the most straightforward explanation of the data, but all available electrophysiological evidence points to an exclusive dendritic location of Ca^{2+} channels (Llinás and Sugimori, 1980a,b, 1982). This conflict could be resolved by postulating a transduction process between the dendritic site of Ca^{2+} entry and the somatic site of K^+ outflux or by assuming a somatic Ca^{2+} conductance weak enough to escape electrophysiological detection, but with a potent effect on the somatic K^+ conductance. A third possibility is that the Ca^{2+}-dependent part of the somatic $[K^+]_0$ response is of dendritic origin and only appears in somatic recordings because of diffusion from the most proximal parts of the dendritic tree. This would imply a much larger $[K^+]_0$ increase in the dendritic region; to date we have not found evidence for this. In spite of these interpretational problems it is clear that $[K^+]_0$ at the level of the soma is influenced by both Ca^{2+}-dependent and voltage-dependent K^+ conductances.

The voltage-dependent component of the somatic $[K^+]_0$ response was studied in isolation by substituting Ba^{2+} for Ca^{2+} in the medium. This eliminated the Ca^{2+} dependent $[K^+]_0$ response since Ba^{2+} readily passes through Ca^{2+} channels, but fails to activate the Ca^{2+}-dependent K^+ conductance (Hagiwara *et al.*, 1974; Gorman and Hermann, 1979). While $[K^+]_0$ reached a steady plateau during a train of Na^+ spikes in normal medium (Figures 6C through 6G, 8A, and 8D) the characteristic long-lasting Ba^{2+} spikes generated a very large initial $[K^+]_0$ increase that declined and reached baseline during the spike (Figure 8B and C). We believe that this is a consequence of the membrane properties of Purkinje cells rather than the $[K^+]_0$ dispersal system. In extracellular records from the soma, the voltage change during Ba^{2+} spikes was predominantly positive, signifying a net outward current (Figures 8E and 8F). The time course of this voltage change and the $[K^+]_0$ change were closely similar to those expected if $[K^+]_0$ was determined by the transmembrane K^+ flux rather than changes in extracellular K^+ clearance. The rapid decline in $[K^+]_0$ accumulation during Ba^{2+} spikes was most likely due to slow inactivation of the voltage-dependent K^+ conductance with continuous depolarization. The alternative explanation is that Ba^{2+} accumulates intracellularly during the spike and blocks the K^+ conductance (Hagiwara *et al.*, 1974); this would imply that $[K^+]_0$ responses should decline with the time in Ba^{2+} medium. This was not seen even after hours in Ba^{2+} medium.

4.3. *Changes in $[K^+]_0$ Recorded outside Glia Cells during Intracellular Current Passage*

We have also used the dual recording regime to explore the membrane properties of glia cells. In the Purkinje cell layer, intracellular

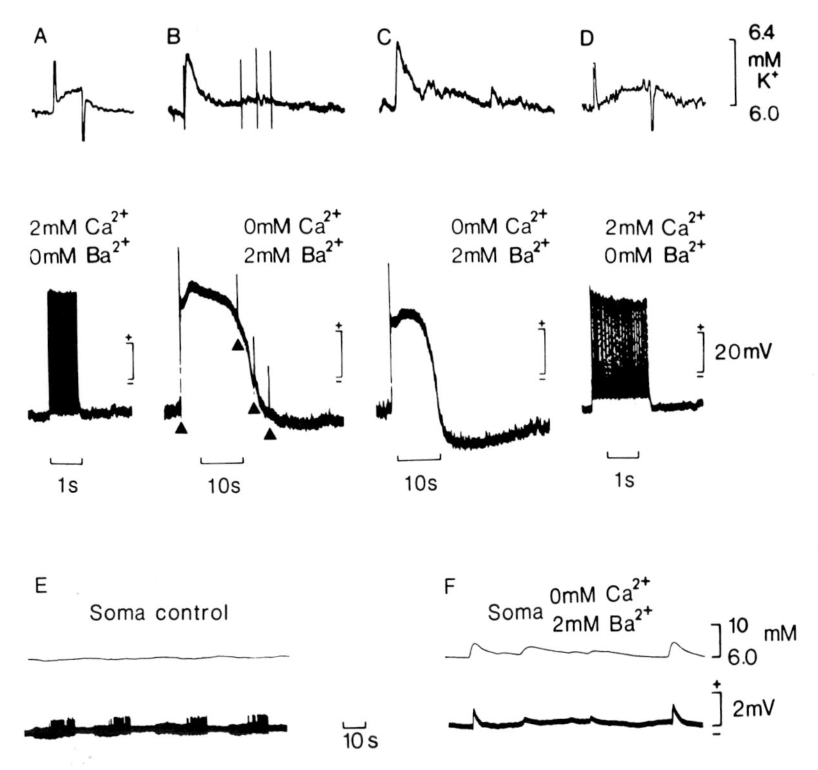

Figure 8. $[K^+]_0$ increase generated by Ba^{2+} spikes evoked by intracellular current injection. Upper record (A–D shows $[K^+]_0$, lower shows intracellular potential. (A,D) controls in normal Ringer; (B,C) Ba^{2+}-Ringer. (B) shows $[K^+]_0$ response associated with a Ba^{2+} spike generated by 200 sec current pulses (arrows). These subsequent pulses failed to generate spikes and $[K^+]_0$ responses. (C) shows $[K^+]_0$ response evoked by a spontaneous Ba^{2+} spike (D) shows response after recovery in normal Ringer. In (E,F), upper traces show extracellular potential, lower traces show $[K^+]_0$. Recording from the cell body of a Purkinje cell in normal medium (E) showed little $[K^+]_0$ associated with spikes (note that the amplification for ion recording was less for all records in this figure than in the other figures). In Ba^{2+} Ringer, the Ba^{2+} spikes are entirely positive and appreciable $[K^+]_0$ signals were seen (F). Ba^{2+} spikes and $[K^+]_0$ responses of at least two different amplitudes could be identified, indicating that more than one Purkinje cell contributed to the records (records A–D from Hounsgaard and Nicholson, 1983).

recordings are often obtained from cells with high membrane potential (75 to 85 mV) and linear membrane characteristics. Morphologically, these cells can be identified as Bergmann glia with intracellular horseradish peroxidase (HRP) injection (M. Sugimori and R. Llinás, personal communication). The $[K^+]_0$ response produced by intracellular current

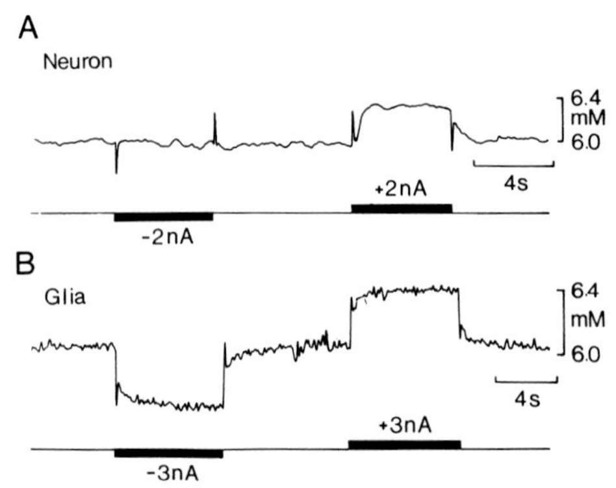

Figure 9. Comparison of $[K^+]_0$ responses to hyperpolarizing and depolarizing intracellular current pulses in a Purkinje cell (A) and a glia cell (B). In the neuron, only depolarization evoked a $[K^+]_0$ response while in the glia cell $[K^+]_0$ decreased during hyperpolarization and increased during depolarization. Note the symmetry of the glia response with changes in current polarity; this result suggests that transmembrane current is predominantly carried by K^+ in glia cells. (See also Figure 6.) Lower records indicate magnitude and duration of intracellular current injection. Modified from Hounsgaard and Nicholson, 1983.

injection clearly differed from the neuronal responses just described. Although measurable $[K^+]_0$ changes in the vicinity of neurons only occurred with suprathreshold depolarizations, both hyperpolarizing and depolarizing current pulses in glia cells resulted in very rapid changes in $[K^+]_0$ (Figure 9). The decrease of $[K^+]_0$ during hyperpolarization was comparable in amplitude to the increase during depolarization. Since K^+ constitutes only a small fraction of the extracellular ionic species, the results provide strong evidence that the current across the glia membrane in the vicinity of the K^+–ISM is almost exclusively carried by K^+. Bergmann glia of the mammalian cerebellum, therefore, have membrane characteristics closely similar to glia cells in invertebrate and lower vertebrate glia cells (Kuffler and Nicholls, 1966).

Mammalian glia cells are usually small and impalement could create a nonspecific leak around the intracellular electrode that may give a falsely low potential change in response to imposed extracellular $[K^+]_0$ changes. In our experiments, this source of error was avoided because the K^+–ISM sampled the ion change several microns away from the site of impalement. Even if a substantial fraction of the current was carried by other ions through the leak, the $[K^+]_0$ recording away from the site of impalement would still reveal the actual K^+-flux at the glia cell membrane.

5. PROSPECTS AND PROBLEMS

We have presented experiments that illustrate new ways to study the brain cell microenvironment with ISMs in cerebellar slices. Our re-

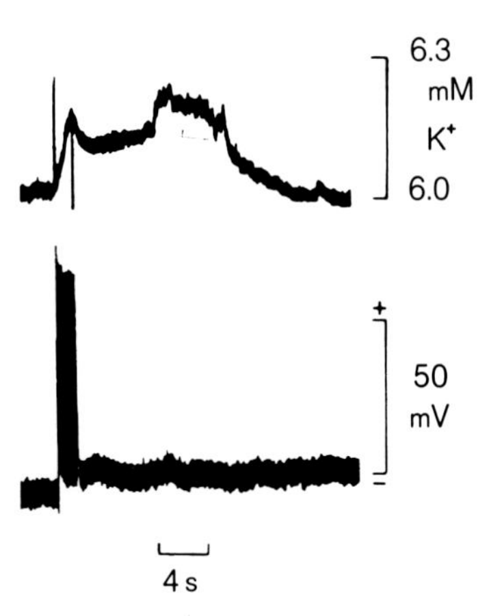

Figure 10. Summation of $[K^+]_0$ responses from two Purkinje cells. Upper record, $[K^+]_0$, lower, intracellular potential in one cell. Activation of one Purkinje cell by injecting current evoked a $[K^+]_0$ response (upper trace) corresponding to the generation of Na^+ spikes. This response continued as a prolonged $[K^+]_0$ plateau and an additional terminal $[K^+]_0$ increase—presumably associated with the activity of a neighboring Purkinje cell (not shown). These latter $[K^+]_0$ excursions are not seen in the intracellular potential recording from the cell originally stimulated. (From Hounsgaard and Nicholson, 1983.)

sults only constitute a modest first step. In order to focus experimental efforts on key questions, it may nevertheless be useful to formulate specific—though speculative—hypotheses about the possible functional significance of interaction between neurons and the brain cell microenvironment.

Classical neurotransmission occurs at specific sites often equipped with mechanisms to eliminate prolonged action and diffusional dispersal of released transmitters. The hypothesis that we stressed at the outset considers an additional mode of neuronal interaction. We suggest that neurons under some conditions release substances to the nonsynaptic extracellular space where they are dispersed predominantly by diffusion. The release sites may be pre- or postsynaptic, restricted to a limited part of the cell (e.g., presynaptic boutons) or widely distributed (e.g., released from the dendritic tree). K^+, released as described above, is an example of such a substance. (For this ion, the dispersal may be governed by mechanisms other than diffusion—see Nicholson, 1980). Substances released from neighboring neurons may add in the extracellular space. This is true for K^+ released from Purkinje cells. Figure 10 shows a $[K^+]_0$ recording during simultaneous intracellular recording from a Purkinje cell. In this case, $[K^+]_0$ stayed elevated after the train of spikes terminated and finally there was an additional increase before $[K^+]_0$ returned to baseline. The prolonged $[K^+]_0$ elevation correlated with the activity of a neighboring neuron monitored with the reference barrel of the K^+–ISM. The terminating $[K^+]_0$ increase occurred during a burst of Ca^{2+} spikes in this neuron and $[K^+]_0$ returned to baseline when firing stopped. Thus, $[K^+]_0$ at the tip of the K^+–ISM in this case was clearly modulated by the activity of the two different Purkinje cells.

Numerous studies have shown that $[K^+]_0$ increases throughout the extracellular space in areas where many neurons are simultaneously active. It seems likely that $[K^+]_0$ in specific areas of the brain is modulated continuously by the varying levels of neuronal activity. The three-dimensional profile of $[K^+]_0$ may thus carry information about the level and spatial distribution of the neuronal activity of an entire ensemble of neurons.

If neuroactive substances are indeeed handled as hypothesized, then the cytoarchitectural diversity in the brain assumes new significance by specifying the geometric distribution of release sites, the pathways of diffusion, and the spatial orientation of the receptive sites of the cells. These factors all add richness to the ways in which the brain cell microenvironment may operate as a communication channel in different areas of the brain, even for simple inorganic ions like K^+. Dendritic bundling (e.g., Scheibel and Scheibel, 1975) is an example of a geometric arrangement that favors direct neuronal interaction via the extracellular space while glia cells at other locations may prevent direct neuronal apposition and so shape the extracellular diffusional pathways.

Let us discuss these ideas specifically by considering the K^+ release from Purkinje cells in the cerebellar cortex. The cerebellar cortex is particularly attractive because of the highly organized cytoarchitecture and synaptic connectivity (Palay and Chan-Palay, 1974). As we have seen, the release of $[K^+]_0$ from Purkinje cells is most intense during Ca^{2+} spikes. In the intact cerebellum, these dendritic spikes probably are predominantly associated with climbing fiber input (Llinás and Sugimori, 1982). Evidence from *in vivo* studies reports the idea that climbing fiber responses strongly affect $[K^+]_0$ (Bruggencate *et al.*, 1976). On the other hand, each climbing fiber usually only innervates a single Purkinje cell in a given region of the cerebellum (Palay and Chan-Palay, 1974) and their firing frequency is so low (Eccles *et al.*, 1967) that the $[K^+]_0$ responses under most conditions will reach a baseline determined by the simple spike activity between climbing fiber responses (CFRs). CFRs may nevertheless be the major factor for $[K^+]_0$ build-up in the cerebellar cortex because of summation of CFR-produced $[K^+]_0$ responses from neighboring neurons. Although speculative, this possibility is supported by the finding *in vivo* that local variations in CFR input produce oscillations in $[K^+]_0$ (Stöckle and Bruggencate, 1980). Equally interesting is the puzzling effect of the CFR on the simple spike activity of Purkinje cells (Ebner and Bloedel, 1981a,b).

The basic prerequisites for neuronal interaction by release and diffusion of substances in the extracellular space thus seem established for K^+ in the cerebellar cortex, but direct evidence that changes in $[K^+]_0$

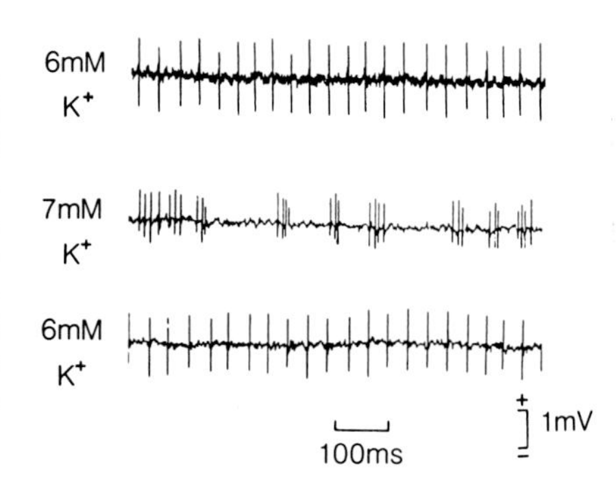

Figure 11. Effect of applied K^+ on spontaneous activity of Purkinje cell. At 6 mM $[K^+]_0$ (normal Ringer), the Purkinje cell fired at a regular rate of 30 Hz (upper trace). Upon introduction of 7 mM $[K^+]_0$ Ringer, the firing rate increased and, in the steady state, the cell fired in repetitive bursts (middle trace). Regular firing pattern returned on reintroduction of 6 mM $[K^+]_0$ Ringer (lower trace). From Hounsgaard and Nicholson, 1983.

significantly affect neuronal function is still lacking. We take this opportunity to point out some of the problems involved in attacking this type of problem. It is instructive to compare the present situation with the criteria that have evolved to identify synaptic transmitters (Werman, 1972).

A neurotransmitter can be identified if the substance is released from the presynaptic terminals by depolarization and if the substance, when applied to the postsynaptic membrane of the target cell, has the same effect as the naturally occurring transmitter. In addition, it should be possible to specifically block the action of the transmitter with a pharmacological agent. Similar strict criteria have to be applied for the kind of neuronal interaction discussed here. Although the release of K^+ from neurons has been amply demonstrated, it is much more difficult to establish that the $[K^+]_0$ changes actually affect neuronal function, a point that was emphasized by Baylor and Nicholls (1969) but almost entirely neglected in mammalian neurophysiology. *In vitro* studies can provide indirect evidence that application of K^+ to the bath, in quantities similar to the experimentally established $[K^+]_0$ increases, does alter neuronal behavior. As shown in Figure 11 a 1-mM $[K^+]_0$ increase clearly changed the firing pattern of spontaneously active Purkinje cells. But, as pointed out by Baylor and Nicholls (1969), the spatial and temporal distribution of experimental $[K^+]_0$ changes of this type is vastly different from that occurring naturally and therefore the two situations are not directly comparable. Also, the site of action of K^+ cannot be determined with any certainty.

In seeking to identify transmitters, specific blockers are often useful. The various K^+ channel blockers available unfortunately change the neuronal firing pattern drastically without preventing all release of K^+ (see Figure 2). Indeed, if a total block of the resting K^+-permeability of the neuronal membrane were to be achieved, it could abolish the resting potential.

Only a few convincing examples of physiological interaction between neurons mediated by $[K^+]_0$ are known. In two cases, synaptic transmission was eliminated by Ca^{2+}-blocking agents (Alkon and Grossman, 1978; Yarom and Spira, 1982). These studies may provide an experimental avenue for persuing physiological actions of $[K^+]_0$ changes in vertebrates but inherently exclude Ca^{2+}-dependent release. Recently, evidence has been presented for K^+-mediated interaction between elements of the cerebellum (Malenka *et al.*, 1981).

It must be emphasized that modulation of extracellular ion concentration is not restricted to K^+. The experimental strategies that have been outlined here are readily applicable to ecluidate the interesting questions that relate to the possible function of $[Ca^{2+}]$ changes—interesting because Ca^{2+} is involved in charge screening, synaptic transmission, and postsynaptic action potentials. New progress in electrode technology promises that even organic compounds (amino acids, peptides, etc) may soon be examined as nonsynaptic transmitter candidates, with techniques that allow real-time monitoring in the brain cell microenvironment.

ACKNOWLEDGMENT. This work was supported by USPHS Program Grant NS-13742 from NINCDS and by the University of Copenhagen.

6. REFERENCES

Alger, B. E. and Teyler, T. J., 1978, Potassium and short-term response plasticity in the hippocampal slice, *Brain Res.* **159**:239–242.

Alkon, D. L. and Grossman, Y., 1978, Evidence for nonsynaptic neuronal interaction, *J. Neurophysiol.* **41**:640–653.

Baylor, D. A. and Nicholls, J. G., 1969, Changes in extracellular potassium concentration produced by neuronal activity in the central nervous system of the leech, *J. Physiol. (London)* **203**:555–569.

Benninger, C., Kadis, J., and Prince, D. A., 1980, Extracellular calcium and potassium changes in hippocampal slices, *Brain Res.* **187**:165–182.

Bruggencate, G. ten, Nicholson, C., and Stockle, H., 1976, Climbing fiber evoked potassium release in cat cerebellum, *Pflügers Arch.* **367**:107–109.

Dietzel, I., Heinemann, U., Hofmeier, G., and Lux, H. D., 1980, Transient changes in the size of the extracellular space in the sensorimotor cortex of cats in relation to stimulus-induced changes in potassium concentration, *Exp. Brain Res.* **40**:432–439.

Ebner, T. J. and Bloedel, J. R., 1981, Role of climbing fiber afferent input in determining responsiveness of Purkinje cells to mossy fiber input, *J. Neurophysiol.* **45**:962–971.

Ebner, T. J. and Bloedel, J. R., 1981, Temporal patterning in simple spike discharge of Purkinje cells and its relationship to climbing fiber activity, *J. Neurophysiol.* **45**:933–947.

Eccles, J. C., Ito, M., and Szentagothai, J., 1967, *The Cerebellum as a Neuronal Machine*, Springer-Verlag, Berlin.

Frankenheuser, B. and Hodgkin, A. L., 1956, The after-effects of impulses in the giant nerve fibers of *loligo*, *J. Physiol.* (*London*) **131**:341–376.

Fritz, L. C. and Gardner-Medwin, A. R., 1976, The effects of synaptic activation on the extracellular potassium concentration in the hippocampal dentate area, *in vitro*, *Brain Res.* **112**:183–187.

Gorman, A. L. F. and Hermann, A., 1979, Internal effects of divalent cations on potassium permeability in molluscan neurons, *J. Physiol.* (*London*) **296**:393–340.

Gutnik, M. J. and Segal, M., 1981, Serotonin and GABA-induced fluctuations in extracellular ion concentration in the hippocampal slice, in: *Ion-Selective Microelectrodes and Their Use in Excitable Tissues* (E. Syková, P. Hník, and L. Vyklický, eds.), Plenum Press, New York, pp. 261–265.

Hagiwara, S., Fukuda, J., and Eaton, D. C., 1974, Membrane currents carried by Ca, Sr, and Ba in barnacle muscle fiber during voltage clamp. *J. Gen. Physiol.* **63**:564–578.

Hansen, A. J. and Olsen, C. E., 1980, Brain extracellular space during spreading depression and ischemia, *Acta Physiol. Scand.* **108**:355–365.

Heinemann, U. and Lux, H. D., 1975, Undershoots following stimulus-induced rises of extracellular potassium concentration in cerebral cortex of cat, *Brain Res.* **93**:63–76.

Hosli, L., Andres, P. F., and Hosli, E., 1978, Neuron-glia interactions: Indirect effects of GABA on cultured glial cells, *Exp. Brain Res.* **33**:425–434.

Hounsgaard, J., 1979, Pacemaker properties of mammalian Purkinje cells, *Acta Physiol. Scand.* **106**:91–92.

Hounsgaard, J. and Nicholson, C., 1983, Potassium accumulation around individual Purkinje cells in cerebella slices, *J. Physiol.* (*London*) **340**:359–388.

Kleine, R. P. and Kupersmith, J., 1982, Effects of extracellular potassium accumulation and sodium pump activation on automatic canine Purkinje fibers, *J. Physiol.* (*London*), **324**:507–533.

King, G. L. and Somjen, G. G., 1981, Extracellular calcium and action potentials of soma and dendrites of hippocampal pyramidal cells, *Brain Res.* **226**:339–343.

Kuffler, S. W. and Nicholls, J. C., 1966, The physiology of neuroglia cells, *Ergeb. Physiol. Biol. Chem. Exp.* **57**:1–90.

Llinás, R. and Sugimori, M., 1980a. Electrophysiological properties of *in vitro* Purkinje cell somata in mammalian cerebellar slices, *J. Physiol.* (*London*) **305**:171–195.

Llinás, R. and Sugimori, M., 1980b, Electrophysiological properties of *in vitro* Purkinje cell dendrites in mammalian cerebellar slices, *J. Physiol.* (*London*) **305**:197–213.

Llinás, R. and Sugimori, M., 1982, Functional significance of the climbing fiber input to Purkinje cells: An *in vitro* study in mammalian cerebellar slices, in: *The Cerebellum—New Vistas* (S. L. Palay and V. Chan-Palay, eds.), Springer-Verlag, Berlin—Heidelberg—New York, pp. 403–411.

Malenka, R. C., Kocsis, J. D., Ransom, B. R., and Waxman, S. G., 1981, Modulation of parallel fiber excitability by postsynaptically mediated changes in extracellular potassium, *Science* **214**:339–341.

Nicholson, C., 1979, Brain cell microenvironment as a communication channel, in: *The Neurosciences: Fourth Study Program* (F. O. Schmitt and F. G. Worden, eds.), MIT Press, Cambridge, Mass.

Nicholson, C., 1980, Dynamics of the brain cell microenvironment, *Neurosci. Res. Prog. Bull.* **18**:177–322.

Nicholson, C. and Hounsgaard, J., 1983, Diffusion in the slice microenvironment and implications for physiological studies, *Fed. Proc.*, **42**:2865–2868.

Nicholson, C. and Phillips, J. M., 1981, Ion diffusion modified by tortuosity and volume fraction in the extracellular microenvironment of the rat cerebellum, *J. Physiol.* (*London*) **321**:225–258.

Nicholson, C., Bruggencate, G. ten, Stöckle, H., and Steinberg, R., 1978, Calcium and potassium changes in extracellular microenvironment of cat cerebellar cortex, *J. Neurophysiol.* **41:**1026–1039.

Nicholson, C., Phillips, J. M., Tobiasz, C., and Kraig, R. P., 1981, Extracellular potassium, calcium and volume profiles during spreading depression, in: *Ion-Selective Microelectrodes and Their Use in Excitable Tissues,* (E. Syková, P. Hník, and L. Vyklický, eds.), Plenum Press, New York, pp. 211–223.

Palay, S. L. and Chan-Palay, V., 1974, *Cerebellar Cortex,* Springer-Verlag, New York.

Phillips, J. M. and Nicholson, C., 1979, Anion permeability in spreading depression investigated with ion sensitive microelectrodes, *Brain Res.* **173:**567–571.

Scheibel, M. E. and Scheibel, A. B., 1975, Dendrites as neuronal couples: The dendrite bundle, in: *Golgi Centennial Symposium Proceedings* (M. Santini, ed.), Raven Press, New York, pp. 347–354.

Schmitt, F. O. and Samson, F. E., Jr., 1969, Brain cell microenvironment, *Neurosci. Res. Program Bull.* **7:**177–417.

Stöckle, H. and Bruggencate, G. ten, 1980. Fluctuation of extracellular potassium and calcium in the cerebellar cortex related to climbing fiber activity, *Neuroscience* **5:**893–901.

Syková, E., Hník, P., and Vyklický, L., eds., 1981, *Ion-Selective Microelectrodes and Their Use in Excitable Tissues.* Plenum Press, New York.

Werman, R., 1972, CNS cellular level: Membranes, *Annu. Rev. Physiol.* **34:**337–374.

Yamamoto, C., 1974, Electrical activity observed *in vitro* in thin sections from guinea pig cerebellum, *Jpn. J. Physiol.* **24:**177–188.

Yamamoto, C. and Chujo, T., 1978, Visualization of central neurons and recording of action potentials, *Exp. Brain Res.* **31:**299–301.

Yarom, Y. and Spira, M. E., 1982, Extracellular potassium ions mediate specific neuronal interaction, *Science* **216:**80–82.

10

Electrophysiological Study of the Neostriatum in Brain Slice Preparation

S. T. KITAI and H. KITA

1. INTRODUCTION

The neostriatum of mammalian species consists of the caudate and the putaman and is a major component of the basal ganglia (Carpenter, 1981). It is a relatively large nuclear mass and is easily dissected from the rest of the brain. In slice preparation, the neostriatum is severed from extrinsic connections. Neuronal responses recorded following electrical stimulation of the neostriatum in slice preparation therefore reflect only the activities of those neuronal elements left intact. These would include the soma, dendrites, and, so long as their course remains within the slice, axons of neostriatal neurons. Added to these are fibers of extrinsic origin and their terminals, isolated from their cell bodies, but remaining electrically excitable. The slice preparation therefore would be well suited to the study of local interactions among striatal neuronal elements and the action of monosynaptic afferents. The slice preparation enables the investigator to manipulate the extracellular milieu in a controlled manner. For this reason, it is also well suited for studying the action of putative neurotransmitters and other agents on the electrical

S. T. KITAI and H. KITA • Department of Anatomy, College of Medicine, The University of Tennessee, Center for the Health Sciences, Memphis, Tennessee 38163.

activities of neurons and characterizing the relationships between these membrane activities and specific electrolytes. Even though the slice technique has been employed in neuropharmacological and biochemical studies of the basal ganglia, laboratories utilizing the slice technique for electrophysiological investigation are limited to only a few (Misgeld and Bak, 1979; Lighthall *et al.*, 1981). In this report, we will describe mainly the technique utilized in our laboratory for the neostriatal slice preparation and a method for intracellular labeling of neurons in the slice preparation by microinjection of horseradish peroxidase (HRP).

2. METHODS

2.1. *Electrode*

Glass microelectrodes are used for both extracellular and intracellular recording. Glass capillaries with outer diameter of 1 to 3 mm are pulled to a fine tip using a pipette puller. For extracellular unitary or field potential recording, glass capillaries are filled with 2 to 4 M NaCl (DC resistance of 5 to 10 MΩ). Intracellular recordings are obtained with glass capillaries filled with 2 M K-citrate, 2 M K-methylsulfate or 3M KCl (DC resistance of 30 to 80 MΩ). For intracellular labeling with HRP, the tip of microelectrode is beveled using either a commercial beveler or a mechanical bumping method (Kitai and Bishop, 1981; Kitai and Wilson, 1982). Beveled glass microelectrodes with a tip diameter of less than 0.5 μm are filled with a 4% solution of HRP in 0.5 M KCl or 0.5 M K-methylsulfate and 0.05 M tris buffer (pH 7.6). Microelectrodes suitable for labeling have DC resistance of 60 to 90 MΩ.

Stimulating electrodes placed near the recording electrode (as close as 0.5 to 1 mm) consist of twisted pairs of wire (i.e., 80-μm diameter nichrome wire) insulated to within 100 μm of the tip and with an intertip distance of 40 to 100 μm. Electrical stimulus pulses are 0.05 to 0.3 msec in duration and 5 to 90 V in amplitude.

2.2. *Slice Chamber*

The slice chamber is constructed of acrylic plastic (Plexiglas). It consists of a recording chamber and a waterbath (Figure 1). The temperature of the waterbath is kept constant at 37°C with a thermostatic heater coil. O_2/CO_2 gas is pumped continuously through the waterbath to create a warm moist environment over the recording chamber. The superfusing Kreb's Ringer passes by gravity feed to the recording chamber via a heat

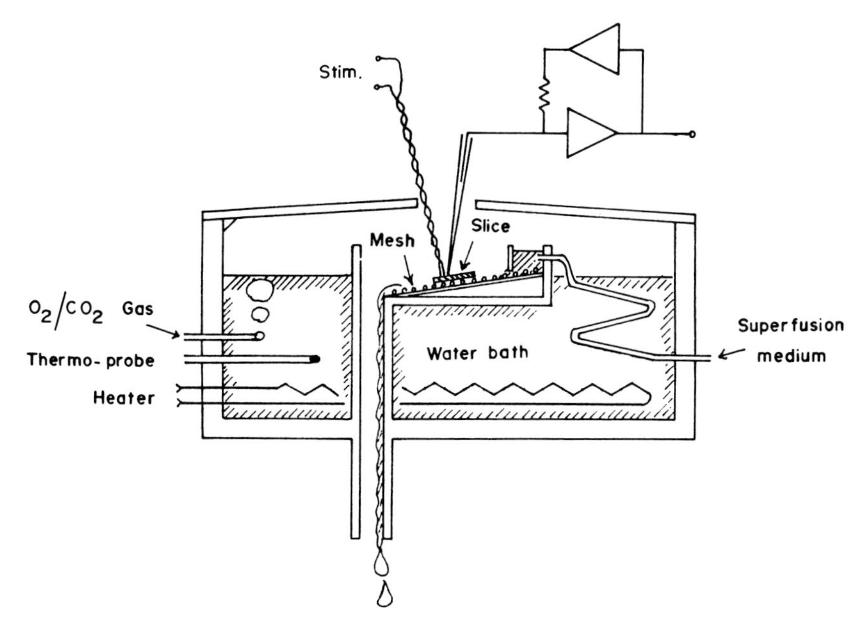

Figure 1. Schematic diagram of the slice chamber. Stim: stimulating electrode.

exchanger in the waterbath. One end of the recording chamber is raised so that the superfusion of Kreb's Ringer can run over the supporting mesh, on which a brain slice is placed, at a constant flow rate of 0.5 to 1.0 ml/min. The slice surface is therefore free of superfusion Kreb's Ringer and is directly exposed to the warm and moist O_2/CO_2 gas mixture. A dissecting microscope is placed over the slice chamber for visualization of stimulating and recording electrode placements.

The superfusion Kreb's Ringer is composed as follows: NaCl 124.0 mM, KCl 5.1 mM, $MgSO_4$ 1.3 mM, CaCl 2.5 mM, KH_2PO_4 1.25 mM, $NaHCO_3$ 26.0 mM, and D-glucose 10.0 mM. The pH of the Kreb's Ringer is maintained at 7.3 to 7.4 by bubbling a gas mixture of 95% O_2/5% CO_2 through the perfusate. The osmolarity of the Kreb's Ringer is checked prior to an experiment with an osmometer and adjusted to 305 ± 5 mOsm. It is noteworthy to mention that the Ca^{2+} concentration of the Kreb's Ringer is a factor in determining patterns of synaptic response in neostriatal slice preparation. Thus, it has been reported that no inhibition exists in rat neostriatal slice preparations, using a Kreb's Ringer of similar composition to that above except with 1.2 mM $CaCl_2$ (Misgeld and Bak, 1979). On the other hand, inhibition was observed when the molarity of $CaCl_2$ was raised to 2.3 to 2.5 mM (Lighthall *et al.*, 1981).

2.3. *Slice Preparation*

The animal (i.e., rat) is decapitated by a small animal guillotine and the brain gently removed from the skull. Neostriatal slices are prepared using either a tissue chopper or a Vibratome for sectioning. With the tissue chopper method, a core containing neostriatum from both hemispheres is initially removed by inserting a coring device (a brass rectangular tube measuring 2.5 mm high by 5.5 mm wide in cross section and 50 mm long) through the brain from one side to the other. The tissue is then extruded from the core device by gently pressing a brass plunger through the tubing onto a small piece of filter paper. The tissue is sectioned with the tissue chopper in parasagittal planes at a thickness of about 350 μm. It is important to use a clean and sharp razor blade in the tissue chopper to ensure better survival of the slice. With the Vibratome sectioning method, the brain, once removed from the skull, is placed immediately in a small petri dish containing cold (~ 5°C) oxygenated Kreb's Ringer. The brain is cut by hand into an appropriately oriented block (i.e., frontal, sagittal, etc.) with a razor blade. The tissue block is then glued to the cutting stage of a Vibratome using a cyanoacrylic cement. The Vibratome bath is filled with cold Kreb's Ringer solution and the block is sectioned to approximately 350 μm thickness. Four to six slices sectioned using a Vibratome or a tissue chopper are stored in a beaker containing continuously oxygenated Kreb's Ringer at room temperature until each slice is ready to be recorded. We have found it usually best to store the tissue in this manner at least 1 hr prior to recording. For the experiments combining intracellular recording and labeling, slices are prepared from ether-anesthetized animals perfused through the heart with a small volume (20 to 50 ml) of cold (5 to 10°C) oxygenated Kreb's Ringer just prior to decapitation. The predecapitation perfusion is helpful in reducing staining of erythrocytes in blood vessels, which often obscures the HRP-labeled neurons. Perfusion prior to decapitation does not seem to alter electrophysiological activities of the slice.

For both tissue chopper and Vibratome sectioning methods, the entire operation from decapitation to placement of the slices in oxygenated Kreb's Ringer should be carried out within 15 to 20 min.

The following procedures are used for an intracellular labeling of neurons with HRP. When the neurons are penetrated and have stable resting potentials of 40 mV or more, HRP is iontophoretically injected using 1 to 3 nA rectangular depolarizing pulses of 150 msec duration, at 3.3 Hz for 3 to 5 min. After the injection is completed, the slice containing the injected neuron is carefully placed on a small piece of lens

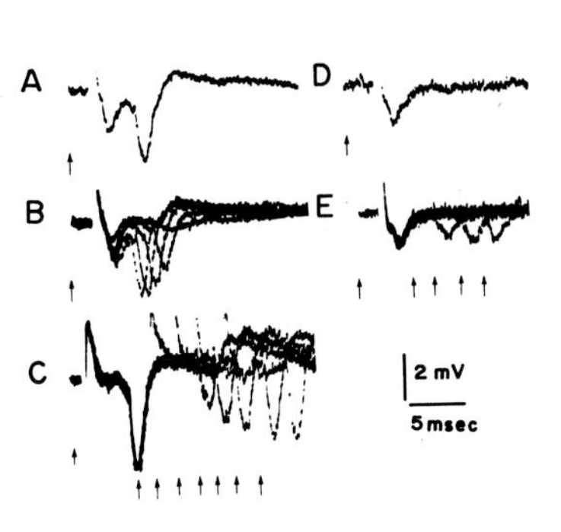

Figure 2. Locally evoked field potential: (A) Two-component (N-1 and N-2) field potential. (B) Changes in stimulation intensity alter N-2 latencies but not N-1. (C) Reduction of the test N-2 response in paired shock at various intrastimulus intervals. Traces are superimposed to show the time course of recovery. (D) Abolishment of the N-2 response in Ca^{2+} free medium. Only the N-1 potential remains. (E) No reduction of the test N-1 response in paired shock at the comparable stimulus intervals during which the N-2 responses are suppressed in (C). Up-going arrows in this and the subsequent figures indicate the onset of stimulation.

paper saturated with the Kreb's Ringer. The paper with the slice is gently floated on fixative solution containing 0.5 to 4.0% formaldehyde and 1.0 to 2.0% glutaraldehyde in standard isotonic buffer for 30 sec to 1 min. The lens paper is then upturned and the slice immersed in the fixative. This procedure of transporting the slice on the lens paper prevents it from curling during fixation. Slices with injected neurons are allowed to remain 3 to 12 hr in the fixative. Following fixation, each slice is sectioned at 50 μm using either a sliding freezing microtome or a Vibratome. For Vibratome sectioning, the slice is embedded in warm agar (2%) and the agar is cut into a block prior to sectioning. The sections are processed according to the standard HRP procedures (see Kitai and Wilson, 1982).

3. RESULTS AND DISCUSSION

3.1. *Field Potentials*

Local electrical stimulation of the neostriatal slice evokes a field potential usually consisting of two negative components: N-1 (latency 0.5 to 2.0 msec) and N-2 (latency 1.5 to 5.0 msec), both ranging in amplitude from approximately 0.5 to 4.0 mV (Figure 2).The N-1 component is unaltered during high-frequency stimulation (i.e., 100 Hz), changes in stimulation intensity, or manipulations of Ca^{2+} in the medium, such as substitution of Ca^{2+} with Mg^{2+} and, therefore, is due to the action currents of presynaptic and/or passing fibers and direct activation of neostriatal neurons. In contrast, the N-2 component is labile to high-frequency stimulation or alteration in Ca^{2+} content (Figure 2) and thus is considered to be a result of synaptically activated responses. These experiments

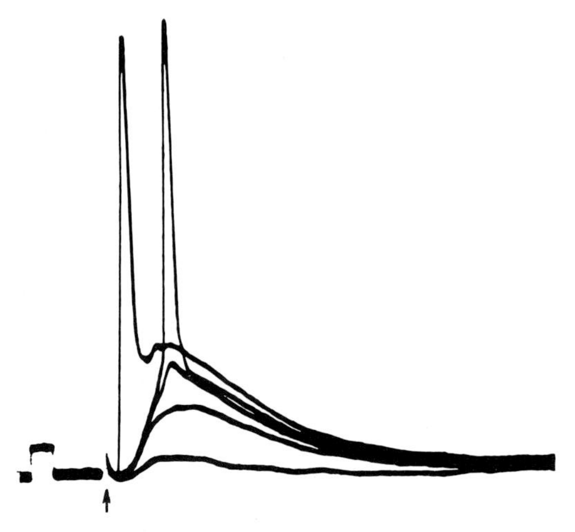

Figure 3. Locally evoked action and synaptic potentials: One trace shows an antidromically activated spike potential accompanied by a depolarizing synaptic potential. Other superimposed traces are synaptic potentials at various stimulus intensity. Note the latency of EPSPs are constant with different stimulus intensities. At a higher stimulus intensity, a spike potential is triggered from the EPSP. Calibration bar = 5 mV, 2 msec.

demonstrate how the components of a compound field potential, in a slice preparation, can be distinguished as pre- or postsynaptic, by manipulations of the medium.

3.2. *Intracellular Recording*

The brain slice presents an ideal preparation for intracellular recording from relatively small neurons since they are virtually free of the pulsations often encountered in *in vivo* preparations. Another advantage is that the depth of penetration is small (i.e., usually less than 400 μm) so that the fine microelectrode tip is less likely to be damaged compared with *in vivo* preparations in which the recording electrode must pass through a long length of brain before it reaches the target area.

A transmembrane potential of more than 60 mV is often recorded as a microelectrode impales neostriatal medium neurons (sizes ranging from 11 to 20 μm). In this pulsation-free condition, a long and stable recording of 1 to 2 hr from a single neuron is very common. Local stimulation can evoke depolarizing potentials from which action potentials are triggered or trigger antidromic spikes followed by synaptic potentials (Figure 3). These depolarizing potentials are considered to be excitatory postsynaptic potentials EPSPs since the potentials increase in amplitude during injection of constant hyperpolarizing current and decrease in amplitude with constant depolarizing current (Figure 4). The latency of the EPSPs are constant with variance in stimulus intensities indicating that it is monosynaptic in nature (Figure 3). The synaptic nature of such

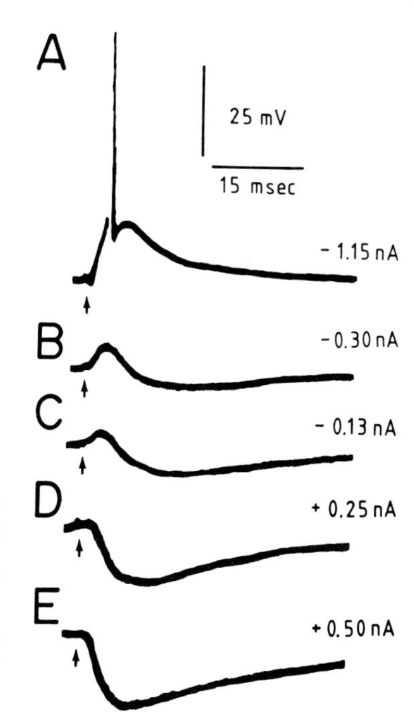

Figure 4. Effect of constant current injection on the amplitude of EPSP–IPSP: (A–C) Increasing hyperpolarizing current eventually elicits an action potential (A). (D–E) The amplitude of IPSP increases with the increase in depolarizing current. Numbers on the right hand side of each trace indicate the intensity of depolarizing (+) and hyperpolarizing (−) current.

depolarizing potentials can be examined by replacement of extracellular Ca^{2+} with Mg^{2+}. The abolishment of the response is one indication that these responses are due to synaptic action. The slice preparation is one method that allows easy manipulation of such procedures. The mean amplitude of action potentials of 86 mV (S.D. ± 11.5 mV) have been recorded from over 70 neostriatal neurons with resting membrane potentials of more than 60 mV. When triggered from EPSPs, an averaged threshold potential is about 5 mV. Though not often, inhibitory postsynaptic potentials (IPSPs) have been observed following EPSPs (Figure 4). The polarity and amplitude of these synaptic potentials could be manipulated by injection of constant current (hyperpolarizing and depolarizing) (Figure 4). The time course of inhibition, measured by the duration of IPSPs (Figure 4), or a reduction in amplitude of test EPSPs in double shock experiments (Figure 5), is found to be approximately 30 msec. These short-duration inhibitions found in the slice preparation is similar to that reported for the recurrent inhibition in the neostriatum (Park *et al.*, 1980), but contrasts to an inhibition of longer duration (200 to 300 msec) observed in the *in vivo* preparation following stimulation of either local structures or extrinsic afferents (Buchwald *et al.*, 1973;

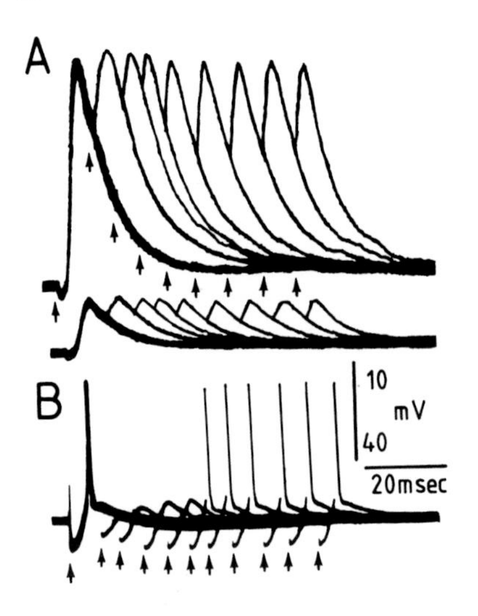

Figure 5. Time course of inhibition of EPSPs and action potentials in paired shock: (A) Conditioned EPSP (first arrow indicates stimulus artifact of conditioning response) reduces last EPSPs at ISIs of 9.5 to 36.5 msec. High gain records are shown in the top traces and low gain in the bottom traces. (B) Action potentials triggered by test EPSPs are blocked by a conditioning action potential.

Kocsis and Kitai, 1977). Wilson *et al.* (1983) have reported that the isolation of the neostriatum from cerebral cortex or thalamus by surgical lesion, abolishes this long-lasting inhibition. This confirms that the long-duration inhibition involves neuronal circuits extrinsic to the neostriatum. The slice preparation in which any extrinsic loop circuits must be open is clearly useful in testing and confirming the above hypotheses.

The passive membrane properties of neostriatal neurons are also relatively easily examined since the stable slice preparation provides the investigator with ample time to manipulate experimental variables. Another advantage of the slice preparation specific for neostriatal neurons is a lack of spontaneous activity. The voltage dependance of the input resistance of neostriatal neurons is studied by determining the current-voltage relationship, with a special care given to the bridge balance of the preamplifier during recording. Hyper- and depolarizing current pulses of known intensity are injected into the cell and potential shifts across the cell membrane are measured. The current-voltage relationship of neostriatal neurons measured demonstrated strong nonlinearity of the input resistance (anomolous rectification) at hyperpolarizing currents higher than 1 nA. A sample record is shown in Figure 6. Depolarizing currents with the value not exceeding the threshold of spike firing and hyperpolarizing currents up to 5 nA were tested. The average input resistance of 52 neostriatal neurons at resting membrane potential is 16.6 MΩ.

Neurons whose electrical activities has been studied can then be intracellularly labeled with HRP iontophoretically injected through the

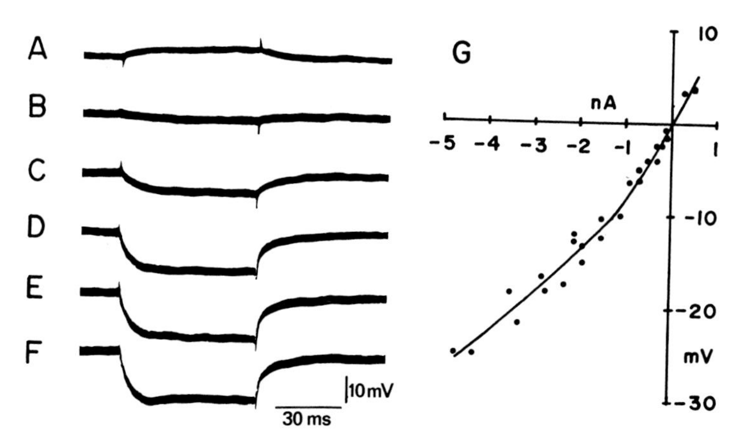

Figure 6. Input resistance determination for one striatal neuron. (G) Current-voltage relation with membrane responses shown in (A to E) for 60-msec depolarizing pulses of 0.3 nA (A) and hyperpolarizing pulses of 0.1 (B), 0.7 (C), 1.5 (D), 1.9 (E), and 2.2 (F), respectively. Note membrane rectification in hyperpolarizing direction.

recording microelectrode. All the neurons so far labeled have been identified as medium spiny neurons, based on their soma-dendritic morphology. The medium spiny neuron is the predominant cell type in the neostriatum and is easily recognizable by its heavy investitude of dendritic spines (Figure 7).

3.3. *Discussion and Summary*

The advantages of slice preparation are clear. (1) The preparation is free of anesthetic effects. (2) The preparation of slice is less time consuming compared with a difficult and long surgical approach one may encounter in *in vivo* preparation. (3) The preparation can expose any structure of the brain and therefore enables a precise placement of recording and stimulating electrodes. (4) The preparation is free of brain pulsations caused either by vascular pulsation or respiratory movements. This is especially critical in intracellular recording. This movement-free preparation promotes a stable and long penetration that allows an investigator to manipulate experimental variables. (5) The neostriatal slice preparation is free of spontaneous activity, most likely due to a severance of tonic extrinsic inputs. The absence of spontaneous activity makes it well-suited to study electrical membrane properties of the neuron. (6) In addition, since the surface of the slice in the recording chamber is exposed to O_2 and CO_2 gas mixture, the electrode capacitance

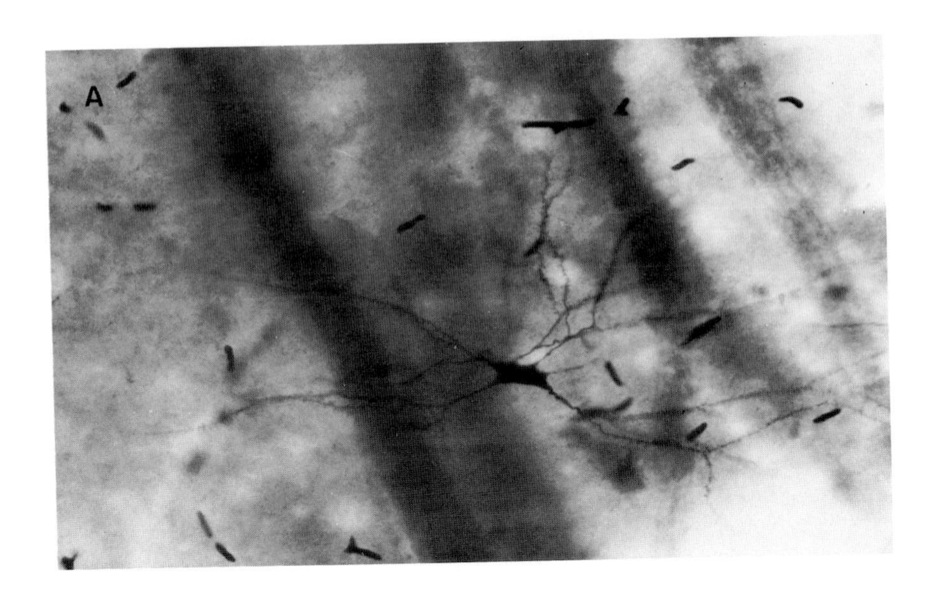

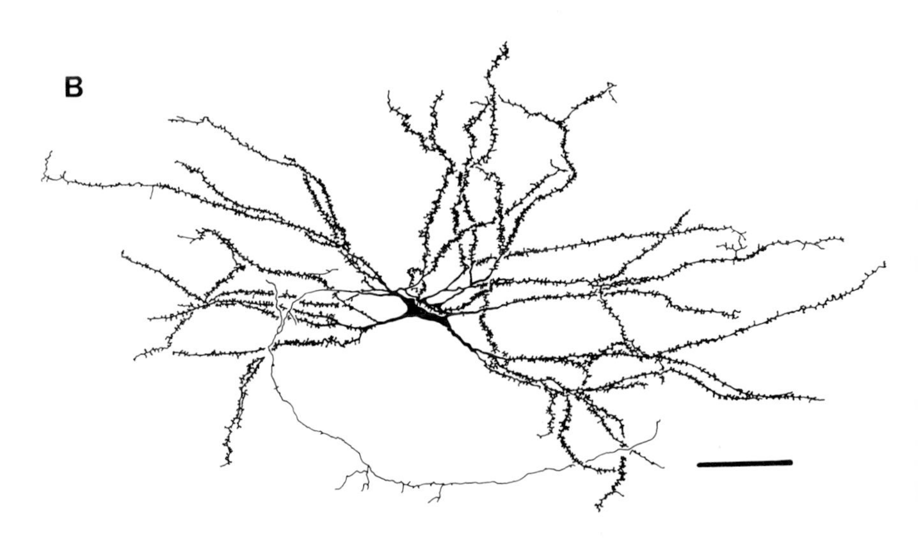

Figure 7. (A) Photomicrograph of a medium spiny neuron intracellularly labeled with HRP. Note that the soma and primary dendrites are spine free while distal dendrites are spine laden. (B) Same neuron as (A) serially reconstructed with the aid of a drawing tube. Calibration bar = 50 μm.

tends to be small and can be readily compensated. This is a clear advantage over an *in vivo* preparation, especially when the recording microelectrode must be inserted for a long distance through the brain tissues to reach a recording site. (7) The slice preparation also allows the experimenter to manipulate the chemical composition of the medium in which the slice is superfused. This preparation is especially suited for neuropharmacological study and electrochemical study of the nerve cell membrane using intracellular recording techniques where the element of time is critical and the precise knowledge of chemical environment of the nerve cell is demanded.

On the other hand, the slice preparation is not without shortcomings. One of the major disadvantages is that the preparation is far from being a normal of *in vivo* situation. The neuronal elements under study are limited by the thickness (usually around 350 to 450 μm) of the tissue. Undoubtedly, many neuronal processes are damaged or destroyed during sectioning. Anatomical studies have demonstrated that the dendrites of medium spiny neurons extend over 300 μm (Fox *et al.*, 1971/72; Kemp and Powell, 1971; DiFiglia, 1976; Sugimori *et al.*, 1978; Preston *et al.*, 1980; Wilson and Groves, 1980). Some of these medium spiny neurons are found to be projection neurons with extensive intranuclear collaterals (Somogyi and Smith, 1979; Preston *et al.*, 1980). However, in our slice preparation, sectioning of the neostriatal tissue by 350 to 400 μm does not seem to affect the electrical properties of the neurons. This statement is based on the fact that there is a very little difference in the resting membrane potentials (i.e., more than 60 mV) or spike potentials (over 70 mV) of neurons in slice preparations and those *in vivo* preparation (Sugimori *et al.*, 1978). In addition, local stimulation easily evokes monosynaptic EPSPs from which spike potentials are triggered. This may mean that either the damaged ends might have been sealed before recording or we were selectively recording from relatively undamaged neurons. All the HRP-labeled neurons, however, are shown to be relatively intact in their dendritic and axon collateral morphology when compared to those labeled in *in vivo* preparations.

In conclusion, care must be taken in the design of an experiment in applying the slice preparations. On the other hand, it is an excellent tool, adding a dimension to the interpretation of *in vivo* data and should provide valuable insights into the working components of the brain.

Acknowledgments. This work was supported by NIH Grant NS 14866 to S.T.K. Thanks are given to Dr. J. Lighthall for the use of collaborative data for a partial construction of some figures.

4. REFERENCES

Buchwald, N. A., Price, D. D., Vernon, L., and Hull, C. D., 1973, Caudate intracellular responses to thalamic and cortical inputs, *Exp. Neurol.* **38:**311–323.

Carpenter, M. B., 1981, Anatomy of the corpus striatum and brain stem integrating systems, in: *Handbook of Physiology*, Section I: *The Nervous System II Part 2* (V. B. Brooks, ed.), American Physiological Society, Bethesda, Maryland, pp. 947–995.

Chang, H. T., Wilson, C. J., and Kitai, S. T., 1982, A Golgi study of rat neostriatal neurons: Light microscopic analysis, *J. Comp. Neurol.* **208:**107–126.

DiFiglia, M., Pasik, P., and Pasik, T., 1976, A Golgi study of neuronal types in the neostriatum of monkeys, *Brain Res.* **114:**245–256.

Fox, C. A., Andrade, A. N., Hillman, D. E., and Schwyn, R. C., 1971/72, The spiny neurons in the primate striatum: A Golgi and electron microscopic study, *J. Hirnforsch.* **13:**181–201.

Kemp, J. M. and Powell, T. P. S., 1971, The structure of the caudate nucleus in the cat: Light and electron microscopy, *Phil. Trans. R. Soc. London, ser. B* **262:**383–401.

Kitai, S. T. and Bishop, G. A., 1981, Intracellular straining of neurons, in: *Neuroanatomical Tract-tracing Methods* (L. Heimer and M. J. RoBards, eds.), Plenum Press, New York, pp. 263–277.

Kitai, S. T. and Wilson, C. J., 1982, Intracellular labeling of neurons in mammalian brain, in: *Cytochemical Methods in Neuroanatomy* (V. Chan-Palay and S. L. Palay, eds.), Alan R. Liss, Inc., New York, pp. 533–549.

Kocsis, J. D. and Kitai, S. T., 1977, Dural excitatory inputs to caudate spiny neurons from substantia nigra stimulation, *Brain Res.* **138:**271–283.

Lighthall, J. W., Park, M. R., and Kitai, S. T., 1981, Inhibition in slices of rat neostriatum, *Brain Res.* **212:**182–187.

Misgeld, U. and Bak, I. J., 1979, Intrinsic excitation in the rat neostriatum mediated by acetylcholine, *Neurosci. Lett.* **12:**277–282.

Park, M. R., Lighthall, J. W., and Kitai, S. T., 1980, Recurrent inhibition in the rat neostriatum, *Brain Res.* **194:**359–369.

Preston, R. J., Bishop, G. A., and Kitai, S. T., 1980, Medium spiny neuron projection from the striatum: An intracellular horseradish peroxidase study, *Brain Res.* **183:**253–263.

Somogyi, P. and Smith, A. D., 1979, Projection of neostriatal spiny neuron to the substantia nigra. Application of a combined Golgi-staining and horseradish peroxidase transport procedure at both light and electron microscopic levels, *Brain Res.* **178:**3–15.

Sugimori, M., Preston, R. J., and Kitai, S. T., 1978, Response properties and electrical constants of caudate nucleus neurons in the cat, *J. Neurophysiol.* **41:**1662–1675.

Wilson, C. J. and Groves, P. M., 1980, Fine structure and synaptic connections of the common spiny neuron of the rat neostriatum: A study employing intracellular injection of horseradish peroxidase, *J. Comp. Neurol.* **194:**599–616.

Wilson, C. J., Chang, H. T., and Kitai, S. T., 1983, Disfacilitation and long-lasting inhibition of neostriatal neurons in the rat, *Exp. Brain Res.* **51:**227–235.

11

Locus Coeruleus Neurons

JOHN WILLIAMS, GRAEME HENDERSON, and ALAN NORTH

1. INTRODUCTION

The nucleus locus coeruleus (LC) is a group of neurons situated close to the floor of the fourth ventricle at the upper border of the pons. The cells of the rat LC are homogeneous by certain criteria. Virtually all the cells contain noradrenaline and its synthesizing enzymes (Dahlström and Fuxe, 1964; Pickel *et al.*, 1979), and, in the anesthetized animal, they all discharge action potentials at a rather regular and slow rate (0.5 to 2 Hz) (Korf *et al.*, 1974).

The LC is a good tissue for electrophysiological studies in the slice preparation for several reasons. First, it is easily identifiable by virtue of its relative translucency to transmitted light, and its position relative to landmarks such as the fourth ventricle, the superior cerebellar peduncle, and the mesencephalic nucleus of the trigeminal nerve. Second, it is a compact nucleus with quite densely packed cells so that, under optimal circumstances, impalements of several neurons can be made in the course of a single electrode track through a 300-μm slice. Third, the homogeneity mentioned above seems to extend to electrophysiological properties and sensitivities to drugs and putative transmitters. Fourth, almost half the noradrenaline (NA)-containing nerve terminals in the

---

JOHN WILLIAMS and ALAN NORTH • Neuropharmacology Laboratory, Department of Nutrition and Food Science, Massachusetts Institute of Technology, Cambridge, Massachusetts 02139. GRAEME HENDERSON • Department of Pharmacology, University of Cambridge, Cambridge, CB2 2QD, England.

central nervous system (CNS) originate from cell bodies in the LC. In view of the possibility that the membrane ion channels at these terminals are similar to those on the cell bodies, intracellular recording from the LC could provide useful information regarding the mechanism of NA release and how this can be manipulated pharmacologically. Indeed, some of the terminals appear to be in the LC itself, where they mediate a "recurrent collateral" form of inhibition (Cedarbaum and Aghajanian, 1976, 1977; Aghajanian *et al.*, 1977). Fifth, there is evidence, either from ligand binding studies, immunohistochemistry, or extracellular recording *in vivo*, that the LC may receive synaptic inputs from cells containing acetylcholine (ACh), NA, 5-hydroxytryptamine, substance P, and enkephalin. The purpose of this review is to describe the electrophysiological properties of the neurons of the LC and the action of some putative neurotransmitters and opiate agonists. From the description of these results, it will be realized that the LC slice technique offers a method to analyze the action of various agents on single neurons that, in combination with results obtained *in vivo*, will provide a better overall understanding of noradrenergic mechanisms in CNS function and drug action.

2. METHODS

The slice was prepared as described by Henderson *et al.* (1982), and is detailed in the Appendix. Briefly, the brain was removed and a block of pons with cerebellum was mounted in a Vibratome. One 300-μm slice containing the LC was chosen for recording. The slice was placed on the net of the recording chamber where it was completely submerged in a warmed (37°C), oxygenated artifical CSF solution that superfused the tissue at 2 ml/min. The volume of the recording chamber was 500 μl. The slice was secured by a titanium electron microscope grid placed over the area of the LC and weighted down with small pieces of platinium wire. Recording electrodes were placed in the region of the LC under visual control. The exact location of the impaled neuron can be determined by intracellular injection of Lucifer Yellow or by histological examination of the slice upon completion of the experiment (Henderson *et al.*, 1982). Since the slices were totally submerged, application of drugs and solutions of differing ionic content by superfusion resulted in effects that were almost immediate in onset with the entry of the drug-containing solution into the bath, and that recovered quickly after returning to the normal Krebs solutions.

Drugs were also applied by pressure ejection from pipettes having tip diameters of 5 to 20 μm. The pipette tip was placed in the superfusion solution (but not touching the slice) close to the recording site. Drugs used for pressure ejection were dissolved in saline at concentrations 100 to 1000 times the threshold concentration used to observe a reproducible effect by perfusion (typically 1 M to 100 μM). The concentration range used in the pressure ejection pipettes was high enough to produce consistent effects that had fast onsets with short (25 to 100 msec) ejection periods, but was not so high as to result in effects caused by leakage or diffusion out of the tip. Control ejection of saline alone failed to produce any response.

3. RESULTS

3.1. *Guinea Pig Locus Coeruleus*

3.1.1. Resting Properties. The resting membrane potential of guinea pig LC cells was -57.7 ± 2.7 mV (mean ± SEM $n = 20$); the input resistance was 58.0 ± 7.6 MΩ ($n = 21$); the membrane time constant was 7.3 ± 1.0 msec ($n = 21$). The voltage–current relationship obtained using pulses of duration sufficient to reach steady state potential displacements was linear for up to 40 mV in the hyperpolarizing direction. The small standard error obtained for measurements of resting properties may be taken to indicate that these neurons are fairly homogeneous.

3.1.2. Active Properties. Depolarization of the neurons using intracellular current injection produced action potentials with amplitudes of 69.9 ± 3.2 mV ($n = 21$) and duration at the point of initiation of 1.38 $\pm$ 0.15 msec ($n = 12$). These action potentials were reversibly eliminated by tetrodotoxin (TTX) (1 μM). Depolarizing pulses up to 500 msec caused repetitive firing without accommodation. Spontaneous action potentials were observed in about 20% of neurons. The origin of the spontaneous activity was probably not synaptic since spontaneous synaptic activity was not observed when the neurons were hyperpolarized to prevent the action potential discharge.

3.1.3. Evoked Potentials. Stimulation with focal glass electrodes in the area surrounding the LC-produced depolarizing potentials having a fast rise time (2.3 to 8.5 msec) and slower decay (13 to 63 msec). The amplitude was dependent on the strength of stimulation, and for a given stimulus the amplitude increased with membrane hyperpolarization. These potentials were completely eliminated by superfusion with calcium-free/6-mM magnesium solution, indicating that they were probably excitatory postsynaptic potentials (EPSPS).

3.2. *Guinea Pig Mesencephalic Nucleus of the Trigeminal Nerve*

Lying immediately adjacent to the lateral border of the LC is the mesencephalic nucleus of the trigeminal nerve (MNV). The neurons contained within this nucleus have large pseudomonopolar cell bodies; their resting and active electrophysiological properties differ markedly from LC neurons and are included here for purposes of comparison. The resting membrane potential of MNV neurons was -51.9 ± 3.6 mV ($n = 13$), the input resistance was 15 ± 1.8 MΩ ($n = 17$), and the membrane time constant was 1.35 ± 0.16 msec ($n = 17$). These cells exhibited strong time-dependent anomalous rectification in response to hyperpolarizing current. MNV neurons were not spontaneously active. Injection of depolarizing current gave rise to an initial burst of firing; with low intensities of current, the cells accommodated, and firing was not maintained throughout the duration of current injection. Action potential amplitude was 59.5 ± 2.9 mV ($n = 16$) with the duration at the point of initiation 0.53 ± 0.04 msec ($n = 8$). Neither evoked nor spontaneous synaptic input could be detected in these neurons.

3.3. *Rat Locus Coeruleus*

3.3.1. Resting Properties. Cells fired spontaneously at frequencies of 0.25 to 5 Hz. The threshold for initiation of the action potential was very constant from cell to cell (-55 mV). When a hyperpolarizing current was applied sufficient to stop the firing, the input resistance measured with hyperpolarizing pulses 300 msec in duration was 201 ± 21 MΩ ($n = 19$) and the membrane time constant was 30 ± 2.9 msec ($n = 15$).

3.3.2. Active Properties. Action potentials were 82.5 mV in amplitude and 1.5 msec duration at their point of initiation. The maximum rate of rise of the action potential was 131 ± 9.4 V/sec; the falling phase had a pronounced shoulder (Figure 1) that was absent when cobalt was added to or calcium removed from the bathing medium. The time course and amplitude of the spontaneous action potentials were identical to those evoked by brief depolarizing pulses. The rise time and amplitude of both spontaneous and evoked action potentials were markedly reduced but not eliminated by TTX (300 nM to 20 μM). The residual spike that persisted in TTX was eliminated by removal of calcium from the superfusing solution or by addition of cobalt or magnesium, suggesting that these spikes were due to calcium entry (Figure 1B). The origin of

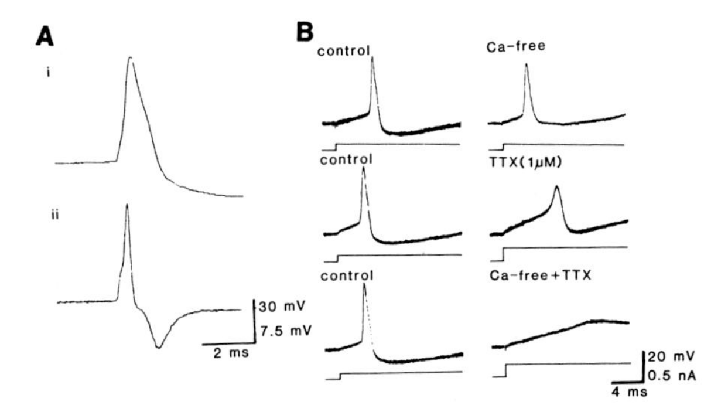

Figure 1. Action potentials of LC neurons. (Ai) Intracellular recording of typical spontaneous action potential (average of 32). (Aii) Extracellular recording of the same spontaneous action potential (average of 32). These records were made by withdrawing the microelectrode from the cell shown in (A) to an immediately extracellular position; the intracellular recording is differentiated by the membrane capacitance. (B) Effects of tetrodotoxin (TTX) and calcium ion removal on the action potential. Top left, control action potential. (The bottom trace in this and other panels is the current pulse used to evoke the action potential.) Top right, after 15 min perfusion with a solution that contained no calcium ions, 5 mM magnesium and 500 μM EGTA. Middle left, control. Middle, after 5-min perfusion with TTX (1 μM). Bottom left, control. Bottom, no action potential evoked after 10-min perfusion with a calcium-free solution that also contained TTX.

the spontaneous firing probably was not synaptic, since spontaneous activity persisted in solutions that abolished evoked synaptic potentials (see Section 3.3.3) and hyperpolarization below the threshold by current injection did not reveal synaptic activity.

3.3.3. *Evoked Potentials.* Stimulation of the tissue with a bipolar tungsten electrode in the area of the LC resulted in a depolarizing potential of about 20 to 100 msec duration, which was usually followed by a hyperpolarization lasting 1 to 10 sec (Figure 2A). The amplitude of these potentials was dependent on the strength of stimulation and both were eliminated in solutions that contained elevated magnesium concentration (10 mM), indicating that they were synaptic in origin. The amplitude of the EPSP usually increased as the cell was hyperpolarized; the range of potentials that could be examined was limited due to membrane rectification, but the extrapolated reversal potential was approximately −10 mV. The inhibitory synaptic potential (IPSP) always decreased in amplitude with hyperpolarization (Figure 2). The reversal potential for the IPSP was −111.5 ± 2.5 mV (n = 8).

3.4. *Pharmacological Studies*

3.4.1. *Opiates.*

3.4.1a. Hyperpolarization. LC neurons of both guinea pig and

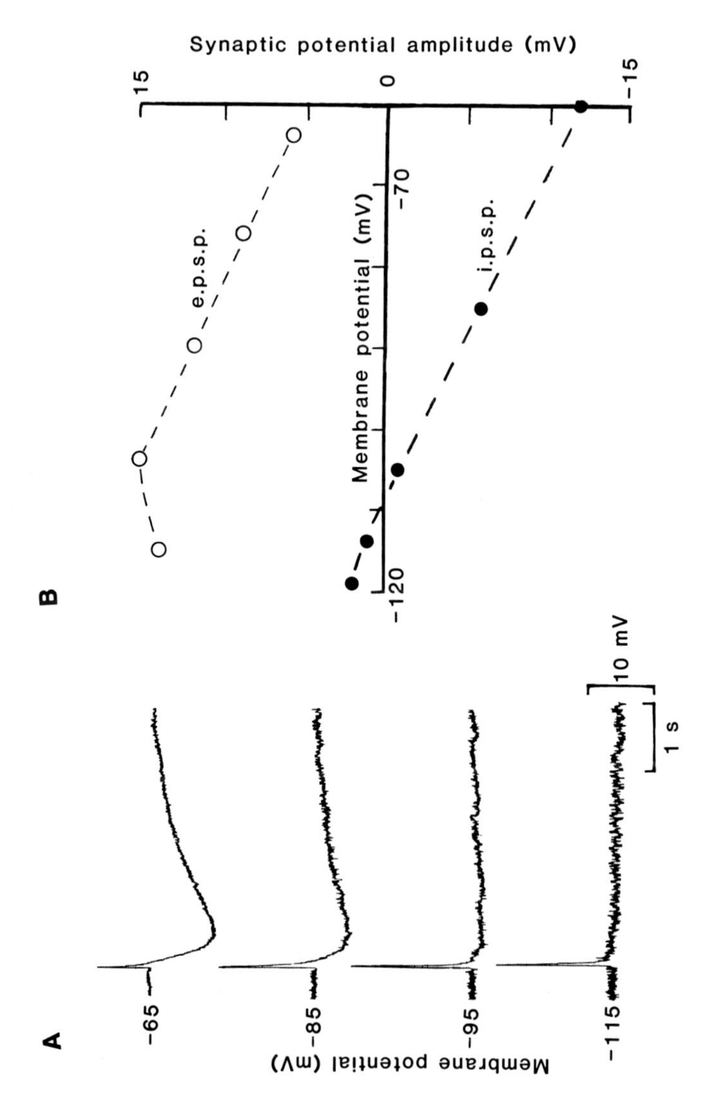

Figure 2. Synaptic potentials recorded in rat LC neurons. (A) Each trace is an average of 4 recordings. Focal electrical stimulation to the slice surface (25 V/300 μsec) evoked an EPSP/IPSP sequence after a latency of about 3 msec. When the cell was hyperpolarized by passage of current, the EPSP became larger and the IPSP smaller. (B) Relation between amplitude of synaptic potential and the membrane potential at which they were evoked, from another neuron. The extrapolated reversal potential for the EPSP was about −10 mV and the IPSP reversed at −110 mV.

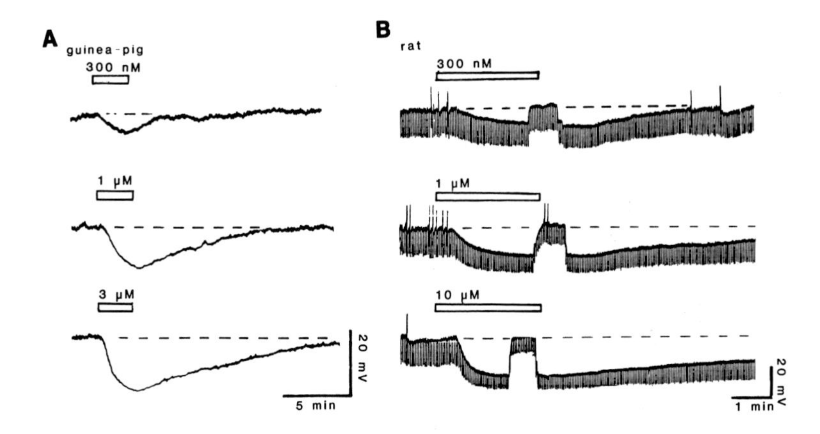

Figure 3. Normorphine hyperpolarizations. The perfusing solution contained normorphine in the concentrations indicated during the periods marked by the bars. (A) Guinea pig LC neuron. Normorphine caused a dose-related hyperpolarization. (B) Rat LC Neuron. The hyperpolarization was accompanied by a dose-related increase in conductance. Upward deflections are occasional spontaneous action potentials (full amplitude not shown). Downward deflections are electrotonic potentials caused by passing current pulses of fixed amplitude. At the peak normorphine hyperpolarization, the membrane potential was temporarily shifted back to the control level so that membrane conductance could be compared. (Part (A) is reproduced, with permission, from Pepper and Henderson, 1980.)

rat were hyperpolarized by opiates and opioid peptide agonists (Figure 3). The effective concentration range was identical in both species and is similar to that found to be effective at peripheral neuroeffector junctions (e.g. myenteric plexus, mouse vas deferens). The hyperpolarization appeared to be a direct effect on the neuron from which the recording was made because it was not changed in solutions in which evoked synaptic activity had been abolished. Repeated application of opiate agonists with pressure ejection resulted in reproducible hyperpolarizations for periods of several hours. Perfusion of opiates for periods of 1 to 2 hr resulted in a prolonged hyperpolarization with no sign of desensitization.

3.4.1b. Opiate Receptor. Naloxone reversed the hyperpolarizing action of opiates on LC neurons. The use of pressure ejection of opiate agonists to produce reproducible dose response curves allowed the demonstration of competitive antagonism by perfusing several known concentrations of naloxone (Figure 4). Equal responses to an agonist were compared and dose-ratios calculated. The slope of the Schild plot (Arunlakshana and Schild, 1959) was close to unity. Naloxone competitively antagonized opiate-induced hyperpolarizations with an equilibrium dis-

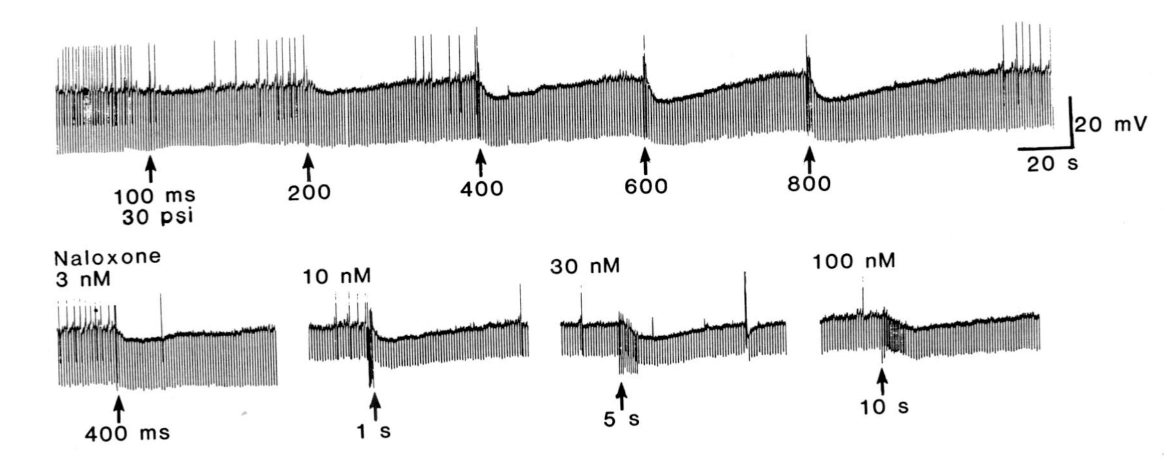

Figure 4. D-Ala^2-D-Leu^5-enkephalin (DADLE) hyperpolarizations in rat LC neurons, and antagonism by naloxone. At the arrows, DADLE was applied by pressure ejection (30 psi) from a pipette tip positioned in the perfusing solution just above the slice surface. Top trace, the amplitude of the hyperpolarization was related to the duration of the pressure pulse applied. Bottom trace, selected records from dose-response curves similar to those shown in the top trace, but repeated in the presence of increasing concentrations of naloxone. Equivalent hyperpolarizations could be obtained in the presence of naloxone when the duration of the DADLE pressure pulse was progressively increased. Downward deflections are electrotonic potentials caused by passing current pulses of fixed amplitude.

sociation constant (K_e) of 1 to 2 nM (Williams *et al.*, 1982). The equilibrium constant is close to that measured in binding on homogenized brain tissue. Future experiments of this kind may permit the distinction among different subtypes of opiate receptor on a single nerve cell.

3.4.1c. Ionic Mechanism. The hyperpolarization was associated with an increase in conductance of the membrane to potassium ions, since it was abolished at the potassium equilibrium potential (E_K). When E_K was changed by altering the extracellular potassium ion content, the reversal potential of the opiate-induced hyperpolarization varied as predicted by the Nernst equation. In addition, extracellular barium (30 to 300 μM) and intracellular cesium administered by diffusion from the recording electrode, two agents that decrease potassium conductance, abolished the hyperpolarizing action of opiate agonists. Indeed, concentrations of barium (100 μM), which had little effect on spike duration, markedly reduced the opiate-induced hyperpolarization; the only other action that barium had at 100 μM was a reduction of the amplitude and duration of the afterpotential following the spike, an effect that was also observed with calcium channel blockers such as cobalt or manganese. Barium may act to block opiate effects at a site inside the cell, since when spontaneous activity was prevented by passing a small hyperpolarizing current (so that barium would not enter during the spikes), the ability of barium to block opiate hyperpolarizations was lost. These findings suggest that opiates may activate a calcium-dependent potassium conductance ($g_{K,Ca}$), as has been previously proposed in the case of myenteric neurons (Tokimasa *et al.*, 1981). The potassium channel opened by opiates certainly seems to be different in some respects from the delayed rectifier ($g_{K,V}$) or the fast potassium current ($g_{K,A}$) since tetraethylammonium (TEA) (10 mM) and 4-aminopyridine (4-AP) (100 μM) prolonged the spike but did not affect the opiate-induced hyperpolarization. Chloride ions probably are not involved at all in the response to opiates. Equal effects were observed using chloride-, acetate-, or citrate-filled electrodes; moreover, hyperpolarizations induced by γ-aminobutryic acid (GABA) were reversed to depolarizations following intracellular chloride injection even when the opioid hyperpolarization was unchanged.

3.4.1d. Site of Action. The reversal potential was estimated in three ways. First, the change in conductance was measured from the amplitude of small hyperpolarizing electrotonic potentials just sufficient in duration to charge the membrane capacitance. The reversal potential was then calculated from

$$\Delta V = (1 - R'/R)\,(E_{rev,R} - E_m) \qquad (1)$$

where ΔV is the amplitude of the hyperpolarization, R is the control input resistance, R' is the resistance during application of opiate (this value was always measured at the same potential as was R), E_m is the resting potential, and $E_{rev,R}$ was this estimate of reversal potential. The value for $E_{rev,R}$ was -135.5 ± 7.4 mV ($n = 11$) when opiates were applied by pressure and -124.0 ± 8.0 mV ($n = 10$) when opiates were applied by superfusion. Second, the reversal potential was also estimated from the change in the time constant of the membrane before (τ) or during (τ') application of opiates according to

$$\Delta V = (1 - \tau'/\tau)\,(E_{rev,\tau} - E_m) \tag{2}$$

(Carlen and Durand, 1981). This value of $E_{rev,\tau}$ was -103.0 ± 5.9 mV ($n = 10$). Third, the reversal potential was measured directly by shifting the membrane potential (E_m) and extrapolating from the linear part of the relation between E_m and hyperpolarization amplitude. This gave a value of -107.7 ± 4.4 mV ($n = 12$) for $E_{rev,E}$. If one assumes an intracellular potassium ion content of 140 mM, then E_K in 2.5 mM external potassium would be -107 mV. Since the value obtained for $E_{rev,R}$ was greater than that obtained for E_K, this would imply that we failed to measure a part of the steady-state conductance increase induced by opiates, perhaps because it was occurring at an electrotonically distant site. The estimate of $E_{rev,E}$ by membrane polarization, although it agrees numerically with E_K, probably is not a useful measure of the site of action in LC neurons. It might be thought that a value of -108 mV indicates a largely somatic site of action; however, a large conductance increase unrelated to opiates occurred with the degree of membrane hyperpolarization used and this would tend to isolate the soma from the electrotonically distant processes.

3.4.1e. Reduction in Action Potential. A second action of opiates on LC neurons was to decrease the amplitude and rate of rise of the action potential. This effect was usually very slight in the case of the normal spike, but was often marked in the case of the smaller and more slowly rising calcium spike recorded in TTX. This action occurred at concentrations of opiates similar to those which activated g_K. Extracellular barium, and cesium diffusion from an intracellular electrode, which blocked the g_K activation, also blocked the depression by opiates of the calcium spike. Therefore, it is considered that the reduction in the calcium spike may result from the potassium activation. If such an action also occurs at nerve terminals, a decreased release of transmitter would result.

3.4.2. Noradrenaline. All LC neurons were hyperpolarized by α-agonists. The concentration range of NA necessary to produce a hyperpolarization was 1 to 30 μM (Figure 5), whereas clonidine was effective at concentrations 1000-fold lower. The hyperpolarization produced by α_2-agonists was also associated with an increase in conductance to potassium ions. The change in conductance agreed quantitatively with the amplitude of the hyperpolarization, assuming that the conductance increase was only to potassium and was located near the recording site (see above). Reversal of the hyperpolarizations was observed at potentials close to E_K. Superfusion with the NA uptake inhibitor, desmethylimipramine (DMI) (100 nM to 10 μM), increased the amplitude and duration of the hyperpolarizations induced by NA applied either by perfusion or by pressure ejection. Superfusion of the slice with the α-receptor adrenoceptor antagonists, phentolamine and yohimbine, antagonized the NA-induced hyperpolarization. DMI also slowed the firing rate and hyperpolarized the cell. Phentolamine also antagonized the DMI-induced hyperpolarization. This implies that tonic release of NA was probably occurring. The K_e for phentolamine against noradrenaline calculated from the Schild plot was 100 nM (Figure 5c).

The IPSP recorded in LC neurons following focal stimulation was also blocked by superfusion with phentolamine or yohimbine at concentrations that blocked the NA hyperpolarization; conversely, the IPSP was augmented by DMI. These findings strongly suggest that the IPSP is caused by the release of noradrenaline, and that this is a "recurrent collateral" IPSP from LC neurons themselves.

4. DISCUSSION

The results presented here serve to illustrate the usefulness of intracellular recording from brain neurons *in vitro* for neuropharmacological studies. The technique offers several advantages. First, it enables the distinction to be made between drug actions on the membrane of the impaled cell and actions secondary to primary effects on other neurons. Second, the concentration of drug at the neuronal membrane is known. Drugs added to the perfusing medium demonstrated steady-state effects upon a cell within a few minutes. These equilibrium concentrations can be compared directly with those measured in the brain following systemic administration of usual doses.

The question of concentration of neurotransmitters is more complicated. It is most unlikely that they act normally under equilibrium conditions, and the profile of their concentration will be markedly affected by processes of degradation and uptake. As we have shown in the case of the NA and IPSP potentiation by DMI, these processes can

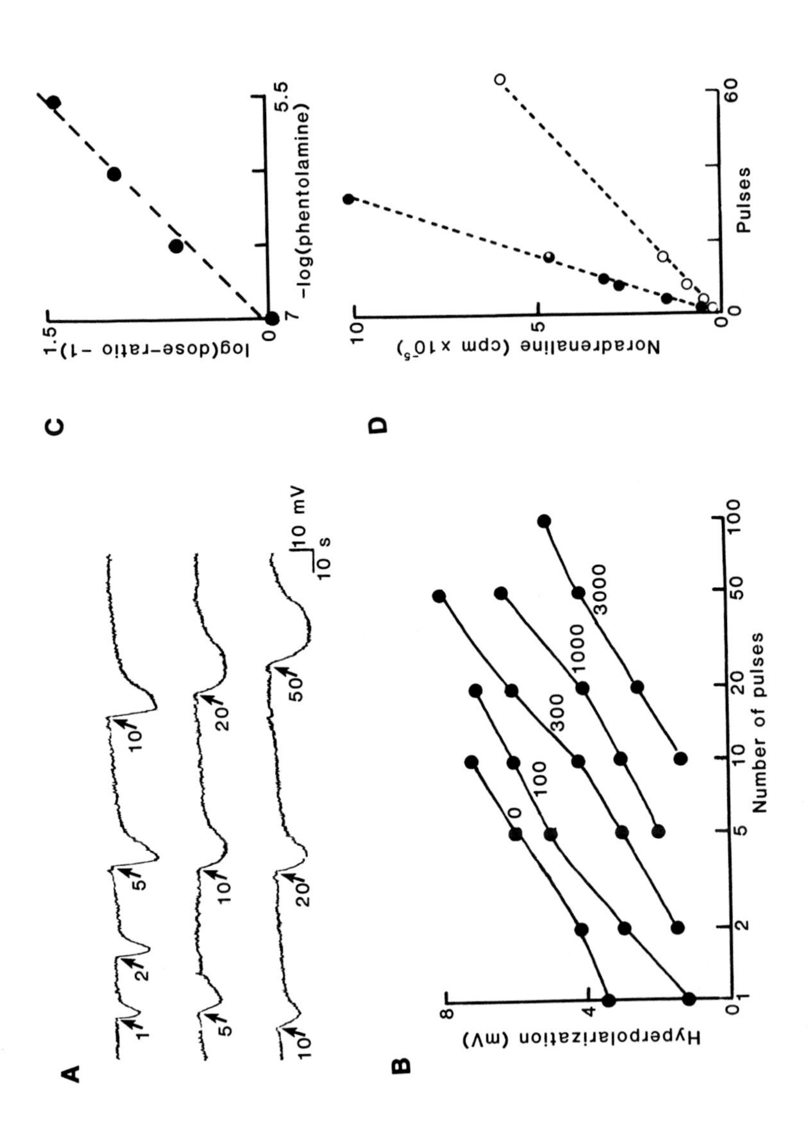
A
1
2
5
10
5
10
20
10
20
50
10 mV
10 s
B
Hyperpolarization (mV)
Number of pulses
0
100
300
1000
3000
C
log(dose-ratio −1)
−log(phentolamine)
1.5
0
7
5.5
D
Noradrenaline (cpm x 10⁻⁵)
Pulses
10
5
0
60

Figure 5. Noradrenaline hyperpolarizations in a rat LC neuron and antagonism by phentolamine. (A) Noradrenaline applied by pressure ejection (arrows) induced membrane hyperpolarizations. The numbers indicate the number of pressure pulses applied (each pulse was 5 psi, 50 msec duration, repeated at 10 Hz). Top trace, in control solution. Middle trace, similar records in the presence of phentolamine (100 nM). Bottom trace, in phentolamine (300 nM). (B) Dose-response curves obtained as described in (A) but from a different cell. The numbers beside each curve refer to the concentration (nM) of phentolamine in the perfusing solution. (C) Schild plot of the data shown in (B), measured at 4 mV response level. The line is straight (r = 0.99) and has a slope of 0.96. The pA_2 for phentolamine was 7.1. (D) Release of ^{3}H-noradrenaline from two typical pressure ejection pipettes. The ejection pulse was 5 psi for 50 msec, and ejections were repeated at 10 Hz. The ejected solution was collected into saline, scintillation fluid added, and the counts are expressed on the ordinate. For these two pipettes, the volumes released per pulse were 30 nl (filled circle, 15 μm tip) and 10 nl (open circles, 7μm tip).

be directly studied in the slice preparation. Furthermore, perfusion of neurotransmitters might be expected to cause desensitization of receptors, which does not normally occur with their brief synaptic release. The pressure application technique offers an approach to brief applications of neurotransmitters, and 10-msec pulses are occasionally effective (Figure 4).

These limitations to the value of knowing precise drug concentrations are largely irrelevant in the case of antagonists. Prolonged perfusion with antagonists ensures equilibrium receptor occupancy, and as most of these compounds are plant alkaloids, it is unlikely that they are significantly taken up or degraded by the tissue. Knowledge of antagonist concentrations permits the use of the null method to assess K_e. For naloxone, the K_e close to 1 nM is close to that found in ligand binding studies; this implies that homogenization procedures and changes in ionic environment commonly used in ligand binding experiments have relatively little effect on the ability of the receptor to bind naloxone. On the other hand, one should be aware that alkaloid antagonists may have actions other than receptor blockade. Phentolamine is known to block catecholamine uptake (Iversen, 1967) and this may increase the agonist concentration at the receptor site. A Schild plot with unit slope for dose-ratios exceeding 10 is likely to indicate competitive antagonism without significant interference from degradation and uptake processes. Larger dose-ratios would be desirable and may be possible with repeated pressure application (Figure 5); unfortunately, although the amount of agonist ejected from the pressure pipette increases linearly with the number of pulses, the time profile of its concentration at the receptor may change.

Third, the ionic mechanism of action of drugs and transmitters can be elucidated by changing the ionic composition of the perfusing solution. Fourth, the site of action of drugs on the neuron can be assessed. Clearly such studies are easier in a structure such as the hippocampus in which a regular organization of cell bodies and dendrites exists. But even in the locus coeruleus, sensitivity of drug effects to polarizing currents and correlations between time constant and resistance changes with observed potential changes can be used to estimate sites of action (see Section 3.4.1d; and Carlen and Durand, 1981).

There is a fifth advantage to neuropharmacological studies on cells such as the locus coeruleus. The terminal projections of these cells release NA at a wide variety of brain sites, and an increasing number of drugs are thought to exert their effects by modifying this release process. Intracellular recording from these terminal projections is not possible, and our knowledge of the mechanism of action of such drugs is limited.

But the calcium action potentials recorded in the cell bodies of the LC neurons may reflect events in the terminals (North and Williams, 1983). Furthermore, the "recurrent collateral" noradrenergic IPSP affords one the possibility to correlate effects on the calcium spike with transmitter release measured directly as the IPSP amplitude or conductance change. Such studies promise to greatly increase our understanding of the function of noradrenergic neurons in the mammalian brain.

5. REFERENCES

Aghajanian, G. K., Cedarbaum, J. M., and Wang, R. Y., 1977, Evidence for norepinephrine-mediated collateral inhibition of locus coeruleus neurons, *Brain Res.* **136**:570–577.

Arunlakshana, O. and Schild, H. O., 1959, Some quantitative use of drug antagonists, *Br. J. Pharmacol.* **14**:48–58.

Carlen, P. J. and Durand, D., 1981, Modelling the location of postsynaptic conductance changes induced by transmitters and drugs, *Neuroscience* **6**:839–846.

Cedarbaum, J. M. and Aghajanian, G. K., 1976, Noradrenergic neurons of the locus coeruleus: Inhibition of epinephrine and activation by the alpha antagonist piperoxane, *Brain Res.* **112**:413–419.

Cedarbaum, J. M. and Aghajanian, G. K., 1977, Catecholamine receptors of locus coeruleus neurons: Pharmacological characterization, *Eur. J. Pharmacol.* **44**:375–385.

Dahlström, A. and Fuxe, K., 1964, Evidence for the existence of monoamine containing neurons in the central nervous system, *Acta Physiol. Scand.* **62**(suppl. 232):1–55.

Henderson, G., Pepper, C. M., and Shefner, S. A., 1982, Electrophysiological properties of neurones contained in the locus coeruleus and mesencephalic nucleus of the trigeminal nerve *in vitro*, *Exp. Brain Res.* **45**:29–37.

Iversen, L. L., 1967, *The Uptake and Storage of Noradrenaline in Sympathetic Nerves*, Cambridge University Press, London.

Korf, J., Bunney, B. S., and Aghajanian, G. K., 1974, Noradrenergic neurons: Morphine inhibition of spontaneous activity, *Eur. J. Pharmacol.* **25**:165–169.

North, R. A. and Williams, J. T., 1983, Opiate activation of potassium conductance inhibits calcium action potentials in rat locus coeruleus neurones, *Br. J. Pharmacol.*, in press.

Pepper, C. and Henderson, G., 1980, Opiates and opioid peptides hyperpolarize locus coeruleus neurons *in vitro*, *Science* **209**:394–396.

Pickel, V. M., Joh, D. H., Reis, B. J., Leeman, S. E., and Miller, R. J., 1979, Electron microscopic localization of substance P and enkephalin in axon terminals related to dendrites of catecholaminergic neurons, *Brain Res.* **160**:387–400.

Tokimasa, T., Morita, K., and North, R. A., 1981, Opiates and clonidine prolong calcium-dependent after-hyperpolarizations, *Nature* (*London*) **292**:162–163.

Williams, J. T., Egan, T. M., and North, R. A., 1982, Enkephalin opens potassium channels on mammalian central neurones, *Nature* (*London*) **299**:74–77.

12

Neocortex
Cellular Properties and Intrinsic Circuitry

BARRY W. CONNORS and MICHAEL J. GUTNICK

1. INTRODUCTION

The neocortex has a striking diversity of cellular morphology. Although histologists have described and catalogued its myriad neuronal forms since the time of Golgi, it is only recently that technical advances in single cell staining have allowed the shape of a cortical cell to be linked with its functional personality (e.g., Kelly and Van Essen, 1974; Christensen and Ebner, 1978; Deschenes *et al.*, 1979; Gilbert and Wiesel, 1979; Lin *et al.*, 1979). These studies are landmark attempts since they represent the convergence of two rich, but basically separate, bodies of knowledge: the structure of single neocortical neurons and their physiological properties. The correlative knowledge to be gained from such an approach will be invaluable to future models of cortical integration (Gilbert and Wiesel, 1981).

As these studies are presently formulated, their definition of neuronal function is restrictive and incomplete, largely due to technical considerations. A single cell is usually characterized by its sensory receptive field properties or a specific form of extrinsic connectivity. Both are vital pieces of information but are removed from a further level of functional diversity, the biophysical properties of the cell's membranes and syn-

BARRY W. CONNORS • Department of Neurology, Stanford University School of Medicine, Stanford, California 94305. MICHAEL J. GUTNICK • Unit of Physiology, Faculty of Health Sciences, Ben Gurion University of the Negev, Beer-Sheva, Israel.

apses. If the spiking pattern of a neuronal axon defines the informational output of that cell, then the cell's biophysical properties together with its shape determine the transformations by which that information was derived. Useful data about the membrane physiology of single neocortical neurons has been difficult to obtain from studies of intact animals, though some notable insights have been achieved. Applied to the neocortex, the brain slice techniques described in this book appear to offer significant advantages for the study of intrinsic membrane properties and synaptic conductances. Although the method as currently used isolates the cortical cell from the identification of its extrinsic purpose, considerable environmental control is achieved. Just as importantly, enough of the neighboring tissue is preserved that meaningful analyses of local cellular interactions can be performed.

Recent anatomical and functional concepts of the workings of neocortex stress the prevalence and importance of vertical interactions between neurons, i.e., the movement of information between cortical layers via a circumscribed subset of neurons that is oriented normal to the cortical surface (Mountcastle, 1978; Szentagothai, 1978). Such a group of neurons has been loosely termed the cortical column. Although horizontal connectivity is recognized as essential, it has taken a secondary role in models of local circuit integration. The current formalizations of cortical organization are built upon the early suggestions of Lorente de Nó (1938), who inferred the functional importance of vertical connections from his observations of Golgi impregnations. His original circuit diagram (Figure 1) is useful as a very schematic representation of one form of *in vitro* neocortical slice preparation, and will be referred to later.

The tremendous utility of brain slice techniques to neurophysiologists is presently unquestionable. Although it is easy, and legitimate, to extole the virtues of this methodology, it is also necessary to bear in mind the limitations it imposes. One constraint is that the neural circuitry of interest be well preserved when separated from the surrounding tissue by two parallel cuts some 0.5 mm apart. This narrow, planar geometry is, fortunately, well suited to the neurons and their local connections in the neocortex, whether derived from mouse or man. In different areas and species, neocortical columns in cross section may take the form of circles, ellipses, slabs, or sheets, and may intersect, interdigitate, and overlap. What is most promising for *in vitro* research is that estimates of the width (or diameter) of various columns indicate that the slice technique yields tissue fragments of very useful proportions. Thus, columns in monkey somatic sensory area are roughly 500 μm wide (Mountcastle, 1957), "barrels" in the somatic sensory cortex of mice are 200 to 300 μm in diameter (Woolsey and Van der Loos, 1970), ocular dominance columns of monkey primary visual cortex are about

Figure 1. Intracortical chains of neurons as schematized by Lorente de Nó (1938). Axons are marked with *a*; synapses are represented by dots and marked with *s*. Note the heavy flow of connectivity in the vertical direction. Inset at right summarizes the diagram at left. Incoming impulses flow through an afferent fiber (*af*), activating a large pyramidal cell both directly and through interneurons (*i1* to *i3*). Impulses leave via efferent fibers (*ef*), while returning some informational flow through recurrent projections.

400 μm wide (Hubel and Wiesel, 1977), and motor columns of monkey precentral motor cortex are 0.5 to 1 mm across (Asanuma and Rosen, 1972). In some conceptions, these "macrocolumns" may be further subdivided into "minicolumns" 30 to 50 μm across (Mountcastle, 1978; Roney *et al.*, 1979), as with the narrow orientation columns of primary visual cortex (Hubel and Wiesel, 1974). The typical *in vitro* slice taken in a plane normal to the cortical surface is roughly the thickness of a macrocolumn, and extends far enough laterally to possibly encompass parts of several adjacent macrocolumns.

2. EARLY USE OF NEOCORTICAL SLICES

Although *in vitro* preparations of hippocampus and olfactory cortex have been most popular in the last decade, electrophysiological recordings from brain slices were first attempted in neocortex. McIlwain (1951) had shown that incubated slices of neocortex displayed the appropriate metabolic changes when electrically stimulated, but under similar conditions he could not demonstrate an electrically manifested response of the tissue (McIlwain and Ochs, 1952). Subsequent studies (Li and McIlwain, 1957; Hillman and McIlwain, 1961; Gibson and McIlwain, 1965) established that large negative resting membrane potentials, and even occasional injury discharges, could be recorded from isolated slices. Unfortunately, synaptically mediated activity remained elusive in brain slices (Hillman *et al.*, 1963) until robust field potentials were finally demonstrated in slices of guinea pig olfactory cortex following lateral olfactory nerve stimulation (Yamamoto and McIlwain, 1966). Synaptic activation of neocortical slices was achieved shortly thereafter (Richards and McIlwain, 1967; Yamamoto and Kawai, 1967).

It is notable that all of the studies cited above utilized slices that were cut tangential to the cortical surface. The majority were comprised of the superficial 200 to 350 μm in a tissue block several millimeters wide. The method isolated primarily the molecular layer and some of layers II and III, with the great disadvantage that most vertical interactions were precluded. This may have contributed to the longstanding silence of the neocortical slice. Most recent investigators of this preparation have turned their blades 90° to allow all cortical layers to be represented in a single living section. The inherent advantages of this approach are only beginning to be exploited.

3. NOTES ON NEOCORTICAL SLICE METHODOLOGY

The same general slicing methods that have proven so successful for other brain areas (reviewed in Teyler, 1980; Dingledine *et al.*, 1981; see also the Appendix in this volume) are directly applicable to study of the neocortex, and will not be further belabored here. Chambers that locate neocortical slices at the gas-liquid interface (Schwartzkroin and Prince, 1976; Stafstrom *et al.*, 1982a; Connors *et al.*, 1982) or submerge them completely (Kato and Ogawa, 1981; Vogt and Gorman, 1982) have yielded intracellular recordings of high quality. Slice temperatures in each case were kept near 37°C during recording, and slice thicknesses were generally 500 μm. Our own studies have given us the distinct impression that slices of 350 μm (the standard for most hippocampal studies) lead to a paucity of impalements in the deepest cortical layers, but not superficially. Since pyramidal cells comprise the majority of neocortical neurons (Winfield *et al.*, 1980) and they are distinctly larger in deeper layers (Ramon y Cajal, 1911), it may be that extensive dendritic amputation renders these cells nonfunctional. On the other hand, surgical separation of apical dendrites from the soma and basal dendrites in hippocampal slices leaves each part viable for several hours (Benardo *et al.*, 1982). Thus, the apparent layer-specific difference in cell survival may also be a matter of low neuronal packing density in deeper layers.

Preparations from a variety of species and neocortical areas have been described, and the utility of the method is clearly not restricted by cortical thickness. When obtained under favorable conditions, even biopsied human neocortex can be sliced and remain electrophysiologically viable for many hours (Kato *et al.*, 1973; Schwartzkroin and Prince, 1976; Prince and Wong, 1981). Slices have also been obtained from sensorimotor areas of the guinea pig, rat, and cat (Gutnick and Prince, 1981; Stafstrom *et al.*, 1982a; Connors *et al.*, 1982, 1983) and cat visual cortex (Komatsu *et al.*, 1981; Ogawa *et al.*, 1981). Rockel *et al.* (1980) have shown that the number of neuronal somata in a 30×25 μm strip through the depth of many different cytoarchitectonic areas is remarkably constant at about 110. Furthermore, this number is conserved in mouse, rat, cat, monkey, and man, the one exception being area 17 of some primates where cell density is 2.5 times higher. This number may approximate the size of the simplest cortical columns (Welt *et al.*, 1967; Hubel and Wiesel, 1974; Mountcastle, 1978), suggesting an evolutionary consistence of the fine structure of neocortical organization. Since cortical thickness may vary by a factor of 3 between mouse and man, the packing density of cell bodies must decrease concurrently. This could affect the probability of successful intracellular impalements in thicker cortices, but increases in the mean size of the cells may compensate for this practical disadvantage.

Vogt and Gorman (1982) have described a very interesting neocortical preparation formed by coronal sections through the rat cingulate cortex. Bilateral sections of cingulate remain functionally joined by the corpus callosum within a single slice, and selective stimulation of this uncontaminated afferent pathway is possible (concurrent, of course, with antidromic activation of some cells). Other forms of neocortical slices do not allow anatomical segregation of the various afferent and efferent subcortical and cortico-cortical pathways. This disadvantage might conceivably be overcome by choosing the proper planes of cut in brains of small animals and using a Vibratome to yield consistent, large slices. Such an approach has yielded slices of mouse brain that retain a patent septohippocampal pathway (S. M. Thompson and D. A. Prince, unpublished experiments).

An exciting application of brain slice technology is the detailed study of developing cells (Schwartzkroin and Altschuler, 1977; Schwartzkroin and Kunkel, 1982). Essentially similar techniques to those used for adults are successful for immature brain. In preliminary efforts, good quality intracellular recordings have been obtained from rat neocortical slices as early as the day of birth (Connors *et al.*, 1983; M. J. Gutnick, L. S. Benardo, B. W. Connors, and D. A. Prince, unpublished experiments). A possible complication arises from the observations of Holtzman *et al.* (1981) that the normal rectal temperatures of very young rat pups are several degrees cooler than 37°C, and that, in fact, seizures may develop if temperatures approach or exceed the normal adult level. Normal body temperatures and seizure thresholds rise during the first 2 weeks of age, leaving the slicer with the dilemma of deciding the proper set point for the chamber thermostat. Although we have never seen epilepticlike activity in young rat neocortical slices maintained at 37°C, we routinely run developmental studies at 35°C regardless of age to help avoid possible metabolic deficits and simplify across-age comparisons.

4. PROPERTIES OF NEOCORTICAL SLICES

4.1. *Electrophysiology of Single Cells*

As the early studies from McIlwain's laboratory demonstrated (Hillman and McIlwain, 1961; Gibson and McIlwain, 1965), the high negative resting potentials of neocortical cells are easily diminished (i.e., membranes depolarize) when extracellular K^+ levels ($[K^+]_0$) are increased, or when the excitatory amino acid, L-glutamate, is applied. Precise quantitative relationships were not attempted at that time, but the results pointed out the advantages of the preparation. Unfortunately, the ionic basis of resting membrane potentials in mammalian central neurons has been investigated only in guinea pig olfactory cortex (Scholfield, 1978) in which the permeability ratio of $Na^+ : K^+$ was about 0.02 and resting Cl^- conductance was apparently quite low. The high resting potentials of olfactory cortical neurons (−74 mV; Scholfield, 1978) are very similar to those of neocortical neurons (−75 mV; Connors *et al.*, 1982), so it may be that their resting permeabilities are similar. These types of experiments are amenable only to isolated preparations, where ionic alterations of the bath may be expected to equilibrate completely with the extracellular fluids of the slice (Benninger *et al.*, 1980). Unfortunately, even in the slice, special efforts are required to eliminate secondary effects. High $[K^+]_0$ is an effective activator of neurotransmitter release, for example (Gage and Quastel, 1965), and the standard methods of blocking release involve addition of divalent cations such as Mg^{2+}, Mn^{2+}, Co^{2+}, and Cd^{2+}, which have nonspecific membrane actions of their own (Hille *et al.*, 1975).

Unique information about the voltage-dependent conductances of neocortical neurons has been obtained under the malleable environment of the slice preparation. In judging the significance of these observations, it should be remembered that the bias of the recording electrode is toward the generally larger, much more common pyramidal neurons, although nonpyramidal cells are occasionally impaled as well (Gutnick and Prince, 1981). However, even classification of a cell as pyramidal would not assure one of consistency; the fast and slow pyramidal tract neurons of cat motor area clearly fit the classical pyramidal form, yet differ distinctly in both physiology (Takahashi, 1965) and morphology (Deschenes *et al.*, 1979). Despite these caveats, both generalities and peculiarities of function have been observed.

When pulses of current are passed into neurons of the guinea pig cortical slice, the thresholds of the spikes are found to be quite high, i.e., an average of 25 mV positive to resting potential (Connors *et al.*, 1982). The voltage region between the rest and threshold levels rarely displays purely passive behavior, however. One method of studying this is to polarize the membrane with very long current pulses while monitoring the voltage deflection (ΔV) produced by a relatively slow (approximately 100 msec) constant amplitude square current pulse (ΔI) (Kandel and Tauc, 1966). The apparent input resistance ($\Delta V/\Delta I$) is then calculated at varying resting potentials. Most neocortical neurons display increases in $\Delta V/\Delta I$ as their membranes are depolarized (Figures 2A and 2B), a process which has been termed "anomalous" rectification because it differs qualitatively from the "delayed" rectification of squid axon (Hodgkin *et al.*, 1952). This process is of extreme interest because it operates in the voltage range below, and perhaps just above (Strafstrom *et al.*, 1982a), the spiking threshold. Synaptically evoked currents would be subject to the same influences of rectification as that of the experimenter's electrode-delivered current, providing a powerful influence on the integrating properties of the single neuron (Kandel and Tauc, 1966; Nelson and Frank, 1967).

The ionic mechanisms of subthreshold rectification have been analyzed pharmacologically (Connors *et al.*, 1981, 1982; Stafstrom *et al.*, 1982a,b). Blockage of active K^+ conductances by intracellular injection of Cs^+ (Benzanilla and Armstrong, 1972) greatly enhanced anomalous rectification, and even uncovered such behavior in neuronal membranes that previously showed linear properties or delayed rectification. Ma-

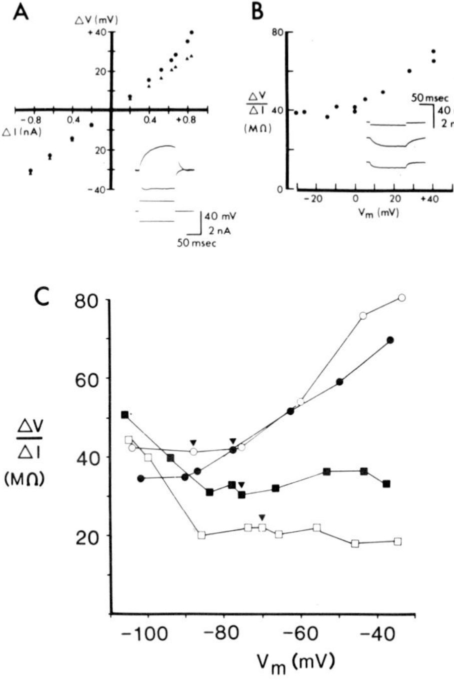

Figure 2. Rectifying properties of neocortical neuronal membranes of guinea pig. (A) Square current pulses of varying amplitude and polarity were passed through the recording microelectrode via a bridge circuit, and voltage deflections were recorded (inset). The graph plots the changes in membrane voltage (ΔV) as a function of current (ΔI) at 25 msec (triangles) and 100 msec (dots) following the onset of the pulse. At longer latencies, this neuron exhibited a marked increase in slope resistance in the depolarizing direction, and a slight decrease in the hyperpolarizing direction. The absolute resting potential was −77 mV. (B) An alternate method was used on the same cell shown in (A) (cf., Kandel and Tauc, 1966). A small, constant amplitude hyperpolarizing pulse (ΔI) was delivered as the resting voltage level (V_m) was altered with tonic current (inset). The amplitude of the voltage deflection (ΔV), normalized to the size of the current pulse, was then plotted as a function of V_m. The depolarizing increase in slope resistance seen in (A) is also quite evident in this measurement. (C) Ionic basis of depolarizing rectification. Measurements of the type shown in (B) were constructed for a different neuron. Anomalous rectification was prominent in the control solution (closed circles), but was eliminated after 4 min (filled squares) and 6 min (open squares) in a solution in which all Na^+ was replaced with choline. Depolarizing rectification returned after the slice was bathed in the control (Na^+-containing) solution (open circles). Resting potentials at each condition are indicated by arrowheads.

nipulations known to block active Na^+ conductances, such as external tetrodotoxin (TTX), internal QX-314, and replacement of external Na^+ with an impermeable substance, abolished the anomolous rectification and usually resulted in delayed rectification (Figure 2C). Efforts to inhibit Ca^{2+} currents by reduction of $[Ca^{2+}]_0$ and addition of Mn^{2+} did not affect subthreshold behavior. Thus, it appears that most neocortical neurons possess voltage-dependent conductances to Na^+ and K^+ that may be activated between resting potential and spike threshold. These currents do not seem to inactivate appreciably over periods of many seconds. The individual physiological character of a given neuron may conceivably depend upon the density and/or spatial distribution of these

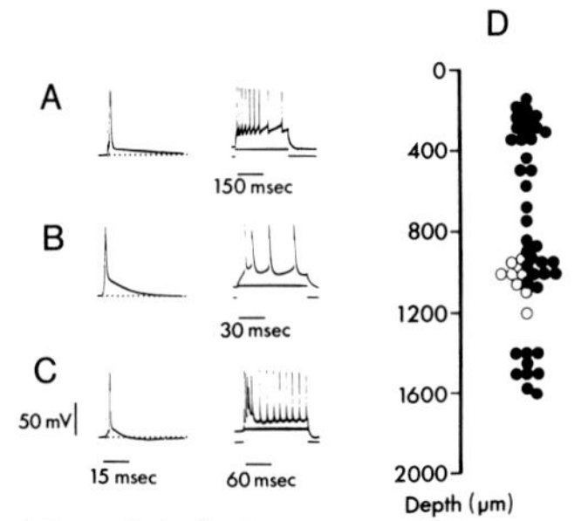

Figure 3. Action potentials of the guinea pig neocortical slice. (A,B) Most cells generated a single spike with a slowly decaying depolarizing afterpotential when activated by a short intracellular current pulse (left). Longer current pulses (right) gave repetitive firing and slow afterhyperpolarizations. (C) Occasional neurons generated all-or-none burst potentials during current injection (right) or synaptic activation. Short intracellular pulses (left) yielded single spikes with both depolarizing and hyperpolarizing afterpotentials. (D) Cortical depth of burst-producing (open circles) and nonburst-producing (closed circles) neurons. Distances were measured from the pia (0 μm). All bursting cells were grouped near layer IV and upper layer V.

conductances. Further, neurotransmitter modulation of subthreshold K^+ currents (Krnjevic, 1974) could alter the dominance of the persistent Na^+ current in various physiological or pathological states.

The spiking characteristics of neocortical neurons are well preserved in the slice (Schwartzkroin and Prince, 1976; Ogawa *et al.*, 1981; Connors *et al.*, 1982). Most impaled cells generate narrow, overshooting action potentials (Figures 3A and 3B) that exhibit marked adaptation during long current pulses, and may show linear increases in firing frequency with current intensity (Ogawa *et al.*, 1981). A unique finding is that occasional recordings in guinea pig sensorimotor area revealed neurons that generated all-or-none burst potentials during either intracellular current injection or, sometimes, synaptic activation (Connors *et al.*, 1982), Figure 3C. When many cells were surveyed through the depth of the cortex, it became apparent that burst-firing neurons could be impaled only at a narrow range of depths around layer IV and upper layer V (Figure 3D). Although the morphological identity of the cell(s) is presently unknown, its rarity, restricted location, and apparent small size (judged by the difficulty involved in stably impaling it with a microelectrode) suggest that it is one of the many types of nonpyramidal local neurons of the neocortex (Feldman and Peters, 1978). The same technical reasons may account for the fact that such intrinsic burst behavior has not been seen *in vivo*.

Ionic analyses of excitability in the more common, nonbursting cells indicate that the spikes are rather conventional, primarily Na^+ dependent, and sensitive to TTX and local anesthetics. However, as with many excitable cells (Hagiwara and Byerly, 1981), active Ca^{2+}-specific currents can be revealed in most cortical neurons by suppression of outward K^+ currents and inward Na^+ currents (Figure 4). Under these condi-

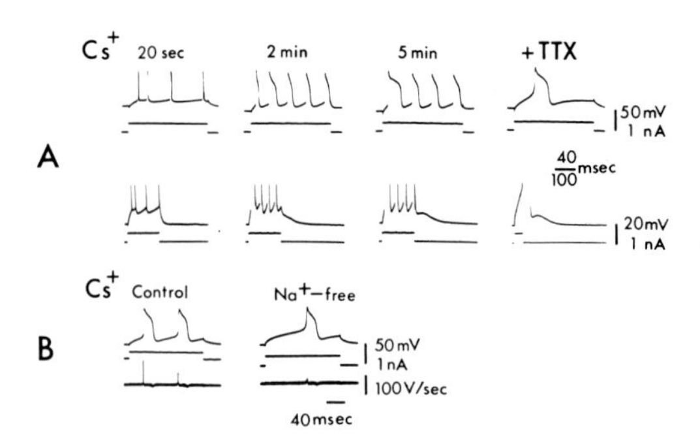

Figure 4. Calcium-dependent action potentials. Recordings obtained with 1 M Cs acetate-filled microelectrodes. (A) Action potential broadening during progressive intracellular injection of Cs^+ [times following impalement shown above; sweeps shown at fast (top) and slow (bottom) speeds]. Note the broad spike shoulders and reversal of the afterpolarization at 5 min. Neuron was exposed to 10^{-6} M TTX (right) to block active Na^+ currents, and a slow, overshooting spike with a depolarizing afterpotential could still be activated. (B) Slow spikes also remained after cells were injected with Cs^+ and perfused with Na^+-free solutions. Note the increase in threshold and greatly reduced rate of rise (bottom trace). In experiments not illustrated, it was demonstrated that nonNa^+-dependent spikes were greatly depressed or blocked by Ca^{2+}-current antagonists such as Mn^{2+} and Co^{2+}.

tions, about 70% of the neurons generate slow, overshooting spikes with a prominent depolarizing afterpotential. The density of Ca^{2+} channels in most neocortical cells would appear to be lower than that of certain other mammalian central neurons such as hippocampal pyramidal cells (Schwartzkroin and Slawsky, 1977), cerebellar Purkinje cells (Llinás and Sugimori, 1980), thalamic neurons (Llinás and Jahnsen, 1982), and inferior olivary cells (Llinás and Yarom, 1981). In the latter cases, regenerative slow spikes could often be evoked following suppression of Na^+ currents, but without blockade of K^+ currents. The alternative explanation is that neocortical cells have more prominent outward K^+ currents than the other cell types, but the conductances associated with hyperpolarizing spike afterpotentials (Connors *et al.*, 1982; Stafstrom *et al.*, 1982b) suggest that slow active K^+ currents actually may be relatively small. Although the described properties are probably most typical of neocortical pyramidal cells, we would speculate that analyses of the middle layer burst-firing cells will reveal a much greater prominence of active Ca^{2+} currents. This is certainly not to imply that there are but two physiological classes of neocortical neuron; the spectrum of subthreshold and spiking behavior argues for a rich diversity of function. Subsequent studies using dye-filled microelectrodes should allow the subtleties of electrical behavior to be correlated with neuron form.

4.2. *Synaptic Behavior of Neocortical Neurons*

One of the most promising uses of slice techniques is to examine local synaptic circuitry. This may be particularly true in neocortex, in which most contemporary theories stress the concept of modular units, yet the specific circuitry of each module appears to be very complex. Since cortical modules (i.e., columns) seem to fit well into the size of a slice, the first order of business is to demonstrate viability of the synaptic mechanisms. The early field potential studies (Richards and McIlwain, 1967; Yamamoto and Kawai, 1967) were promising, but subsequent intracellular measurements have provided more detailed data (Schwartzkroin and Prince, 1976; Prince and Wong, 1981; Connors *et al.*, 1982; Vogt and Gorman, 1982). Although stably impaled neurons in the cortical slice do not display spontaneous spiking, there is detectable background synaptic activity in virtually every cell. Such activity is exclusively depolarizing, and is almost entirely abolished with TTX. This seems to indicate that there is spontaneous impulse activity in a class of cells too small to reliably record. Spontaneous hyperpolarizing potentials (presumed inhibitory postsynaptic potentials (IPSPs)) are revealed only during depolarization of membranes, and only in 10% of the neurons. However, in layer V cingulate neurons, spontaneous synaptic activity is greatly enhanced after intracellular injection of $C1^-$, probably because the $C1^-$ equilibrium potential is shifted positively (L.S. Benardo, unpublished observations). Polarization experiments (Connors *et al.*, 1982) suggest that spontaneous inhibitory synaptic potentials may have reversal potentials slightly positive to rest under normal recording conditions.

Stimulation of the white matter in a coronal section of guinea pig sensorimotor cortex (at site *a* in Lorente de Nó's diagram; Figure 1) would be expected to activate a variety of local pathways. Pyramidal cells (marked with 2, 4, or 8) are the most likely sites of intracellular recording, and would be activated synaptically via afferent axons, recurrent collaterals of other pyramidal cells, and secondarily by local neurons, both excitatory and inhibitory. These synaptic arcs are shown schematically on the right side of Figure 1. Microelectrode recordings in fact show that all neurons can be activated synaptically following white matter stimulation (Figure 5A). The most typical pattern is the development of excitatory postsynaptic potentials (EPSPs) at relatively low stimulus strengths, without obvious IPSPs. At higher strengths, a long-latency, prolonged hyperpolarization commonly follows the excitation and is characterized by a large conductance increase (Figure 5B) and a reversal potential 10 to 20 mV negative to rest (Figure 5C). The current under-

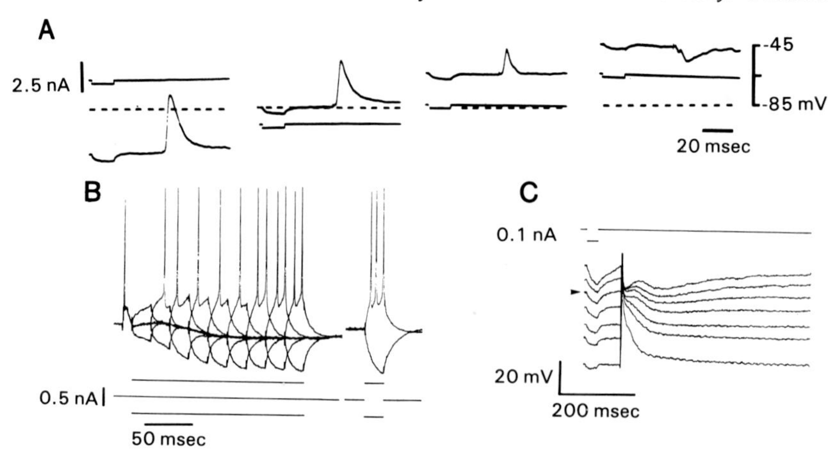

Figure 5. Postsynaptic potentials (PSPs) activated by local circuit stimulation. (A) Stimulation of the white matter elicited a large amplitude depolarizing PSP that never undershot the resting membrane potential (dashed line; second frame). When membrane was polarized from rest, large inhibitory components were revealed and resulted in complete polarity reversal at −40 mV (fourth frame). (B) In a different cell from (A), stimulation at higher intensities and low frequency (0.1 Hz) generated a much longer lasting, hyperpolarizing PSP that inhibited spike generation and was accompanied by a significant conductance increase (monitored by short hyperpolarizing current pulses). (C) Polarization of the long IPSP shows a reversal level some 15 to 20 mV negative to resting potential (indicated by arrow). Records for (B) and (C) were digitized and graphically superimposed.

standing of neocortical synaptic mechanisms is extremely sketchy; however, several interesting aspects are emerging. The early excitation seems, in most cells, to be accompanied by simultaneous inhibition. Thus, membrane depolarization often reverses the composite PSPs at relatively negative potentials and apparent reversed unitary IPSPs can be observed during membrane depolarization (Connors *et al.*, 1982). Focal application of γ-aminobutyric acid (GABA), a substance very likely to be an important inhibitory cortical transmitter (Krnjevic, 1974), causes multiphasic depolarizations (Figure 6) rather than the hyperpolarizations described *in vivo* (Krnjevic and Schwartz, 1967).

These data lead us to a tentative scheme of cortical inhibitory synaptology. The unusually high resting potentials (−75 mV) in slice neurons, together with a Cl^- equilibrium potential slightly positive to rest, result in GABA-mediated IPSPs that are slightly depolarizing and inhibitory by virtue of their large conductance increases. By analogy with the hippocampal pyramidal cell (Alger and Nicoll, 1979), there is probably another distinct GABA receptor that is coupled to an ionic mechanism with a significantly more positive equilibrium potential. The spatial distribution of the two GABA systems appear to be different (B. W.

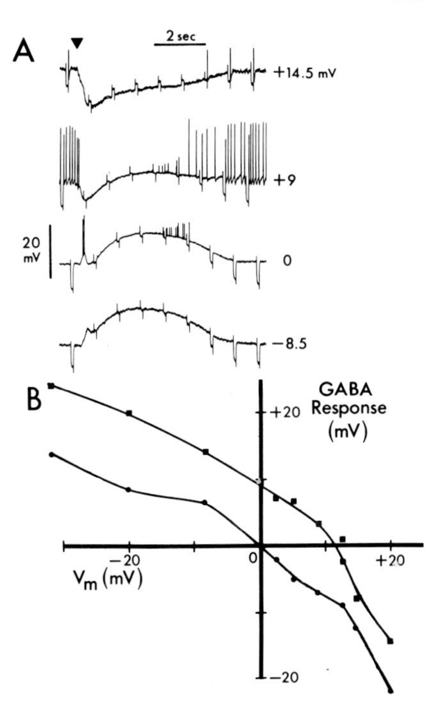

Figure 6. GABA effects on neocortical neuron. GABA (5 mM) was applied by pressure ejection from a micropipette (2 μm tip diameter, 100 msec duration pulse) at a distance of about 100 μm from the recording site. (A) Chart recordings of GABA response at varying membrane potentials. Small hyperpolarizing current pulses (100 msec, 1 Hz) were delivered to monitor input resistance, and tonic current was injected to vary membrane potential. GABA pulse was applied at arrow (top). Resting membrane potential (0 mV in figure) was −75 mV relative to the bath. Action potentials are truncated by the chart recorder, and bridge was slightly imbalanced at +14.5 mV. (B) Amplitudes of GABA response versus membrane potential (V_m). Response was measured at 0.4 sec (circles) and 2.5 sec (squares) after GABA pulse.

Connors, unpublished observations) and it is unclear whether the second type is related to specific synaptic contacts. Finally, the long-duration hyperpolarizing IPSP must work by yet another ionic mechanism with an equilibrium potential that is very negative. The most probable candidate for this is a K^+ channel, and again, by analogy with the hippocampal pyramidal cell, this IPSP may reflect the action of Ca^{2+} which enters through excitatory synaptic channels (Nicoll and Alger, 1981). Alternatively, a variety of suspected cortical transmitters may have some role in directly activating inhibitory K^+ channels. These hypotheses are admittedly quite tenuous, but hint at the rich and perhaps unanticipated group of postsynaptic mechanisms at work in cortical circuits.

As if the complexity of chemical synapses in cerebral cortex were not enough of an experimental challenge, both electron micrographic evidence and physiological data from slice work seem to argue for the presence of some electrical communication. The accepted morphological substrate of intercellular coupling, the gap junction, has so far been observed between adult monkey (Sloper and Powell, 1978) and rat (Peters, 1980) neocortical neurons. With the timely advent of the highly flourescent, membrane-impermeable, gap junction-permeable marker,

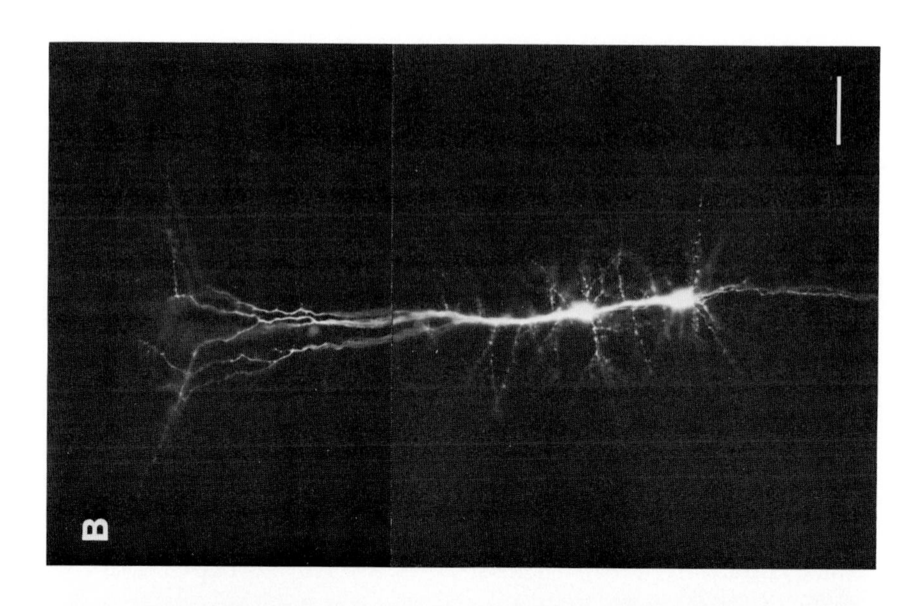
B

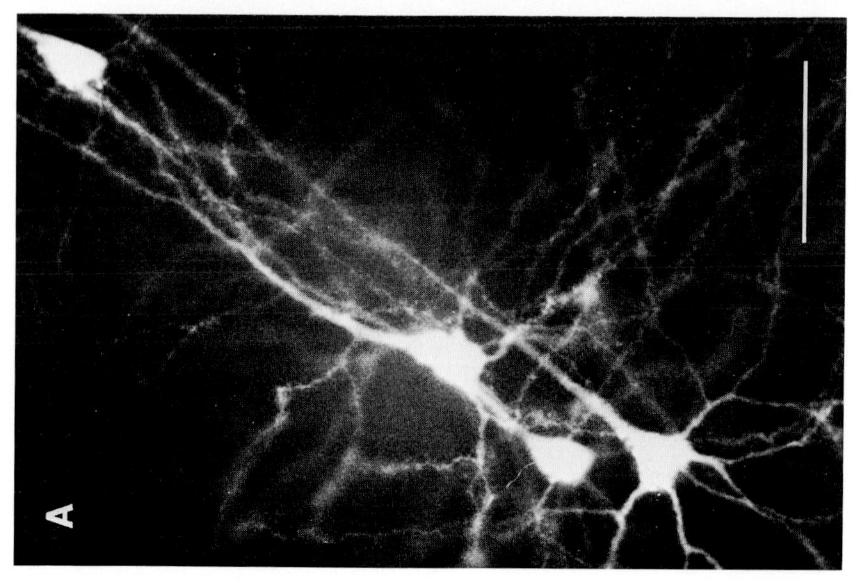
A

Figure 7. Dye-coupling between neocortical neurons. (A) Staining of column of four pyramidal cells in the adult guinea pig neocortex following Lucifer Yellow injection into a single cell. Apical dendrites point toward the pial surface (top right), and the deepest neuron soma lies about 450 μm from the pia (from Gutnick and Prince, 1981). (B) Dye-coupled pyramidal cells of 6-day-old rat neocortex. Two neurons arranged radially; apical dendrites (top) and axons (bottom) run in parallel courses. Pial surface is at top; photograph is a composite from two planes of focus (L. S. Benardo and B. W. Connors, unpublished observations). Calibration bars = 50 μm.

Lucifer Yellow CH (Stewart, 1978), combined with brain slice techniques, apparent observations of neuronal coupling in the mammalian CNS have burgeoned (reviewed in Dudek *et al.*, 1983). Intraneuronal injections of Lucifer Yellow into the guinea pig neocortical slice often reveals extensive spread of the dye into adjacent neurons (Gutnick and Prince, 1981; Connors *et al.*, 1983), (Figure 7). The prevalence of dye-coupling is greater in superficial cortical layers, but is occasionally seen in deeper cells as well. Both pyramidal and nonpyramidal cells can be coupled. The extent of coupling also varies greatly between species, being much lower in the adult rat than in guinea pig, and between cortical areas within a species, being much higher in parietal than occipital cortex of guinea pig (M. J. Gutnick, unpublished observations). Neuronal coupling also changes during development. Rat cortex from 1 to 4 days following birth yields neuronal dye-coupling rates of 70%, with aggregates of 3 to 7 neurons being relatively common (Connors *et al.*, 1983). The coupling rates fall to 30 to 40% by 10 to 15 days and are about 20% in the mature rat, where more than two cells per coupled group is very rare. Further evidence for neuronal coupling comes from experiments in which chemical synaptic activity has been blocked by reduction of bathing Ca^{2+} levels and the addition of Mn^{2+}. Antidromic activation by stimulation of the white matter is then capable of eliciting small, subthreshold depolarizations in many of the cells where dye-coupling is prevalent (Gutnick and Prince, 1981; Connors *et al.*, 1983). The relative voltage-insensitivity, lack of collision with orthodromic spikes, and all-or-none components of these potentials are properties consistent with the idea that they are electronically conducted spikes from coupled cells (Bennett, 1977).

The impressive specificity of dye-coupling in different layers, areas, species, and ages of cerebral cortex is striking, yet there is no ready explanation of its significance. Most commonly, dye-coupled cortical cells have a predominantly radial organization of pyramidal somata with apically extending, intermingling dendrites reminiscent of the bundles described in Golgi studies (Roney *et al.*, 1979). It is tempting to offer this as fodder for the columnar theory of cortical organization. In particular, the arrangement of the coupled cells is consistent with an important role for electrical communication within a minicolumn (Mountcastle, 1978). The developmental correlates suggest an alternative and/or concurrent purpose. In the rat, a large decrease in coupling precedes the formation of most extrinsic cortical connections (Wise and Jones, 1978), precluding any behaviorally relevant role at this young age. However, the ability of gap junctions to transfer small organic substances as well as inorganic ions makes them attractive as potential mediators of developmental signals (Bennett *et al.*, 1981), although the identity and even existence of these signals is still questionable. Morphologically, the columnar arrangement of neocortical cells is nowhere as clear as during early development, when neurons migrate long distances by guiding along the extended radial processes of glial cells (Rakic, 1972). The pillar of neurons associated with each radial process may constitute a single clone (Meller and Tetzlaff, 1975), and it is conceivable that the neurons of coupled aggregates represent some subset of this clone.

It should be pointed out that there is still some question as to the validity of dye-coupling as an indicator of physiologically relevant electrotonic coupling in the mammalian CNS (cf., Knowles *et al.*, 1982; Dudek *et al.*, 1983). Several potential sources of artifact have not been completely ruled out. These fall into two basic categories: (1) the possibility that the act of slicing the tissue causes spurious coupling, perhaps by fusing neuronal membranes or by inducing *de novo* synthesis of gap junctions, and (2) the possibility that the microelectrode causes damage leading to leakage of dye into adjacent cells. In rat hippocampal pyramidal cells, dye-coupling has been observed following intracellular injections *in vivo* (MacVicar *et al.*, 1982); comparable experiments have not yet been done in neocortex. The second source of artifact is perhaps more insidious, and further experimentation will be necessary to determine its significance. However, the relative specificities of neuronal dye coupling in the neocortex suggest that it is low. Further, injection of the dye into one astrocyte yields a spherical constellation of dye-coupled glia, but not neurons (Gutnick *et al.*, 1981).

4.3. *Intrinsic Connectivity of Neocortical Slices*

In the introduction to this chapter, we pointed out the enormous potential of slice methods for detailed physiological analyses of neocortical circuitry. Unfortunately, we must reveal that such analyses have only barely begun, though there is ample justification for our enthusiasm. The difficulties *in vivo* of combining single cell recording, especially intracellular, with precisely placed intracortical stimulation are clear from the study of Asanuma and Rosen (1973). Stability was very limited, precluding changes of stimulus position during a given cellular recording, and electrode placements had to be reconstructed histologically after termination of the experiment. Simultaneous intracellular recordings in intact neocortex have never been reported, to our knowledge. Tremendous advantages are bestowed by an *in vitro* approach.

Shaw and Teyler (1982) have recently described the evoked extracellular potentials of coronal neocortical slices of rat. The methodology allowed measured lateral and radial placements of electrodes while stimulating the white matter. The evoked potentials were similar to those seen *in situ*, and consisted of fast, afferent fiber-related waves and slower, mono- and polysynaptic waves. Consistent with concepts of columnar cortical organization, the lateral spread of activity was much more limited than that in the radial direction. With a similar experimental arrangement, Komatsu *et al.* (1981) showed marked long-term potentiation in slices of kitten visual cortex. Evidence was presented that potentiation occurred at both geniculocortical and intrinsic synaptic contacts and, remarkably, that potentiation occurred only during a narrow developmental window. The latter study employed current source-density analysis to localize the layers of the potentiated synapses. A similar approach in two dimensions, with multiple, interactive stimulus sites, could give a very detailed view of intracortical synaptic relationships at a neuron population level.

Mountcastle (1978) has pointed out that different areas of neocortex can probably be defined by three simultaneously varying qualities: a distinctive cytoarchitecture, a specific pattern of extrinsic connections and a physiologic "function." He speculates that a fourth and very poorly understood variable, intrinsic microconnectivity, may also serve to distinguish a cortical area. The reason for the relative experimental neglect of this parameter is clearly technical. Golgi impregnations or horseradish peroxidase (HRP) injections may allow tracing of some intrinsic axonal arborizations, but synaptic relationships are difficult and tedious to work out. The possibility that quantitative differences in connectivities are of primary physiologic importance greatly compounds the problem. The pattern of degenerating axons following punctate lesions is enlightening (e.g., Gould and Ebner, 1978), but not very specific. However, since the anatomic arrangement of local circuitry will weigh heavily on the functional characteristics of an area, it may be feasible to employ sensitive electrophysiologic criteria in the mapping of neocortical regions. Thus, just as a histologist isolates a thin section of brain tissue for visual examination, the physiologist might employ a living slice for functional categorization. In this context, the application of voltage-sensitive dyes to neocortical brain slices (Grinvald and Segal, this volume) might be especially useful in charting the flow of activity. Ultimately, variations in microconnectivity might be understood in light of the specific duties of a given cortical area.

Slice preparations have recently been used to examine a pathology of intrinsic cortical circuitry, the generation of focal epileptic discharges. Application of GABA antagonists such as bicuculline or penicillin appear to reversibly block some of the local inhibitory circuits of the cortex in a dose-dependent manner. Above a certain threshold concentration, the result of such treatment is not a simple augmentation of EPSPs but the activation of a wholly unique electrical event, the depolarization shift (DS). The DS is a well-characterized process in *in situ* models of epilepsy (Matsumoto and Ajmone-Marsan, 1964; Prince, 1968), and is the intracellular manifestation of the interictal spike of the electroencephalogram. In both intact and sliced neocortex, the DS is a large, slow membrane depolarization with an associated burst of action potentials. It occurs as a distinctly all-or-none event with a very long refractory period. In cortical slices, as *in vivo*, DSs follow very low intensity stimuli with long and shifting latencies (Courtney and Prince, 1977; Gutnick *et al.*, 1982a) and under some conditions may occur spontaneously (B. W. Connors, unpublished observations). Unit recordings *in vivo* suggested that occurrence of a DS in one neuron was mirrored in every other neuron of a local epileptic population (Matsumoto and Ajmone-Marsan, 1964). This fact distinguishes the DS as an event critically dependent upon neuronal synchronizing mechanisms, the most obvious of which are EPSPs mediated via local axons. By suppressing outward K^+ currents with intracellularly injected Cs^+, it is indeed possible to demonstrate that the DS-associated slow wave has underlying synaptic conductances with a reversal potential similar to nonDS-associated EPSPs (Gutnick *et al.*, 1982a; Figure 8).

A study of the spatial and temporal character of DS expression might give clues to the means by which this event is triggered and propagated (Connors and Gutnick, in press). Based on an investigation of penicillin foci in intact cat visual cortex, Gabor *et al.* (1979) suggested that the minimal epileptogenic aggregate is a cortical column, i.e., a neuron population with similar orientation preference and ocular dominance. This exciting conclusion further supports the concept of strong radial interactions in neocortex. It also leads to the possibility that an epileptic focus may be thought of as an aggregate of modules, each capable of producing DS activity, but coordinating this activity via horizontally mediated excitation. To address some of these problems, coronal neocortical slices were treated with the convulsant agent, picrotoxin, and further subdivided with cuts 400 to 500 μm apart, either tangential (separating layers) or radial (separating columns) (Gutnick *et al.*, 1982b). Both isolated columns and layers could generate DSs when stimulated; however, tenfold higher concentrations of picrotoxin were required and the DS duration was roughly correlated with the narrowness of the slab. Thus, no particular layer (and perhaps no particular cell type except the pyramidal cell) is necessary for the initiation of the DS, and horizontal cortical connections appear to contribute to the intensity of a DS in adjacent

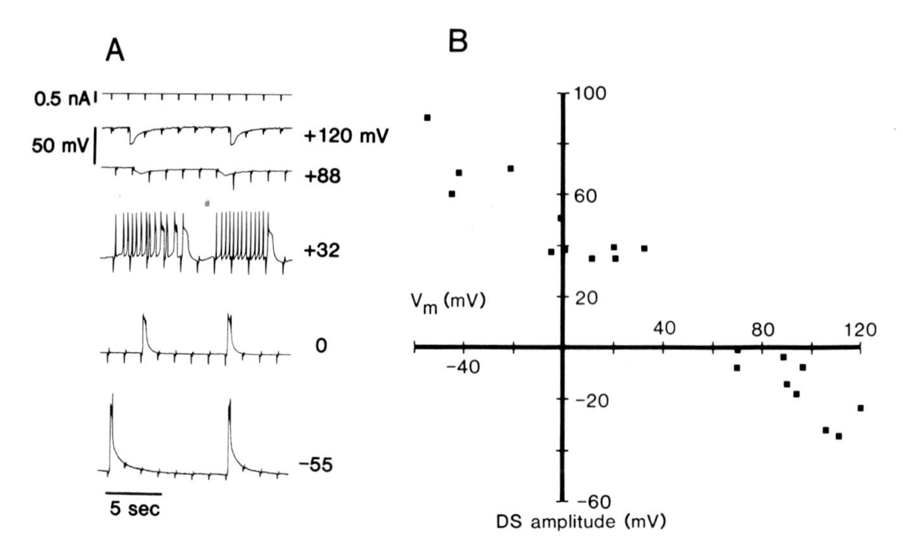

Figure 8. Effects of membrane potential on neocortical DS. Slices were bathed in bicuculline (5×10^{-5} M) and neuron was injected intracellularly with Cs^+ to increase input resistance. (A) Stimuli were applied to layer I (0.67 Hz), and DS was generated after every 5th to 7th stimulus. Small hyperpolarizing current pulses preceded each shock (downward deflections). Cell was polarized from resting potential (0 mV in figure, −65 mV absolute) to the levels indicated at right. DS-associated slow waves reversed polarity at very depolarized levels (+88 and +120 mV) and grew in amplitude at hyperpolarized levels (−55 mV). (B) Plot of amplitude of DS-associated slow wave as a function of membrane potential. Reversal potential is about 60 to 70 mV positive to the resting level. Note that measurements of the DS were impossible between +35 and +65 mV because of membrane potential instability.

areas. These studies do not pinpoint the layers of DS initiation when vertical connections are intact, but studies *in situ* have shown that the region between laminae III and V is most sensitive to focally applied penicillin (Lockton and Holmes, 1980).

An alternative approach to the task of mapping paroxysmal activity is the use of ion-selective microelectrodes (see Hounsgaard and Nicholson, this volume). The release of K^+ to the extracellular space is a necessary consequence of enhanced neural activity (Somjen, 1979) and provides a convenient measure of the temporal and spatial properties of the DS. Single stimuli applied to the white matter normally elicit a small change in $[K^+]_o$, which peaks in deeper layers near the stimulation site (Figure 9A, closed symbols). When penicillin is added, however, single weak stimuli elicit an all-or-none DS field potential with an as-

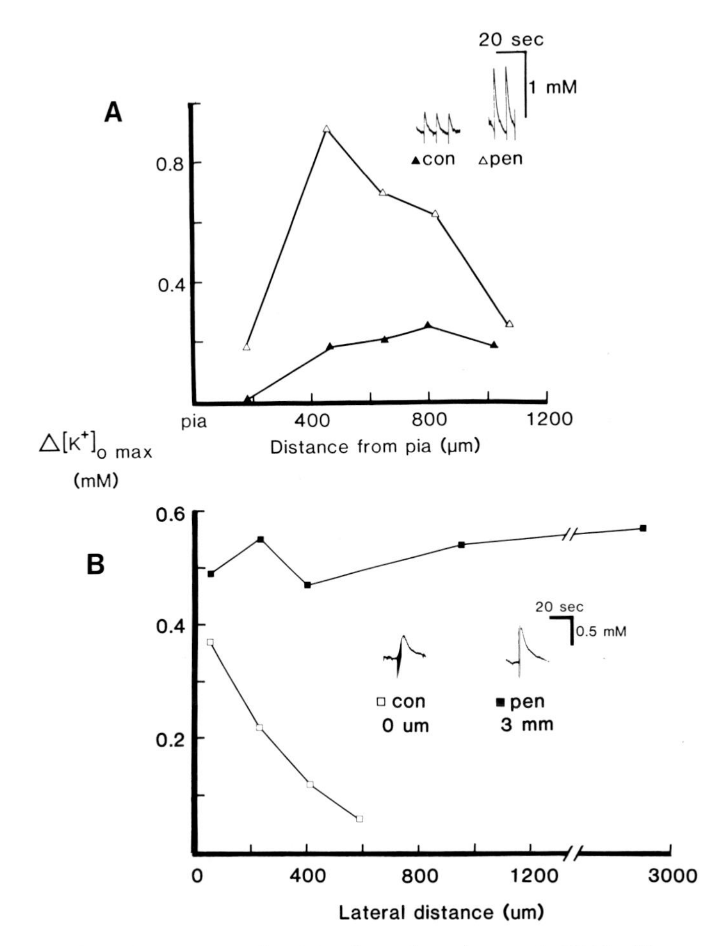

Figure 9. Spatial properties of activity-dependent changes in $[K^+]_0$. Slices were bathed in either normal solution or 3mM penicillin-containing solution, and stimuli were applied to the white matter. All data were obtained from 14-day-old rat neocortex. (A) Radial profile of the change in $[K^+]_0$ ($\Delta[K^+]_0$) evoked by a single strong shock (con) in normal solution, and by single, weak, DS-initiating shocks in penicillin (pen). K^+-sensitive microelectrode was advanced 100 μm into the slice at varying distances from the pia. (B) Lateral profile of $\Delta[K^+]_0$ measured 500 μm from the pia. Inset shows response to a short stimulus train (5 Hz, 3 sec) in control solution at the point directly in line with the stimulus site (0 μm), and response in penicillin following a single DS-initiating shock recorded 3 mm lateral to the stimulus site.

sociated large, transient change in $[K^+]_0$, which has a peak amplitude in the middle layers (Figure 9A, open symbols). When measured tangentially in the same layer, $[K^+]_0$ changes are spatially restricted in the control state, presumably due to a limited spread of activity (cf., Shaw and Teyler, 1982). With the addition of penicillin, the DS-associated change in $[K^+]_0$ is seen to propagate several millimeters to the farthest reaches of the slice, often undiminished in size (Figure 9B). The laminar distribution of $\Delta[K^+]_0$ following the DS is quite similar to that recorded *in vivo* (Futamachi and Pedley, 1976), and indicates a redistribution of activity following the convulsant. The horizontal extent of paroxysmal activity reiterates the intensity of lateral excitatory connections. It would be of great interest to apply the convulsant agent focally, rather than uniformly as above, and map the DS propagation laterally and radially. This model might more closely resemble an *in vivo* focus, and allow the intrinsic mechanisms that limit the spread of activity to interact with the paroxysmal discharge. The advantages of the slice preparation make it the ideal system for such an inquiry.

5. CONCLUSIONS

During the relative short period that *in vitro* brain slice techniques have been applied to the study of neocortical physiology, the approach has yielded some unique insights. These include information about the intrinsic membrane conductances of cortical neurons, the presence and patterns of specific coupling between both neurons and glia, aspects of neuronal development and a detailed look at the epileptic discharge. All of these investigations were greatly facilitated by the technical advantages of the *in vitro* preparation, and many would have been unfeasible by other available means. Prospects for continued advances, using slices, in these and most questions of neocortical function are undoubtedly excellent. A perusal of the earlier chapters of this book will provide an inkling of the experimental possibilities as they have been applied to the hippocampus.

Enthusiasm must be tempered with caution at this point. The neocortical slice, as with all brain slices, is a traumatized tissue subject to its many biological and physical reactions to such insults. The physiological properties, though often remarkably similar to what is measured in the living organism, can be distinctly different. A further annoyance is the fact that functional properties may vary over time, never quite reaching that convenient steady state that facilitates data collection and interpretation. The experimental environment and technique are unfailingly different from laboratory to laboratory, making data comparison difficult. To be fair, all of these problems are present to some degree in all invasive experimental procedures. Of greatest immediate concern is the lack of understanding of precisely what changes do occur, and the relative lack of interest in tackling this admittedly pedestrian, though vital, question.

If we could lend our bias to the prospectus for neocortical brain slice studies, the emphasis would be toward an unraveling of the physiology of local cortical circuitry. Active contemporary interest in the anatomical groundwork, extensive information on the functional organization and a testable conceptual framework have arisen from the more classical methodologies. With high-resolution, quantitative physiological techniques applied cleverly to *in vitro* cortical circuits, the details of cortical function at the cellular level may be illuminated.

ACKNOWLEDGEMENTS. We are very grateful to L. S. Benardo, F. E. Dudek, D. A. Prince, C. E. Stafstrom, S. M. Thompson, and B. A. Vogt for sharing preprints and unpublished data with us. We also acknowledge the excellent technical support of G. Dennison and J. Kadis and manuscript preparation by C. Joo. The authors' studies were supported by NIH grant NS 06477 to Dr. D. A. Prince, a Lennox Fellowship from the American Epilepsy Society (BWC), and the United States–Israel Binational Science Foundation (MJG).

6. REFERENCES

Alger, B. E. and Nicoll, R. A., 1979, GABA-mediated biphasic inhibitory response in hippocampus, *Nature (London)* **281**:315–317.

Asanuma, H. and Rosen, I., 1972, Topographical organization of cortical efferent zones projecting to distal forelimb muscles in the monkey, *Exp. Brain Res.* **14**:243–256.

Asanuma, H. and Rosen, I., 1973, Spread of mono- and polysynaptic connections within cat's motor cortex, *Exp. Brain Res.* **16**:507–520.

Benardo, L. S., Masukawa, L. M., and Prince, D. A., 1982, Electrophysiology of isolated hippocampal pyramidal dendrites, *J. Neurosci.* **2**:1614–1622.

Bennett, M. V. L., 1977, Electrical transmission: A functional analysis and comparison with chemical transmission, in: *Handbook of Physiology,* Section I: *The Nervous System,* Volume 1: *Cellular Biology of Neurons* (E. R. Kandel, ed.), American Physiological Society, Bethesda, pp. 357–416.

Bennett, M. V. L., Spray, D. C., and Harris, A. L., 1981, Electrical coupling in development, *Am. Zool.* **21**:413–427.

Benninger, C., Kadis, J., and Prince, D. A., 1980, Extracellular calcium and potassium changes in hippocampal slices, *Brain Res.* **187**:165–182.

Benzanilla, F. and Armstrong, C. M., 1972, Negative conductance caused by entry of sodium and cesium into potassium channels of squid axons, *J. Gen. Physiol.* **60**:588–608.

Christensen, B. N. and Ebner, F. F., 1978, The synaptic architecture of neurons in opossum somatic sensory-motor cortex: A combined anatomical and physiological study, *J. Neurocytol.* **7**:39–60.

Connors, B. W. and Gutnick, M. J., Cellular mechanisms of neocortical epileptogenesis in an acute experimental model, in: *Electrophysiology of Epilepsy* (P. A. Schwartzkroin and H. Wheal, eds.), Academic Press, in press.

Connors, B. W., Gutnick, M. J., and Prince, D. A., 1981, Electrophysiological properties of neocortical neurons maintained *in vitro, Soc. Neurosci. Abstr.* **7**:593.

Connors, B. W., Gutnick, M. J., and Prince, D. A., 1982, Electrophysiological properties of neocortical neurons *in vitro, J. Neurophysiol.* **48**:1302–1320.

Connors, B. W., Benardo, L. S., and Prince, D. A., 1983, Coupling between neurons of the developing rat neocortex, *J. Neurosci.* **3**:773–782.

Courtney, K. R. and Prince, D. A., 1977, Epileptogenesis in neocortical slices, *Brain Res.* **127**:191–196.

Deschenes, M., LaBelle, A., and Landry, P., 1979, Morphological characterization of slow and fast pyramidal tract cells in the cat, *Brain Res.* **178**:251–274.

Dingledine, R., Dodd, J., and Kelly, J. S., 1981, The *in vitro* brain slice as a useful neurophysiological preparation for intracellular recording, *J. Neurosci. Methods* **2**:323–362.

Dudek, F. E., Andrew, R. D., MacVicar, B. A., Snow, R. W., and Taylor, C. P., Recent evidence for and possible significance of gap junctions and electrotonic synapses in the mammalian brain, in: *Basic Mechanisms of Neuronal Hyperexcitability* (H. H. Jasper and N. M. van Gelder, eds.), Alan R. Liss, Inc., New York, in press.

Feldman, M. L. and Peters, A., 1978, The forms of non-pyramidal neurons in the visual cortex of the rat, *J. Comp. Neurol.* **179**:761–794.

Futamachi, K. J. and Pedley, T. A., 1976, Glial cells and extracellular potassium: Their relationship in mammalian cortex, *Brain Res.* **109**:311–322.

Gabor, A. J., Scobey, R. P., and Wehrli, C. J., 1979, Relationship of epileptogenicity to cortical organization, *J. Neurophysiol.* **42**:1609–1625.

Gage, P. W., and Quastel, D. M. J., 1965, Dual effect of potassium on transmitter release, *Nature* **206**:625–626.

Gibson, I. M. and McIlwain, H., 1965, Continuous recording of changes in membrane potential in mammalian cerebral tissues *in vitro*: Recovery after depolarization by added substances, *J. Physiol. (London)* **176**:261–283.

Gilbert, C. D. and Wiesel, T. N., 1979, Morphology and intracortical projections of functionally characterised neurones in the cat visual cortex, *Nature (London)* **280**:120–125.

Gilbert, C. D. and Wiesel, T. N., 1981, Laminar specialization and intracortical connections in cat primary visual cortex, in: *The Organization of the Cerebral Cortex* (F. O. Schmitt, F. G. Worden, G. Adelman, and S. G. Dennis, eds.), MIT Press, Cambridge, pp. 163–191.

Gould, H. J. and Ebner, F. F., 1978, Interlaminar connections of the visual cortex in the hedgehog (*Paraechinus hypomelas*), *J. Comp. Neurol.* **177**:503–518.

Gutnick, M. J. and Prince, D. A., 1981, Dye-coupling and possible electrotonic coupling in the guinea pig neocortical slice, *Science* **211**:67–70.

Gutnick, M. J., Connors, B. W., and Ransom, B. R., 1981, Dye-coupling between glial cells in the guinea pig neocortical slice, *Brain Res.* **213**:486–492.

Gutnick, M. J., Connors, B. W., and Prince, D. A., 1982a, Mechanisms of neocortical epileptogenesis *in vitro, J. Neurophysiol* **48**:1321–1335.

Gutnick, M. J., Grossman, Y., and Carlen, P., 1982b, Epileptogenesis in subdivided neocortical slices, *Neurosci. Lett. Supplement* **10**:5226.

Hagiwara, S. and Byerly, L., 1981, Calcium channel, *Annu. Rev. Neurosci.* **4**:69–125.

Hille, B., Woodhull, A. M., and Shapiro, B. J., 1975, Negative surface charge near sodium channels of nerve: Divalent ions, monovalent ions and pH, *Philos. Trans. R. Soc. London* **270**:301–318.

Hillman, H. H. and McIlwain, J., 1961, Membrane potentials in mammalian cerebral tissues *in vitro*: Dependence on ionic environment, *J. Physiol.* (*London*) **157**:263–278.

Hillman, H. H., Campbell, W. L., and McIlwain, H., 1963, Membrane potential in isolated and electrically stimulated mammalian cerebral cortex: Effects of chlorpromazine, cocaine, phenobarbitone, and protamine on the tissue's electrical and chemical responses to stimulation, *J. Neurochem.* **10**:325–339.

Hodgkin, A. L., Huxley, A. F., and Katz, B., 1952, Measurement of current-voltage relations in the membrane of the giant axon of *Loligo, J. Physiol.* (*London*) **116**:424–448.

Holtzman, D., Obana, K., and Olson, J., 1981, Hyperthermia-induced seizures in the rat pup: A model for febrile convulsions in children, *Science* **213**:1034–1036.

Hubel, D. H. and Wiesel, T. N., 1974, Sequence regularity and geometry of orientation columns in the monkey striate cortex, *J. Comp. Neurol.* **158**:267–294.

Hubel, D. H. and Wiesel, T. N., 1977, Functional architecture of macaque monkey visual cortex, *Proc. R. Soc. London, Ser. B.* **198**:1–59.

Kandel, E. R. and Tauc, L., 1966, Anomalous rectification in the metacerebral giant cells and its consequences for synaptic transmission, *J. Physiol.* (*London*) **183**:287–304.

Kato, H. and Ogawa, T., 1981, A technique for preparing *in vitro* slices of cat's visual cortex for electrophysiological experiments, *J. Neurosci. Methods* **4**:33–38.

Kato, H., Ito, Z., Matsuoka, S., and Sakurai, Y., 1973, Electrical activities of neurons in the sliced human cortex *in vitro, EEG Clin. Neurophysiol.* **35**:457–462.

Kelly, J. P. and Van Essen, D. C., 1974, Cell structure and function in the visual cortex of the cat, *J. Physiol.* (*London*) **238**:515–547.

Knowles, W. D., Funch, P. G., and Schwartzkroin, P. A., 1982, Electrotonic and dye coupling in hippocampal CA1 pyramidal cells *in vitro, Neuroscience* **7**:1713–1722.

Komatsu, Y., Toyama, K., Maeda, J., and Sakaguchi, H., 1981, Long-term potentiation investigated in a slice preparation of striate cortex of young kittens, *Neurosci. Lett.* **26**:269–274.

Krnjevic, K., 1974, Chemical nature of synaptic transmission in vertebrates, *Physiol. Rev.* **54**:418–540.

Krnjevic, K. and Schwartz, S., 1967, The action of gamma-aminobutyric acid on cortical neurones, *Exp. Brain Res.* **3**:320–336.

Li, C. H. and McIlwain, H., 1957, Maintenance of resting membrane potentials in slices of mammalian cerebral cortex and other tissues *in vitro, J. Physiol.* (*London*) **139**:178–190.

Lin, C. S., Friedlander, J., and Sherman, S. M., 1979, Morphology of physiologically identified neurons in the visual cortex of the cat, *Brain Res.* **172**:344–348.

Llinás, R. and Jahnsen, H., 1982, Electrophysiology of mammalian thalamic neurones *in vitro, Nature* (*London*) **297**:406–408.

Llinás, R. and Sugimori, M., 1980, Electrophysiological properties of *in vitro* Purkinje cell somata in mammalian cerebellar slices, *J. Physiol.* (*London*) **305**:171–195.

Llinás, R. and Yarom, Y., 1981, Electrophysiology of mammalian inferior olivary neurones *in vitro*. Different types of voltage-dependent ionic conductances, *J. Physiol.* (*London*) **315**:549–567.

Lockton, J. W. and Holmes, O., 1980, Site of the initiation of penicillin-induced epilepsy in the cortex cerebri of the rat, *Brain Res.* **190**:301–304.

Lorente de Nó, R., 1938, The cerebral cortex: Architecture, intracortical connections and motor projections, in: *Physiology of the Nervous System* (J. F. Fulton, ed.), Oxford University Press, Oxford, pp. 291–339.

MacVicar, B. A., Ropert, N., and Krnjevic, K., 1982, Dye-coupling between pyramidal cells of the rat hippocampus *in vivo, Brain Res.* **238**:239–244.
Matsumoto, H. and Ajmone-Marsan, C., 1964, Cortical cellular phenomena in experimental epilepsy: Interictal manifestations, *Exp. Neurol.* **9**:286–304.
McIlwain, H., 1951, Metabolic response *in vitro* to electrical stimulation of sections of mammalian brain, *Biochem. J.* **49**:382–393.
McIlwain, H. and Ochs, S., 1952, Absence of electrical responses of brain slices on *in vitro* stimulation, *Am. J. Physiol.* **171**:128–133.
Meller, K. and Tetzlaff, W., 1975, Neuronal migration during the early development of the cerebral cortex: A scanning electron microscopic study, *Cell Tissue Res.* **163**:313–325.
Mountcastle, V. B., 1957, Modality and topographic properties of single neurons of the cat's somatic sensory cortex, *J. Neurophysiol.* **20**:408–434.
Mountcastle, V. B., 1978, An organizing principle for cerebral function: The unit module and the distributed system, in: *The Mindful Brain* (G. M. Edelman and V. B. Mountcastle, eds.), MIT Press, Cambridge, pp. 7–50.
Nelson, P. G. and Frank, K., 1967, Anomalous rectification in cat spinal motoneurons and effect of polarizing currents on excitatory postsynaptic potential, *J. Neurophysiol.* **30**:1097–1113.
Nicoll, R. A. and Alger, B. E., 1981, Synaptic excitation may activate a calcium-dependent potassium conductance in hippocampal pyramidal cells, *Science* **212**:957–959.
Ogawa, T., Ito, S., and Kato, H., 1981, Membrane characteristics of visual cortical neurons in *in vitro* slices, *Brain Res.* **226**:315–319.
Peters, A., 1980, Morphological correlates of epilepsy: Cells in the cerebral cortex, in: *Antiepileptic Drugs: Mechanisms of Action* (G. H. Glaser, J. K. Penry, and D. M. Woodbury, eds.), Raven Press, New York, pp. 21–48.
Prince, D. A., 1968, The depolarization shift in "epileptic" neurons, *Exp. Neurol.* **21**:467–485.
Prince, D. A. and Wong, R. K. S., 1981, Human epileptic neurons studied *in vitro, Brain Res.* **210**:323–333.
Rakic, P., 1972, Mode of cell migration to the superficial layers of fetal monkey neocortex, *J. Comp. Neurol.* **145**:61–84.
Ramón y Cajal, S., 1911, *Histologie due Système Nerveux de l'homme et des Vertèbrés,* Volume 2 (translated by L. Azoulay), Maloine, Paris.
Richards, C. D., and McIlwain, H., 1967, Electrical responses in brain samples, *Nature (London)* **215**:704–707.
Rockel, A. J., Hiorns, R. W., and Powell, T. P. S., 1980, The basic uniformity in structure of the neocortex, *Brain* **103**:221–244.
Roney, K. J., Scheibel, A. B., and Shaw, G. L., 1979, Dendritic bundles: Survey of anatomical experiments and physiological theories, *Brain Res. Rev.* **1**:225–271.
Scholfield, C. N., 1978, Electrical properties of neurones in the olfactory cortex slice *in vitro, J. Physiol. (London)* **275**:535–546.
Schwartzkroin, P. A. and Altschuler, R. J., 1977, Development of kitten hippocampal neurons, *Brain Res.* **134**:429–444.
Schwartzkroin, P. A. and Kunkel, D. D., 1982, Electrophysiology and morphology of the developing hippocampus of fetal rabbits, *J. Neurosci.* **2**:448–462.
Schwartzkroin, P. A. and Prince D. A., 1976, Microphysiology of human cerebral cortex studied *in vitro, Brain Res.* **115**:497–500.
Schwartzkroin, P. A. and Slawsky, M., 1977, Probable calcium spikes in hippocampal neurons, *Brain Res.* **135**:157–161.

Shaw, C. and Teyler, T. J., 1982, The neural circuitry of the neocortex examined in the *in vitro* brain slice preparation, *Brain Res.* **243**:35–47.

Sloper, J. J. and Powell, T. P. S., 1978, Gap junctions between dendrites and somata of neurones in the primate sensori-motor cortex, *Proc. R. Soc. London, Ser. B* **203**:39–47.

Somjen, G. G., 1979, Extracellular potassium in the mammalian central nervous system, *Annu. Rev. Physiol.* **4**:159–177.

Stafstrom, C. E., Schwindt, P. C., and Crill, W. E., 1982a, Negative slope conductance due to a persistent subthreshold sodium current in cat neocortical neurons *in vitro, Brain Res.* **236**:221–226.

Stafstrom, C. E., Schwindt, P. C., Crill, W. E., and Flatman, J. A., 1982b, Membrane currents in cat neocortical neurons *in vitro, Soc. Neurosci. Abstr.* **8**:413.

Stewart, W. W., 1978, Functional connections between cells as revealed by dye-coupling with a highly fluorescent naphthalimide tracer, *Cell* **14**:741–759.

Szentagothai, J., 1978, The neuron network of the cerebral cortex: A functional interpretation, *Proc. R. Soc. London, Ser. B* **201**:219–248.

Takahashi, K., 1965, Slow and fast groups of pyramidal tract cells and their respective membrane properties, *J. Neurophysiol.* **28**:908–924.

Teyler, T. J., 1980, Brain slice preparation: Hippocampus, *Brain Res. Bull.* **5**:391–403.

Vogt, B. A. and Gorman, A. L. F., 1982, Responses of cortical neurons to stimulation of the corpus callosum *in vitro, J. Neurophysiol.* **48**:1257–1273.

Welt, C., Aschoff, J. C., Kameda, K., and Brooks, V. B., 1967, Intracortical organization of cat's motosensory neurons, in: *Symposium on Neurophysiological Basis of Normal and Abnormal Motor Activities* (M. D. Yahr and D. P. Purpura, eds), Raven Press, New York, pp. 255–293.

Winfield, D. A., Gatter, K. G., and Powell, T. P. S., 1980, An electron microscopic study of the types and proportions of neurons in the cortex of the motor and visual areas of the cat and rat, *Brain* **103**:245–258.

Wise, S. P. and Jones, E. J., 1978, Developmental studies of thalamocortical and commisural connections in the rat somatic sensory cortex, *J. Comp. Neurol.* **178**:187–208.

Woolsey, T. A. and Van der Loos, H., 1970, The structural organization of layer IV in the somatosensory region (SI) of mouse cerebral cortex. The description of a cortical field composed of discrete cytoarchitectural units, *Brain Res.* **17**:205–242.

Yamamoto, C. and Kawai, N., 1967, Origin of the direct cortical response as studied *in vitro* in thin cortical sections, *Experientia* **23**:821–822.

Yamamoto, C. and McIlwain, H., 1966, Electrical activities in thin sections from the mammalian brain maintained in chemically defined media *in vitro, J. Neurochem.* **13**:1333–1343.

13
Hypothalamic Neurobiology

GLENN I. HATTON

1. INTRODUCTION

Reviewed in this chapter are the contributions made to our current understanding of hypothalamic neurobiology by studies involving the use of brain slices. These contributions must be viewed in the context of what questions were perceived to be important at the time that the hypothalamic *in vitro* slice became a reality. With this in mind, an attempt has been made to give a brief, but informative background sketch of each field of hypothalamic function to which slice technology has been successfully applied. My expectation is that the reader will agree that the picture of neural function in the hypothalamus, viewed as it were, through the slice, is somewhat clearer than it was heretofore. Some of our prior notions, gleaned from *in vivo* studies, have been confirmed; others have been modified beyond recognition. A few extremely well-kept hypothalamic secrets have been revealed, though many await our further arduous sleuthing activities.

2. THE DEVELOPMENT OF THE HYPOTHALAMIC SLICE PREPARATION

2.1. *Motivating Factors*

The notion that a hypothalamic slice preparation might be feasible and useful in physiological studies was first proposed to me by Dr. Gary

GLENN I. HATTON • Department of Psychology, and the Neuroscience Program, Michigan State University, East Lansing, Michigan 48824.

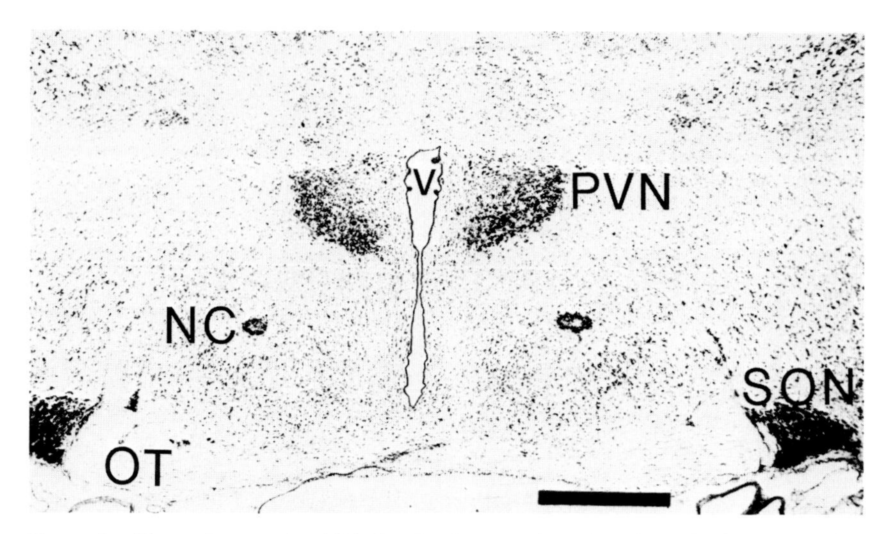

Figure 1. Photomicrograph of Nissl-stained, coronal section through the anterior hypothalamus of rat. NC: nucleus circularis; OT: optic tract; PVN: paraventricular nucleus; SON: supraoptic nucleus; V: third ventricle. Bar = 500 μm.

S. Lynch in 1975. It was during that same year that I visited Dr. Lynch's laboratory, where hippocampal slice work was routinely done, in order to see if one could indeed record spontaneous electrical activity from single neurons in slices of hypothalamus. Our initial attempts were successful, as far as they went.

Of particular interest to us at that time was a little known group of cells, called the nucleus circularis (Peterson, 1966), which lies in the anterior hypothalamus about midway between the supraoptic nucleus (SON) and paraventricular nucleus (PVN) (see Figure 1). We had hypothesized on the basis of primarily morphological work (Hatton, 1976; Tweedle and Hatton, 1976) as well as some lesion studies in the literature (Blass and Epstein, 1971; Peck and Novin, 1971; Almli and Weiss, 1974), that this cell group may contain osmosensitive neurons and may, therefore, play a role in water regulation in the rat. It was, however, too small a nucleus (~ 275 cells) to allow one to study it successfully using the blind man's buff approach commonly referred to as the stereotaxic method. Since one of the singular advantages of brain slice methodology is placement of recording and stimulating electrodes under direct visual guidance, we reasoned that perhaps this tiny nucleus might at least occasionally be visible in 400- to 500-μm-thick slices of tissue. This proved to be only too true: nucleus circularis was *occasionally* visible in such preparations. At about that same time, interest in the other, more extensively investigated magnocellular hypothalamic nuclear groups, SON and PVN, was further increasing because of new developments and findings with quantitative cytoarchitectonic (Hatton *et al.*, 1976), autoradiographic (Conrad and Pfaff, 1976), horseradish peroxidase (HRP) (Sherlock *et al.*, 1975), and immunocytochemical methods (Swaab *et al.*, 1975; Vandesande and Dierickx, 1975). Since these cell groups were usually in the same hypothalamic slices as the smaller nucleus, were composed of similar cell types, and were much easier to locate visually, it was both convenient and interesting to study them.

In addition to visual guidance of electrodes, an advantage of *in vitro* methodology that contributed to our development of the hypothalamic slice was the possibility of finely controlling the osmotic pressure of the medium bathing the tissue. Obviously, if one were going to be able to demonstrate osmoreceptivity in a group of neurons, the parameters of the osmotic stimulus would have to be precisely known and controlled. It was the lack of such knowledge and control of the actual osmotic stimulus as it impinged on putative receptors that had been a major stumbling block in the location of central osmoreceptors using *in vivo* techniques. That is, one cannot, by presently known methods, change the osmotic pressure of a mammal's blood and determine with any degree of accuracy the osmotic stimulus that reaches the various brain areas that may contain osmoreceptive elements. The *in vitro* system allows one to know more precisely both the stimulus and its temporal characteristics.

Thus, two main factors provided the impetus for the development of the hypothalamic slice: visual guidance of electrodes and precise knowledge and control of the extracellular fluid surrounding the cells of interest. Other factors worthy of mention have been and continue to be important in making progress toward understanding hypothalamic neurobiology through the use of slices. The stability of the *in vitro* preparation is a particularly important advantage because of the relatively rich vascularization of the hypothalamus, which creates pulsating nightmares for electrophysiologists who have the temerity to attempt *in vivo* recording. Another factor that will make possible the eventual identification of cell groups that respond to particular changes in their chemical environments, is the ability, *in vitro,* to completely and reversibly block chemical synaptic transmission. This, of course, allows one to determine whether observed cellular responses are primary responses to applied stimuli or secondary, synaptically mediated responses. This is rarely if ever possible to determine in *in vivo* studies of central neural mechanisms.

2.2. *Early Problems Encountered*

The cutting of fresh tissue into slices of uniform thickness varies in difficulty directly with the size of the tissue block face to be cut. Bilateral hypothalamic slices are approximately double the size of hippocampal slices. They also contain tissues that are of nonuniform density and composition (e.g., large myelinated fiber tracts, such as the fornix and optic chiasm, next to cellular areas) and discontinuities (i.e., the third ventricle). These features, along with the fact of no well-known neural circuitry available for use in testing the viability of hypothalamic slices, made this tissue considerably more difficult than was the hippocampus in investigate during experimentation. To some extent, these difficulties, stemming from the need to have a large tissue slice and a lack of hippocampal-like organized redundancy, continue to impede progress in understanding hypothalamic neurobiology. Rather than overcoming these problems, it seems fair to say at this time that progress has been made despite them.

3. HYPOTHALAMIC SLICE PREPARATIONS AND THEIR USES

Compared to hippocampal slice electrophysiology, hypothalamic slice work is in its infancy. Most of the work and the majority of the papers published to date relate to the magnocellular peptidergic areas of the anterior hypothalamus. This is a logical extension of the earlier studies, since we also knew most about these areas from the large amount of *in vivo* work that had been done over many years. Therefore, this section will deal first with experimental investigations of the SON and PVN, followed by a discussion of those few reports of studies made of the predominantly parvocellular hypothalamus.

A few words on the method of producing hypothalamic slices are in order before describing the findings of studies that have employed them. The usual procedures are as follows. The animal is decapitated and the brain is quickly removed.* It is then blocked in a manner similar to that shown in Figure 2. This block of tissue is either placed directly on the stage of a tissue chopper or glued (with cyanoacrylate cement) to the mounting chuck of a vibrating microtome. Orientation of the block

* A common practice in brain slice work is to stun the animal with a sharp blow to the neck before decapitation. Since this procedure often results in hemorrhage of the blood vessels at the base of the brain (e.g., the Circle of Willis and its tributaries) with formation of clots that obstruct one's view of hypothalamic surface features, stunning is undesirable and generally to be avoided when working with this brain area.

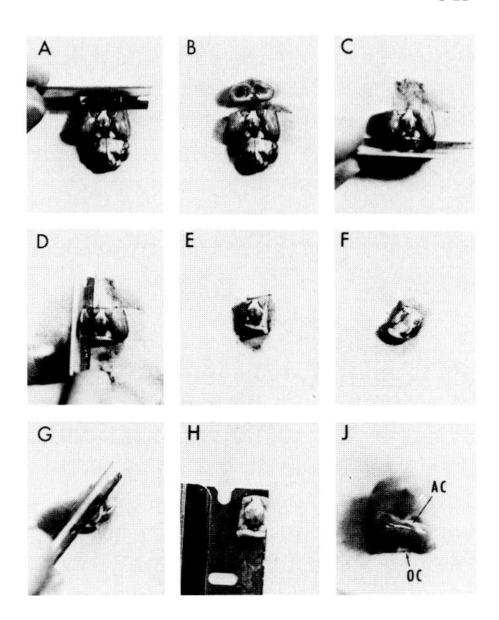

Figure 2. (A–H) Photographs of sequence of cuts used in blocking the hypothalamus in preparation for cutting slices on a tissue chopper. In (F), the block shown in (E) has been placed on its side (ventral is to the right) so that a horizontal cut [shown in (G)] at the level of the anterior commissure can be made. The tissue may then be picked up on the blocking blade [as in (H)] and transferred to a chopper or Vibratome stage. (J) A slightly magnified view of the anterior face of the block showing anterior commissure (AC) and the optic chiasm (OC) (from Hatton *et al.*, 1980).

on the stage or chuck, of course, determines the plane in which the slices will be cut (i.e., coronal, sagittal, horizontal, or oblique). Slices of hypothalamic tissue are usually cut at 400 to 500 μm and put into a buffered, balanced salt solution that has been gassed with a 95% O_2/5% CO_2 mixture. The slices to be studied are then transferred to a supporting substrate (such as a nylon net) in a chamber in which they are maintained at or near body temperatures in an oxygenated atmosphere. For a complete description of one method used with hypothalamic slices, see Hatton *et al.* (1980).

Electrophysiological recordings are generally not attempted until the brain slices have been in the chamber for 1 to 2 hr, because during this time there is little or no activity and membrane potentials are near zero. Although the reasons for this depression are not completely understood, it is probably due in large part to the extreme elevation in extracellular K^+ that follows rapid onset cerebral ischemia.

3.1. *Magnocellular Areas*

The peptide hormones vasopressin (VP)* and oxytocin (OX), are known to participate in a variety of physiological functions including body water regulation, control of vascular smooth muscle, the uterine

* Also sometimes called antidiuretic hormone.

contractions of parturition, the milk ejection reflex during lactation, and discharge of epididymal semen (see Forsling, 1977; Roberts, 1977, for reviews). Also, VP is released during pain and cold stimulation, is in turn a releaser of adrenocorticotropic hormone, and thus, may be important in the stress response (Defendini and Zimmerman, 1978). In recent years, direct actions of these peptides, particularly VP, on the brain have been inferred from studies of learning and memory in animals (see de Wied, 1980, for review).

It has long been known that the cells responsible for the manufacture of VP and OX are located in the magnocellular nuclei of the hypothalamus (Bargmann and Scharrer, 1951). Most, but not all, of these cells terminate in the perivascular spaces surrounding fenestrated capillaries in the neurohypophysis (or neural lobe) of the pituitary gland (Palay, 1957), their hormones being transported there by axoplasmic flow. When appropriately activated, hormone (along with a "carrier protein," neurophysin) is released probably by exocytosis into the perivascular space and enters the general circulation. The elements of these hypothalamic nuclear groups that do not send terminating projections to the neural lobe, project to diverse brain areas including the median eminence (Antunes *et al.*, 1977; Wiegand and Price, 1980), the brain stem and spinal cord (Saper *et al.*, 1976; Swanson, 1977), as well as to other forebrain structures (Conrad and Pfaff, 1976; Silverman *et al.*, 1981).

Adequate evidence exists supporting the notion that neurosecretion is coupled, albeit in complex ways, to the electrical activity of the hormone-containing cell (for reviews see Finlayson and Osborne, 1975; Hayward, 1977; Mason and Bern, 1977). In the rat hypothalamoneurohypophysial system, studies of dehydration, hemorrhage, and suckling (the three most used stimulus complexes) have relatively consistently suggested that increases in hormone release are accompanied by specific alterations in the patterns of neuronal firing observed at the level of the cell bodies in the hypothalamus. Further, these alterations are different for presumed OX-containing and VP-containing cells, even when, as with dehydration, both cell types are activated by the same set of stimulus conditions (see Figure 3). In the unstimulated state (Figures 3A and 3B), cells of the SON, for instance, that have been antidromically identified as ones having terminals in the neural lobe, generally exhibit slow (0.5 to 1 Hz), unpatterned activity (Poulain *et al.*, 1977; Wakerley *et al.*, 1978; and many others reviewed in Poulain and Wakerley, 1982). When stimulated osmotically, two activated patterns have been seen to emerge and predominate (as in Figures 3C and 3D), though not all cells are activated by any one set of conditions. One major cell type increases its firing rate to ~ 10 to 20 Hz and fires in a "fast-continuous" manner;

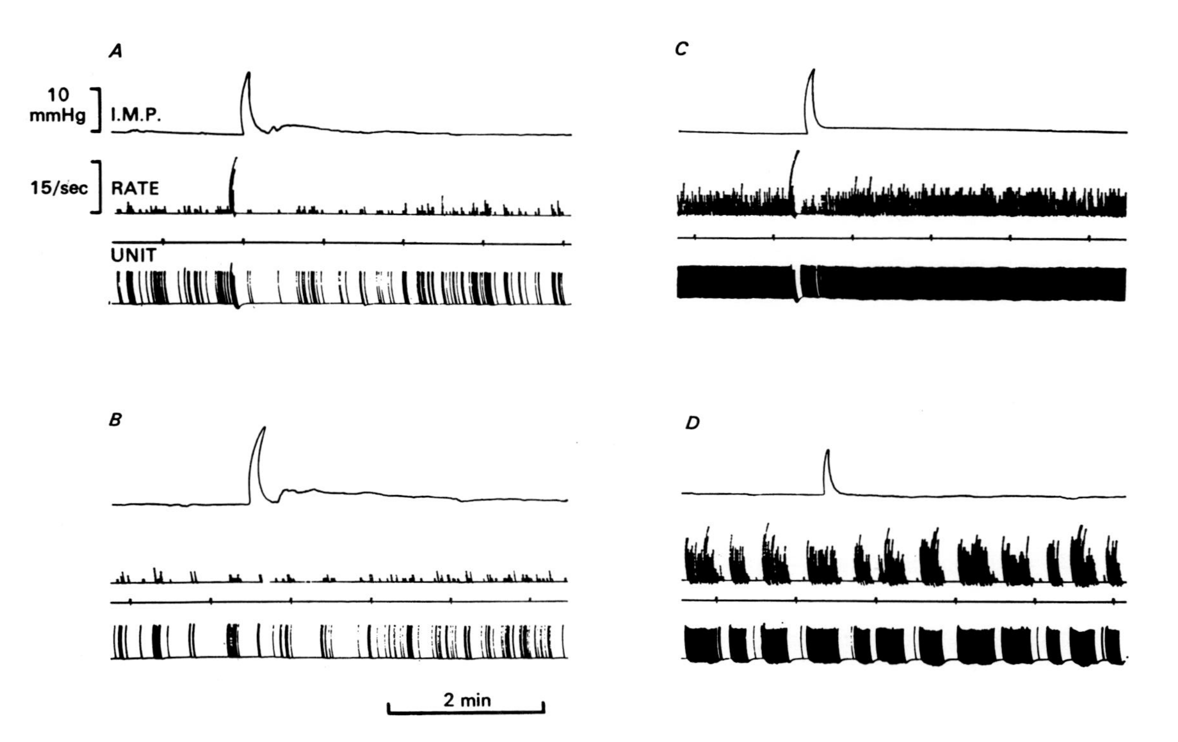

Figure 3. Polygraph recordings showing examples of the firing patterns of cells encountered in the supraoptic nucleus. Slow irregular firing is seen in both oxytocin (A) and vasopressin (B) neurons under control conditions. When stimulated osmotically, oxytocin cells usually fire continuously (C), but vasopressin cells fire phasically (D). In each panel, the upper trace is intramammary pressure (IMP), the middle trace a ratemeter record (RATE) and, on the lower trace, each pen deflection represents one spike (UNIT) (from Brimble and Dyball, 1977).

and, in lactating animals at least, occasional high-frequency bursts (60 to 80 Hz) are superimposed upon the already fast rate. The other major type of stimulated pattern is one described as "phasic," in which the cell may fire a prolonged burst of 30- to 60-sec duration and then remain silent for a variable time interval after which this general pattern repeats. Observations made in the lactating rat (Wakerley and Lincoln, 1973; also see Mason and Bern, 1977) have generated a strong hypothesis that one of these cell types represents the OX-containing cells, while the other is thought to contain VP. Immediately (~ 15 sec) preceding the rise in intramammary pressure that accompanies the milk ejection reflex, the fast-continuous type of cell emits its burst. The bursts of nearly all such cells are phase-locked to the intramammary pressure rise, whereas neither the bursts nor the silent periods of the phasic bursting cells bear any consistent relationship to intramammary pressure changes. From such observations, it has been hypothesized that OX-containing cells fire synchronously to produce the requisite "pulse" of OX necessary to obtain the milk ejection reflex. Further, it was generally assumed (from a lack of data) that this synchrony was due to synaptic driving of OX-containing cells as a direct result of suckling. The phasic bursting activity of putative VP cells has long been thought to be controlled by recurrent synaptic inhibition. This subject is covered in some detail in Section 3.1.2.

3.1.1. Osmosensitivity.

3.1.1a. Extracellular Studies. In our initial attempt at determining the viability of hypothalamic slices and the feasibility of recording electrophysiologically from the nucleus circularis, we also manipulated the osmolality of the medium. We found that the addition of small amounts of 3M NaCl to the bathing medium occasionally appeared to influence the rates and/or patterns of firing of cells in the region of the nucleus. This result was reported as an unpublished observation (Hatton, 1976), however, these manipulations were crude and many important measurements were not made (e.g., changes in medium osmolality due to evaporation). The results of a more carefully controlled set of experiments, in which neurons from both PVN and nucleus circularis were studied, were presented in preliminary form (Hatton *et al.*, 1977) and as part of a complete paper (Hatton *et al.*, 1978).

In these studies, extracellular recordings of spontaneously occurring action potentials were made primarily from neurons of nucleus circularis and the PVN. Samples of activity were also taken from cells in the nucleus of the diagonal band of Broca, periventricular regions of the anterior hypothalamus, preoptic area and the suprachiasmatic nucleus to determine whether these areas, too, were spontaneously active in the slice. The main objectives of these early studies were to establish the hypothalamic slice as a useful investigative tool by determining whether: (1) the activity one could record resembled the patterns, firing rates, etc that had been recorded *in vivo*, and (2) changes in activity could be effected by applying physiological stimuli. Activity rates were quite similar to those already in the literature and magnocellular neurons in the slice frequently showed either fast continuous or phasic firing patterns characteristic of SON and PVN cells. This was important, of course, because such patterns are thought to be associated with OX- and VP-containing cells, respectively. Further, there was some evidence, even in these early experiments, for responsiveness to physiological stimuli. Some slowly discharging neurons were observed to increase their firing rates and others to become phasic when the osmolality of the medium was increased by approximately 20 mOsm/kg, which was well within the physiological range. Also, many more phasically active cells were encountered in slices bathed in high (320 mOsm/kg) than in low (290 mOsm/kg) osmolality medium.

These initial results did not, however, go unchallenged. In a series of papers (Brimble *et al.*, 1978; Haller *et al.*, 1978; Haller and Wakerley, 1980), osmotic stimuli were consistently reported to be ineffective in producing changes in either rates or patterns of neuronal discharge from units in the SON, PVN, and nearby anterior hypothalmic areas of slices. A problem with these studies was that the rates of spontaneous activity recorded *in vitro* by Wakerley and colleagues were far below those usually recorded from similar areas either *in vivo* or *in vitro*. Also, at least half of the cells they recorded were silent, only responding during mechanical deformation of the tissue (e.g., as when the electrode was advanced) or to extracellular applications of monosodium glutamate. As with all negative results, it is difficult or impossible to assign causation. Brain slices in which spontaneous activity is abundant and firing rates are in the range of those recorded *in vivo* (e.g., Hatton *et al.*, 1977, 1978) must differ in a number of ways from those in which all activity is depressed. One of these ways appears to responsiveness to osmotic stimuli. In our report (Hatton *et al.*, 1978), we attributed the negative results of Wakerley and co-workers to their inability at that time to produce optimal preparations. That this guess was probably not too far off the mark is suggested by the fact that more recent reports have confirmed the osmotic responsivity of magnocellular neurons in hypothalamic slices (Mason, 1980; Noble and Wakerley, 1982) and in explant preparations (Bourque and Renaud, 1981). Bourque and Renaud (1981) using the isolated perfused basal hypothalamus *in vitro*, recorded extracellularly from SON neurons both under control conditions and with

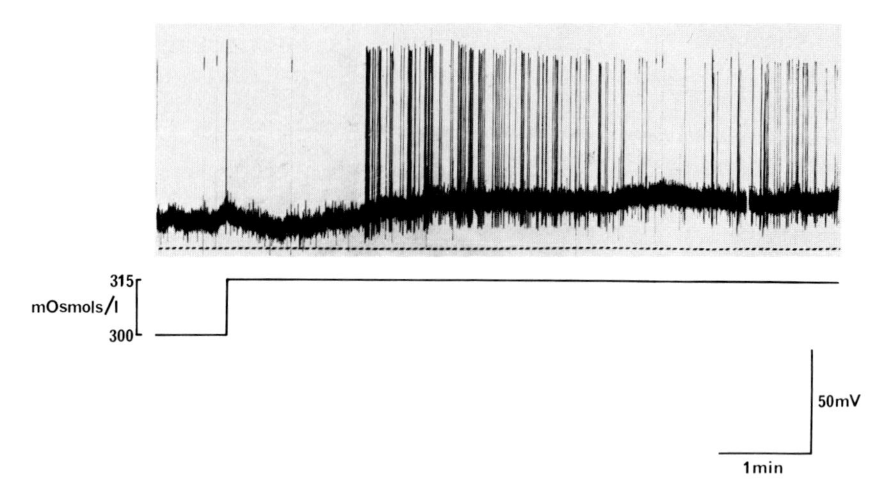

Figure 4. Intracellularly recorded response from a neuron in the SON of the rat hypothalamus, recorded from a 400-μm slice maintained *in vitro*. The perfusing medium was changed where indicated to one of higher osmolarity by addition of mannitol [from Mason, W. T., 1980, Supraoptic neurones of rat hypothalamus are osmosensitive, *Nature* (*London*) **287**:154–157. Reprinted by permission from Macmillan Journals, Ltd].

synaptic transmission blocked by 12 mM Mg^{2+}. (There was, however, no independent assessment of the completeness of this blockade reported.) Over 70% of SON neurons responded to the application of osmotic stimuli (in the range 315 to 354 mOsm/liter); NaCl, sucrose and mannitol were used. The responses observed were the ones typically seen *in vivo* when osmotic stimuli are applied: induction of firing in silent cells, increased frequency of firing, emergence of phasic bursting patterns, increases in burst duration and rate within bursts, and decreases in interburst intervals for phasic cells. Similar results were obtained in the presence and the absence of synaptic transmission, supporting the idea that SON neurons are themselves osmosensitive.

3.1.1b. Intracellular Studies. The work of Mason (1980) certainly comes close to providing a definite, affirmative answer to the question of whether magnocellular hypothalamic neurons are capable of responding to osmotic stimuli. He recorded intracellularly from SON neurons in hypothalmic slices and found that osmolality increases of 15 mOsm/kg were sufficient to trigger bursts of activity in many cells (see Figure 4). Then, with synaptic transmission apparently blocked by 15 mM Mg^{2+}, he recorded osmotically stimulated membrane depolarizations in these neurons. These depolarizations were not, however, accompanied by action potentials (Figure 5). This result may have been due to a need for synaptic activation, as Mason hypothesized at the time, or it may be that the high Mg^{2+} concentration simply raised the spike threshold of these neurons so that even the relatively large (20 mV) depolarization did not trigger a spike train. Work by Abe and Ogata (1982), however, indicates that SON neurons both depolarize and fire rapidly in response to hypertonic solutions applied in the presence of 0mM Ca^{2+} and 12 mM Mg^{2+}.

Therefore, it appears that within the 400- to 500-μm brain slice there reside elements or complexes of elements that are capable of responding to alterations of the osmotic pressure of the extracellular fluid. This is not yet to conclude too firmly that the magnocellular neurons of the SON and PVN are themselves osmoreceptors, for other possibilities remain. Although synaptic connections within some areas of the slice are perhaps not crucial, nonneuronal elements may be involved in osmoreception (evidence reviewed 3.1.2c).

3.1.2. Phasic Bursting Activity. Recent intra- and extracellular studies of phasically firing (putative VP-containing) neurons in the SON have shown that this type of activity can be triggered synaptically by electrically stimulating nearby cholinergic neurons (Hatton *et al.*, 1983a). Acetycholine applied by micropressure injection in the vicinity of the SON produced a similar effect. This phasic bursting is apparently mediated by nicotinic receptors, since the synaptic triggering of such bursts was blocked reversibly by either hexamethonium bromide or d-tubocurarine chloride. The firing rates of putative oxytocin-containing neurons were not altered either by stimulation of these cholinergic cells or by applications of acetylcholine. This newly discovered pathway is probably the one that mediates the osmotic input to vasopressin neurons, since it has been shown that osmotic stimulation of hypothalamo-neurohypophysial explants fail to release vasopressin if the cholinergic receptors are blocked by hexamethonium (Sladek and Joynt, 1979). Further research is needed to determine whether these cholinergic neurons, which lie just dorsolateral to the SON, are indeed themselves osmosensitive.

Evidence that at least some magnocellular hypothalamic neurons are characterized by recurrent synaptic inhibition was first reported by Kandel (1964) in his studies of the goldfish preoptic nucleus. Kandel's evidence was unequivocal in that he was able to show inhibitory postsynaptic potentials in the cell soma following antidromic activation of the axon by electrical stimulation. These potentials were enhanced by injection of depolarizing current through the recording electrode. Ko-

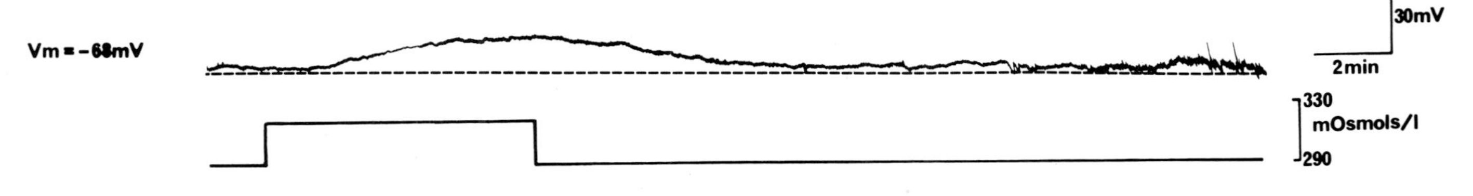

Figure 5. The effect of increased osmolarity on the membrane potential of a supraoptic neuron [from Mason, W. T., 1980, Supraoptic neurones of rat hypothalamus are osmosensitive, *Nature* (*London*) **287**:154–157. Reprinted by permission from Macmillan Journals, Ltd].

izumi and Yamashita (1972) observed prolonged periods of hyperpolarization in SON cells following posterior pituitary stimulation in dogs and cats. They interpreted these long afterhyperpolarizations to indicate the presence of recurrent synaptic inhibition. It is important to note that in neither of these two early reports were the cells that appeared to show recurrent inhibition identified as phasically active neurons. Nevertheless, it was these two reports that formed the foundation for the hypothesis that phasic bursting in putative VP neurons was due to recurrent chemical synaptic inhibition (Dyball, 1971; Dreifuss and Kelly, 1972). It was generally thought that this inhibition was either via a recurrent axon collateral from the magnocellular neuron itself, or that a small local interneuron, similar to a spinal Renshaw cell, was involved. In either case, no definite evidence for synapses containing neurosecretory granules of the type found in SON and PVN magnocellular neurons has ever been reported to be presynaptic to these neurons. There have been many electron microscopic studies of these neurosecretory nuclei, some specifically focusing on synaptic morphology (e.g., Theodosis *et al.*, 1981; Hatton and Tweedle, 1982), so that ample opportunity was there to observe presynaptic terminals containing VP-size dense core granules if they existed within the SON or PVN proper. There is some evidence, from Golgi impregnated material, for axon collaterals arising within the PVN (van den Pol, 1982). Furthermore, recent evidence from studies combining the uses of retrograde tracers, immunocytochemistry and electrophysiology give firm support to the possibility that axon collaterals may be given off and synapse at some distance from the nuclear borders of the PVN and SON (Cobbett *et al.*, 1983; Hatton *et al.*, 1983b). This possibility is, of course, in line with the idea of feedback through interneurons, the cell bodies of which may lie outside of the confines of the magnocellular nuclei.

Dreifuss *et al.* (1976) found that antidromic activation by a train of stimuli presented during the silent period of phasically firing neurons could trigger the next burst. This finding argued against recurrent collateral activity as the terminator of bursting activity. Pittman *et al.* (1981) specifically studied phasic neurons and found no compelling evidence for recurrent inhibition. Still, it was not possible to approach this problem directly *in vivo* since the direct approach requires blocking of chemical synaptic transmission.

3.1.2a. Bursting without Synaptic Transmission. Such a blockade is, of course, readily possible *in vitro* and the phenomenon of phasic bursting has now been studied in the hypothalmic slice (Hatton, 1982). In order to be certain of the effectiveness of the procedures used to block synaptic transmission, slices of hypothalmus were placed in an *in vitro*

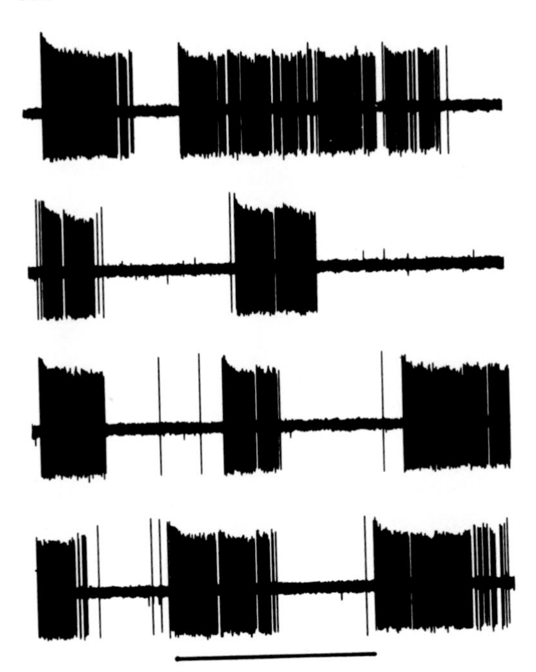

Figure 6. Phasically firing cell from lateral portion of PVN recorded in medium containing 18.7 mM Mg^{2+} and 0.05 mM Ca^{2+}. Firing rate at start of each burst ~ 20 to 30/sec. Bar = 30 sec. From Hatton, 1982.

chamber with hippocampal slices. The synaptic response in the CA1 cell layer from Schaffer collateral stimulation was monitored before, during, and after synaptic transmission was blocked by superfusion of medium containing high Mg^{2+} (either 18.7 or 9.3 mM) and low Ca^{2+} (0.05 mM). This well-studied pathway was chosen as an assay of synaptic blockade because hypothalamic circuitry is relatively unknown. The electrical activity of 22 phasic bursting neurons in the lateral portion of the PVN (as defined by Armstrong *et al.*, 1980) was recorded. This portion is rich in VP-containing cells (Swaab *et al.*, 1975). Nineteen of 22 phasic PVN neurons were recorded only after synaptic transmission was blocked. The remaining three cells were firing phasically in standard medium when first encountered and continued to display phasic bursting activity for up to 1.25 hr after synaptic blockade. Active cells in nearby hypothalamic areas did not show phasic bursting patterns either before or after synaptic transmission was blocked. The phasic bursting activity of the PVN neurons in this study and that of previously reported PVN cells *in vivo* were similar in: (1) firing rate within bursts, (2) burst length, and (3) silent period duration (see Figures 6 and 7). It was concluded that phasic bursting in PVN magnocellular neuropeptidergic cells is not dependent upon synaptically mediated excitation or upon recurrent in-

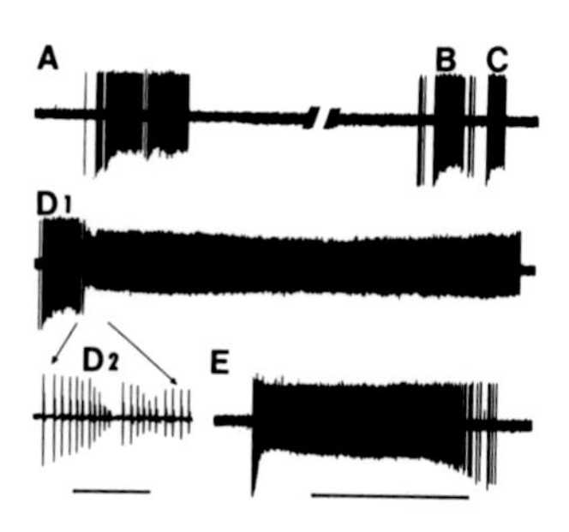

Figure 7. Samples of activity of PVN cell beginning 4.5 hr after synaptic blockade (medium: 9.3 mM Mg^{2+}, 0.05 mM Ca^{2+}). (A–C) Relatively brief bursts showing slight spike inactivation. (D) Prolonged burst with brief period of total spike inactivation shown at faster speed in (D2). (E) Later prolonged burst with initial spike inactivation which recovers as burst ends. Bar in (D2) = 2 sec. Bar in (E) = 30 sec; (A–D1) are at the same speed as (E). From Hatton, 1982.

hibition as had been hypothesized earlier. Both the onset and termination of the bursts occur without chemical synaptic input. It may be that such cells are of the "pacemaker" type described by Gähwiler and Dreifuss (1979) for cultured SON-area neurons. In my study, one phasically bursting nucleus circularis cell was recorded which was of the "follower cell" type, in that it ceased to be spontaneously active when synaptic transmission was blocked, though it continued to show antidromic activation. In a preliminary study of SON neurons in explant preparations containing preoptic area, SON, and posterior pituitary (W. E. Armstrong and G. I. Hatton, unpublished observations), three of three phasic bursting SON cells, shut off in a high Mg^{2+}, low Ca^{2+} medium. This is a small sample of SON and nucleus circularis neurons, but it suggests that PVN may have a relatively larger proportion of cells that can generate phasic bursting activity without chemical synaptic inputs. However, Bourque and Renaud (1981) continued to observe phasically firing SON cells in their preparations after blocking synaptic transmission, so it appears that at least some of these cells are similar to PVN neurons.

3.1.2b. Evidence for Endogenous Membrane Currents. If recurrent chemical synaptic inhibition is not involved in phasic bursting of these neurons, then what mechanisms are responsible? Several possibilities exist and have been proposed (Dreifuss *et al.*, 1976; Hatton, 1982, 1983), and since they are not mutually exclusive, they may all be subtly and simultaneously at play. The first of these, initially proposed by Dreifuss *et al.* (1976), is that bursting is an endogenous property of VP neurons. Evidence in support of this hypothesis comes from intracellular studies of hypothalamic slices in two independent laboratories. R. D. Andrew and F. E. Dudek (unpublished observations) found that current-evoked bursts in SON and PVN cells are followed by an afterhyperpolarization lasting as long as several hundred msec (see Figure 8). Since such afterhyperpolarizations were not blocked by tetrodotoxin (TTX) but

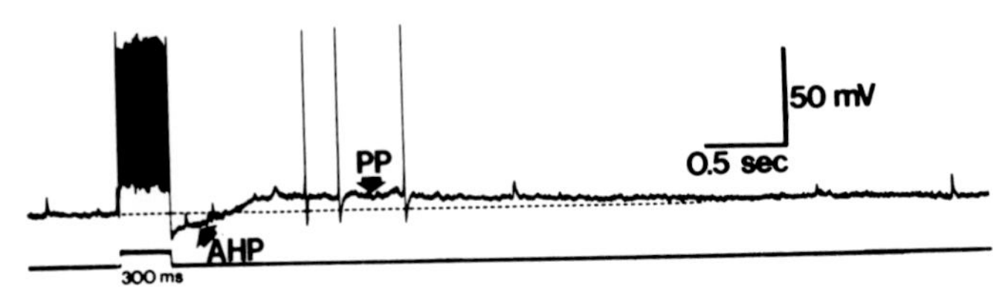

Figure 8. Afterhyperpolarization and plateau potential of silent neuron in SON. A spike train was elicited upon injection of 0.4 nA of depolarizing current through the recording electrode. An afterhyperpolarization (AHP), which probably results from a Ca^{2+}-activated K^+ conductance, interrupts the plateau potential (PP) that usually follows low-frequency firing. Data courtesy of R. D. Andrew and F. E. Dudek, Tulane University.

were blocked by intracellular injection of the Ca^{2+} chelator, EGTA, it appears that the results point to the presence of a calcium-activated K^+ conductance. As shown in our earlier work (Dudek *et al.*, 1980), some spontaneously occurring bursts terminate not with hyperpolarizations, but with depolarizations of the membrane (see Figure 9). Such bursts are also frequently inferrable from extracellular recordings of bursts ending in spike inactivation.

The results of injecting EGTA are also consistent with the hypothesis that the action potentials of magnocellular neurons have a large calcium component. MacVicar *et al.* (1982) suggested this on the basis of the duration and shape of the action potentials of magnocellular neurons compared to those of hippocampal CA1 cells (see Figure 10). With action potentials equated with respect to spike height and durations measured at one-third amplitude, the mean (± SD) duration in msec

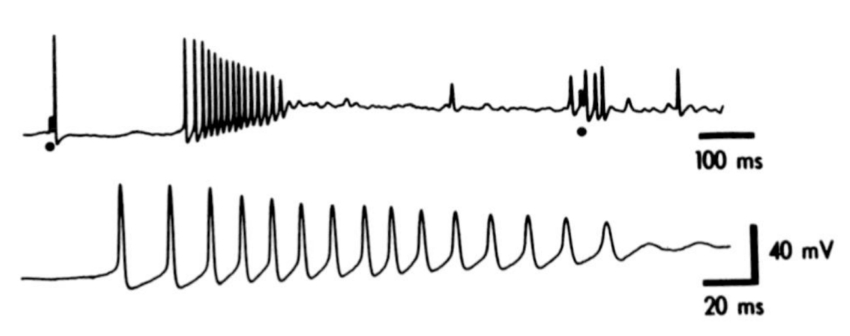

Figure 9. (Upper trace) High frequency burst of action potentials with a progressive decrease in amplitude and followed by maintained depolarization lasting several seconds. (Lower trace) Same burst shown at faster time scale to illustrate the decrease in amplitude and increase in duration of the action potentials during the burst. Within 7 sec after the burst, the cell fired action potentials with waveforms indistinguishable from the one at left of the upper trace. Filled circles mark the occurrence of 5 msec, 10 mV calibration pulses. From Dudek *et al.*, 1980.

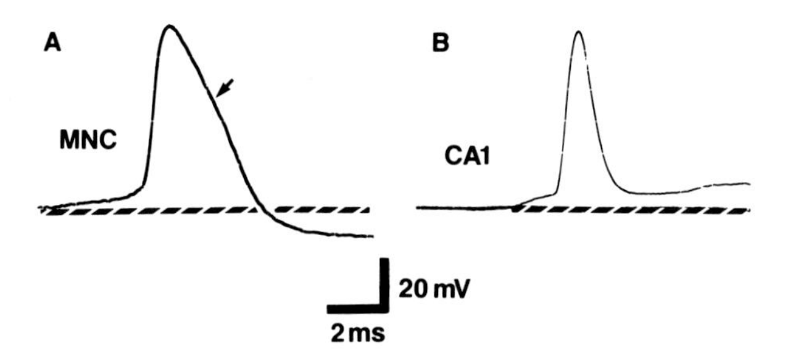

Figure 10. Action potentials of magnocellular neurosecretory cell (MNC) from hypothalamus and CA1 pyramidal cell from hippocampus. The typical shape of a prolonged MNC action potential is illustrated in (A). A hump is apparent on its falling phase (arrow) and an afterhyperpolarization follows the spike. The broker line indicates resting potential. However, an action potential of similar amplitude from a CA1 pyramidal cell (B) shows a relatively rapid falling phase. Pyramidal cell spikes did not have afterhyperpolarizations but often had depolarizing after potentials (from MacVicar *et al.*, 1982).

was 2.06 ± 0.6 for magnocellular neurons and 1.17 ± 0.29 for CA1 cells. This difference was highly significant ($p < 0.001$). Such differences between peptidergic and nonpeptidergic neurons were expected, of course, since this is generally the case in all systems, vertebrate and invertebrate, that have been studied thus far (see Finlayson and Osborne, 1975, for review).

Andrew and Dudek also observed oscillations in the resting membrane potential of phasically firing cells, with bursts occurring at the peak depolarization of the oscillation. This same observation has been made in my laboratory by P. Cobbett. Both groups of workers report that the membrane potential oscillations are independent of chemical synaptic input (Figure 11). So, it seems that the idea of endogenous membrane currents as factors involved in phasic bursting of magnocellular hypothalmic neurons is supported by these very recent experiments.

3.1.2c. Extracellular K^+ Fluxes. A second factor of possible importance in phasic firing of these cells is extracellular K^+ concentration. If the fluid concentration in the space immediately surrounding the magnocellular neurons were to rise just a few millimolar, several effects might be observed. The excitability of the cells would increase, making initiation of bursts more probable. This should not affect the balance of excitation and inhibition on the cells, since both types of presynaptic terminals should be equally influenced by K^+ alterations. Continued increases in K^+ concentration during bursting, particularly if neighbor-

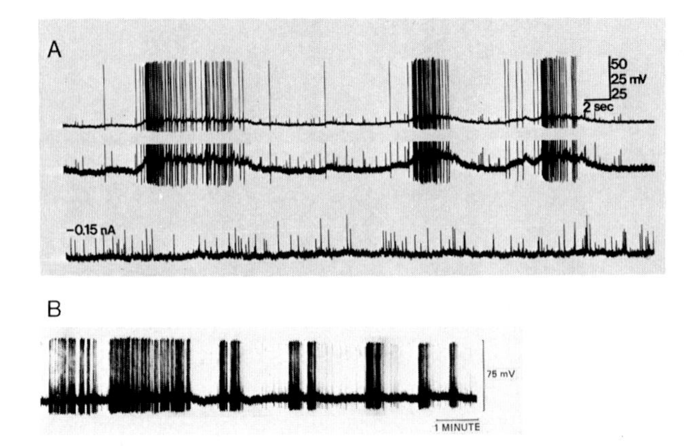

Figure 11. Phasic neurons in SON. Note the slow depolarization that underlies each burst of action potentials. (A) The upper and middle traces show the same recording at lower and higher gains respectively. When the neuron was hyperpolarized (lowest trace), the oscillation in membrane potential disappeared and unpatterned postsynaptic potentials were revealed. Both observations suggest an endogenous mechanism underlying the phasic bursting of this cell (courtesy of R. D. Andrew and F. E. Dudek, Tulane University). (B) Similarly oscillating SON neuron recorded by W. T. Mason, Babraham, Cambridge.

ing cells are also active, might even depolarize the membrane of the bursting neuron sufficiently to terminate a burst (as in Figure 9).

By what mechanism might such changes in extracellular K^+ occur? Are not the astrocytic glial cells surrounding these neurons functioning as K^+ wicks as they do in other systems (Orkand, 1977), and, thus, would they not prevent rises in the concentration of K^+? Also, is the quantity of K^+ efflux into the extracellular space during normal neuronal activity sufficient to really become a factor? Evidence bearing on these questions comes from several domains. Ultrastructural studies have shown that under conditions, such as dehydration, which produce increases in the numbers of phasically firing cells as well as in the length of the bursts, there is glial cell retraction from between and among magnocellular neurons (Gregory *et al.*, 1980; Tweedle and Hatton, 1976, 1977, 1982). This retraction allows the neurosecretory neurons to form close appositions (intermembrane distances of ~ 6 to 8 nm; see Figure 12) with one another, appositions involving large areas of membrane (as judged from two-dimensional profiles). With clefts of this size, and without intervening glial processes, it is likely that neuronal activity would result in significant local elevations of extracellular K^+ (for review, see Hertz, 1977).

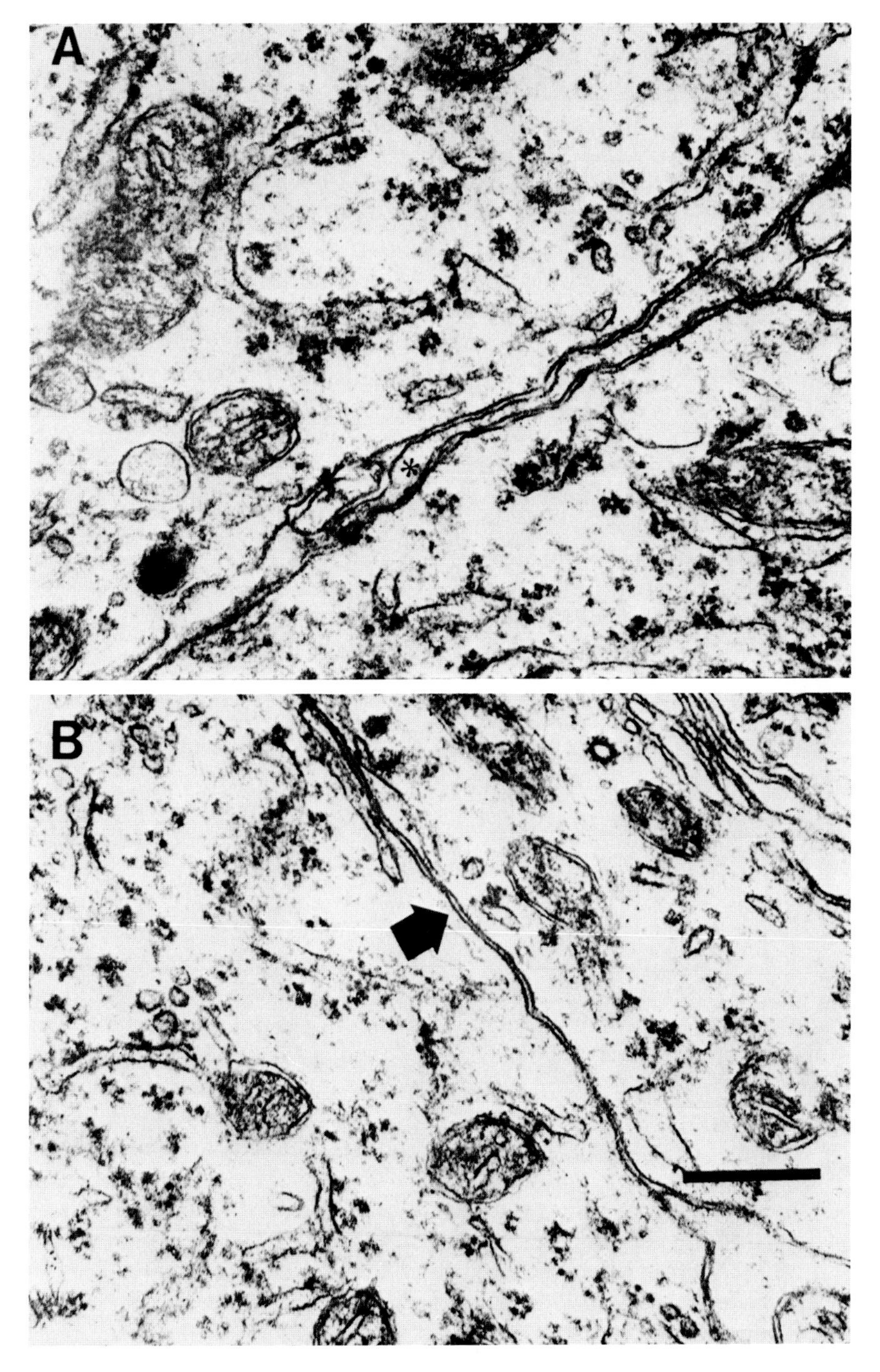

Figure 12. High-power electron micrographs of SON cells. (A) Two cell bodies are separated by a fine glial process (*). (B) Two neurons with direct membrane appositions (arrow). Bar = 300 nm.

Direct evidence that the amounts of K^+ measurable at or near the membrane are sufficient to influence nearby cells comes from the work of Hounsgaard and Nicholson (1981, 1983; Nicholson and Hounsgaard, this volume). Using ion-sensitive microelectrodes, these investigators measured the changes in K^+ concentration adjacent to individual, intermittently active Purkinje cells in slices of guinea pig cerebellum. They observed 1 to 4 mM elevations in K^+ activity during bursts of action potentials. Also bearing on the question of cell–cell interactions between closely apposed neurons via extracellular K^+ concentrations, is a study by Yarom and Spira (1982). They recorded intracellularly from two closely apposed (intermembrane distance: 7 to 10 nm) cockroach neurons and found that an action potential in one cell resulted in an excitatory postsynapticlike potential of >1 mV in the other. The cells were not electrotonically coupled, and injection of tetraethylammonium (TEA) prolonged the action potential of the cell into which it was injected while reducing the amplitude of the potential in the other cell. The conclusion drawn from these results was that the induced potentials were due to "localized increases in extracellular potassium concentrations as a consequence of firing in one neuron" (Yarom and Spira, 1982, p. 80). Thus, it seems entirely possible that extracellular potassium changes accompanying the activity of closely apposed SON and PVN neurons could be influencial in initiating, prolonging, and/or terminating bursts.

3.1.2d. Electrotonic Coupling. A third factor that may be involved in the control of phasic activity is electrotonic coupling among magnocellular neurons. A completely unexpected finding from our studies in which we injected the fluorescent dye, Lucifer Yellow (Stewart, 1978), into cells in order to correlate morphology and function, was that some magnocellular neurons in both PVN and SON were dye-coupled to as many as three others (Andrew *et al.*, 1981). The coupling was found to be either soma-somatic, soma-dendritic, or dendro-dendritic. Since this dye crosses gap junctions, we also looked for and found evidence for these junctional complexes in the SON using freeze-fracture methods. Mason (1982) has confirmed the presence of dye-coupling in SON cells in hypothalamic slices. These two independent lines of evidence strongly suggest that many SON and PVN cells are electrotonically coupled, since gap junctions are thought to be the morphological substrate of electrotonic conduction (Bennett, 1978).

3.1.2e. Immunocytochemical Identification of Coupled Cells. Preliminary evidence from a study combining Lucifer Yellow injections and immunocytochemical identification of the hormone in dye-coupled cells indicates that, at least, VP-containing cells are coupled to one another (Cobbett *et al.*, 1982). OX-containing cells have been injected and immunostained, but at this writing, there is no positive evidence available that OX-containing cells are coupled. Such coupling would not, of course, be an unexpected outcome in future studies. Certainly, electrotonic conduction within groups of magnocellular neurons would influence excitability levels of the coupled cells independently of chemical synaptic inputs and could therefore make an important contribution to the patterned activity of phasically firing cells.

These three factors (endogenous membrane currents, extracellular K^+ fluxes, and electrotonic conduction) taken together may be sufficient to produce the complex phenomenon of phasic bursting in the magnocellular peptidergic neurons of the hypothalamus. Future research should determine the relative contributions of each of these factors as well as others yet to be uncovered.

3.1.3. *Intrinsic Connections.* The problem of local circuits in the hypothalamus has only begun to be investigated physiologically. Data so far obtained are indeed meager in amount. There is a good deal of synaptic activity that exists in the 400- to 500-μm coronal slice, even hours after preparation (Dudek *et al.*, 1980; MacVicar *et al.*, 1982). Figures 13 to 15 illustrate intracellular records of excitatory and inhibitory postsynaptic potentials, both spontaneously occurring and evoked by extracellular stimulation within the slice. At least some of these potentials may represent the activity of local circuit neurons. In particular, it is unlikely that the spontaneous postsynaptic potentials arise from cut axons in which the cells of origin lie outside of the slice, although potentials evoked by extracellular stimulation could be from such sources as well as from intact neurons within the slice. In any case, these intrinsic connections only suggest the possibility of local circuits.

Tribollet and Dreifuss (1981) found that injections of HRP into the PVN consistently labeled, retrogradely, neurons in nucleus circularis and the anterior commissural nucleus, another small group of magnocellular neurons lying ~ 250 μm rostral to the PVN. Evidence supporting the existence of these connections comes from the following observations: (1) that extracellular stimulation in the area of the nucleus circularis in coronally cut slices excited orthodromically some spontaneously active PVN neurons and (2) that such stimulation rostral to the PVN in horizontally prepared slices was also excitatory to some PVN cells (G. I. Hatton, unpublished observations). Of course, the actual elements activated by extracellularly applied stimulation cannot be known with certainty but, taken together with the anatomical findings, it appears that there are at least two intrinsic connections involving magnocellular neurons. This is puzzling, given the absence of observed neurosecretory

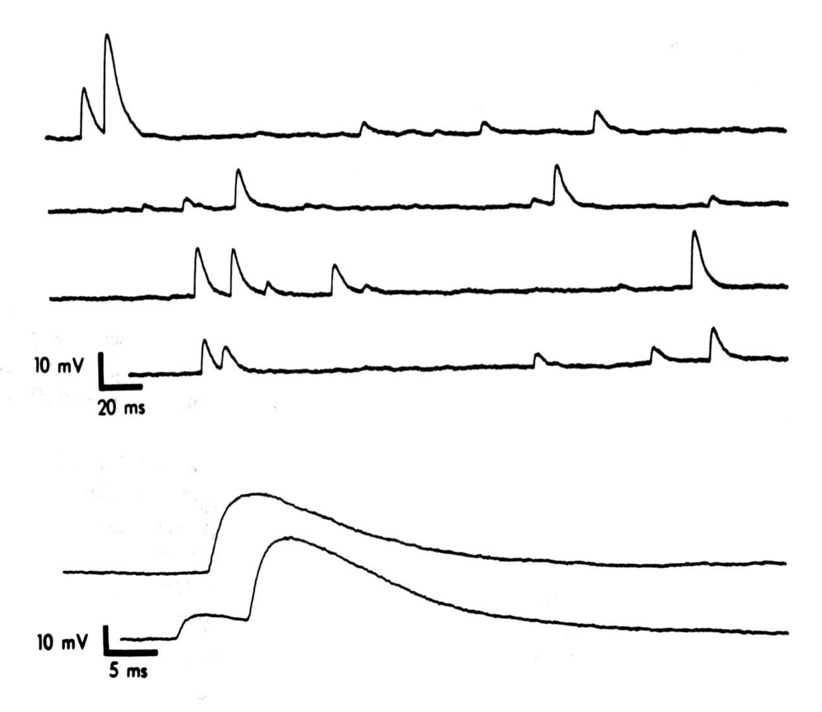

Figure 13. Spontaneously occurring depolarizations recorded from silent PVN neurons *in vitro*. Upper four traces from one cell; lower two traces are from a different cell and shown at faster time scale to reveal the waveforms of these postsynaptic potentials (from Dudek *et al.*, 1980).

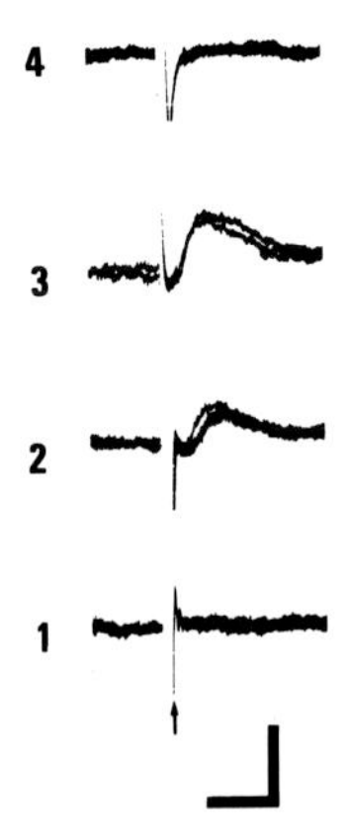

Figure 14. Intracellular response of an identified magnocellular neuron in PVN to extracellular stimulation in a coronally cut slice. Excitatory postsynaptic potentials were evoked by stimulation at two sites ventrolateral to the fornix, but not by stimulation near SON or dorsal to the fornix. Calibration: 10 mV, 10 msec (from MacVicar *et al.*, 1982).

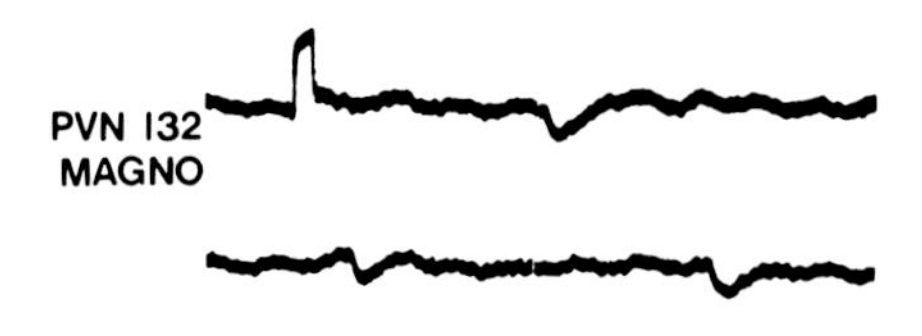

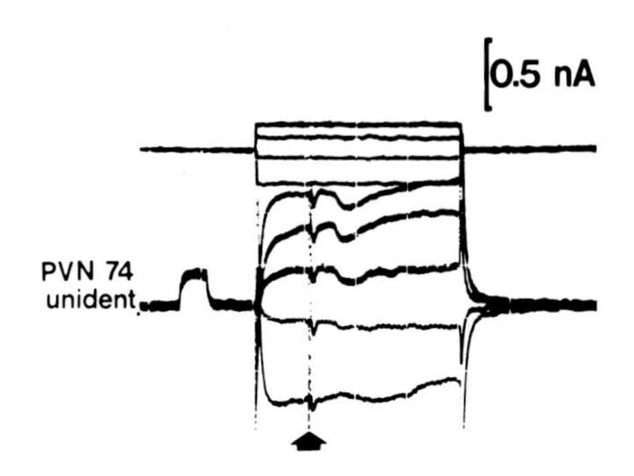

Figure 15. Spontaneous (above) and extracellularly evoked inhibitory postsynaptic potentials (below, arrow) recorded from two different PVN cells with Lucifer Yellow-filled electrodes. The evoked inhibitory postsynaptic potential was reduced and then slightly reversed with increasing hyperpolarizing current.

granule-containing synapses in the PVN and compels one to hypothesize the existence of nonOX- , nonVP-containing interneurons in these pathways. Further physiological work needs to be done in order to determine the presence of such interneurons.

Intracellular injections of Lucifer Yellow into PVN magnocellular neurons has produced one unequivocal example of an axon collateral which may be indicative of local circuit feedback (Cobbett *et al.*, 1983). Figure 16 is a drawing made from a fluorescence photomontage of this neuron. Its axon collateral forms a terminal arborization in the relatively cell-poor area between the lateral subdivision of the PVN and the fornix. Immunocytochemical analysis showed that this neuron was immunoreactive with neurophysin II antiserum and thus is probably vasopressinergic. Although we cannot be certain which cells receive inputs from these collaterals, it is interesting to note that this perifornical area contains cells that show met-enkephalinlike immunoreactivity (Salm and Hatton, 1982). As diagrammed in Figure 16, these enkephalin-containing cells appear to send processes back into the medial PVN. PVN neurons have been shown to be inhibited by met-enkephalin analogues (Pittman *et al.*, 1980). Since our results are preliminary, we are treating them as observations to generate working hypotheses about local circuitry, rather than as hard evidence for such circuits at this level.

3.2. *Parvocellular Areas*

The term *parvocellular* is used here simply to distinguish from the magnocellular areas already discussed, the groups of cells in the anterior

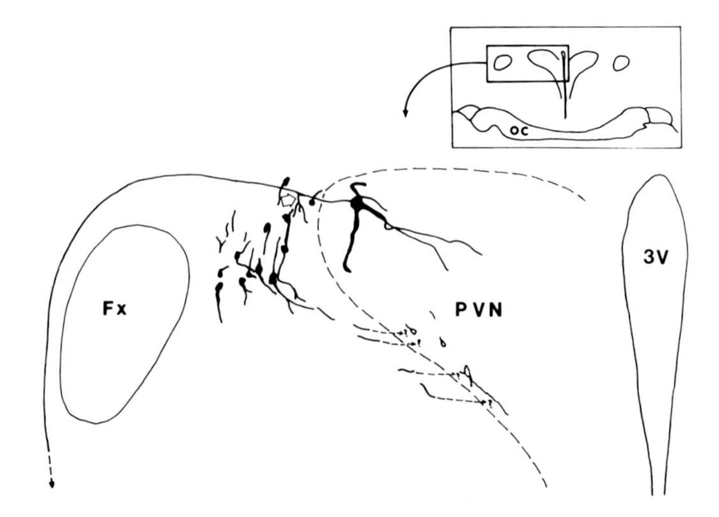

Figure 16. Diagram made by superimposing two hypothalamic preparations. Inset shows area enlarged. Magnocellular neuron from a 400-μm slice was injected intracellularly with Lucifer Yellow and photographed in whole mount. Smaller filled-in cells between the PVN and Fx, and unfilled cells in PVN are from a 15-μm section stained for met-enkephalinlike immunoreactivity. About 100 to 200 μm after leaving the nuclear boundary, the axon of the PVN cell thickened slightly and projected a single arborizing collateral (open arrow), which appeared to terminate in the area of the cells that densely immunostained with met-enkephalin antiserum. These cells, in turn, appear to send processes back toward or into the more ventral areas of the PVN. Unfilled cells in the nucleus were lightly immunostained, filled cells lateral to PVN were densely stained. 3V: third ventricle; Fx: fornix; OC: optic chiasm; PVN: paraventricular nucleus.

hypothalamus consisting of generally smaller neurons than those found in the SON or PVN. To be sure, the largest cells in, for instance, the lateral preoptic area are at least equal in size to the smaller cells of the SON. Other than their general location in the brain and their apparent functional roles in body homeostasis (in which they inevitably interact), there is little that can be said that would apply to all of these areas. Two of these parvocellular areas have been studied physiologically, in some detail, using brain slice methodology: the preoptic area-anterior hypothalmus (POA-AH) (temperature regulation), and the suprachiasmatic nucleus (SCN) (circadian rhythms). See Figure 17 for the locations of these hypothalamic entities.

3.2.1. The POA-AH and Thermoregulation. Temperature-sensitive neurons have been known for many years to reside in this brain area (see Boulant, 1980, for review). *In vivo* studies had firmly established

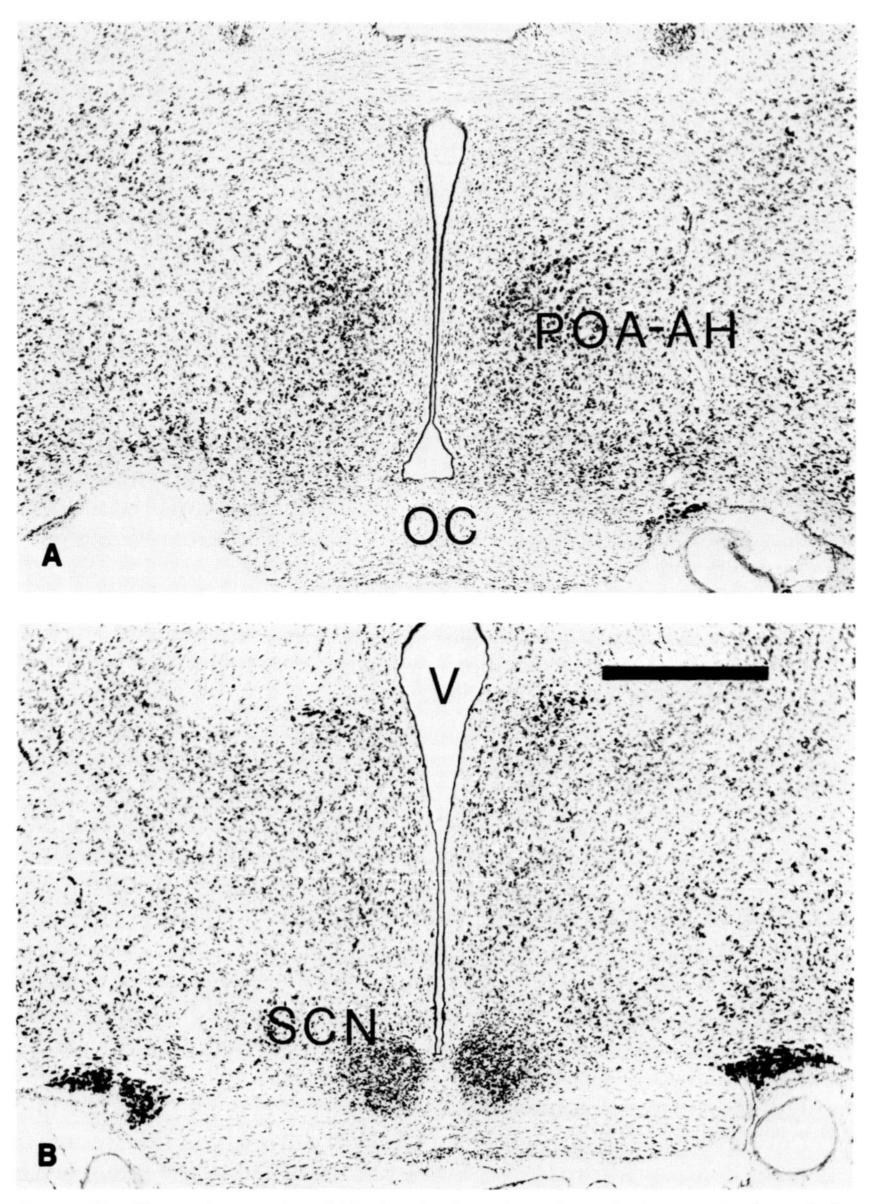

Figure 17. Photomicrographs of Nissl-stained sections through the anterior hypothalamus of the rat. (A) Area where temperature sensitive cells have been recorded. (B) Area containing cells maintaining a circadian rhythm *in vitro*. OC: optic chiasm; POA-AH: preoptic area-anterior hypothalamus; SCN: suprachiasmatic nucleus; V: third ventricle. Bar = 500 μm for (A) and (B).

the presence of warm-sensitive and cold-sensitive cells as measured by their firing rates in response to changes in skin, core, or brain temperature. In general, it was found that somewhat less than one-half of the neurons recorded in this brain area showed any temperature sensitivity. Of those that were sensitive, about three-quarters were warm sensitive and the remainder were cold sensitive. The warm-sensitive neurons were further divisible into two subpopulations: those that appeared to receive a large amount of excitatory afferent input from the periphery, with accordingly higher basal discharge rates, and those that did not, these latter cells displaying relatively low rates of activity.

Studies of *in vivo* preparations could not determine whether these POA-AH neurons were able to respond directly to changes in brain temperature because these are always confounded with alterations in vasoconstriction, blood flow and peripheral inputs in the intact animal. Any of these "secondary" alterations in response to temperature changes could, of course, affect POA-AH cell firing. Also, it is difficult to totally eliminate peripheral influences on central neuronal activity in the whole animal without producing untold compensatory side effects.

These interesting problems, along with a host of theoretical considerations about the mechanisms of thermoregulation (which are beyond the scope of this chapter), led two independent research teams to employ hypothalamic slice techniques (Hori *et al.*, 1980; Kelso *et al.*, 1982). Although there were some minor differences between the results of these studies, the major results were quite similar. Changes in temperature of the medium bathing the slices resulted in repeatable alterations in neuronal firing rates. Both warm-sensitive and cold-sensitive neurons were observed. The proportions of these as well as of temperature insensitive neurons were similar to those commonly reported from *in vivo* preparations, i.e., 60% insensitive, 30% warm-sensitive, and 10% cold-sensitive.

As would be predicted when the excitatory peripheral inputs were removed, the basal firing rates were low (< 10 Hz) in the slices. Furthermore, only one population of warm-sensitive units was observed, consistent with the idea that the two subpopulations exist *in vivo* primarily because of differential afferent input. Kelso *et al.* (1982), in the more extensive study of the two, were able to conclude that thermosensitivity was an intrinsic property of POA-AH cellular networks, if not the individual neurons themselves.

In a further study, Kelso and Boulant (1982) found that warm-sensitive neurons retained their thermosensitivity when the perfusion medium bathing the slices contained 0.3 mM Ca^{2+} and 9 mM Mg^{2+}, which solution was presumed to block chemical synaptic transmission. In addition some neurons, that were temperature-insensitive in normal medium, showed increased warm sensitivity, and all cold-sensitive neurons became temperature insensitive in the low Ca^{2+}, high Mg^{2+} medium. From these results the authors concluded that warm sensitivity was an independent property of some POA-AH neurons, while neurons that appear to be cold-sensitive, depend upon synaptic input from warm-sensitive cells for their temperature information.

3.2.2. The SCN and Circadian Rhythms. In rodents, at least, the SCN has a firmly established role in circadian rhythms of a number of functions, including locomotor activity, feeding, drinking, sleep-waking cycles and plasma corticosterone levels (see Rusak and Zucker, 1979, for review). This nucleus receives primary retinal projections as well as inputs relayed from other visual target areas. It also has a rather complex intrinsic structure (van den Pol, 1980). The SCN is probably an important integrative area for the entrainment of physiological functions and behaviors to light–dark cycles. When the SCN is destroyed bilaterally, rats show no circadian rhythms, though they can still make light–dark discriminations. Whether the SCN contains a primary oscillator, capable of sustaining independently a 24-hr rhythm, or merely receives instructions from some other bodily mechanism(s) is a question that has intrigued investigators in circadian rhythms research for some time. This question seemed unanswerable *in vivo*.

The earliest reported attempt to broach this question using brain slice methods appears to be that of Groos and Hendriks (1979). They found 16 out of 98 cells whose firing frequencies were quite regular during the time that they were recorded. For example, one cell fired at ~ 7 Hz for 18 min. Little significance was attached to this observation except that such cells were only rarely encountered in *in vivo* recordings from the SCN.

A second study by Kita *et al.* (1981) took the problem a step further. Extracellular recordings were made from single neurons in the SCN of slices from rats taken out of the dark or the light part of the light–dark cycle. They found that the firing rates of cells with relatively stable firing frequencies were high (~ 7 to 8 Hz) during the light and lower (~ 3 to 4 Hz) during the dark. These results indicated that entrained SCN cells would continue, at least for a few hours, to reflect the firing frequencies being generated at the time the rat was killed. This, however, still left open the question of whether the SCN was a primary oscillator.

An affirmative answer to that question is suggested by the results obtained by Green and Gillette (1982). These workers recorded single unit activity of SCN cells in slices of rat hypothalamus for 24 to 36 hr. Cells from the same slices were, therefore, recorded over at least one complete diurnal cycle, though no single cell was recorded for that long a time. In some cases, slices from animals 12 hr out-of-phase were placed in the same recording chamber in order to control for possible unknown environmental influences. The results were striking (see Figure 18). The cells of this nucleus continued to show a circadian rhythm in firing rates after being totally isolated from all sources of visual and humoral, and most other sources of neural input. As observed by Kita *et al.* (1981), the rates were highest during the light phase and lowest during the dark phase of the cycle. Remarkably, even the actual peak and trough firing rates were highly similar between the two studies.

Taken together, these findings with the slice are consistent with the notion that the area of the SCN contains an oscillator capable of generating and sustaining a rhythm in the relative absence of external inputs. Further research is needed to determine which of the several elements within this area are responsible for this capability.

4. SUMMARY AND CONCLUSIONS

The hypothalamic slice preparation has only recently been developed. It apparently has the potential, if appropriately exploited, to reveal a great many of the well-guarded secrets of the hypothalamus. Already, the use of this preparation has produced findings that are leading to a reconceptualization of the functioning of the magnocellular hypothalamoneurohypophysial system. There is evidence that cells of the supraoptic and paraventricular nuclei are osmosensitive without external inputs. Other evidence indicates that the phasic bursting patterns of activity associated with presumed vasopressinergic neurons are endogenously generated. Experiments with slices have revealed the presence of endogenous membrane currents and the ability of phasically firing neurons to maintain their patterns of activity in the absence of chemical synaptic transmission. These results make untenable the formerly held idea that phasic bursting in magnocellular neurons is due to recurrent synaptic inhibition. Cells of this magnocellular system have been found to be dye-coupled through gap junctions, strongly suggesting electrotonic coupling. Although this phenomenon is of considerable theoretical importance in understanding functional relationships, the crucial information concerning the extent of this coupling and the hormone types involved have only begun to be investigated. Intrinsic connections of this magnocellular elements with nearby hypothalamic cells have also begun to be investigated in the slice. What evidence there is to date

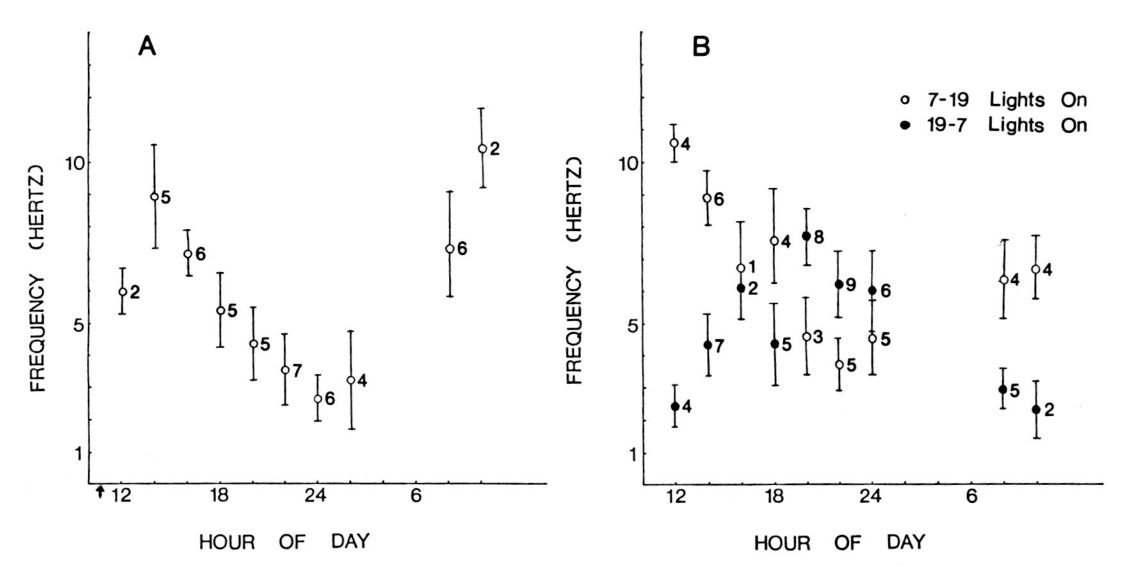

Figure 18. (A) Frequency of 48 units recorded at various times over 24 hr. The rat was sacrificed at the arrow. Lights on 0700. (B) Filled circles show average frequency of units from three rats on reversed light cycle (lights on 1900). Open circles are average frequencies of units from three rats with lights on 0700. Each data point for (A) and (B) contains units recorded starting at that hour and for the next 2 hr. Error bars are standard error of the mean. Numbers next to circles indicate the number of cells recorded in that time period. 12:12 light/dark cycle. From Green and Gillette, 1982.

suggests the presence of local axon collaterals and possibly of local circuits in both PVN and SON.

Only two of the many parvocellular areas of the hypothalamus have been studied physiologically in slices and in sufficient detail to be reviewed here. Investigations of the preoptic–anterior hypothalamic area have shown that some 40% of the cells are temperature sensitive, 30% warm sensitive, and 10% cold sensitive. This proportion is similar to what is typically found *in vivo*. Thus, thermosensitivity appears to be an intrinsic property of the cellular networks in this hypothalamic area. Studies of the suprachiasmatic nucleus in the brain slice indicate that certain regularly firing cells are capable of maintaining a circadian rhythm of activity for up to 36 hr (the longest time tested) after being removed from the animal. This rhythm consisted of relatively rapid firing rates (7 to 10 Hz) during the light part of the cycle and slow rates (3 to 5 Hz) during the dark phase. Such results raise the possibility that this hypothalamic nucleus contains a primary oscillator responsible for the generation of at least some circadian rhythms.

ACKNOWLEDGMENTS. The support of the National Institutes of Health through research grants NS 09140 and NS 16942 and Fogarty Senior International Fellowship TW 00690, is gratefully acknowledged. I thank P. J. R. Cobbett, A. A. Nuñez, and A. K. Salm for helpful comments on an earlier draft of the manuscript, and K. G. Smithson and J. Harper for technical and typing assistance.

5. REFERENCES

Abe, H. and Ogata, N., 1982, Ionic mechanism for the osmotically-induced depolarization in neurons of the guinea-pig supraoptic nucleus *in vitro*, *J. Physiol.* (*London*) **327:**157–171.

Almli, C. R. and Weiss, C. S., 1974, Drinking behaviours: Effects of lateral preoptic and lateral hypothalamic destruction, *Physiol. Behav.* **13:**527–538.

Andrew, R. D., MacVicar, B. A., Dudek, F. E., and Hatton, G. I., 1981, Dye transfer through gap junctions between neuroendocrine cells of rat hypothalamus, *Science* **211:**1187–1189.

Antunes, J. L., Carmel, P. W., and Zimmerman, E. A., 1977, Projections from the paraventricular nucleus to the zona externa of the median eminence of the rhesus monkey: An immunohistochemical study, *Brain Res.* **137:**1–10.

Armstrong, W. E., Warach, S., Hatton, G. I., and McNeill, T. H., 1980, Subnuclei in the rat paraventricular nucleus: A cytoarchitectural, horseradish peroxidase and immunocytochemical analysis, *Neuroscience* **5:**1931–1958.

Bargmann, W. and Scharrer, E., 1951, The site of origin of the hormones of the posterior pituitary, *Am. Scientist* **39:**255–259.

Bennett, M. V. L., 1978, Junctional permeability, in: *Intercellular Junctions and Synapses* (J. Feldman, N. B. Gilula, and J. D. Pitts, eds.), Halsted Press, New York, pp. 23–36.

Blass, E. M. and Epstein, A. N., 1971, A lateral preoptic osmosensitive zone for thirst in the rat, *J. Comp. Physiol. Psychol.* **76**:378–394.

Brimble, M. J. and Dyball, R. E. J., 1977, Characterization of the responses of oxytocin- and vasopressin-secreting neurones in the supraoptic nucleus to osmotic stimulation, *J. Physiol.* (*London*) **271**:253–272.

Brimble, M. J., Haller, E. W., and Wakerley, J. B., 1978, Supraoptic and paraventricular units in hypothalamic slices incubated in iso- or hypertonic medium, *J. Physiol.* (*London*) **278**:38P–39P.

Boulant, J. A., 1980, Hypothalamic control of thermoregulation: Neurophysiological basis, in: *Handbook of the Hypothalamus,* Volume 3, Part A (P. J. Morgane and J. Panksepp, eds.), Marcel Dekker, New York, pp. 1–82.

Bourque, C. and Renaud, L. P., 1981, Osmosensitivity of rat supraoptic neurosecretory neurons in the isolated and perfused basal hypothalamus, *Can. Physiol.* **12**:93.

Cobbett, P., Smithson, K. G., and Hatton, G. I., 1982, Immunoreactivity to neurophysins I and II in dye-coupled magnocellular hypothalamic neurons, *Soc. Neurosci. Abstr.* **8**:531.

Cobbett, P., Hatton, G. I. and Salm, A. K., 1983, Evidence for local circuits in the paraventricular nucleus of the rat hypothalamus, *J. Physiol.* (*London*) **338**:43P.

Conrad, L. C. and Pfaff, D. W., 1976, Efferents from medial basal forebrain and hypothalamus in the rat—II. An autoradiographic study of the anterior hypothalamus, *J. Comp. Neurol.* **169**:221–262.

Defendini, R. and Zimmerman, E. A., 1978, The magnocellular neurosecretory system of the mammalian hypothalamus, in: *The Hypothalamus* (S. Reichlin, R. J. Baldessarini, and J. B. Martin, eds.), Raven Press, New York, pp. 137–152.

de Wied, D., 1980, Behavioral actions of neurohypophysial peptides, *Proc. R. Soc. London Ser. B* **210**:183–195.

Dreifuss, J. J. and Kelly, J. S., 1972, The activity of identified supraoptic neurones and their response to acetylcholine applied by iontophoresis, *J. Physiol.* (*London*) **220**:105–118.

Dreifuss, J. J., Tribollet, E., Baertchi, A. J., and Lincoln, D. W., 1976, Mammalian endocrine neurons: Control of phasic activity by antidromic action potentials, *Neurosci. Lett.* **3**:281–286.

Dudek, F. E., Hatton, G. I., and MacVicar, B. A., 1980, Intracellular recordings from the paraventricular nucleus in slices of rat hypothalamus, *J. Physiol.* (*London*) **301**:101–114.

Dyball, R. E. J., 1971, Oxytocin and ADH secretion in relation to electrical activity in antidromically identified supraoptic and paraventricular units, *J. Physiol.* (*London*) **214**:245–256.

Finlayson, L. H. and Osborne, M. P., 1975, Secretory activity of neurons and related electrical activity, *Adv. Comp. Physiol. Biochem.* **6**:165–258.

Forsling, M. L., 1977, *Anti-Diuretic Hormone,* Volume 2, Eden Press, Montreal.

Gähwiler, B. H. and Dreifuss, J. J., 1979, Phasically firing neurons in long-term cultures of rat hypothalamic supraoptic area: Pacemaker and follower cells, *Brain Res.* **177**:95–103.

Green, D. J. and Gillette, R., 1982, Circadian rhythm of firing rate recorded from single cells in the rat suprachiasmatic brain slice, *Brain Res.* **245**:198–200.

Gregory, W. A., Tweedle, C. D., and Hatton, G. I., 1980, Ultrastructure of neurons in the paraventricular nucleus of normal, dehydrated and rehydrated rats, *Brain Res. Bull.* **5**:301–306.

Groos, G. A. and Hendriks, J., 1979, Regularly firing neurones in the rat suprachiasmatic nucleus, *Experientia* **35:**1597–1598.

Haller, E. W. and Wakerley, J. B., 1980, Electrophysiological studies of paraventricular and supraoptic neurones recorded *in vitro* from slices of rat hypothalamus, *J. Physiol.* (*London*) **302:**347–362.

Haller, E. W., Brimble, M. J., and Wakerley, J. B., 1978, Phasic discharge in supraoptic neurones recorded from hypothalamic slices, *Exp. Brain Res.* **33:**131–134.

Hatton, G. I., 1976, Nucleus circularis: Is it an osmoreceptor in the brain?, *Brain Res. Bull.* **1:**123–131.

Hatton, G. I., 1982, Phasic bursting activity of rat paraventricular neurones in the absence of synaptic transmission, *J. Physiol.* (*London*) **327:**273–284.

Hatton, G. I., 1983, Some well-kept hypothalamic secrets revealed, *Fed. Proc.* **42:**2869–2874.

Hatton, G. I. and Tweedle, C. D., 1982, Magnocellular neuropeptidergic neurons in hypothalamus: Increases in membrane apposition and number of specialized synapses from pregnancy to lactation, *Brain Res. Bull.* **8:**197–204.

Hatton, G. I., Hutton, U. E., Hoblitzell, E. R., and Armstrong, W. E., 1976, Morphological evidence for two populations of magnocellular elements in the rat paraventricular nucleus, *Brain Res.* **108:**187–193.

Hatton, G. I., Gregory, W. A., and Armstrong, W. E., 1977, Bursting activity in rat hypothalamic neurosecretory cells *in vitro* in response to osmotic stimulation, *Soc. Neurosci. Abstr.* **345:** in press.

Hatton, G. I., Armstrong, W. E., and Gregory, W. A., 1978, Spontaneous and osmotically-stimulated activity in slices of rat hypothalamus, *Brain Res. Bull.* **3:**497–508.

Hatton, G. I., Doran, A. D., Salm, A. K., and Tweedle, C. D., 1980, Brain slice preparation: Hypothalamus, *Brain Res. Bull.* **5:**405–414.

Hatton, G. I., Ho, Y. W., and Mason, W. T., 1983a, Synaptic activation of phasic bursting in rat supraoptic nucleus neurons recorded in hypothalamic slices, *J. Physiol* (*London*), **345:** in press.

Hatton, G. I., Ho, Y. W., and Mason, W. T., 1983b, Rat supraoptic (SON) neurones have axon collaterals: anatomical and electrophysiological evidence, *J. Physiol* (*London*), in press.

Hayward, J. N., 1977, Functional and morphological aspects of hypothalamic neurons, *Physiol. Rev.* **57:**574–658.

Hertz, L., 1977, Drug-induced alterations of ion distribution at the cellular level of the central nervous system, *Pharmacol. Rev.* **29:**35–65.

Hori, T., Nakashima, T., Hori, N., and Kiyohara, T., 1980, Thermo-sensitive neurons in hypothalamic tissue slices *in vitro*, *Brain Res.* **186:**203–207.

Hounsgaard, J. and Nicholson, C., 1981, Dendritic and somatic action potentials in single Purkinje cells change $[K^+]_0$ and $[Ca^{2+}]_0$, *Soc. Neurosci. Abstr.* **7:**225.

Kandel, E. R., 1964, Electrical properties of hypothalamic neuroendocrine cells, *J. Gen. Physiol.* **47:**691–717.

Kelso, S. R. and Boulant, J. A., 1982, Effect of synaptic blockade on thermosensitive neurons in hypothalamic tissue slices, *Amer. J. Physiol.* **243:**480–490.

Kelso, S. R., Perlmutter, M. N., and Boulant, J. A., 1982, Thermosensitive single-unit activity of *in vitro* hypothalamic slices, *Am. J. Physiol.* **242:**R77–R84.

Kita, H., Shibata, S., and Oomura, Y., 1981, Circadian rhythmic changes of SCN neuronal activity in the rat hypothalamic slice, *Soc. Neurosci. Abstr.* **7:**858.

Koizumi, K. and Yamashita, H., 1972, Studies of antidromically identified neurosecretory cells of the hypothalamus by intracellular and extracellular recordings, *J. Physiol.* (*London*) **221:**683–705.

MacVicar, B. A., Andrew, R. D., Dudek, F. E., and Hatton, G. I., 1982, Synaptic inputs and action potentials of magnocellular neuropeptidergic cells: Intracellular recording and staining in slices of rat hypothalamus, *Brain Res. Bull.* **8**:87–93.

Mason, W. T., 1980, Supraoptic neurones of rat hypothalamus are osmosensitive, *Nature* (*London*) **287**:154–157.

Mason, W. T., 1982, Dye coupling, gap junctions and synchronous unit activity in the rat supraoptic nucleus (SON), *J. Physiol.* (*London*) **327**:44P.

Mason, C. A. and Bern, H. A., 1977, Cellular biology of the neurosecretory neuron, in: *Handbook of Physiology,* Section I: The Nervous System, Volume 1 Cellular Biology of Neurons, Part 2 (E. R. Kandel, ed.), American Physiological Society, Bethesda, pp. 651–689.

Nicholson, C. and Hounsgaard, J., 1983, Diffusion in the slice microenvironment and implications for physiological studies, *Fed. Proc.* **42**:2865–2868.

Orkand, R. K., 1977, Glial cells, in: *Handbook of Physiology, Section I: The Nervous System, Volume I Cellular Biology of Neurons, Part 2.* (E. R. Kandel, ed.), American Physiological Society, Bethesda, pp. 855–875.

Noble, R. and Wakerley, J. B., 1982, Behavior of phasically active supraoptic neurones *in vitro* during osmotic challenge with sodium chloride or mannitol, *J. Physiol.* (*London*) **327**:41P.

Palay, S. L., 1957, The fine structure of the neurohypophysis, in: *Ultrastructure and Cellular Chemistry of Neural Tissue* (H. Welsch, ed.), Hoeber Press, New York, pp. 31–49.

Peck, J. W. and Novin, D., 1971, Evidence that osmoreceptors mediating drinking in rabbits are in the lateral preoptic area, *J. Comp. Physiol. Psychol.* **74**:134–147.

Peterson, R. P., 1966, Magnocellular neurosecretory centers in the rat hypothalamus, *J. Comp. Neurol.* **128**:181–190.

Pittman, Q. J., Hatton, J. D., and Bloom, F. E., 1980, Morphine and opioid peptides reduce paraventricular activity: Studies on the rat hypothalamic slice preparation, *Proc. Natl. Acad. Sci. USA* **77**:5527–5531.

Pittman, Q. J., Blume, H. W., and Renaud, L. P., 1981, Connections of the hypothalamic paraventricular nucleus with the neurohypophysis, median eminence, amygdala, lateral septum and midbrain periaqueductal gray: An electrophysiological study in the rat, *Brain Res.* **215**:15–28.

Poulain, D. A. and Wakerley, J. B., 1982, Electrophysiology of hypothalamic magnocellular neurones secreting oxytocin and vasopressin, *Neuroscience* **7**:773–808.

Poulain, D. A., Wakerley, J. B., and Dyball, R. E. J., 1977, Electrophysiological differentiation of oxytocin- and vasopressin-secreting neurons, *Proc. R. Soc. London* **196**:367–384.

Roberts, J. S., 1977, *Oxytocin,* Volume 1, Eden Press, Montreal.

Rusak, B. and Zucker, I., 1979, Neural regulation of circadian rhythms, *Physiol. Rev.* **59**:449–526.

Salm, A. K. and Hatton, G. I., 1982, Distribution of met-enkephalin, oxytocin and vasopressin-like immunoreactivities in the rat paraventricular nucleus (PVN), *Anat. Rec.* **202**:166A.

Saper, C. B., Lowey, A. D., Swanson, L. W., and Cowan, W. M., 1976, Direct hypothalamo-autonomic connections, *Brain Res.* **117**:305–312.

Sherlock, D. A., Field, P. M., and Raisman, G., 1975, Retrograde transport of horseradish peroxidase in the magnocellular neurosecretory system of the rat, *Brain Res.* **88**:403–414.

Silverman, A. J., Hoffman, D., and Zimmerman, E. A., 1981, The descending afferent connections of the paraventricular nucleus of the hypothalamus (PVN), *Brain Res. Bull.* **6**:47–61.

Sladek, C. D. and Joynt, R. J., 1979, Characterization of cholinergic control of vasopressin release by the organ-cultured rat hypothalamo-neurohypophyseal system, *Endocrinology* **104:**659–663.

Stewart, W. W., 1978, Functional connections between cells as revealed by dye-coupling with a highly fluorescent naphthalimide tracer, *Cell* **14:**741–759.

Swaab, D. F., Pool, C. W., and Nijveldt, F., 1975, Immunofluorescence of vasopressin and oxytocin in the rat hypothalamo-neurohypophysial system, *J. Neural Trans.* **36:**195–216.

Swanson, L. W., 1977, Immunohistochemical evidence for a neurophysin-containing autonomic pathway arising in the paraventricular nucleus of the hypothalamus, *Brain Res.* **128:**346–353.

Theodosis, D. T., Poulain, D. A., and Vincent, J.-D., 1981, Possible morphological bases for synchronisation of neural firing in the rat supraoptic nucleus during lactation, *Neuroscience* **6:**919–929.

Tribollet, E. and Dreifuss, J. J., 1981, Localization of neurones projecting to the hypothalamic paraventricular nucleus area of the rat: A horseradish peroxidase study, *Neuroscience* **6:**1315–1328.

Tweedle, C. D. and Hatton, G. I., 1976, Ultrastructural comparisons of neurons of supraoptic and circularis nuclei in normal and dehydrated rats, *Brain Res. Bull.* **1:**103–121.

Tweedle, C. D. and Hatton, G. I., 1977, Ultrastructural changes in rat hypothalamic neurosecretory cells and their associated glia during minimal dehydration and rehydration, *Cell Tissue Res.* **181:**59–72.

Tweedle, C. D. and Hatton, G. I., 1982, Specialized synapses and direct cell–cell apposition in rat supraoptic nucleus: Increases with chronic physiological stimulation, *Soc. Neurosci. Abstr.* **8:**745.

van den Pol, A. N., 1980, The hypothalamic suprachiasmatic nucleus of the rat: Intrinsic anatomy, *J. Comp. Neurol.* **191:**661–702.

van den Pol, A. N., 1982, The magnocellular and parvocellular paraventricular nucleus of rat: Intrinsic organization, *J. Comp. Neurol.* **206:**317–345.

Vandesande, F. and Dierickx, K., 1975, Identification of the vasopressin producing neurons in the hypothalamic magnocellular neurosecretory system of the rat, *Cell Tissue Res.* **165:**153–162.

Wakerley, J. B. and Lincoln, D. W., 1973, The milk ejection reflex of the rat: A 20- to 40-fold acceleration in the firing of paraventricular neurones during oxytocin release, *J. Endocrinol.* **57:**477–493.

Wakerley, J. B., Poulain, D. A., and Brown, D., 1978, Comparison of firing patterns in oxytocin- and vasopressin-releasing neurones during progressive dehydration, *Brain Res.* **148:**425–440.

Wiegand, S. J. and Price, J. L., 1980, The cells of origin of the afferent fibers to the median eminence in the rat, *J. Comp. Neurol.* **192:**1–9.

Yarom, Y. and Spira, M. E., 1982, Extracellular potassium ions mediate specific neuronal interaction, *Science* **216:**80–82.

14

Brain Slice Work
Some Prospects

PER ANDERSEN

1. INTRODUCTION

In writing this chapter on future prospects of brain slice work, I realized some of the difficulties in undertaking such a project. For example, the opinions expressed in such a chapter must necessarily be highly biased and may well have a modest predictive value. In Scandinavia, we are fond of citing the famous Danish humorist and philosopher Storm-Pedersen, who has a cartoon running: "To predict is difficult, the future in particular."

However, since science depends on innovation, it may still be helpful to ponder the possibilities for further brain slice development.

2. OPTIMAL CONDITIONS

Because of the opportunities offered by brain slices, the last decade has witnessed the enthusiastic adoption of the brain slice method. Most investigators have been so interested in using the slices for their particular problem that, so far, little effort has been given to the study of the optimal conditions under which the slices work. There is a great necessity for parametric studies.

PER ANDERSEN • Institute of Neurophysiology, University of Oslo, Oslo 1, Norway.

Until recently, there were remarkably few studies on the optimal level of oxygen supply and of nutrients. The chapter by Lipton and Whittingham in the present volume is, therefore, very valuable. Still, we are far from understanding the limitations of the technique and knowing which parameters are the essential ones. The viability of the perforant path/granule cell activation is dependent upon the glucose concentration (Bachelard and Cox, 1981). A reduction from 10 to 5 mM glucose already showed a clear change in the rate of rise of the extracellular excitatory postsynaptic potential (EPSP) and of the population spike amplitude. However, since substitution of pyruvate plus malate for glucose strongly depressed both the EPSP and the population spike, the behavior of the granule cells in the presence of lowered glucose did not correlate with the energy state of the tissue. Clearly, the limiting factor is not simply ATP synthesis by the cerebral mitochondria.

Further, it would be useful to know the exact requirement for oxygen concentration. How important are the different buffers used? If a pure oxygen/Hepes or oxygen/Tris system could be used, the expensive carbogen mixture could be dispensed with. How important are trace materials like amino acids and other nutrients? Is it possible to extend the life of the slices by adding antibiotics or other measures against infection to supplant the artificial cerebrospinal fluid medium with tissue culture material? Such experiments have been proposed, but it is uncertain how essential the changes are (Lynch *et al.*, 1978).

Similarly, new technical advances with regard to cutting may be of great importance. The interesting studies by Garthwaite *et al.* (1979) showed that cerebellar slices that were manually cut preserved their morphology much better than slices made with a tissue chopper.

We also need to know more about the temperature of the tissue during dissection and cutting. A reduced temperature seems to have protective value for the survival of the slices. On the other hand, will not prolonged cooling reduce their viability and quality? What is the reason for the early postslicing recovery? The initial interpretation by Tower (1960) that this period is used for reactivation of the membrane pump to restore the potassium/sodium distribution remains a likely explanation. However, the restoration is far from complete, as the remarkable physiological resilience of the tissue could lead one to believe. The striking discrepancy between the often marked pathological appearance of the tissue and the very active tissue as seen with physiological, biochemical, and pharamacological methods calls for new correlative studies.

Finally, the development of better chambers is going to change the field considerably. We have already witnessed a tendency to simplification of the original large chambers (Haas *et al.*, 1979; Nicoll and Alger, 1981; Koerner and Cotman, 1983).

3. SLICES FROM NEW REGIONS

Slices have now been made from a great number of different areas of the brain, in particular from cortical tissue. However, there are several regions in which the difficulties have been considerable. This applies particularly to certain parts of the brain stem and the spinal cord. The resilience of the cortical tissue has been partly due to the existence of large bundles of myelinated fibers that resist proper cutting. However, with vibrating knives, this difficulty has been largely overcome. Still, the ventral part of the brain stem and large areas of the spinal cord (perhaps with exception of the dorsal horn) have been largely refractory. Ventral horn slices, for example, have shown motoneuronal response in neonatal rats only (Takahashi, 1978). In the brain stem, the dorsal part is said to survive better than the ventral portions (Fukuda and Loeschke, 1977). Specific regions of the brain stem nuclei have been relatively difficult to study, with the exception of hypothalamic tissue. The main difficulty has been to make the slices survive. Nevertheless, several interesting studies have been made of this tissue, including correlations of physiological and biochemical activity and also the interesting relations between the morphology and the hormonal content of nerve cells (Kelly *et al.*, 1979). The lower brain stem and the spinal cord, therefore, represent a real challenge for investigators wanting to use the slice technique.

Another possibility would be to remodel some slices to include afferent and efferent connections. For example, in some sensory systems, it should be possible to retain the sense organs with the brain slice. Furthermore, it might be possible to retain the connections to certain effector organs and maintain the brain slice and the organs in the same bath. In this way, system slices could possibly be developed. A retina–optic nerve/tract–lateral geniculate slice might be possible as well as an olfactory mucosa–nerve-bulb preparation. The development of the slice technique should go through a transitory phase, in which the brains and organs of lower animals are used, before mammalian preparations are developed.

Another approach would be large-scale rearrangements of the brain before the sections are cut. For example, with many parts of the brain, it is possible to straighten previously curved structures before sectioning. In this way, large thin sections may be produced, for example, using the curved structure of the thalamocortical fibers, an initial cut is made following the line of the thalamocortical radiation. The tissue is then flattened by laying this cut surface against a glass plate after which a thin section may be cut parallel to the plate containing both the thalamic nucleus with the cells of origin, the axons of these cells in the radiation, and the receiving cortical tissue. This procedure is most easily performed for the somatosensory parts and in small mammals like mice. However, the visual system in the rat should also lend itself to this procedure.

4. NEW USES OF SLICES

A possible development will be an enhancement of the already considerable collaboration between different types of neurobiologists who use slices to obtain information. A good example of this collaboration is the combined approach in which cells identified by physiologists are later studied by histological and biochemical techniques (Lynch and Schubert, 1980).

Another role for the slice technique would be to test some of the earlier data from intact animals. For example, many of the pharmacological data obtained with systemic and iontophoretic applications should be compared with relevant intracellular tests in slices. In this way, a comparision between previously used techniques and approaches in slices may give fruitful results (Ryall and Kelly, 1978). In particular, intracellular studies on the effect of drugs applied in the immediate vicinity of the impaled cell must be of great importance. The advantage of delivering the drugs to known positions of identified neurons is particularly valuable. Obviously, one should not overestimate the value of a single technique. As always, a combination of techniques gives greater insight in a problem than sticking to a single approach.

5. NEW APPROACHES IN SLICE EXPERIMENTS

Other advances in slice work could come through modification of the tissue before or after the sectioning. By previous surgical, radiation, or chemical treatment of the tissue, one can produce a modified nervous system. Slices from such systems could be even easier to handle than those presently used. Along the same line of thought, modification after sectioning may also be of great value. For example, microsurgery has already been proven to be of great assistance in certain areas. In invertebrates, it has also been possible to load individual nerve cells with

Lucifer Yellow and kill them with a light pulse of ultraviolet. In a similar way, it might be possible to treat slices in various ways to knock out groups of cells that share a chemical, morphological, or functional trait. In this way, a further dissection of the slice into simpler components could appear possible.

Another promising approach is to combine slices with tissue culture. A very interesting technique has been developed in which slices are placed in roller tube culture (Gähwiler, 1981). The cells survive for several months and migrate from their original position to form a single or double layer of neurons some distance away. Interestingly, some of the neurons remain attached to afferent fibers that seem to retain their functional capabilities. Because of the separation, it is possible to penetrate visually identified cells that have contact with at least some of their afferent fibers. The technique represents a hybrid between the explanted culture technique and the slice with its preformed connectivity. In the former case, new and functional connections may be established (LaVail and Wolf, 1972), but the type and position of the synapses may not be the same as those occurring *in vivo*. On the other hand, the slice shows the standard system obtained in normal adult or young animals as an end product after much modification during the development. The Gähwiler technique allows stable recording from visually identifiable individual cells. In addition, the preparation clears itself of the cells that were damaged during the slicing procedure, assuming that the remainders are in excellent condition.

6. THE NEED FOR CORRELATION

I would like to end by stressing the need for correlation. We must always remember that the brain slice represents an artificial simplification of the normal state of the nervous system. For this reason, it is imperative that the data obtained from slices are correlated with results from intact preparations and from other techniques. Only then will the results from slice work acquire the attention the enthusiasts feel they deserve.

7. REFERENCES

Bachelard, H. S. and Cox, D. W. G., 1981, Effects of low glucose, pyruvate/malate and sodium fluoride on the components of field potentials recorded *in vitro* from granule cells in the dentate gyrus, *J. Physiol.* (*London*) **317**:63–64P.

Fukuda, Y. and Loeschke, H. H., 1977, Effect of H^+ on spontaneous neuronal activity in the rat medulla oblongata *in vitro, Pflügers Arch.* **371**:125–134.
Garthwaite, J., Woodhams, P. L., Collins, M. J., and Balazs, R., 1979, On the preparation of brain slices: Morphology and cyclic nucleotides, *Brain Res.* **173**:373–377.
Gähwiler, B. H., 1981, Organotypical monolayer cultures of nervous tissue, *J. Neurosci. Methods* **4**:329–342.
Haas, H. L., Schaerer, B., and Vosmansky, M., 1979, A simple perfusion chamber for the study of nervous tissue slices *in vitro, J. Neurosci. Methods* **1**:323–325.
LaVail, J. H. and Wolf, M. K., 1972, Postnatal development of the mouse dentate gyrus in organotypic cultures of the hippocampal formation, *Am. J. Anat.* **137**:47–66.
Lynch, G. and Schubert, P., 1980, The use of *in vitro* brain slices for multi-disciplinary studies of synaptic function, *Ann. Rev. Neurosci.* **3**:1–22.
Nicoll , R. A. and Alger, B. E., 1981, A simple chamber for recording from submerged brain slices *J. Neurosci. Methods* **4**:153–156.
Kelly, M. J., Kuhnt, U., and Wuttke, W., 1979, Morphological features of physiologically identified hypothalamic neurons as revealed by intracellular marking, *Exp. Brain Res.* **34**:107–116.
Koerner, J. F. and Cotman, C. W., 1983, A microperfusion chamber for brain slice pharmacology, *J. Neuroscience Methods* **7**:243–251.
Ryall, R. W. and Kelly, J. S., 1978, Iontophoresis and transmitter mechanisms in the mammalian central nervous system, Elsevier/ North Holland, Amsterdam-New York.
Takahashi, T., 1978, Intracellular recording from visually identified motoneurons in rat spinal cord slices, *Proc. R. Soc. London, Ser B* **202**:417–421.
Tower, D. B., 1960, *Neurochemistry of Epilepsy,* Charles C. Thomas, Springfield, Illinois.

Appendix

Brain Slice Methods

BRADLEY E. ALGER, S. S. DHANJAL, RAYMOND DINGLEDINE,* JOHN GARTHWAITE, GRAEME HENDERSON, GREGORY L. KING, PETER LIPTON, ALAN NORTH, PHILIP A. SCHWARTZKROIN, T. A. SEARS, M. SEGAL, TIM S. WHITTINGHAM, and JOHN WILLIAMS

1. INTRODUCTION

The purpose of this Appendix is to provide the prospective slicer with a convenient package of methods; where choices exist among a number

BRADLEY E. ALGER • Department of Physiology, University of Maryland School of Medicine, Baltimore, Maryland 21201. S. S. DHANJAL and T. A. SEARS • Sobell Department of Neurophysiology, Institute of Neurology, The National Hospital, University of London, London, England. RAYMOND DINGLEDINE and GREGORY L. KING • Department of Pharmacology, University of North Carolina, Chapel Hill, North Carolina 27514. JOHN GARTHWAITE • Department of Veterinary Physiology and Pharmacology, The University of Liverpool, Liverpool, England. GRAEME HENDERSON • Department of Pharmacology, University of Cambridge, Cambridge CB2 2QD, England. PETER LIPTON • Department of Physiology, University of Wisconsin, Madison, Wisconsin 53706. ALAN NORTH and JOHN WILLIAMS • Neuropharmacology Laboratory, Department of Nutrition and Food Science, Massachusetts Institute of Technology, Cambridge, Massachusetts 02139. PHILIP A. SCHWARTZKROIN • Department of Neurological Surgery, University of Washington, Seattle, Washington, 98195. M. SEGAL • The Weizmann Institute of Science, Rehovot 76100, Israel. TIM S. WHITTINGHAM • Laboratory of Neurochemistry, National Institute of Neurological and Communicative Disorders and Stroke, National Institutes of Health, Bethesda, Maryland 20205.
* To whom correspondance should be directed.

of approaches, this section also provides a discussion of those areas of agreement and disagreement among current practitioners. By collecting this information into one narrative, it should be easier for the newcomer to choose a set of procedures appropriate to slicing his favorite brain region. As is the case for all experimental preparations, the brain slice should be chosen only if it is an appropriate way to approach particular problems. The prospective slicer should be warned that it is unrealistic to consider sliced tissue as "normal," no matter how skillful and careful one is with the preparation. Slices are isolated tissue (without normal inputs) immersed in an artificial environment. This environment can be manipulated to "bias" the tissue however the investigator desires, but it is certainly never "normal." Investigators must weigh the potential advantages of the preparation (e.g., technical accessibility) with the obvious disadvantages (e.g., loss of normal input pathways, shearing of dendritic processes and axons) as they apply to a particular problem. For supplemental reading, a number of reviews on brain slice methodology are available (Dingledine *et al.*, 1980; Hatton *et al.*, 1980; Lynch and Schubert, 1980; Teyler, 1980; Kerkut and Wheal, 1981).

2. PREPARATION OF SLICES

2.1. *Slicing the Brain*

It is perhaps on this topic more than others where myths, unfounded dogmas, and notions based on intuition or anecdotal evidence tend to influence the choice of method. The following comments reflect our own experiences and are not meant to discourage the enterprising investigator from testing alternatives. Regardless of the method employed, the goal is to prepare a slice of tissue having those neurons, glia, and synapses that are important to the experiment in a viable condition. Although the procedure used to dissect the brain will depend somewhat on whether a tissue chopper or Vibratome is used to prepare the slice, some general comments can be made. A gentle brain dissection may be the single most important factor in producing healthy slices. It appears that younger animals (e.g., guinea pigs less than 400 g, rats less than 175 g) produce better results than older animals. This may be in part because their thinner skulls are easier to remove, allowing a less traumatic dissection. If female animals are used, it is well to keep in mind the findings of Teyler *et al.* (1980), who showed sex-linked differences, and differences across the estrous cycle, in the responses of hippocampal neurons to estradiol and testosterone.

In general, the speed of dissection seems not nearly so important as the care taken in removing and slicing the tissue. The method of slicing the tissue is undoubtedly important as documented by Garthwaite *et al.*, (1979). They found that careful chopping and careful hand-cutting of cerebellum did not appear to differ outwardly in the "trauma" inflicted on the tissue, yet there was a huge difference in the ultimate quality of the slices, as judged by electrophysiological and histological criteria. These findings point to the actual cutting of the tissue as being a critical step.

Brains taken from lightly anesthetized rats are more often mushy or sticky during the slicing procedure. Animals that are deeply anesthetized with ether may yield better slices than marginally anesthetized ones, possibly due to the protective effects of ether anesthesia against spreading depression (van Harreveld and Stamm, 1953). In place of ether anesthesia, decapitation or killing by a blow to the neck has also been employed. Tissue in contact with stagnated blood (or deoxygenated blood) may suffer, not so much because of the hypoxia but because of effects of the blood itself. This impression is supported by the following comparison (P. A. Schwartzkroin, unpublished superstition): if one wants to make slices from adult rabbit or cat hippocampus (or cortex), one can either (1) anesthetize the animal, expose the relevant structure, and excise a piece for study, or (2) decapitate the animal and then do the dissection. Decapitation tends to be slower but less bloody, and it seems to yield superior results.

2.1.1. *Hippocampus.*

2.1.1a. The Tissue Chopper. A variety of methods can be employed to dissect out the hippocampus; the following is one technique found to be reliable. After removal from the skull case, the brain is placed into a beaker of cold oxygenated artificial CSF (ACSF) for a minute or so. The brain is then bisected sagittally with a razor blade, and one hemisphere placed on a flat surface with its medial face up. The tissue should be kept wet with chilled ACSF during dissection. The cerebellum and lower brainstem are lifted with forceps to expose the choroid fissure. Fine forceps can then be used to gently strip away the bulk of the tough choroid plexus. At this point, the ventral surface of the hippocampal formation is visible as a curved structure. Spatula cuts are made to free the septal and temporal ends of the hippocampus, and the spatula then inserted into the ventricular slit to gently roll the hippocampus out of the surrounding tissue. It is sometimes helpful to free the fimbria with forceps cuts as the hippocampus is rolled out. The entorhinal area is then trimmed away with a minimum of spatula cuts to completely free the hippocampus, which has the appearance of a small cashew nut. The

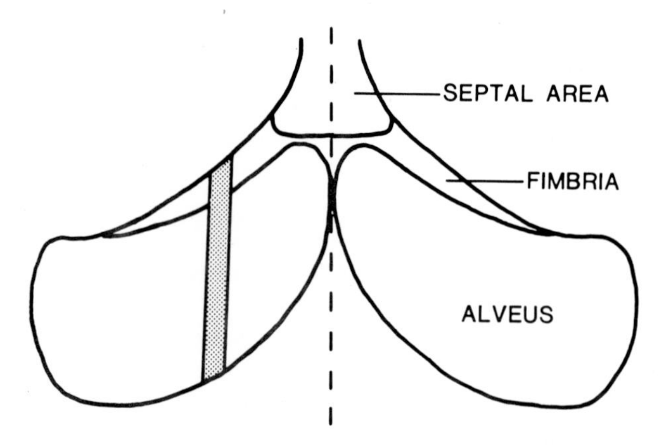

Figure 1. Dorsal view of the rat hippocampal formation. The ventricular surface (alveus) is exposed by removing the overlying neocortex. The hatched bar represents the approximate orientation for preparing hippocampal slices.

hippocampus can be stored in fresh, chilled ACSF while its counterpart from the other half of the brain is removed.

Either or both hippocampi are placed on a layer of moistened filter paper glued to the chopping block, and the block rotated to adjust the tissue to the desired orientation for slicing (see hatched area in Figure 1). It is usually possible to view striations on the alvear surface with oblique lighting; these striations can be conveniently placed parallel to the blade. Excitatory pathways are better preserved when an angle of 15 to 30° from the transverse axis is used (Andersen *et al.*, 1971; Rawlins and Green, 1977). Inhibitory pathways might be better preserved by a more longitudinal orientation (Struble *et al.*, 1978; Dingledine and Langmoen, 1980), although this is not the impression of Schwartzkroin and co-workers in studying interneurons. It seems prudent to clean the chopper blade with ether to dissolve oils, etc. that are added by the manufacturer for lubrication and rust prevention, although no obvious difficulties have arisen when this step was occasionally forgotten.

Slices of 350 to 500 μm are serially cut and gently removed in a rolling motion with a fine sable brush into a Petri dish containing chilled oxygenated ACSF. The slices are placed individually in the recording chamber (or in a holding chamber) with a wide bore glass tube and suction bulb. The transfer pipette can be conveniently made by flaring and polishing the end of a broken Pasteur pipette in a Bunsen flame.

Several different tissue choppers have been used successfully. Commercial models include the Sorvall and McIlwain choppers. Of these two, the Sorvall, although 3 to 4 times more expensive, is easier to use since it allows fine adjustment of the blade height and blade angle, as well as rotation of the chopping block to orient the tissue. Duffy and Teyler (1975) have described a robust chopper, the blade of which falls by gravity into the tissue block. This design can be modified so that a solenoid with adjustable force is used to drive the chopping arm. Whatever machine is used, it appears important that the blade have minimal lateral movement and that the vertical position of the blade be adjusted so that it does not slam into the surface of the chopping block. A blade height that just dimples the surface of the wet filter paper on the chopping blocks works well.

2.1.1b. The Vibratome. Currently two vibrating knives are available commercially: the Oxford Vibratome (Oxford Laboratories, U.S.A.) and the Vibroslice Oscillating Tissue Slicer designed by Jefferys (1981) (available from Campden Instruments, U.K. and Frederick Haer, U.S.A.).

This method requires more time and is less convenient than chopping, although it may be more versatile. The tissue dissection differs from that described above in that the hippocampus need not be completely removed from the surrounding tissue before slicing. The brain is removed and blocked by parasagittal razor blade cuts to obtain a 3- to 5-mm thick slab that contains the hippocampus in the correct orientation for sectioning (Figure 1). The cerebellum and neocortex are removed with iridectomy scissors and fine forceps, and the resulting piece of tissue is fixed to a brass or aluminum block with a thin layer of cyanoacrylate glue. Small blocks of agar (2 to 5% made up in saline) are similarly glued around the tissue to provide lateral support; alternatively a barely viscous (39 to 40°C) agar solution can be poured over the tissue, which is then quickly immersed in ice-cold medium for a minute or so. The assembly is clamped in the Vibratome chuck so that the alvear surface faces the blade. Gillette "superthin" blades work well. The tissue and blade are covered with ice-cold, preoxygenated bathing medium. The rate of advance and vibration amplitude of the blade are best set at the maximum values that will permit rapid cutting without compressing or "pushing" the tissue. Often this requires a relatively slow advance with only moderate amplitude vibrations. The slices usually need to be trimmed away from the surrounding tissue (septum, entorhinal cortex, thalamus) with fine scissors and forceps. The choroid plexus occasionally presents a problem because of its relative toughness, and so it should be gently stripped away when possible.

2.1.2. Cerebellum. The cerebellum is removed simply and rapidly from a severed head by making lateral scissor-cuts, about 2 cm long, through the occipital and parietal bones. The upper skull is levered off carefully and bent back, exposing the cerebellum, which is then excised by displacing it gently away from the inferior colliculi (scissors or spatula) and cutting the peduncles together with any other adhering tissue.

2.1.2a. The Tissue Chopper. Mechanical chopping has proved to be one of the methods that can be used successfully to obtain viable slices of hippocampus or hypothalamus. This is not the case for adult rat cerebellum. In Garthwaite's laboratory, the vast majority of neurons in cerebellar slices prepared using a McIlwain chopper were in advanced stages of necrosis (Garthwaite *et al.*, 1979). Poor biochemical responses to transmitters matched the poor structural preservation.

Because of the potential usefulness of this technique, particularly for experiments in which slices are required in large number, several modifications of the chopping method were tested. These included anesthetizing the animals before decapitation (ether, urethane, or barbiturate), preimmersing the tissue for various times in chilled medium (various temperatures between 4°C and 20°C), precoating the tissue in agar (2 to 4% in saline) with the intention of reducing distortion during chopping, removing meninges, reducing the tissue to smaller blocks (1 to 4 mm high) by slicing off the tough pial surface and/or base of the cerebellum before chopping, and combinations of the above. Furthermore, a different chopper (Sorvall) and different ways of removing the cut slices were tested, but none of these variations led to significant improvements. Reasons for the differences between the cerebellum and hippocampus or hypothalamus in this respect are unclear at present.

Unlike the adult, however, well-preserved slices of immature (14 days or less) rat cerebellum can readily be prepared using a tissue chopper (Garthwaite *et al.*, 1980). Even in tissue "prisms" (chopped in two directions, 400 μm by 400 μm), the cellular preservation is remarkably good and damage at the cut edge is minimal.

2.1.2b. Hand-Slicing. This method for preparing brain slices was developed by McIlwain (1975) and is still widely used for various areas including olfactory cortex (Richards and Sercombe, 1968; Scholfield, 1978), interpeduncular nucleus (Brown and Halliwell, 1979; Ogata, 1979) and cerebellum (Okamoto and Quastel, 1973; Garthwaite *et al.*, 1979; Crepel *et al.*, 1981).

The apparatus required is a bow cutter, a guide, and a stable elevated cutting platform. Examples of these are shown in Figure 2. The frame for the cutter is made from a single length of steel wire (3 mm in diameter) bent into the shape illustrated. The blade is the cutting edge of a ribbon from a Gillette Techmatic disposable razor head. This blade is much thinner (40 μm) than other blades commonly used. A suitable

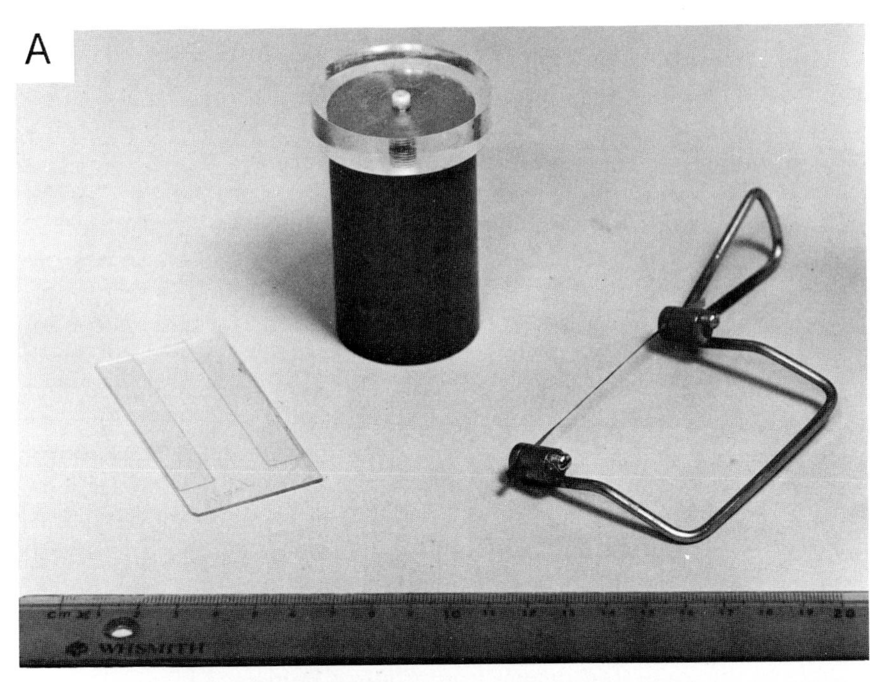

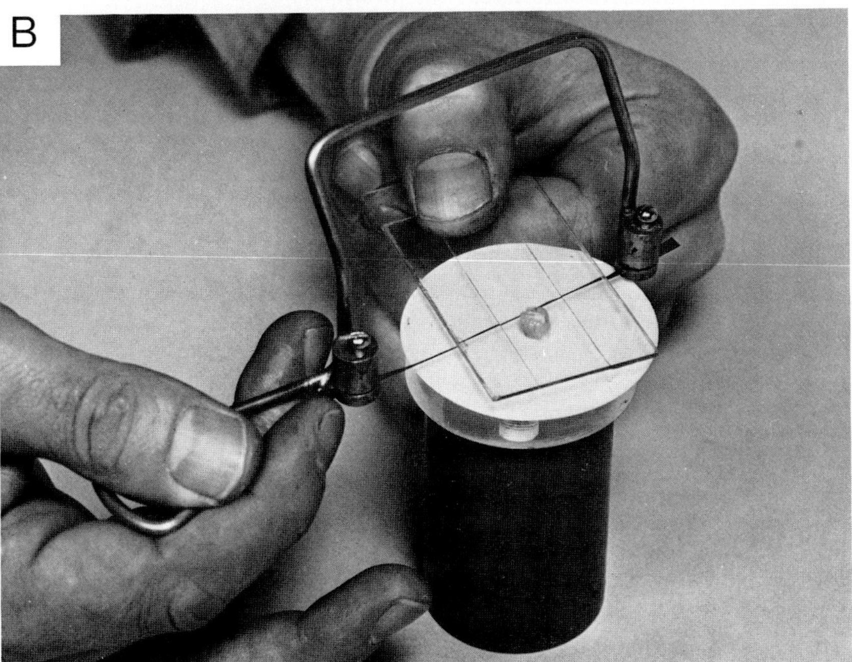

Figure 2. Bow cutter, glass guide, and cutting table used for preparing brain slices manually (Garthwaite).

length of the ribbon is trimmed to a width of about 1 mm and is secured under tension by the screw fittings of the frame. Any jagged edges are smoothed using a fine stone and the blade is degreased before use. Glass guides are made from microscope slides (76 by 38 mm), on the sides of which are glued half cover slips (22 by 64 mm). These are built up as required to form a central recess, the depth of which determines the slice thickness. The cutting table has a cylindrical brass base (6.5 cm high, 4.1 cm in diameter) onto which is secured, by means of a central screw, a circular polished perspex top (1.2 cm high, 5 cm in diameter).

To prepare slices using this method, the tissue is removed and placed on hardened filter paper, premoistened with ACSF, on the cutting table. The blade is wetted with ACSF while the guide is used dry. Blade and guide are positioned as shown in Figure 2B. The middle of the guide is lowered so as to contact and slightly compress the tissue. Slices are cut using a side-to-side sawing motion (1 to 2 mm amplitude) keeping the guide still (the hand holding the guide may be steadied against the cutting table). Slight pressure of the blade on the guide is maintained in order to ensure even slice thickness. The cut slices should adhere to the guide but detach from it when immersed in ACSF (e.g., in a Petri dish).

The advantages of this method are that the apparatus required is very simple and inexpensive to make and that the slices produced show good preservation. Surface (pial) slices of cerebellum cut in this manner have been used in pharmacological, biochemical, and morphological analyses (Okamoto and Quastel, 1973; Garthwaite, 1982; Garthwaite and Wilkin, 1982), although slices cut in the sagittal plane (in order to preserve the main afferent and efferent pathways) have been used to study synaptic excitation of Purkinje cells (Crepel *et al.*, 1981). The main disadvantage is that some skill and practice are necessary to produce good slices reproducibly. Slices cut by the inexperienced are often either rather squashed or uneven. Also, the number of slices that can be obtained from one animal are rather small (2 to 3 usually).

2.1.2c. The Vibratome. Purkinje cells and other neuronal elements have been shown to survive in thin (80 μm) slices of guinea pig cerebellum cut in the sagittal plane using an Oxford Vibratome (Chujo *et al.*, 1975). Antidromic and climbing fiber activation of Purkinje cells following white matter stimulation could be demonstrated in 200 μm thick slices prepared in a similar way (Llinás and Sugimori, 1980). Thicker (400 μm) slices of cerebellum cut with the new Vibroslice have also been studied recently. On electrophysiological (Crepel and Dhanjal, 1981) and morphological (J. Garthwaite, unpublished observations)

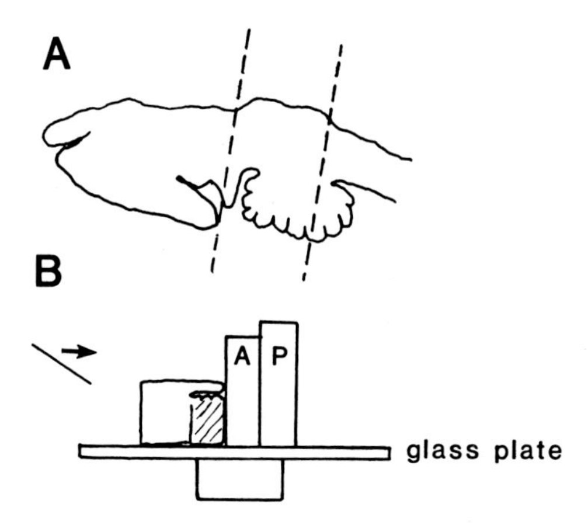

Figure 3. Slicing the brainstem with the Vibratome. (A) A sagittal drawing of the rat brain, showing the approximate position of the razor blade cuts to block the brain. (B) Mounting of brainstem block on the Vibratome chuck. P represents a Plexiglas block, and A, an agar slab that supports the brainstem. The arrow indicates the angle and direction of Vibratome blade motion (Williams, Henderson, and North).

grounds, the slices compare favorably with hand-cut slices, as do those of hippocampus (Jefferys, 1981).

A block of cerebellar vermis (3 to 5 mm high) is glued onto a Teflon stage that is then positioned in a chamber containing ACSF (80 ml capacity). The assembly is advanced by a hand-operated screw towards a blade that vibrates horizontally at a speed up to 3000 rpm at fixed amplitude (*ca.* 1 mm). The slicing arm moves in the vertical plane to adjust for the desired slice thickness. A particularly advantageous feature of the design is the direct manual control over the rate of slicing; thus, where the tissue is tough (connective tissue) and/or sticky (white matter), a slow rate of advance is used while grey matter can be cut quite quickly.

2.1.3. Brainstem. The procedure described in this section may be applied to other brainstem nuclei, but was specifically developed for the locus coeruleus (LC) (Henderson *et al.*, 1982). Isolation of the brainstem with the least amount of damage is most easily done by removal of the brain rostal end first. Care must be taken to remove completely the dura overlying the cerebellum before lifting the brain out of the cranium. The brain is then placed on a dissection plate with the ventral surface up and cold ACSF is poured over it to cool it, wash away debris, and facilitate clean separation of the pons from the rest of the brain stem with a single-edged razor blade. Best results are obtained if a wide margin is left between the rough cuts made at this point and the position of the LC (Figure 3A). The section of pons is trimmed of pial tissue and dorsal cerebellum, and then glued with cyanoacrylate adhesive on a glass plate with its rostral surface uppermost (Figure 3B). This is placed in a Vibratome containing cool ACSF. The temperature of the ACSF in the

Vibratome seems to be critical for obtaining good sections. The ease of cutting seems to be inversely related to the temperature; i.e., the brainstem cuts cleanly without compression or squashing when the temperature is less than 4°C. However, the most viable slices are obtained if the cutting temperature is between 4 and 8°C. After a suitable slice has been cut, it is immediately placed in the recording chamber. The entire procedure takes 8 to 12 min. Use of the Vibratome offers several advantages over the tissue chopper. The first is that the brainstem is soft and tends to be squashed by the chopper. Another advantage is the ease of identification of the LC which is contained within only two 300-μm slices of the pons. Careful visual inspection is necessary to select a slice that contains the nucleus. Finally, with the use of the Vibratome handling the slice is reduced to a minimum. From the Vibratome bath, the slice of choice is lifted with a spatula and placed directly onto the net of the recording chamber.

2.1.4. Spinal cord. Recently, viable spinal cord slices have been prepared from adult rats (Dhanjal and Sears, 1980a) from which preliminary electrophysiological studies of the dorsal horn have been made (Dhanjal and Sears, 1980b). These authors have also had some encouraging results from the ventral horn (S. S. Dhanjal and T. A. Sears, unpublished observations). One of the problems encountered in preparing spinal cord slices, as with neocortical slices, is cutting through the pia without compressing the tissue when a guillotine-type chopper is employed. This problem has been resolved by using an angled, silicone-coated Gillette Valet blade that in its cutting action passes through a polystyrene block on which the cord rests in a groove. The use of an angled blade achieves a shearing as opposed to a chopping action, thus cutting through the tough pia relatively easily.

This method is satisfactory for preparing transverse slices of spinal cord, but for longitudinal slices a number of other maneuvers are first required to obtain greater cord stability. Viable longitudinal slices have been obtained by using the Vibroslice vibrating-blade cutter. The method used is as follows: upon isolation of the cord from the anesthetized animal, a 1-cm piece of cord is first hemisected and one (or both) pieces glued down with cyanoacrylate cement, placing the cut surface on the cutting platform. Following immersion of the fixed cord in freshly oxygenated ACSF at room temperature, the blade is first positioned to remove the lateral 200 to 400 μm of the cord. The next cut is aimed at a thickness of 600 to 800 μm to yield a longitudinal, parasagittal slice. The slice is then transferred to a holding chamber containing ACSF at room temperature, bubbled continuously with 95%

O_2/5% CO_2. The slice remains in this bath for 30 min to 1 hr before being transferred to the recording bath, maintained at 32°C.

2.1.5. Other Areas. Many other brain regions have been successfully studied in slice preparations. The hypothalamic, striatal, and neocortical slice are treated in detail elsewhere in this volume. The art of slicing neocortex, particularly when dealing with large animals (cats, monkeys, humans) is somewhat more difficult than cutting hippocampus. First, it is important that the tissue sample be trimmed to a reasonable size so that (1) the chopper arm does not bang into it (i.e., the tissue set on the stage is no higher than the exposed blade width) and (2) the slice can be manipulated with a brush without pulling or tearing. Second, as with spinal cord, one must deal with the problem of how to cut through the pia (which is often relatively tough) without squashing the tissue sample. Guillotine-type choppers cannot easily cut through tissues surrounded by a pial sheath; the falling blade does not cut the capsule before the tissue is mashed. This problem can sometimes be avoided by setting the pial surface of the tissue down on the chopper stage. As with cerebellum, the neocortex might be cut best with a Vibratome.

2.2. Slice Chambers

The environment that a chamber provides for slices must include appropriate oxygenation, osmolarity, and temperature. In addition, it is necessary to have good visual control. Although some investigators use a static-pool chamber design (Teyler, 1980), the most commonly used chamber allows for superfusion of ACSF across the slice. Constant superfusion demands good mechanical stability, preferably with small dead space if drugs are to be added.

In superfusion chamber designs, the slices either rest on a net at the gas–liquid interface (an "interface chamber"), or are totally submerged (a "submersion chamber"). Different chamber designs provide particular advantages, and the "best" design depends on experimental goals and objectives. For instance, chambers with small volumes that allow fast exchange rates are good for certain pharmacologic studies but may sacrifice some mechanical stability. Submersion designs may optimize diffusion of ions and drugs into the tissue (diffusion occurs from both surfaces of the slice), but visualization of the electrodes may suffer if a water immersion lens is not used on the microscope. Each investigator will undoubtedly want to modify a chamber design so as to make the chamber most efficient for his own work.

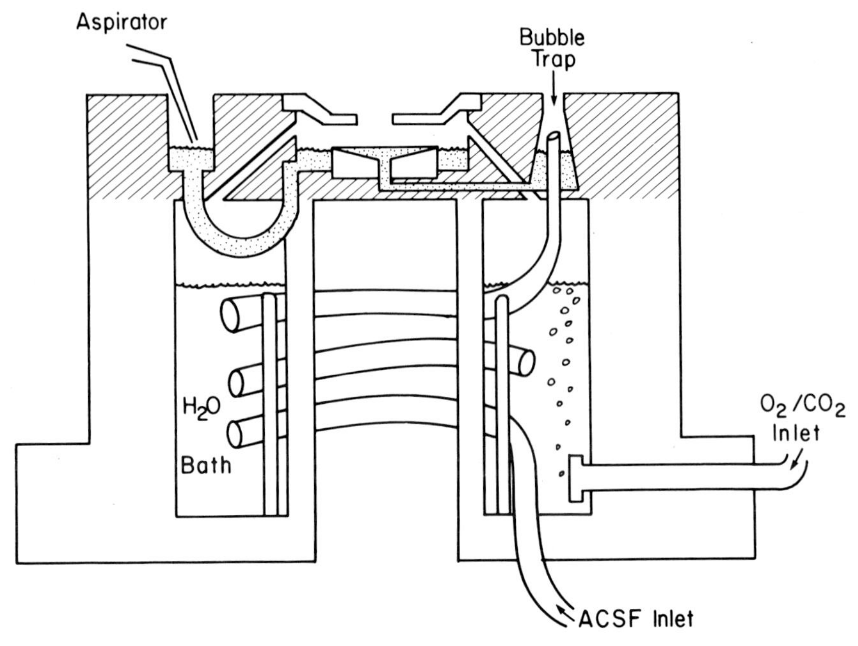

Figure 4. The Oslo interface chamber. This chamber is constructed from lucite, and features a bubble trap and separate aspiration well to prevent drying out of the slices. ACSF (dotted regions) flows through the bubble trap, under and around the slices, and spills over into a surrounding moat, from which it is drained into the aspiration well. Up to 25 slices rest on a piece of Kodak lens tissue, which is draped over a taut nylon net (bridal veil works well) that is glued to a slightly countersunk lucite cylinder. The angled lid above the center well directs the flow of warmed, humidified gas over the slices and also allows for the placement of electrodes under stereomicroscopic control. The chamber is approximately 6 inches in diameter and 4 inches high (King and Dingledine).

2.2.1. Slice Chamber Design. Figure 4 is a diagram of an interface chamber used by many investigators, sometimes called the "Oslo chamber" from its place of origin in Andersen's laboratory (Schwartzkroin, 1975). In this chamber, which was modified from an earlier design by McIlwain (Li and McIlwain, 1957), the slices rest on a nylon net affixed to a central pedestal. They are exposed to 95% O_2/5% CO_2 that is humidified by bubbling the gas through fritted stones in an outer warm bath filled with water or saline. The slices rest on a circle of lens paper, which reduces the surface tension of the ACSF and distributes it evenly around the slices. The ACSF flows through a bubble trap (Figure 4) and is introduced from below the slices, where it flows over and around them to drain into a concentric "moat." The fluid height is controlled by aspiration. It should be noted that placing the aspirator into the same well as the slices increases the risk of drawing laboratory air over the slices, thus cooling them and drying them out. To avoid this problem, the aspiration well can be connected to the moat by a fluid-filled tube (Figure 4).

The Oslo chamber design, with slices at the gas–liquid interface, permits adequate oxygenation of 300 to 500-μm-thick slices for many horus at 35 to 37°C. To maintain adequate mechanical stability, however, the flow rate is kept rather low, typically less than about 1 ml/min. In addition, the upper surface of each slice receives ACSF only by capillary flow and diffusion of substances into the slice is thereby slowed. For example, at a flow rate of 0.5 ml/min, most drugs require 12 to 20 min to reach peak effect, while the fluid surrounding the slices is completely exchanged within 3 to 5 min. To circumvent problems related to slow flow rates and fluctuating fluid level, other chambers have been designed.

Several modifications have been made to Oslo-type designs. An alternative and much simpler chamber to construct, which was developed by Haas *et al.* (1979), allows the ACSF to drain across the slice and drip into a collecting dish. The Haas chamber reduces mechanical switching artifacts and the formation of air bubbles by providing a fluid reservoir open to the air between the perfusion line and the slice. This reservoir acts as a baffle to dampen the changes in fluid level that may occur during switching. Although drying of slice surfaces can be a problem with all interface chambers, this seems to be especially troublesome with the Haas design.

In the submersion chamber developed by White *et al.* (1978), ACSF is directed downward through the slice, which rests on a support, into a collection cone and through an outflow tube. The ACSF flows circumferentially from all directions, which may help stabilize the slice and minimize the possibility of tissue or electrode displacement. Like the Oslo chamber, there is an outer bath compartment that provides humidified O_2/CO_2. The inner compartment, however, contains several smaller chambers, each of which holds an individual slice. This allows slices to be superfused separately and not all exposed to a particular drug at one time. The chamber in use is connected to a pressurized inflow system, while medium is recirculated through the others with a peristaltic pump. Each compartment volume is about 100 μl, and, with a 2.5 ml/min flow rate, is replaced every 5 sec.

A modified Oslo chamber is also used for maintaining spinal cord (Dhanjal and Sears, 1980 a,b) and cerebellar (Crepel *et al.*, 1981, 1982) slices. The major modification made to this chamber is an elevation of

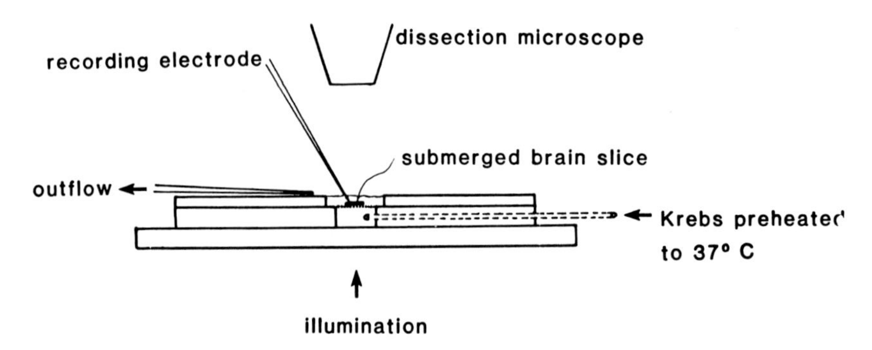

Figure 5. The Scottish submersion chamber. (Williams, Henderson, and North).

the gas inlet portholes in order to submerge the slices by raising the fluid level. In addition, the formation of bubbles in the recording bath is virtually eliminated by pregassing the incoming incubating medium in a small vessel fitted within the warm body of the Oslo chamber.

Lipton and Whittingham use a simple submersion chamber with a volume of about 5 ml in which the slices are placed on nylon bolting cloth and pinned down at one end by a lucite bar. ACSF flows from a reservoir maintained at about 41°C by a Haake water circulator. The flow rate is rapid, about 25 ml/min via a gravity feed, and the ACSF is pumped out of the chamber by a peristaltic pump and recirculated to the reservoir. In spite of the fast flow rate, mechanical stability is reported to be adequate for extracellular recordings. The chamber has the virtue of being extremely simple and the rapid flow offers two important features: first, a constant temperature in the chamber (± 0.5°C) in spite of changes in ambient temperature, and second, the ability to change the composition of the ACSF in the chamber completely within 30 sec with little effect on tissue position. This allows satisfactory biochemical studies.

The Scottish chamber (Figure 5), was named for its designer (G.H.) and because of its inexpensive and simple construction. It is constructed of three pieces of Plexiglas that are glued together. The ACSF is warmed by a heating jacket to 37°C and enters the bath through the bottom. The slice sits on the net in the middle of the bath and the solution is removed from the top. One or two titanium electron microscopy grids are placed on top of the slice, and these are immobilized either by a small platinum wire pressure foot mounted on a manipulator or by placing small platinum "logs" on the grid as weights. Since the volume of the bath is small, precise and even control of the bath temperature is obtained with a flow rate of 1.5 to 2.5 ml/min. With the rapid turnover time of the

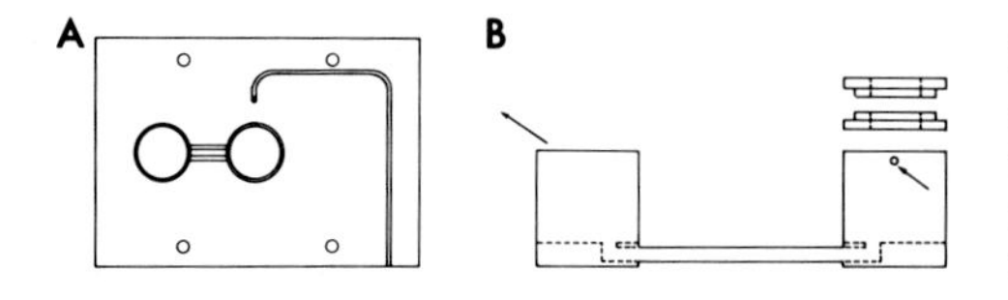

Figure 6. The Nicoll/Alger submersion chamber. (A) A view of the bottom of the brass block, showing the milled groove in which lays the perfusion tubing, and the two holes drilled out to hold the recording and aspirating wells. The brass block is firmly attached to a Peltier thermoelectric unit by nylon screws through the four small outermost holes. (B) An enlarged side view of the two wells, the tube connecting the wells, and the two nets positioned above the recording well. Arrows indicate inflow (right) and outflow (left). The recording and aspirating wells, and the rings holding the nylon nets, are constructed from plastic syringe parts. From Nicoll and Alger, 1981.

bath solution, and the fact that the slice is completely submerged, perfusion of drugs provides reproducible and dependable responses with fast onset and washout (see Williams *et al.*, this volume).

The submersion chamber (Figure 6) developed by Nicoll and Alger (1981) is similar to that used by Scholfield (1978) and differs from the chambers described above in that it incorporates a Peltier thermoelectric heating/cooling module (Cambion Thermionics model 806-7243-01) in place of the water filled Plexiglas reservoirs. Heat transfer from the thermoelectric unit to the experimental chamber occurs through a brass plate. A bottom view of the brass plate is provided in Figure 6A. The assembly shown in Figure 6B fits snugly into the holes in the brass plate. Basically, this assembly consists of two cylindrical wells connected by a tube. The well in the center of the brass plate is the experimental chamber. ACSF flows into the top of the experimental chamber (arrow), through the tube, and is aspirated out the back well. The bath is grounded with a reference electrode in the back well. Total dead space (volume of the system from the switching point to the top of the experimental well) is about 1 ml, and solution exchange is complete within about 5 min. Switching between different experimental media can be controlled by a latching solenoid valve (General Valve, Inc.). A single slice is sandwiched between two nylon nets, each held taut by a plastic and rubber frame. The slice is held in this sandwich about 1 mm below the surface of the medium. Illumination is from above, and contrast in the slice is enhanced by placing black tape beneath the experimental well on the bottom of the brass plate.

The advantages of this system include: (1) Good temperature control. Temperature changes in the chamber can be effected in a few, rather than tens of, minutes. (2) Size. The thermoelectric unit is much smaller than the Oslo reservoir, which makes the chamber less cumbersome.

This, plus the fact that the lip of the experimental chamber is not recessed with respect to the top of the brass place, makes the chamber possibly more accessible to a complicated array of electrodes. (3) Affordability. The chamber is inexpensive and simple to use. A final point of departure of the Nicoll/Alger chamber from previous designs is the use of a separate incubation chamber to maintain a batch of slices. Their holding chamber consists of a Plexiglas box and hinged top with about 2 cm of distilled water or ACSF in the bottom of the box. A pedestal is attached to the bottom of the box such that a shallow dish of medium rests on top of the pedestal 1 cm above the distilled water. Slices sit on a moist piece of filter paper that just touches the liquid in the shallow dish. A 95% O_2/5% CO_2 gas is bubbled through the liquid. A holding chamber is useful since not only does it keep the slices "naive" with respect to different experimental conditions, but it also allows slices to be transferred easily from a central spot to any setup in the laboratory. Finally, holding slices at room temperature may slow their rate of deterioration.

2.2.2. *Oxygenation.* If flexible tubing such as Tygon is used in the perfusion line, O_2 and CO_2 escape by diffusion as the perfusate travels from the prebubbled reservoir to the chamber. The amount of loss will depend on the flow rate and the size of the tubing, among other factors. For example, over a 3-ft length of Tygon (1/32 inch wall, 1/32 inch ID), the O_2 pressure falls from 600 to 380 Torr when the flow rate is 0.5 ml/min, but hardly at all when the flow rate is 10 ml/min (S. Ji and R. Dingledine, unpublished observations). For chambers in which slices are not submerged but perfused at a slow flow rate, this loss is presumably made up by oxygen uptake from humidified gas.

2.2.3. *Mechanical Stability.* The mechanical stability of the slices is a function of, among other factors, the inflow and aspiration rates and the balance achieved between them. Drainage of fluid, at the proper rate, appears to be a most difficult and critical factor in obtaining stability. For tissue supported on a mesh at a fluid–gas interface, breaking the fluid's surface tension at the level of the mesh is an initial critical step, and may be accomplished by covering the mesh with lens paper before the slices are placed down. Since lens paper may interfere with precise visualization (especially with slices illuminated from below), an alternative is to drape tissue over the edge of the mesh to form a wick.

In addition to surface tension problems, mechanical stability can be disrupted by gas bubbles in either the inflow or outflow lines, or by abruptly switching from one solution to another, or from one gas mixture to another. Bubbles of gas that come out of solution, or that enter the line inadvertently when the medium is switched, can be diminished by good overall temperature control and a "bubble-trap" in the perfusate line. A bubble-trap can be constructed by allowing the ACSF to flow into a well open to the atmosphere before entering the inner chamber.

In all chambers, however, bubbles can form by gas condensing out of the ACSF onto supporting nets, chamber walls, etc. The problem of bubbles may be eliminated completely if all solutions are vigorously oxygenated at the same temperature as the chambers. An additional trick for controlling occasional bubbles is to attach a broken micropipette (tip $\sim$ 20 μm) to a suction line. The pipette can be inserted between the holes in the support net and individual bubbles sucked up. If the composition of gases in the slice chambers is changed, mechanical artifacts can be avoided by using a common metering valve.

2.2.4. Osmolality. The maintenance of constant ACSF osmolality in slice chambers has received little systematic attention, with one exception. Investigators who use hypothalamic slices obviously need to monitor osmolality. Only when a static-pool chamber design was used did a considerable, rapid change in medium osmolality occur due to evaporation (see Teyler, 1980, for details and preventative techniques). Hatton *et al.* (1978) have noted, however, that evaporative loss may occur even in a flowing chamber. They found that, in some experiments, the osmolality of their ACSF increased to 390 mOsm within a few hours.

2.3. *Bathing Medium Composition*

2.3.1. Ions. Comparison of the ionic composition of ACSF with reported *in situ* values (see Tables I and II) shows that the most significant differences are in the K^+ and Ca^{2+} concentrations. Most investigators use slightly higher K^+ and Ca^{2+} concentrations in their ACSF than are found in CSF. These ions exert profound effects on neurons so the differences may be important.

A recent NRP workshop (Nicholson, 1980) summarized the specific neuronal effects of raised $[K^+]_0$, which include depolarization of cells and increased activity of the Na/K pump. In slices of olfactory cortex, increasing $[K^+]_0$ from 3 to 6 mM depolarized the large pyramidal neurons by about 8 mV (Scholfield, 1978); there was an associated increase in the evoked population spike of between 50 and 100% (Voskuyl and ter Keurs, 1981). Changes in the size of population spikes and synaptic potentials in the hippocampal slice have been reported when $[K^+]_0$ was varied within its normal range (Lipton and Whittingham, 1979; Hablitz and Lundervold, 1981; King and Somjen, 1981). The metabolic effects of raising $[K^+]_0$ are numerous. Increasing the concentration of this cation from 3 to 6 mM increases the activity of brain ATPases by approximately 25% (Kimelberg *et al.*, 1978). This procedure also causes an increase in

Table I. Composition of CSF (in mM) for Selected Species

Na^+	K^+	Mg^{2+}	Ca^{2+}	Cl^-	HCO_3^-	Glucose	pH	mOsm	Species	Reference designations[a]
158	2.69–3.28	1.33–1.47	1.50–1.67	138–144	18.3–25.6[b]	3.68	—	314.8	Cat	*a,b,d*
149	2.90	0.87	1.24	130	22.0	5.35	7.27	305.2	Rabbit	*d*
149–152	2.95–4.10	0.79–0.80	1.00–1.60	127–132	25.8	4.15	7.42	304.4–305.2	Dog	*b,d,f,i,l*
139–147	2.82–2.84	0.97–1.12	1.14–1.32	113–123	23.3	3.30	7.31	289	Human	*c,e,g,h,j,k*

[a] *a*, Ames *et al.*, 1964; *b*, Bito and Davson, 1966; *c*, Bradley and Semple, 1962; *d*, Davson, 1967; *e*, Fremont-Smith *et al.*, 1931; *f*, Friedman *et al.*, 1963; *g*, Hendry, 1962; *h*, Hunter and Smith, 1960; *i*, Kemény *et al.*, 1961; *j*, Salminen and Luomanmäki, 1962; *k*, McGale *et al.*, 1977; *l*, Oppelt *et al.*, 1963a,b.
[b] Calculated rather than measured.

Table II. Composition of ACSF (in mM) Used for Brain Slices

Na^+	K^+	Mg^{2+}	Ca^{2+}	Cl^-	HCO_3^-	Glucose	H_2PO_4	mOsm[a]	Structure	Species	Reference designations[b]
150–152	3.5–6.2	1.3–2.4	1.5–2.5	132–136	24–26	4–10	1.20–1.40	302–311	Hippocampus	Rat	*b,c,d,g,i,m,n*
143–150	5.4–6.2	1.3–2.0	2.0–2.5	127–133	26–26.2	10–11	0.92–1.25	294–307	Hippocampus	Guinea pig	*a,l*
150	6.4	1.3	2.5	134	26.0	10	1.25	304	Neostriatum	Guinea pig	*h*
151	5.0	2.0	2.0	133	26.0	10	1.25	307	Neocortex	Guinea pig	*e*
150	6.2	1.3	2.4	134	26	10	1.20–1.24	307	Cerebellum	Guinea pig	*f,j*
150	6.2	1.3	2.4	134	26	10	1.24	307	Hypothalamus	Guinea pig	*n*
153	5.0	1.3	2.4	136	26	10	1.20	308	Locus coeruleus	Guinea pig	*k*

[a] Calculated.
[b] *a*, Alger and Nicoll, 1980; *b*, Brown *et al.*, 1979; *c*, Dingledine, 1981; *d*, Dunwiddie and Lynch, 1979; *e*, Gutnick and Prince, 1981; *f*, Hounsgaard and Yamamoto, 1979; *g*, Lee *et al.*, 1981; *h*, Lighthall *et al.*, 1981; *i*, Lipton and Whittingham, 1979; *j*, Llinaś and Sugimori, 1980; *k*, Pepper and Henderson, 1980; *l*, Schwartzkroin, 1975; *m*, Segal, 1981; *n*, Yamamoto, 1973.

cell K^+ of about 15 mM and an equal decrease in Na^+ (Grisar and Franck, 1981; Lipton and Robacker, 1982), while levels of high-energy phosphates are not affected (Lipton and Heimbach, 1978). Finally, increasing $[K^+]_0$ from 3 to 6 mM increases the rate of protein synthesis in brain slices from guinea pigs by about 40% (Lipton and Heimbach, 1977).

The elevated Ca^{2+} and Mg^{2+} levels used in *in vitro* studies may well affect the physiological properties of slices in at least two obvious ways. First, the stabilizing effect of divalent cations on excitable membranes is well documented, and it is possible that elevated Ca^{2+} and Mg^{2+} concentrations raise action potential thresholds in slices. Consistent with this idea are reports that reducing the $[Ca^{2+}]_0$ to 0.75 mM in hypothalamic slices (Pittman *et al.*, 1981), or 0.2 mM in hippocampal slices (Jefferys and Haas, 1982), induced spontaneous activity, and raising $[Ca^{2+}]_0$ from 2.4 to 4.8 mM reduced population spike amplitude in hippocampal slices by about 25% (P. Lipton and D. Korol, unpublished observations). Second, synaptic transmission is quite sensitive to relatively small changes in $[Ca^{2+}]_0$. This has been shown in slices of olfactory cortex (Richards and Sercombe, 1970), cerebellum (Hackett, 1976) and hippocampus (Dingledine and Somjen, 1980). In all these tissues, the magnitude of postsynaptic field potentials fell as $[Ca^{2+}]_0$ was reduced below 2.4 mM, and in the hippocampus the efficiency of synaptic transfer was sharply decreased as $[Ca^{2+}]_0$ was lowered from 1.2 mM. It is worthwhile noting that free Ca^{2+} in most physiological buffers is below total Ca^{2+} because of complexing with bicarbonate and phosphate anions (Schaer, 1974).

Although most experimenters attempt to use ACSF that approximates the ionic composition of CSF, this goal is clearly not attainable *in extremis* and may not even be desirable in certain situations. It must be acknowledged that there is no "ideal" ionic composition for slice ACSF. An experimenter's final choice will be dictated primarily by what biological problem is under investigation. Ionic manipulations may be used to bias activity in slices as discussed above with the understanding that multiple physiological effects are invariably produced when $[K^+]_0$ and $[Ca^{2+}]_0$ are altered.

2.3.2. Organic Compounds. The primary energy substrate for the CNS in glucose. Perry *et al.* (1975) and McGale *et al.* (1977) have reported the amino acid composition of human CSF, and although only glutamine appears in substantial amounts (0.6 mM), the total concentration of amino acids (10 mM) may add a substantial amount of metabolic intermediates to CSF. It is very improbable, however, that these other compounds act as energy substrates, for in the brain *in situ* 95% or more of the oxygen consumption can be accounted for by the breakdown of glucose to H_2O and CO_2 (Siesjo, 1978). The high glucose level does not appear necessary, at least to obtain satisfactory electrophysiological responses in the hippocampal slice. Thus, P. Lipton (unpublished observations) sees no attenuation of the population spike in the guinea pig dentate gyrus until glucose falls to 0.2 mM. Requirements for glucose in order to observe long-term potentiation (LTP) or other multiple response-dependent phenomena have not been strictly determined, but 4 mM glucose, which is the normal blood level, supports LTP in rat hippocampal slices (I. S. Kass, unpublished observations). To date, there have been no systematic studies of the potential benefits that might accrue from using either alternative or additional energy substrates, although one might predict that the inclusion of glutamine would help maintain synaptic transmission in fibers using glutamate (Hamberger *et al.*, 1978). Two of us (PAS and RD) have tried Eagle's minimum essential medium, which contains a variety of amino acids and vitamins, for some slice experiments. We noticed no appreciable improvement over our usual blanced salt solutions. However, Schwartzkroin has found that addition of 10% calf serum to slice ACSF can improve the condition and longevity of hippocampal slices.

2.3.3. *Osmolality and pH.* The range of values found for both osmolality and pH of true CSF, 289 to 315 mOsm and 7.27 to 7.42, respectively, are well approximated by slice investigators. Although most investigators employ bicarbonate as a buffer, Pittman *et al.* (1981) have reported the use of Hepes, and Llinás and Sugimori (1980) Hepes and tris. There have been few reports concerning the possible effects of changes in pH on slice physiology, although in other preparations, changes in pH affect gap junction permeability (Spray *et al.*, 1981) and membrane potential (Marshall and Engberg, 1980). Reducing extracellular pH from 7.4 to 7.1 by lowering $HCO^-{}_3$ reduced synaptic transmission in the hippocampal slice by 50% (Lipton and Korol, 1981). It is obviously important to ensure pH stability when additions to the ACSF are made. This is especially true when Ca^{2+} or its substitutes are added since these divalent cations can complex phosphate and bicarbonate.

2.3.4. *Oxygen.* As might be expected, brain slice preparations are quite sensitive to oxygen levels; this presents a potential problem since the method of preparation produces a brief hypoxic period from which most slices eventually recover. With intracellular recordings during hypoxic episodes, Schwartzkroin has observed reduced synaptic activity [especially inhibitory postsynaptic potentials (IPSPs)], low input resistance, difficulty in holding cells, and spontaneous epileptiform bouts. *In vivo*, periods of hypoxia or anoxia produce the following: (1) increased intracellular lactate levels and resultant decreased pH, (2) massive exchange of extracellular for intracellular ions, (3) tissue edema, (4) diminished oxidative metabolism (5) failure of neurotransmitter synthesis (Gibson *et al.*, 1981), and (6) failure of synaptic transmission (Eccles *et al.*, 1966). Hypoxia is also one of the predisposing factors that make brain tissue more susceptible to spreading depression, which involves a massive release of intracellular K^+ (see Nicholson and Kraig, 1981). Most of these phenomena have been seen, to some degree, in slice preparations.

Many of the early slice investigators reported severe alterations in the structure (i.e., swelling) and metabolism of slices, which was a likely consequence of hypoxic conditions (Franck, 1972). The action of hypoxia on hippocampal slice function and its role in the observed differences between slice and *in situ* metabolism are discussed more fully in the chapter by Lipton and Whittingham (see Chapter 5, this volume). Some effects of hypoxia are summarized here. In both guinea pig and rat slices, 5 min of almost complete anoxia at 36°C produces no irreversible damage to slice electrophysiology or metabolism; this casts doubt on explanations of abnormalities in slice function that assume they necessarily result from the hypoxia during the isolation procedure. Although slices can recover from short periods of almost complete anoxia, they are rapidly affected *during* the period of lowered oxygen tension; the population spike is completely, but reversibly, abolished after 3 to $3\frac{1}{2}$ min of anoxia at 35°C (Lipton and Whittingham, 1982). There are also significant changes in monovalent cations and cell water during hypoxia. Thus, hypoxia produces the same exchange of intracellular K^+ for extracellular Na^+ and Cl^- seen *in vivo* (van Harreveld, 1966; Vyskocil *et al.*, 1972). After 10 min of anoxia in the rat hippocampal slice, there is a 60 mM increase in cell Na^+ and an equal but opposite shift in cell K^+ (Kass and Lipton, 1982). Cell volume increases within 30 sec of the onset of anoxia, as measured by brain slice reflectance. The volume increase is reversible if anoxia lasts 4 min or less, but is largely irreversible if anoxia lasts for 10 min (Lipton, 1973; Kass and Lipton, 1982).

Brain slices and ACSF are typically humidified and bubbled, respectively, with 95% O_2/5% CO_2. The resultant Po_2 of the gas at the slice surface should be about 675 mm Hg. There may be a reduction of Po_2 at the gas liquid interface (cf., Bingmann and Kolde, 1982) and, as mentioned earlier, there may be diffusion of O_2 out of the input line. Thus, the slices are probably exposed to a Po_2 less than 675 mm Hg. These estimates are consistent with the data of Fujii *et al.* (1981, 1982) who found U-shaped depth profiles for Po_2 in guinea pig olfactory slices maintained in an interface chamber at 37°C. Their slices (320 to 480 μm thick) showed similar Po_2 at both surfaces (ranging from 252 to 432 mm

Hg); in slices thicker than 430 μm, there was an anoxic central core (Po_2 = 0 mm Hg). G. L. King *et al.* (unpublished observations) have observed a similar range for the Po_2 values of hippocampal slices in an interface chamber, using oxygen polarimetry. Whether submerged slices have a different Po_2 profile has not been reported.

Toxicity due to high oxygen tensions may be a contributing factor to the rationale for recording deep within the slice. Toxic effects of O_2 in the respiration of brain tissue occur at 1 to 3 atmospheres pressure (for review, see Davies and Davies, 1965) and Po_2 values of 400 mm Hg are equivalent to 1.7 atmospheres of pressure. King and Parmentier (1983) have recently observed that synaptic transmission in the hippocampal slice preparation is depressed by 100% O_2 at even 1.5 atm. If such toxicity occurs, it may add to the damage induced by the tissue dissection.

2.3.5. Temperature. The temperature at which slices have been successfully maintained ranges from room temperature (25°C) to normal animal body temperature (38 to 39°C for small animals like the rat or guinea pig). Although, intuitively, it would seem best to maintain brain tissue at or near normal body temperature, investigators have generally found that the preparation survives longer, and in a healthier state, at lower temperatures (30 to 35°C). In addition, small temperature fluctuations around 38 to 39°C, particularly in the warmer direction, can result in large changes in slice excitability. Epileptiform discharges become common at 40 to 41°C.

There is, of course, a real conceptual problem in maintaining slices at lower temperatures. Within the temperature range of most slice experiments, electrophysiological responses are neither uniform nor linear. This is expected since many cell processes are temperature dependent. A good example is the effect of low chamber temperature on field potential amplitudes and action potential durations. Action potentials become increasingly broad with lowering temperature. The magnitude of the temperature effect is in dispute. In hippocampus, the experience of both Schwartzkroin and Dingledine is that the effect can be extremely dramatic, with 5-msec action potentials not unusual at 30°C. Alger, on the other hand, reports more typical values (2 to 2.5 msec) at 30°C, and doesn't see marked broadening until 22 to 25°C. Broad action potentials may result in increased transmitter release at terminals, and extremely large excitatory postsynaptic potentials (EPSPs). If one records only field potentials, the evoked potentials may look normal, or even "healthier" than normal. Although this effect may be advantageous for those studying synaptic responses, the investigator must again weigh the technical advantages of this manipulation against the interpretational problems it subsequently presents.

2.4. *Slice Thickness*

There are three main considerations in deciding the thickness of slice to be used: visibility, viability, and preservation of neuronal circuits. The thickness of sliced tissue is limited at the upper extreme by diffusional restrictions, and at the lower extreme by slicer-induced surface damage. Determination of how thick to cut slices must be made, in part, on the basis of what questions are being asked and what techniques are to be used. If one uses Nomarksi or Hoffman optics to visualize cells (Chujo *et al.*, 1975; Llinás and Sugimori, 1980), or does experiments that deal with single cell properties, it is clearly desirable to use thin slices (less than 200 μm). If experiments deal with, or depend on intact circuitry, however, thicker slices will provide better results. Also, thicker slices are useful when the tissue is very fragile (e.g., unmyelinated immature brain), since slices fray less.

Harvey *et al.* (1974) calculated for olfactory cortex slices incubated in an atmosphere of 95% oxygen that the maximum slice thickness that could be tolerated at 37°C without anoxia was 690 μm (unstimulated slice), 600 μm (lateral olfactory tract stimulation, 100 Hz) or 530 μm ("trans-slice" general stimulation, 100 Hz). The corresponding limiting thickness at 25°C were 55% greater. From observations on the population spikes, these authors considered that thicker (700 μm) slices maintained at 25°C were beneficial for more complete preservation of neuronal circuits without compromising viability. Using the same preparation, Fujii *et al.* (1982) reported a limiting thickness of 430 μm based on direct measurements of tissue P_{O_2}.

Bak *et al.* (1980) have noted that neostriatal slices of different thickness differ appreciably in their biochemical, histological, and electrophysiological properties. In the central zone of 300-μm slices incubated at 36°C, the cells displayed well-preserved morphology and electrical properties. In 700-μm slices incubated either at 36°C or at room temperature, swelling and necrosis was observed throughout most of the slice, even in regions not expected to suffer directly from hypoxia, i.e., 100 to 200 μm in from the cut edge.

Morphological studies on rat cerebellar slices (Garthwaite) indicate that the limiting thickness of an unstimulated slice is 450 μm at 37°C. Regions of the slices that were thicker than this value exhibited centrally-located necrotic cells, suggestive of hypoxic damage. It is noteworthy that Purkinje cells adjacent to, but outside, the central hypoxic zone also tended to be necrotic.

At present, the favored thickness for spinal cord slices of 600 to 800 μm is based largely on macroscopical observations that show generally good structural integrity of the tissue that tends to decrease with decreasing thickness (S. S. Dhanjal and T. A. Sears, unpublished observations). However, the preparation is maintained at 32°C in the hope of compensating for greater metabolic requirements of the thicker slice. It should be remembered that there is a relatively greater proportion of white matter in the cord compared with most brain slice preparations, and therefore, the possibility of sustaining thicker slices may be greater.

3. EVALUATION OF SLICE DATA

3.1. *General Remarks*

Many of us are drawn to slices from our earlier *in vivo* studies. The feasibility of making a detailed *in vitro* study of cellular phenomena originally pursued in the whole animal provides a strong attraction. The point has been made repeatedly, however, that the relevance of data collected in slices to the *in vivo* situation is dependent on how well the important *in vivo* characteristics can be preserved *in vitro*. Of course, which *in vivo* characteristics are "important" depends on the type of question asked. For example, studies of the pharmacological responsiveness of individual neurons, or of the properties of most individual ionic conductances, would seem relatively immune to serious problems. At the other end of the spectrum are investigations requiring statistical comparisons among slices prepared from different animals, which are fraught with difficulties of interpretation. Some of the more interesting neurobiological questions fall into the latter category (e.g., are membrane or synaptic properties changed during the aging process or in animals made chronically epileptic?).

Despite worries about the suitability of slices for certain studies, and the problems that an isolated *in vitro* preparation introduces into interpretation of data, investigators should remember that virtually all experimental preparations—*in vivo* as well as *in vitro*, chronic as well as acute—introduce interpretational difficulties. The most satisfying "validation" of slice phenomena has been the general finding that *in vitro* findings are similar to *in vivo* findings. Clearly, this reasoning is somewhat circular, and our criteria for useful data are invariably arbitrarily set.

Certain laminated structures such as the cerebellum and hippocampus are useful for making comparisons to the *in vivo* situation since their cellular physiology has been well characterized during a detailed series of *in vivo* experiments carried out over the past 20 years. Other structures such as the locus coeruleus and certain hypothalamic nuclei have not been so thoroughly characterized by *in vivo* experiments, and in these cases slices are providing completely new information about cellular properties. The provision that slice data should conform to whole animal data cannot be accepted blindly. For example, the increased mechanical stability provided by *in vitro* conditions undoubtedly permits certain neuronal properties (e.g., membrane time constant, input resistance, spike amplitude, and threshold) to be measured with much more fidelity and reliability than is possible in the whole animal. Better recording stability is also likely to influence the shape of synaptic potentials, the prominence of active calcium conductances, etc. However, it must also be considered that the *in vitro* nature of the preparation may remove many tonic synaptic influences, which could lead to artifactual values for certain "resting" parameters.

The following discussion is divided into electrophysiological, histological, and biochemical approaches to evaluating the suitability of slices.

3.2. *Electrophysiology*

3.2.1. Hippocampal Field Potentials. In the hippocampus and dentate gyrus of healthy animals, characteristic synaptic and antidromic field potentials can be recorded in response to the appropriate stimuli (see Andersen, 1975, for review). The essential features of these evoked potentials are reproduced in slices, as shown in Figure 7. The dendritic field evoked by afferent stimulation (S1–R2 in each part of Figure 7) reflects mainly excitatory postsynaptic currents flowing across pyramidal and granule cell membranes. The sharp negativity recorded in the cell layers (S1–R1) is a result of the synchronous firing of the principle cells in each region, and is termed the "population spike." Focal stimulation of the axons of pyramidal cells (alveus for CA1 regions, Schaffer collaterals for CA3) or granule cells (mossy fibers) results in a typical antidromic population spike (S2–R1).

There is one component of the synaptic field potentials that can be recorded in slices routinely but is not prominent in whole animal recordings (but see Andersen *et al.*, 1971, Figure 3C; also Leung, 1979,

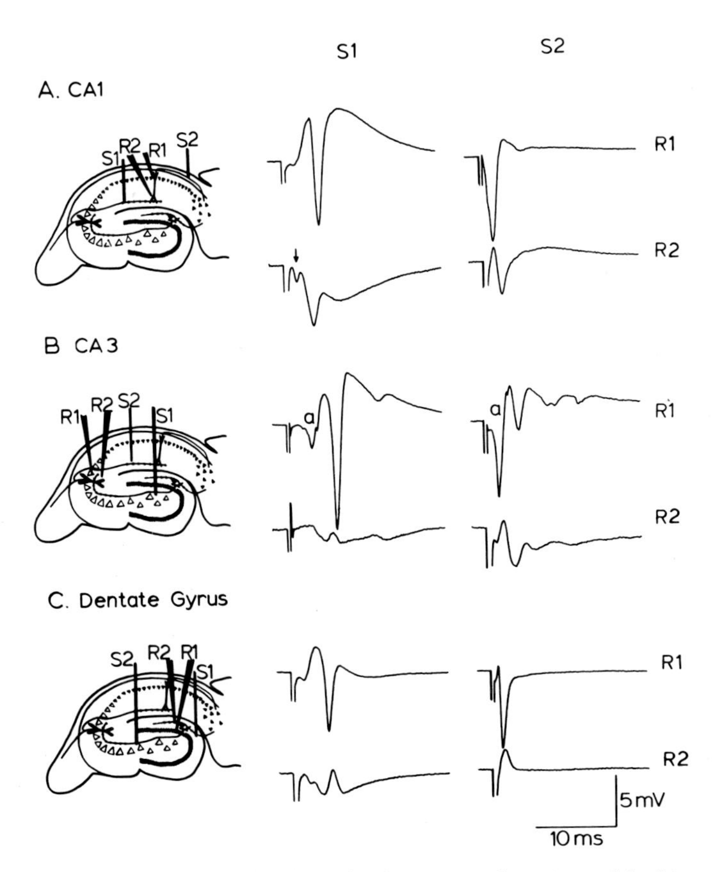

Figure 7. Field potentials recorded from the three principle regions of the hippocampal slice preparation. In each diagram and associated set of recordings, S1 represents an orthodromic stimulus and S2 an antidromic stimulus; R1 is a recording electrode placed in the cell layer and R2 a recording electrode placed in the dendritic receiving zone for orthodromic activation. (A) In the CA1 region, S1 is placed in the s. radiatum to activate Schaffer/ commissural inputs, and S2 in the alveus to activate pyramidal cell axons. The arrow in S1–R2 designates the presynaptic fiber volley that is recorded only from the dendritic electrode. (B) In the CA3 region, S1 is placed among the mossy fibers and S2 in the Schaffer collateral fiber pathway. "a" represents an antidromic component of both stimulating lines, which is more prominent when Schaffer collaterals are stimulated. (C) In the dentate gyrus, S1 is placed among the perforant path fibers and S2 in the mossy fiber zone (R. Dingledine and G. L. King, unpublished observations).

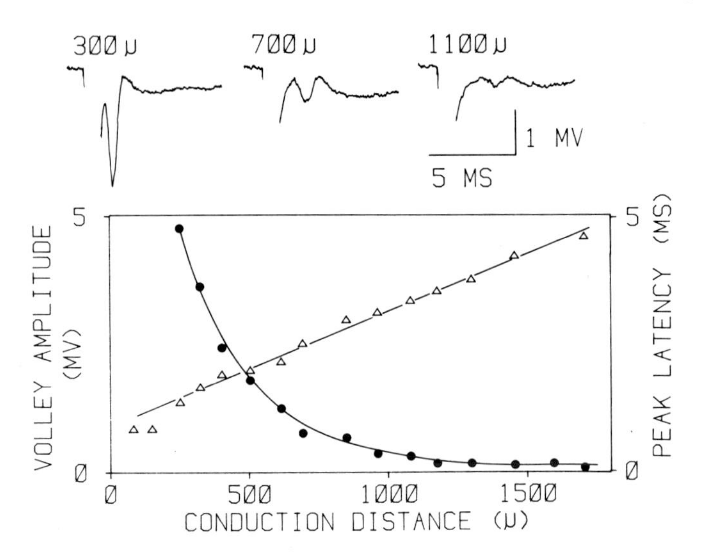

Figure 8. Impulse propagation in unmyelinated fibers within the hippocampal CA1 region. A stimulating electrode was placed in s. radiatum, and recordings were made at successively increasing distances along a line parallel to the cell body layer. At each recording site, 8 to 64 responses were averaged. Representative averages at three distances are shown at the top. In the graph, the peak-to-peak amplitude of the volley (circles) and the peak latency (triangles) are plotted as a function of distance. The propagation velocity of these fibers was 0.40 m/sec. The volley half-width also increased regularly with increasing distance (not plotted). (R. Dingledine, unpublished observations.)

Figure 3C). This is the presynaptic fiber volley, marked with an arrow in Figure 7A, which results from the extracellular currents surrounding the synchronously activated unmyelinated fibers running in the dendritic layers (Andersen *et al.*, 1978). Two explanations for this apparent discrepancy between *in vitro* and *in vivo* results can be noted. First, it is our experience that field potentials, including fiber volleys, tend to be larger in interface chambers when compared to submersion chambers, presumably due to the more restricted paths of current flow (thus higher extracellular current density) in exposed slices. This factor would be expected to contribute to a reduced opportunity for observing the afferent volleys in whole animal experiments. Second, and likely of greater importance, the presynaptic fiber volley is most conspicuous when the stimulating and recording electrodes are quite close to each other, a condition not easily achieved in whole animal experiments. Figure 8 demonstrates that as the propagation distance is increased, the volley amplitude decreases very sharply, while the peak latency increases as expected. At distances greater than 1 mm, the fiber volley is usually undetectable. These observations suggest that the Schaffer/commissural fibers may fan out considerably as they course within the stratum (s.) radiatum (cf., Andersen *et al.*, 1980).

In addition to the excitatory phenomena described above, the hippocampus is noted for displaying pronounced synaptic inhibition and a range of synaptic potentiations following tetanic stimulation. The recurrent inhibitory pathway, well characterized in whole animal experiments (Kandel *et al.*, 1961; Andersen *et al.*, 1964), is also functional in hippocampal slices (Dingledine and Langmoen, 1980; Lee *et al.*, 1980). Although recurrent IPSPs can be elicited in slices, they do not appear as prominent as they are *in vivo*. *In vitro*, the IPSP timecourse is approximately half that reported *in vivo* (50 to 150 msec compared to 80 to 600 msec), and the IPSP conductance change seems less intense than that reported in the *in vivo* literature. It is known that the barbiturate anesthesia used in whole animal studies greatly prolongs and intensifies hippocampal IPSPs (Nicoll *et al.*, 1975), which accounts for a good deal of the difference. The more negative resting potential recorded *in vitro* additionally contributes to the difference in IPSP (and EPSP) amplitude.

A multitude of criteria have been proposed to judge the "healthiness" of hippocampal slices. There are often two separate questions under review: "How normal, with respect to *in vivo* conditions, is the slice?" and "How good, in a technical sense, is the intracellular penetration?" Field potentials are convenient for assessing the overall state of the slice, or at least of small regions within a slice. Several tests have been put forward, such as the maximum amplitude of population spikes, the minimum stimulus current needed to evoke a population spike, or the prominence of the dipole positivity in the CA1 or dentate cell layers. These measures are greatly influenced by electrode position and the "wetness" of the slice, however, and so are unsuitable for general use.

An initial visual inspection can be useful in rejecting obviously damaged slices. Damaged slices are often distinguished by one of the following characteristics: a ripped or frayed alveus, excessive translucency, and an appearance of "melting" into the supporting mesh or filter paper. Good slices typically have smooth, well-formed borders, and a firm consistency. The configuration of the field potentials evoked by weak orthodromic stimulation does seem to be a good indicator of some aspects of slice physiology—primarily synaptic transmission. In a healthy CA1 or dentate gyrus, as one increases the stimulus current, the population spike gradually increases in size and decreases in latency. Typically, only a single spike is evoked, even at near maximal stimulus currents; although in a sick or anoxic preparation, two or more spikes are elicited by only moderate stimulating currents. On the other hand, one can record from apparently healthy pyramidal cells in slices with poor field potentials, and, conversely, high-amplitude field potentials can also be associated with cellular "abnormalities"—e.g., broad action potentials (slice too cool) or hyperexcitability (high K^+).

3.2.2. *Hippocampal Intracellular Potentials.* A strict comparison of *in vitro* with *in vivo* results here seems unrewarding. Although resting membrane properties are higher *in vitro,* is is not clear whether most *in vivo* records are of "abnormal" poor quality, or whether most *in vitro* records are biased by the relative lack of tonic synaptic conductances. Nevertheless, a general consensus is developing about what constitutes a good intracellular penetration, at least for CA1 and CA3 pyramidal cells. A well-penetrated cell typically shows a stable membrane potential, spike height and input resistance over long periods of time, in the absence of steady hyperpolarizing current. In response to threshold orthodromic stimulation, a single spike of at least 80 mV amplitude appears at the crest of a well-defined EPSP. A weak depolarizing current pulse (i.e., 0.2 to 0.25 nA) triggers a single spike at a long and reproducible latency; as the current is increased the spike latency decreases and more spikes are recruited in the train. A poorly penetrated cell generally is unstable with respect to one or more of the above properties. Cell spontaneous activity also provides information regarding cell integrity; pyramidal cells *in vitro* normally have a relatively low spontaneous rate (0 to 2/sec), whereas interneurons *normally* have high rates (20 to 50/sec). These criteria have been elaborated in several previous publications (Schwartzkroin, 1975, 1977; Langmoen and Andersen, 1981).

Which of the above measures provides the most sensitive indicators of cell health? Little meaning can normally be attached to the absolute values reported for resting potentials measured with high-resistance micropipettes. Possible changes in tip potential as the tip moves from an extracellular to an intracellular environment are difficult to quantitate satisfactorily. Membrane input resistance and time constant (which should be maximized for the particular cell population under study) may provide useful measures of the quality of the seal around the microelectrode, although misinterpretations can arise if the cell population is not truly homogeneous. Overshooting action potentials are considered a necessity by all.

It is Schwartzkroin's impression that resting potential, spike amplitude, and even recording stability can be relatively poor indicators of good penetrations. In his opinion (recording from hippocampal and neocortical neurons), the most sensitive measure of cell health is the cell's ability to produce a regular, rhythmic train of action potentials in response to moderate depolarization. Injured neurons will often respond with a single action potential at the onset of a depolarizing current pulse; the "injured" response pattern can be seen in cells with high resting potentials and good spike amplitude.

Unquestionably, what one sees with an intracellular electrode may be very strange. High resting potentials (−75 to −85 mV) may be accompanied by poor input resistance (<15 MΩ), but reasonable action potentials, or penetrations showing high input resistance, may also show small spikes. The first of these might reflect a dual penetration of neuron and glial cell, the second a dendritic impalement. However, unless the cell and electrode tip can be observed directly, it is impossible to know for sure what's going on.

Even with the above criteria for accepting a penetration as apparently healthy, cells show quite a wide range in properties. In one laboratory, for example, given the criteria of a steady resting potential, typical orthodromic responses, and a spike amplitude >80 mV, a sample of 113 CA1 pyramidal cells gave the following ranges: spike amplitude: 80 to 128 mV; input resistance: 9 to 64 MΩ; membrane potential: −52 to −87 mV. An attempt was made to determine whether any of these measurements were interdependent; for example, whether cells with lower input resistance also have relatively smaller spikes. As shown in Figure 9, however, such correlations are weak or absent for most parameters. The spike amplitude was only slightly correlated with either membrane potential or input resistance (Figure 9A and 9C). In both cases, only 10% of the variance could be accounted for by the correlation, although this small degree of correlation reached significance at the $p<0.01$ level. There was no correlation between membrane potential and input resistance ($p>0.10$).

Perhaps the most interesting finding was a moderately strong correlation ($p<0.001$, accounting for 30% of the variance) between the input resistance and electrode resistance (Figure 9D). Higher resistance electrodes, which presumably had smaller tips, tended to yield cells with higher input resistance. This correlation holds even when one eliminates the results of penetrations with electrodes of resistance > 150 MΩ, i.e., cases in which bridge balance is more difficult to judge. These data are compatible with high-resistance electrodes penetrating smaller cells more often, or doing less damage to a population of rather uniformly sized pyramidal cells. The increasing practice of going to low resistance electrodes in order to achieve greater control over the neuron should be viewed with these considerations in mind.

3.2.3. Brainstem. Evaluation of viability of the LC is more difficult than in hippocampus or olfactory cortex, highly organized structures in

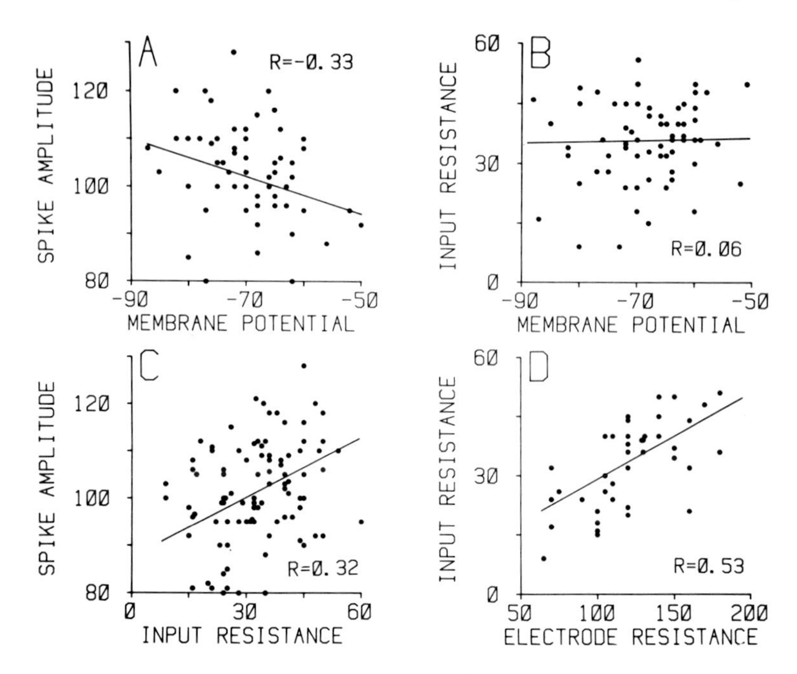

Figure 9. Correlations among measured pyramidal cell properties in the hippocampal slice. The data were gathered over a period of a few months from a population of 113 CA1 pyramidal cells in which satisfactory penetrations were obtained. For each graph, the linear regression line is drawn and the correlation coefficient (R) is displayed. A reasonably high correlation was found only between membrane input resistance and electrode resistance. There is a wide scatter of measured values for all parameters, even though all cells met the criterion of a steady resting potential and spike height greater than 80 mV. (R. Dingledine, unpublished observations.)

which field potentials are a good measure. There are no structurally identifiable afferents or efferents of the LC in the coronal slice, so that recording of field potentials is difficult. The viability of a given slice from rats can be assessed qualitatively by the appearance of large amounts of extracellular spike activity as the electrode tip is advanced through the tissue. In slices from guinea pig, most cells are quiescent, but unfortunately no *in vivo* data exist for comparison.

The appearance in the slice of spontaneous firing identical in rate and pattern with extracellular recording *in vivo* may indicate that recordings *in vitro* are comparable to those in normal conditions. LC neurons exhibit this firing for up to 14 hr after isolation. The threshold for spike activation and the spike amplitude are quite constant from cell to cell (−55 mV and 80 mV respectively for over 200 cells). The consistency of recordings made in the slice preparation indicate that physiological and pharmacological results obtained *in vitro* may be relevant to LC neurons *in vivo*.

3.2.4. Neocortex. One difficult problem in evaluating neocortical potentials is the heterogeneity of cell types. In contrast to a relatively homogeneous hippocampal cell population, the neocortex presents almost every imaginable cell type. Since one rarely knows at the time of penetration what cell type has been penetrated, it is difficult to judge the impalement on the basis of predetermined criteria. The problem of electrode selectivity is also aggravated by cellular heterogeneity. Electrodes will, in general, record most easily from the largest cellular elements they can find. One thus records infrequently from small cells, even if they make up a significant part of the cortical population.

3.2.5. Spinal Cord. A number of the problems encountered in evaluating the viability of brain stem and neocortical slices mentioned above are also common to spinal cord slices. Some properties of extracellular unitary field potentials recorded from the dorsal horn *in vitro* appear to match those recorded *in vivo* but with much greater stability (S. S. Dhanjal and T. A. Sears, unpublished observations). As with the brainstem slice preparation, the best assessment of viability is the appearance of large numbers of "spontaneous" extracellular spikes upon advancing the electrode through the tissue. Another method has been to demonstrate the reversible nature of a postsynaptic potential following a short period of anoxia. This is shown in Figure 10 as an "N" wave recorded in the dorsal horn upon stimulation of the dorsal column in a longitudinal spinal cord slice preparation (S. S. Dhanjal and T. A. Sears, unpublished observations).

3.3. *Histology*

To many, even nonhistologists, the most obvious way to investigate the cellular and subcellular integrity of a tissue is to examine it under the microscope. Until recently, however, there have been relatively few such studies on brain slices. Instead, many earlier workers in this area chose to assess the health of slices by measuring water "spaces". Undoubtedly, this approach was adopted in part because of justifiable uncertainties regarding artifacts arising from the processing of the tissue for microscopy. These artifacts may well have contributed to the rather poor microscopic appearance of slices in earlier studies (Gerschenfeld *et al.*, 1959; Cohen and Hartmann, 1964; Wanko and Tower, 1964). With modern methods, however, most of these problems have been circum-

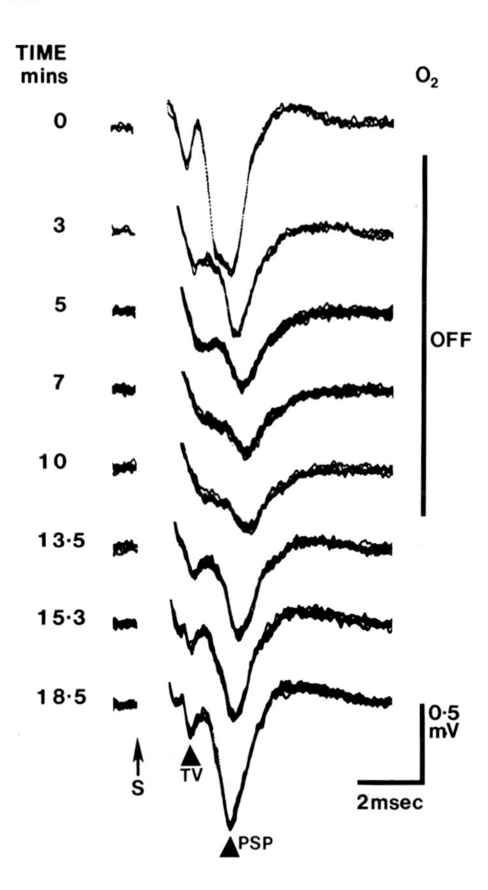

Figure 10. Effect of an anoxic period on evoked responses in rat dorsal horn. A longitudinal, parasagittal slice was prepared. Each trace consists of several superimposed sweeps of the response to stimulation of the dorsal column. The period of anoxia is marked by the vertical bar on the right, and time during the experiment is shown on the left. S, Stimulus artifact; TV, tract volley; PSP, postsynaptic field potential. (S. S. Dhanjal and T. A. Sears, unpublished observations.)

vented. Morphology must be considered an effective tool for evaluating the quality of brain slices and a particularly important one for properly interpreting the results of experiments in which other objective parameters of slice viability are not measured.

Histological studies at the light level, particularly in experiments that have included dye-injection of neurons through intracellular pipettes, have revealed normal-appearing neurons (including fine processes). There have been relatively few reports regarding the ultrastructure of incubated slice tissue. This is due, at least in part, to the fact that electron microscopic examination of incubated slices often shows considerable ultrastructural damage. In Schwartzkroin's experience, hippocampal slices initially (i.e., soon after chopping) appear relatively normal ultrastructurally with small patches of clearly abnormal tissue (dead cells, large extracellular space, large vacuoles in cells). These abnormal patches become larger with longer periods of incubation. This description does not coincide with Garthwaite's observations on cerebellar slices. The procedures he has developed to examine brain slices histologically are detailed below.

3.3.1. Processing Brain Slices for Microscopy. Slices should be incubated for at least 1½ hr; if less, it is sometimes difficult to decide if cells are dying or recovering. They are then immersed into fixative (5 to 8 ml) in small pots. An appropriate fixative for brain slices is a mixture of 2.5% glutaraldehyde and 4% paraformaldehyde in 0.1 M phosphate buffer (pH 7.4). A comparison of cerebellar slices fixed at room temperature or in the cold (4 to 6°C) suggested the former to be preferable, probably because of the more rapid penetration by the aldehydes under these conditions (J. Garthwaite, unpublished observations). After 30 to 45 min fixation, the slices are washed (5 min) in buffer and then postfixed in 1% osmium tetroxide (in buffer) for 1 hr. After a further buffer wash (5 min), the slices are dehydrated through a graded series of alcohols (30, 50, 70, 90, and 100% ethanol; 2 × 5 min each), gradually impregnated with Spurr's (or other) resin (0, 25, 50, 75, and 100% resin in the appropriate solvent; 20 min each). The slices are then transferred into molds filled with 100% resin and hardened. For light microscopy, semithin (0.5 to 1.0 μm) sections are cut and stained with Toluidine Blue (0.2 to 1% in 1% borax, pH 11). Ultrathin sections are contrasted with uranyl acetate and lead citrate for electron microscopy.

3.3.2. Light Microscopy. The overall quality of a slice can be assessed most objectively by examining semithin sections through the light microscope. Despite this, relatively few authors investigating the morphological preservation of brain slices have included light micrographs in their publications (Garthwaite *et al.*, 1979, 1980; Bak *et al.*, 1980; Frotscher *et al.*, 1981).

To evaluate the uniformity of the slice, it should be cross-sectioned in at least two planes. Larger sections cut in the plane of the slice are technically more difficult to obtain but can be instructive especially for preparations in which the cellular composition varies from region to region, such as in hippocampal slices. The microscopic appearance should be compared with that of the brain area fixed *in situ* by intracardiac perfusion. Immersion fixation (i.e., dissecting out the brain area and immersing it, or sections of it, into fixative) is not used as it produces artifacts due to hypoxia (which are presumed to reverse on incubation).

Damage to cut edges characteristically appears as a band of tissue that may stain more lightly than intact regions (due to swelling) and in which the cells are necrotic. In the case of cerebellar slices, the extent of damage at the cut edge can be estimated from the depth over which

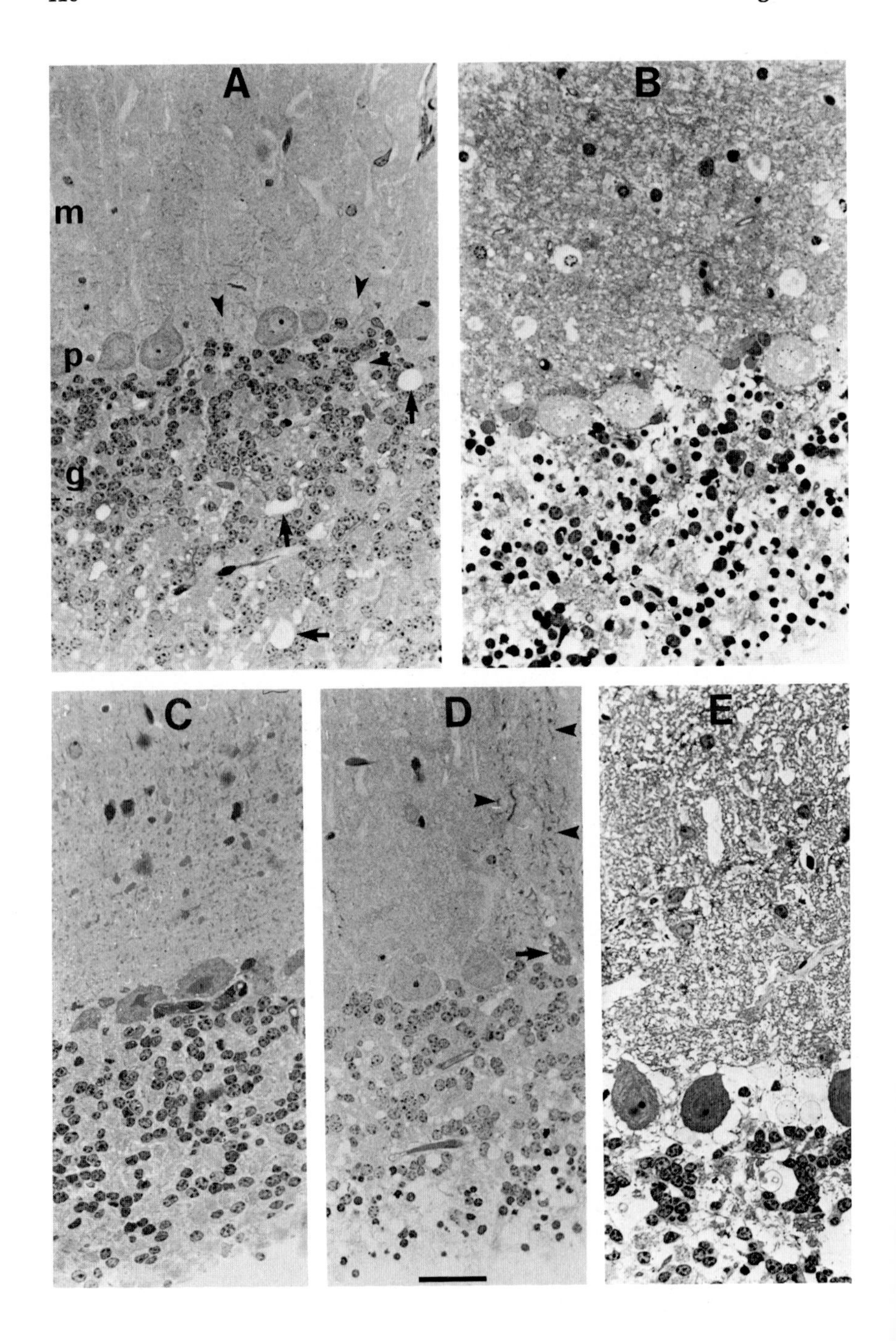
A
B
C
D
E
m
p
g

granule cell nuclei are pyknotic, i.e., shrunk and darkly-staining. This is usually about 50 μm in the cerebellar slices but is likely to vary with different areas. For example, in the CA3 pyramidal cell regions of hippocampal slices cut in a similar way, damage at each of the cut edges extends 100 μm or so (Frotscher *et al.*, 1981). A relatively small number of cells in the center of the slice survive, while in the granule cell layers of the dentate gyrus and in CA1 pyramidal layer the cut edge damage is much less (J. Garthwaite *et al.*, unpublished observations).

Other areas of necrosis can be recognized easily. This is perhaps best illustrated by comparing "good" and "bad" slices under higher power. Figure 11A shows the different cell layers in a hand-cut cerebellar slice, in which the morphological preservation is broadly comparable with the tissue fixed *in situ*; Figure 11B is the corresponding view of a slice prepared using a mechanical chopper and shows considerable necrosis. In most cases, degenerating large neurons tend to stain weakly and numerous vacuoles are seen in their cytoplasm (see Purkinje cells in Figure 11B). In more advanced stages of necrosis, only large vacuoles containing a few lightly-staining fragments can be seen in their place. In other cases, however, such cells may be shrunk with darkly-staining nuclei and cytoplasm (Figure 11C); in extreme cases, the cell somata become very shrunk, their cytoplasm vacuolated and dark dendrites can be seen in the molecular layer (Figure 11D). Necrotic cells are rare in surface cerebellar slices but when they exist, they are often spotted amidst other normal-looking cells. This pattern of necrosis is observed in a number of pathological conditions *in vivo* (see Schwob *et al.*, 1980) and is evident in slice preparations of striatum (Bak *et al.*, 1980) and hippocampus (Frotscher *et al.*, 1981). Small neurons (e.g., granule cells in cerebellar or hippocampal slices), by contrast, tend to display shrun-

Figure 11. Examples of pathological changes found in cerebellar slices. (A) shows the good preservation commonly observed in surface slices of rat cerebellum prepared by hand-slicing. Note the compact molecular layer (m) and the normal light staining perikarya of Purkinje cells (p), granule cells (g), and glial cells (arrowheads). A few vacuoles (arrows) in the granule cell layer are usually evident. In contrast, massive disruption in a mechanically chopped cerebellar slice is apparent in (B). Purkinje cells are badly swollen and vacuolated; granule cells and some cells in the molecular layer show condensed, pyknotic nuclei and expanded cytoplasm; and the neuropil contains many swollen structures. Glial cell perikarya in the Purkinje cell layer appear intact but slightly shrunken. (C) and (D) illustrate different degrees of shrinkage of large neurons that can sometimes be seen. In (D), in addition to the shrinkage of Purkinje cells soma (arrow), dark, shrunken dendrites can be seen extending up into the molecular layer (arrowheads). (E) Swelling of glial cell somata and processes elicited by exposing a slice to 3 mM L-glutamate for 1 hr. Calibration = 40 μm (J. Garthwaite, unpublished observation).

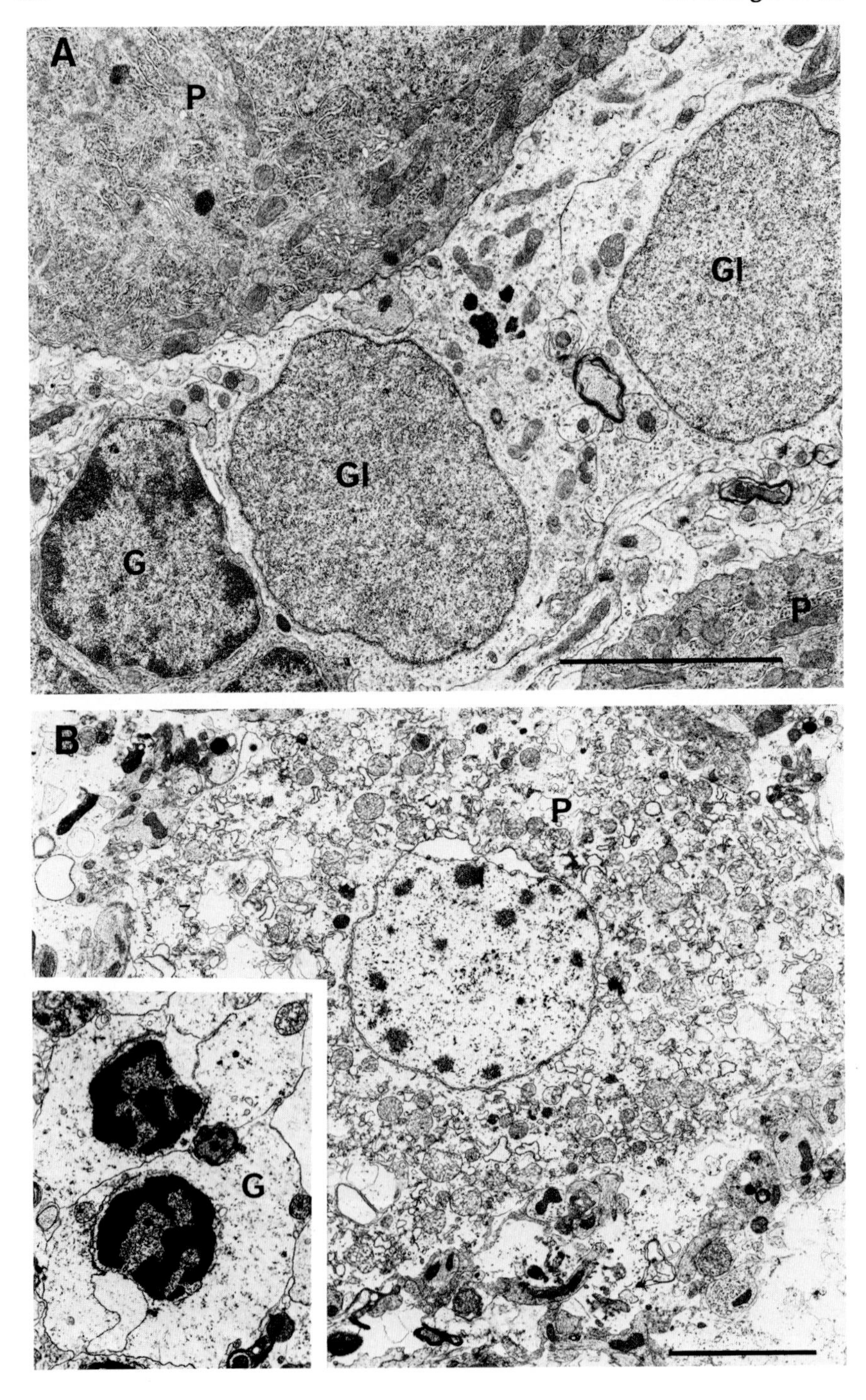
A
P
Gl
Gl
G
P
B
P
G

ken, pyknotic nuclei, surrounded by empty, highly swollen cytoplasm when necrotic, the so-called "bull's eye" profile (Figure 11B).

Damaged neuropil is recognized by its being peppered with vacuoles of varying size (Figure 11B) in contrast to its rather evenly-staining appearance when well preserved (Figure 11A). Glial cell bodies can also be recognized under the light microscope (Figure 11A). These are not normally swollen in well-preserved slices, and even in heavily disrupted tissue (Figure 11B), the glial somata appear to be more prone to shrinkage rather than swelling. For comparison, massive glial swelling induced experimentally by exposing slices to L-glutamate is illustrated in Figure 11E.

3.3.3. *Electron Microscopy.* In general, electron miscroscopy confirms the impression gained from light microscopy. Thus, those areas that appear compact and evenly staining usually appear normal at the ultrastructural level. For example, Purkinje, granule, and glial cell bodies in the slice illustrated in Figure 11A are seen under the electron microscope (Figure 12A) to have well-preserved internal organelles (mitochondria, endoplasmic reticulum, Golgi apparatus and nucleus) and there is no clear evidence of any pathological alterations. The degenerating Purkinje cells noted in Figure 11B, on the other hand, display massive disruption of internal structure and, apparently, discontinuous plasma membranes under the electron microscope (Figure 12B); necrotic granule cells with shrunken, pyknotic nuclei and highly expanded cytoplasm are also illustrated (Figure 12B inset). Purkinje and other cells appearing shrunk and dark at the light level show electron dense cytoplasm and expansion of their endoplasmic reticulum at the ultrastructural level (see Frotscher *et al.*, 1981; Garthwaite and Wilkin, 1982).

Not evident under the light microscope is the condition of neuronal processes such as dendrites and nerve terminals. In the molecular layer of well-preserved slices (Figure 13A), the majority of these elements appear normal. Occasionally, condensed, dark terminals with tightly packed vesicles can be seen. Large mossy fiber terminals in the granule cell layer showing similar features are also found in small numbers (Figure 13B); this has been noted in other slice preparations (Frotscher *et*

Figure 12. Electron micrographs of cell bodies in cerebellar slices. (A) From the hand-cut slice shown in Figure 11A to show that the ultrastructural preservation is comparable to that seen in the cerebellum *in vivo*. P, Purkinje cell somata; Gl, glial cells; G, granule cell. (B) At the other extreme, a heavily vacuolated Purkinje cell, necrotic granule cells, and swollen glial processes are seen in a slice prepared by mechanical chopping. Calibration bars = 5 μm. (J. Garthwaite, unpublished observation.)

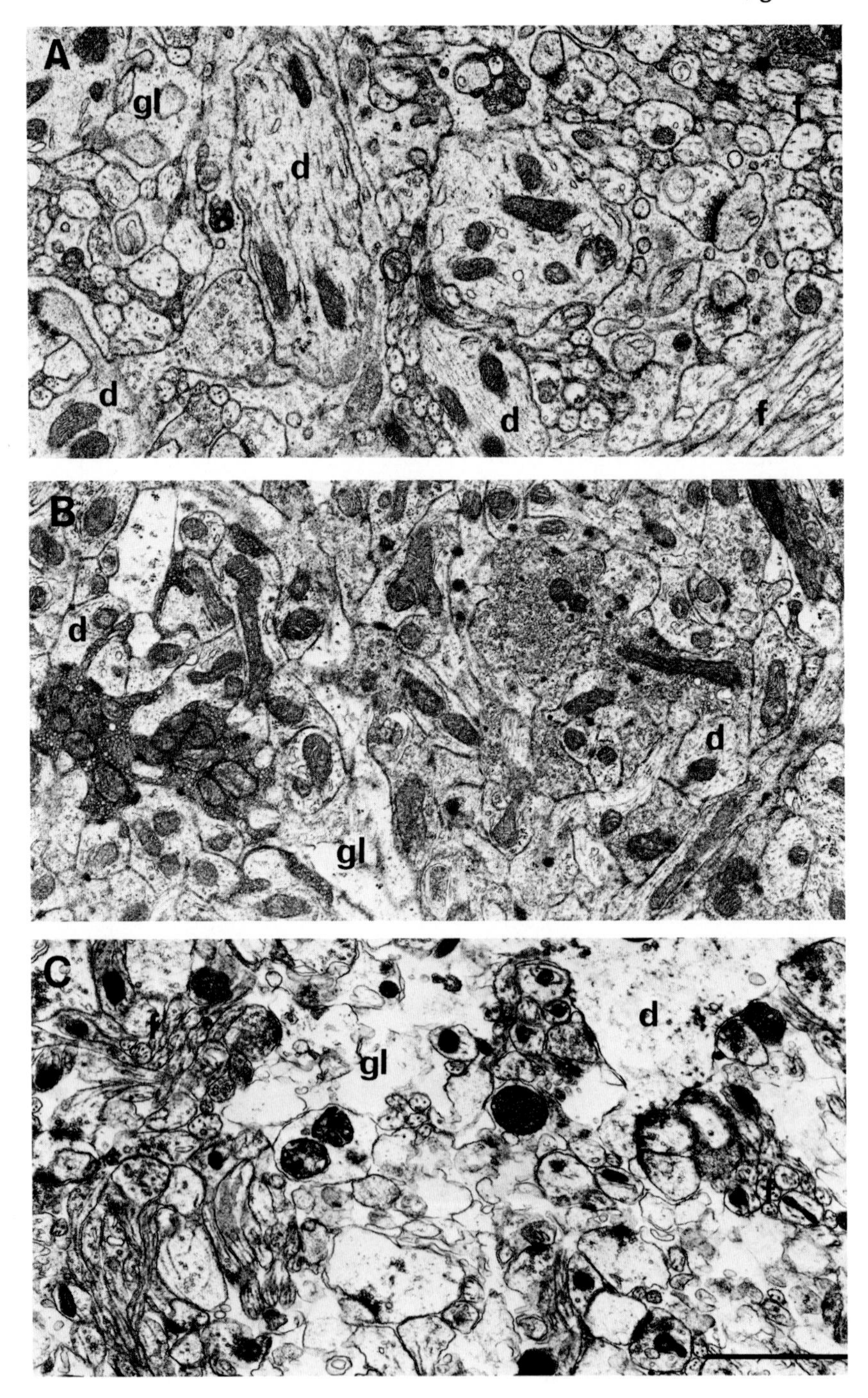
A
gl
d
d
d
f
f
B
d
d
gl
C
f
gl
d

al., 1981). In badly damaged neuropil, postsynaptic elements and glial processes are highly swollen and unmyelinated fibers are expanded (Figure 13C). Synaptic terminals, on the other hand, retain many of their normal morphological features. In view of the fact that nerve terminals survive homogenization and other insults during the preparation of synaptosomes, that is not altogether surprising.

3.3.4. *Implications.* Most of the pathological changes occurring in incubated brain slices can be detected readily using light microscopy of semithin, plastic-embedded sections. It is really only when a slice preparation has been evaluated on this basis that the more selective impression given by the electron microscope is of significant value. Whether a slice is judged satisfactory or not depends, of course, on what it is being used for. For instance, those studying certain properties of nerve terminals (e.g., transmitter uptake or release) might prefer to use slices in which these structures are preserved most selectively; others working on individual cells, or groups of cells, may be content with small areas of slices in good condition. Certain studies, however, require each slice to resemble its *in vivo* counterpart as reproducibly and as closely as possible. It is clear from morphological analyses that substantial differences in cellular and subcellular preservation can exist between slices of different brain areas prepared using the same method, between slices of the same area prepared using the same method but from animals of different ages, and between different cell types in the same slice (Garthwaite *et al.*, 1979, 1980). Inadequate checks, therefore, greatly restrict the interpretation that can be placed on many experimental results.

Despite these reservations, it is still possible to carry out light and electron microscopic investigations of either physiologically identified neurons (e.g., cells filled with horseradish peroxidase HRP) or slices subjected to certain treatments (e.g., long-term potentiation). It is critical, however, that sufficient baseline data be collected to enable the experimenter to distinguish treatment-induced from slice-induced phe-

---

Figure 13. Electron micrographs of synaptic regions in cerebellar slices. The appearance of dendrites (d), unmyelinated fibers (f), glial processes (gl), and synapses are illustrated. (A) A well-preserved molecular layer in a hand-cut surface slice. (B) Synaptic glomeruli in the internal granule cell layer of a hand-cut slice. The glomerulus on the right shows normal features while an adjacent one (left) contains a dark, condensed mossy fiber terminal. Other elements (dendrites and inhibitory Golgi terminals) appear normal. (C) Molecular layer of mechanically chopped slice displaying highly swollen dendrites, variable expansion of fibers but also several nerve terminals in reasonable condition. Calibration bar = 2 μm. (J. Garthwaite, unpublished observation.)

nomena (e.g., HRP-fill vs. dark, dying neuron; stimulation-induced dendritic changes vs. incubation-induced swelling).

3.4. *Metabolism*

Elsewhere in this volume, Lipton and Whittingham (see Chapter 5, this volume) discuss some biochemical characteristics of the hippocampal slice, particularly in regard to energy metabolism and the evaluation of metabolic data from the slice. In summary, there are quite large differences in the values of many high energy metabolites and in rates of oxidative phosphorylation between the slice and the *in situ* brain. The basis for these differences is not known. The "healthy slice" or, in other terms, the slice on which electrophysiological measurements are generally carried out, is one in which ATP levels and 0_2 consumption are 50 to 67% of their *in situ* values and in which intracellular pH is about 0.3 to 0.4 units more alkaline. The intracellular K/Na ratio is 75% of normal.

Two parameters, energy charge* and the phosphocreatine (PCr) to ATP ratio, are markers for the energetic viability of the slice. These are both ratios and are therefore independent of protein content, which is beneficial because much of the reduction in brain slice metabolite levels may be accounted for by the presence of protein associated with destroyed cells along the cut edges of the slice. Thus, the ratios will indicate the energetic viability of the relatively undamaged areas of the interior portion of the slice. Steady-state energy charge is an indication of the percentage of adenylate content which is in the form of ATP and, both *in situ* and *in vitro*, lies in the range 0.85 to 0.95. The PCr/ATP ratio, which is around 1.5 *in situ*, is 2.0 to 2.6 in slices. This ratio is a sensitive indicator of compromised energy metabolism, as PCr is rapidly used in energetically stressful situations to maintain ATP levels. In addition to these ratios, the total creatine content (creatine + PCr) can be used to indicate the degree of tissue damage. An increased steady-state concentration of total creatine is generally accompanied by higher levels of PCr, ATP, and total adenylates, indicating a greater amount of unperturbed tissue in the brain slice. In addition, total creatine content is dependent on slice thickness. It is generally in the range of 45 to 55 nmole/mg protein for immersed hippocampal slices (T. Whittingham, unpublished observations). All these values, except total creatine, are very sensitive to anoxia or to inhibition of glycolysis. For example, hippocampal slice ATP is reduced by about 20% if glucose is replaced by

* Energy charge is defined as (ATP + 0.5 ADP)/(ATP + ADP + AMP).

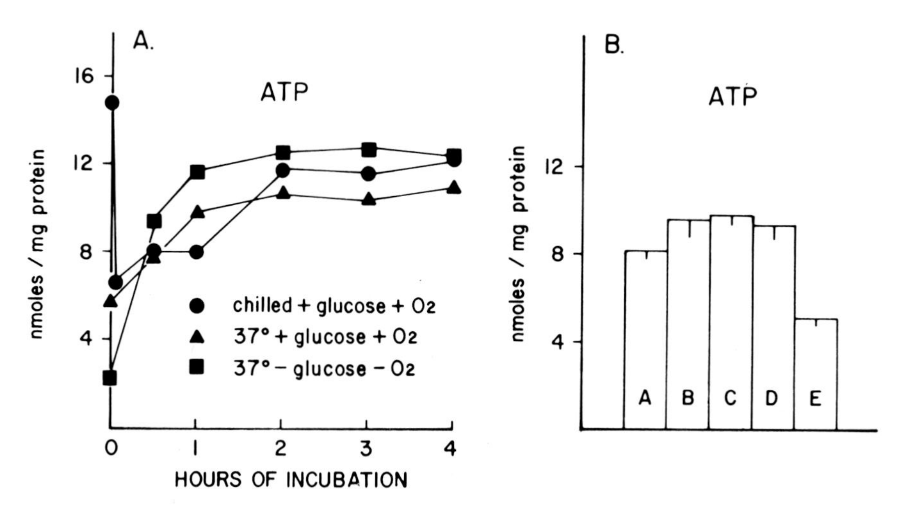

Figure 14. ATP content of hippocampal slices prepared in different ways. (A) Recovery of ATP content in 500-μm-thick slices prepared in different media. Slices were prepared during a 7-min period while maintaining the tissue at either 37°C or 4°C (chilled) and in the presence or absence of glucose and oxygen. They were subsequently perfused in a submersion chamber for the times indicated. (B) Effect of slice thickness on ATP content. Slices were prepared using a Sorvall chopper at 37°C, with both glucose and oxygen present in the ACSF. The tissue was incubated for 7 hr in a submersion chamber prior to fixation and metabolic analysis. Slice thickness: A = 250 μm, B = 500 μm, C = 750 μm, D = 1000 μm, E = 2000 μm. (T. S. Whittingham, unpublished observations.)

pyruvate as an energy source; the evoked population spike in CA1 is reduced by about 25% and the intracellular K/Na ratio reduced by about 25%. Other energy substrates, such as β-OH butyrate, glycerol, or endogenous stores, are unable to support any synaptic transmission in the hippocampal slice and, in addition, reduce ATP levels to below 50% of normal (Lipton and Robacker, 1983).

Figure 14B demonstrates the effect that slice thickness has on ATP levels. Slices of thickness up to 1000 μm exhibit similar metabolic values. The 2000-μm-thick slices appear to be metabolically compromised, probably from the presence of a significantly large anoxic core. The relationship between metabolic and electrophysiological measures is questionable, since 1000-μm-thick hippocampal slices, immersed at 37°C, would not be expected to retain normal responses. When slices were incubated for up to 8 hr, there was no indication of metabolic failure, the 8-hr metabolite levels generally resembling *in vivo* values more closely than levels at 4 hr (T. Whittingham, unpublished observations).

Metabolite levels have not been measured in brain slices maintained for periods longer than 8 hr.

How quickly do slices approach a metabolic steady state after preparation? Steady-state levels of ATP and PCr are reached within 2 to 4 hr of incubation (the major portion of the recovery occurs within the first hour), and appear to be independent of the temperature or availability of glucose and oxygen during the preparation procedure. Figure 14A illustrates the time course of recovery to steady-state level for ATP following the preparation of 500-μm slices in three media. The circles represent slices prepared at 4°C in ACSF containing glucose and oxygen, the triangles, slices cut at 37°C with glucose and oxygen; and the squares, slices cut at 37°C in the absence of glucose and oxygen. The timecourse for recovery of PCr, energy charge and intracellular pH were similar to that for ATP. Although metabolic recovery was similar in all three groups, the maximum evoked population spike in the dentate gyrus in slices prepared at 37°C in the absence of glucose and oxygen was only 30% of that obtained in the other two groups (T. Whittingham, unpublished observations). This is a remarkable finding and is at the same time unsettling. Perhaps the metabolic measures chosen are not tightly coupled to the electrophysiological "health" of the slices. Alternatively microdissected subregions within a slice may show better correlations with electrophysiology than does the whole slice (Lipton and Whittingham, 1982).

The inability of workers to produce sliced tissue in which concentrations of ions and energy metabolites come close to their *in situ* values is perhaps one of the major engimas in slice studies. Certainly, determination of the basis for the metabolic differences between slices and *in situ* brain tissue and the difference between metabolic and electrophysiological indices of slice health are essential if we are to gain further confidence in the slice as a model for *in vivo* events. Elimination of these differences is an even more important goal and, perhaps paradoxically, probably more easily attained. It has not yet been possible to eliminate the differences with supplements to ACSF, which suggests that the absence of some substance in the extracellular fluid is probably not an explanation. A recent abstract suggests that treating living animals before the brain is removed might be a way to approach the problem; thus, elevated slice ATP and K/Na ratios were achieved by administering ascorbate (2 g/kg) to rats 30 min prior to decapitation (Mishra and Kovachich, 1982).

There appears to be little metabolic evidence for an "anoxic core" in well-prepared thin tissue slices (Lipton and Whittingham, Chapter 5, this volume) so that oxygen availability should not be considered a major source of the problem. Thus, one is left with the possibility that the trauma of isolation may produce the metabolic deficits. Some lines of research that might be fruitful in this regard would be (1) efforts to reduce the physical trauma of the isolation process by studying different slicing methods, (2) efforts to decrease the biochemical trauma of preparation using antioxidants, antiperoxidants, or other agents, such as indomethacin, which might inhibit formation of potentially toxic substances.

3.5. *Spreading Depression*

Under certain conditions, hippocampal slices demonstrate a phenomenon akin to or identical with spreading depression (SD) as seen *in vivo* (Bureš *et al.*, 1974; Nicholson and Kraig, 1981). SD can be triggered in the slice in a number of ways; for example, by depositing an iontophoretic bolus of an excitatory amino acid into the slice, by a short bout of anoxia, by perfusing with high $[K^+]_o$, by inadvertent prodding with a stimulating electrode, or by high-frequency orthodromic stimulation (10 to 60 Hz). When high frequency stimulation is employed, a longer than usual pulse width (1 msec instead of 0.05 to 0.1 msec) is especially effective. Whatever the trigger, the hallmark of SD is a rather sudden, negative DC shift in the extracellular potential that reaches 15 to 30 mV in amplitude, accompanied by cessation of all evoked responses. During intracellular recording the expected large depolarization and precipitous fall in input resistance is seen (R. Dingledine, unpublished observations). As demonstrated first *in vivo* (see Nicholson and Kraig, 1981, for references), the interstitial $[K^+]_o$ rises and $[Ca^{2+}]_o$ falls (Figure 15B). All responses return to near normal within several minutes. When triggered by high-frequency stimulation, SD is often preceded by a period of synchronous afterdischarge bursts (arrows in Figure 15A) that lasts for several seconds. It is our impression that SD can be elicited only with great difficulty in otherwise healthy slices, and not at all in frankly sick tissue. Some ill-defined middle ground seems optimal, in concert with the prevailing idea that brain tissue requires "conditioning" to undergo SD.

4. METHODS OF DRUG APPLICATION

Among the more important advantages of slice preparations is the accessibility of the recorded neurons to application of known amounts of drugs and the possible control of the ionic environment of the cells. There are four usual modes of drug application in slices. Each has some

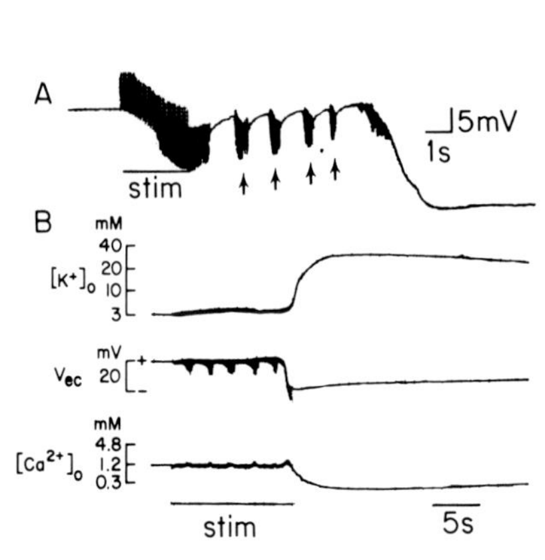

Figure 15. Spreading depression in hippocampal slice. (A) A DC recording of potential (referenced to a distant ground) in the extracellular space in s. pyramidale, during a tetanic train (60 Hz, 1 msec pulse duration) of stimuli applied to fibers in s. radiatum. The stimulus train caused a negative shift in the extracellular potential, and following the train synchronous, spontaneous bursts of population spikes were observed (arrows), the last of which was terminated by spreading depression. (B) Simultaneous measurement of $[Ca^{2+}]_o$ and $[K^+]_o$ with ion-selective microelectrodes, as well as of extracellular potential (V_{ec}), during a stimulus-evoked bout of spreading depression. (A) From R. Dingledine (unpublished observations). (B) From Somjen *et al.* (1981).

advantages over the others yet each may constitute a source of variability and artifacts. A combination of different modes of drug application should be selected, depending on the drug and the tissue studied. The following discussion is relevant mostly to intracellular recordings. Extracellular recordings are associated with less stringent concerns and will not be dealt with here.

4.1. *Superfusion*

Ions and drugs can be applied via the superfusion medium in known concentrations. Several factors determine the efficiency of the superfusion: (1) The volume and shape of the recording chamber. The larger the chamber and the further away the slice is from the mainstream of flow of the medium, the slower the exchange of the old by the new medium. (2) The rate of superfusion. Obviously the faster the superfusion, the faster the exchange. (3) The mode of anchoring the slice to the floor of the chamber. These last two factors determine the stability of recording during the superfusion; a stable recording situation is crucial during intracellular experiments. The stability is also determined by the manner in which the waste fluid is drawn out of the recording chamber. Fluid can be drained by gravity or by suction. Either method can be associated with drastic changes in fluid level during superfusion, if care is not taken in the design of the slice chamber.

Advantages: The superfusion method is the only one in which the precise extracellular concentration of drugs of ions can be known. It allows ultimately a homogeneous distribution of drug within the slice.

Disadvantages: There are several problems with the superfusion method that prohibit its universality for neuropharmacology. First, the concentration of the added drug grows gradually during the perfusion, and does not reach a plateau level within the slice until some time after the bath concentration has reached its maximum. Unless the intraslice concentration of the drug is measured, which is not a common practice, it is difficult to predict the drug concentration within the tissue at any given time after the onset of superfusion. The magnitude of this problem will depend on the thickness of the slice, the type of drug studied, and the design of the slice chamber. Interface chambers, in which slices are exposed to the atmosphere, seem particularly prone to slow exchange rates within the slice. In both the Scottish chamber and the Nicoll/Alger chamber, which are submersion designs, the onset of drug effects approaches the exchange time of the extraslice space. Second, continuous exposure of brain slices to certain drugs, especially putative neurotransmitters, may desensitize the receptors or change the equilibrium potentials of the associated ions. Thus, the true time course and maximal extent of the reponse may not be revealed. Likewise, due to long washout times, the construction of full dose–response curves in a single slice is not often practical.

4.2. *The Nanodrop*

The nanodrop technique allows one to apply minute (<1 to ~ 50 nl) volumes of drug-containing media onto the surface of the slice. Usually a broken micropipette with a tip diameter of 10 to 30 μm, which contains the chemical dissolved in ACSF, is held in the air above the recorded slice. The pipette is connected to a microsyringe by a fluid-filled tube before use. Gentle pressure is applied and a small droplet of fluid is formed, the size of which can be estimated with an eyepiece micrometer or in relation to known landmarks on the slice. The pipette is then lowered until the drop touches the surface of the slice.

Advantages: Application of the droplet is nearly instantaneous and therefore fast responses can be separated from slow ones in the event that a given agonist produces a mixed response when acting on different types of receptors. The speed of application also reduces the effects of desensitization of receptors. Furthermore, the size of the droplet can be adjusted such that a given drop can activate regions of the cell near the soma, apical or basal dendrites, or alternatively the entire cell surface. Thus, responses that might not be seen with the use of other techniques may show up when using the nanodrop. Rapid successive application

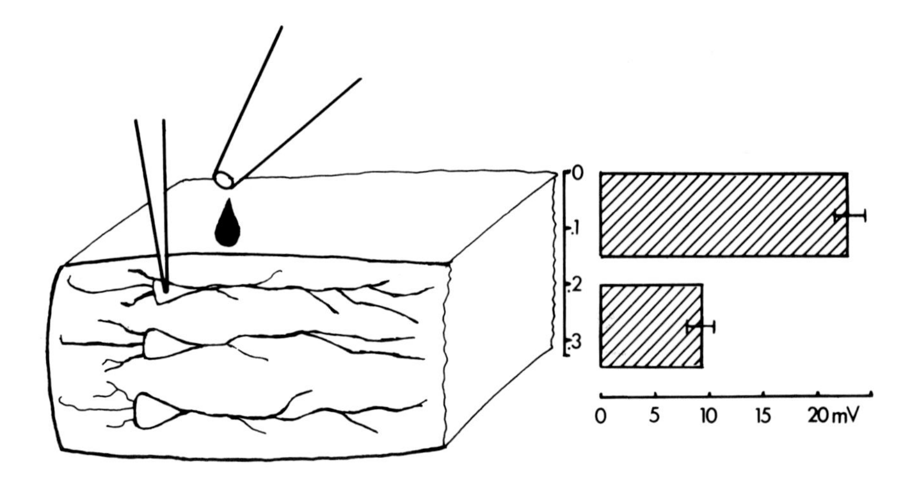

Figure 16. Magnitude of depolarizing response of CA1 cells to droplet application of L-glutamate as a function of cell depth within the slice. The cells were divided into shallow (0 to 175 μm) and deep (175 to 350 μm), as measured from the surface of the slice. A constant droplet of L-glutamate (10 mM, approx. 20 nl) produced a much larger response in shallow than in deep cells. This difference is partly due to the efficiency of the tissue in terminating the actions of applied glutamate, which is taken up by the tissue so that less reaches deeper cells. Such a large gradient is not seen with drugs that are not taken up or rapidly metabolized by the tissue. (M. Segal, unpublished observation.)

of known drug concentrations can be easily achieved, although the precise drug concentration achieved at its active site is unknown.

Disadvantages: Although this is a fast and reproducible method, there are several problems with the nanodrop technique. First, the application of a drop of fluid on the surface of the slice can shunt stimulus currents or produce mechanical vibration or temperature fluctuations, strong enough to cause a potential or resistance change. Mechanical artifacts are usually easily distinguishable from real biological effects and are typically over in 2 to 4 sec, but the possibility that the droplet produces a local cooling requires separate controls. Second, the duration of exposure of the slice to the applied substance cannot be controlled. The action of the substance can be terminated only when it diffuses away or is removed by uptake into the slice. Third, the distance of the recorded cell from the surface of the slice is extremely important for the development of a response (Figure 16). Some chemicals are taken up rapidly into the slice and, when the cell is positioned deep within the slice, it might be exposed to a smaller amount of that chemical. This is a source of variation that should be taken into account. Finally, the nanodrop technique cannot be used with submersion chambers.

4.3. *Iontophoresis*

Microiontophoresis is a common method for drug application, in both *in vitro* and *in vivo* studies. Several reviews have dealt critically with iontophoresis methodology (Kelly *et al.*, 1975; Purves, 1981) and so it will be discussed here only briefly. Several-barrelled (3-, 5-, or 7-) micropipettes are commonly used. They are pulled together to form a tip of 3-to-8-μm diameter. Concentrated drug solutions, prepared usually in distilled water or saline, are placed in the barrels of these pipettes. DC current is passed through the pipette to a reference electrode in the bath; charged molecules are thereby ejected from the pipette and diffuse toward the recorded cell. Compared to other techniques, only a small amount of the compound is released and only a small region of the cell membrane is normally exposed to the compound. This situation may give rise to cases of "false-negatives," in which no responses are registered when in fact the iontophoresis pipette was not properly positioned.

Advantages: Iontophoresis is clearly superior to the other methods in its precision of temporal and spatial control over the drug administration. The onset is relatively rapid, the duration and termination are more easily controlled. The amount of drug released is proportional, to a certain degree, to the amount of current ejected. The detection of response "hot spots," and the analysis of the spatial distribution of receptors, is possible in a laminated tissue.

Disadvantages: First, it can sometimes be difficult to differentiate drug from current effects, especially with intracellular DC recording. Application of 10 to 1000 nA through the iontophoresis pipette can produce a voltage deflection in the nearby recording pipette that may mislead the unwary. This problem can often be overcome by the use of a balancing pipette through which current is returned. This does not provide foolproof protection, however, and the effect of iontophoresis on tip and ground potentials should always be measured with the recording pipette situated just outside the recorded neuron. Second, the concentration of the drug at its receptors is unknown. There are usually numerous neural and nonneural elements between the pipette and the receptors, which can, by sequestering the drug, cause a much steeper concentration gradient than is predicted by diffusion alone. Even the amount released from the pipette normally can only be estimated. The transport number, which describes the relation between applied current and amount of drug released, is different for each drug and for each pipette. The transport number is seldom measured in order to calculate the amount of drug released. Occasionally, a pipette "plugs" and does not release the expected amount of drug. This condition is often use-dependent, but is not always readily detectable, and may be misinterpreted as desensitization of receptors. A basic requirement for iontophoresis is that the drug is a highly soluble, charged molecule. This is not the case for many drugs of interest. Finally, iontophoresis really comes into its own when the impaled cell can be visualized, which is usually not possible in most slice chambers.

4.4. *The Pressure Pipette*

The pressure micropipette was designed to overcome some of the problems associated with the iontophoresis method. Essentially, a micropipette with a tip diameter of 1 to 5 μm is connected to a high-pressure system which, when activated, can release drug solution from the pipette. No current is used and therefore no current artifacts are expected. The pipette is lowered into the slice to the desired depth and is used to apply a known concentration of the drug at a known location.

Advantages: The pressure pipette has several obvious advantages over the iontophoresis method, but it also shares some advantages with the former. Drugs can be applied at a given location and for a specified duration. It is not necessary that the drug be charged, although solubility is still an important factor. Quantitation of released drug is usually better than with iontophoresis.

Disadvantages: First, drugs may leak out of the pipette in the intervals between pressure ejection. There is no easy way to prevent this leakage, unless the pipette is situated above the slice between applications. Second, the tissue or impurities in the drug solution can block the pressure pipette. Obviously, if only minute quantities are released into the depths of the tissue, it is not easy to monitor this visually. If the experimental design permits, the pipette tip can be withdrawn into air and tested for blockage by visually checking for the presence of a small droplet at the tip. Third, heterogeneity of pipette tip diameter, or of tip shape, is a more crucial variable here than with the iontophoresis method. The larger the tip, or the larger the cone angle at the tip, the larger the volume released at a given pressure. With routine use, these variables can be partly controlled. Indeed, the micropressure pipette has replaced the iontophoresis pipette in many laboratories in recent years.

Another form of pressure ejection uses pipettes with larger tip diameters (5 to 20 μm) and lower pressure (2 to 20 psi). Ejection volumes using this method are generally larger (5 to 20 nl). The pipette is placed above slices that are completely submerged. Since the tip is in solution above the slice, the effect of leakage of drugs between ejection times is limited. Very brief pressure pulses are effective, and drug release is often linear with an increasing number of pulses. Results can therefore be treated quantitatively so long as absolute concentrations are not required.

4.5. *Summary*

In most studies, it is advantageous to use a combination of methods of drug application. In many cases, the largest physiological effects can be detected with the nanodrop technique, and this method is appropriate when the quantitative aspects of drug action are not important. When repeated applications are planned, iontophoresis, or for uncharged molecules, the micropressure pipette, is recommended. For ion substitution experiments, or for administration of compounds such as receptor antagonists, which are not expected to produce direct effects on neuronal properties, the perfusion technique may be best. To assure that a certain effect of a drug is not an artifact, it should be applied in at least two different modes of application.

5. CONCLUDING STATEMENT

In this document, we, as a group, have attempted to discuss the technical aspects of brain slice methodology, and in so doing, reflect our unpublished opinions and considered prejudices. In the process, a good deal of spirited discussion developed among us, especially on the topics of slice chamber design and the proper means of evaluating slice data. The areas of best overall agreement were in the preparation of slices and in the composition of ACSF, perhaps because such a wide range of techniques seems to work.

Over a period of approximately 7 years, the isolated brain slice has gained increasing acceptance among neurobiologists as an appropriate tool to investigate the mammalian brain at the cellular level. The mechanical stability, direct visual access to brain structures, and control of fluid environment are advantages inherent in slice experiments that continue to attract workers to these preparations. A recent explosion of publications dealing with slices, and an increasingly refined and critical concept of what constitutes a "healthy" slice, indicate that the use of brain slices is approaching maturity as a technique. Nevertheless, many opportunities remain for the enterprising investigator to improve matters. Mutually beneficial interactions are becoming more widespread among electrophysiologists, histologists, and neurochemists. This development probably more than any other will continue to advance the usefulness of brain slices in neurobiology.

ACKNOWLEDGMENTS. The preparation of this manuscript was supported in part by NIDA grant DA-02360 and NIH grant NS-17771.

6. REFERENCES

Alger, B. E., and Nicoll, R. A., 1980, Epileptiform burst afterhyperpolarization: Calcium-dependent potassium potential in hippocampal CA1 pyramidal cells, *Science* **210:**1122–1124.

Ames, A., III, Sakanolle, M. and Endo, S., 1964, Na, K, Ca, Mg and Cl concentrations in choroid plexus fluid and cisternal fluid compared with plasma ultrafiltrate, *J. Neurophysiol.* **27:**672–681.

Andersen, P., 1975, Organization of hippocampal neurons and their interconnections, in: *The Hippocampus* Volume I, (R. L. Isaacson and K. H. Pribram, eds.), Plenum Press, New York, pp. 155–175.

Andersen, P., Bliss, T. V. P., and Skrede, K. K., 1971, Lamellar organization of hippocampal excitatory pathways, *Exp. Brain Res.* **13:**208–221.

Andersen, P., Eccles, J. C., and Løyning, Y., 1964, Location of postsynaptic inhibitory synapses on hippocampal pyramids, *J. Neurophysiol.* **27:**592–607.

Andersen, P., Silfvenius, H., Sundberg, S. H., and Sveen, O., 1980, A comparison of distal and proximal dendritic synapses on CA1 pyramids in guinea-pig hippocampal slices *in vitro, J. Physiol. (London)* **307:**273–299.

Andersen, P., Silfvenius, H., Sundberg, S. H., Sveen, O., and Wigström, H., 1978, Functional characteristics of unmyelinated fibers in the hippocampal cortex, *Brain Res.* **144:**11–18.

Bak, I. J., Misgeld, U., Weiler, M., and Morgan, E., 1980, The preservation of nerve cells in rat neostriatal slices maintained *in vitro*: A morphological study, *Brain Res.* **197:**341–353.

Bingmann, D. and Kolde, G., 1982, PO_2-profiles in hippocampal slices of the guinea pig, *Exp. Brain Res.* **48:**89–96.

Bito, L. Z. and Davson, H., 1966, Local variations in cerebrospinal fluid composition and its relationship to the composition of the extracellular fluid of the cortex, *Exp. Neurol.* **14:**264–280.

Bradley, R. D. and Semple, S. J. G., 1962, A comparison of certain acid-base characteristics of arterial blood, jugular venous blood and cerebrospinal fluid in man and the effect on them of some acute and chronic acid-base disturbances, *J. Physiol. (London)* **160:**381–391.

Brown, D. A. and Halliwell, J. V., 1979, Neuronal responses from the rat interpeduncular nucleus *in vitro, J. Physiol. (London)* **292:**9–10P.

Brown, T. H., Wong, R. K. S., and Prince, D., 1979, Spontaneous miniature synaptic potentials in hippocampal neurons, *Brain Res.* **177:**194–199.

Bureš, J., Burešova, O., and Zacharova, D., 1974, *The Mechanism and Applications of Leao's Spreading Depression of Encephalographic Activity*, Academic Press, New York.

Chujo, T., Yamada, Y., and Yamamoto, C., 1975, Sensitivity of Purkinje cell dendrites to glutamic acid, *Exp. Brain Res.* **23:**293–300.

Cohen, M. M. and Hartmann, J. F., 1964, Biochemical and ultrastructural correlates of cerebral cortex slices metabolizing *in vitro*, in: *Morphological and Biochemical Correlates of Neural Activity*, (M. M. Cohen and R. S. Snider, eds.), Harper and Row, New York, pp. 57–74.

Crepel, F. and Dhanjal, S. S., 1981, Sensitivity of Purkinje cell dendrites to glutamate and aspartate in cerebellar slices maintained *in vitro*. *J. Physiol.* (*London*) **320:**54P.

Crepel, F., Dhanjal, S. S., and Garthwaite, J., 1981, Morphological and electrophysiological characteristics of rat cerebellar slices maintained *in vitro*, *J. Physiol.* (*London*) **316:**127–138.

Crepel, F., Dhanjal, S. S., and Sears, T. A., 1982, Effect of glutamate, aspartate and related derivatives on cerebellar Purkinje cell dendrites in the rat. An *in vitro* study, *J. Physiol.* (*London*) **329:**297–317.

Davies, H. C. and Davies, R. E., 1965, Biochemical aspects of oxygen poisoning, in: *Handbook of Physiology. Section 3: Respiration, Vol. II*, (W. O. Fenn and H. Rahn, eds.), American Physiological Society, Bethesda, pp. 1047–1058.

Davson, H., 1967, *Physiology of the Cerebrospinal Fluid*, J. and A. Churchill, London.

Dhanjal, S. S. and Sears, T. A., 1980a, An *in vitro* slice preparation of adult mammalian spinal cord, *J. Physiol.* (*London*) **312:**12–13P.

Dhanjal, S. S. and Sears, T. A., 1980b, Electrical activity of rat substantia gelatinosa Rolandi studied *in vitro*, *J. Physiol.* (*London*) **312:**19P.

Dingledine, R., 1981, Possible mechanisms of enkephalin action on hippocampal CA1 pyramidal neurons, *J. Neurosci.* **1:**1022–1035.

Dingledine, R. and Langmoen, I. A., 1980, Conductance changes and inhibitory actions of hippocampal recurrent IPSPs, *Brain Res.* **185:**277–287.

Dingledine, R., and Somjen, G., 1981, Calcium dependence of synaptic transmission in the hippocampal slice, *Brain Res.* **207:**218–222.

Dingledine, R., Dodd, J., and Kelly, J. S., 1980, The *in vitro* brain slice as a useful neurophysiological preparation for intracellular recording, *J. Neurosci. Methods* **2:**323–362.

Duffy, C. J. and Teyler, T. J., 1975, A simple tissue slicer, *Physiol. Behav.* **14:**525–526.

Dunwiddie, T. V. and Lynch. G., 1979, The relationship between extracellular calcium concentration and the induction of hippocampal long-term potentiation, *Brain Res.* **169:**103–110.

Eccles, R. M., Løyning, Y., and Oshima, T., 1966, Effects of hypoxia on the monosynaptic reflex pathway in the cat spinal cord, *J. Neurophysiol.* **29:**315–332.

Franck, G., 1972, Brain slices, in: *The Structure and Function of Nervous Tissue*, Volume VI. Structure and Physiology (G. H. Bourne, ed.), Academic Press, New York, pp. 417–465.

Fremont-Smith, F., Dailey, M. E., Merritt, H. H., Carroll, M. P., and Thomas, G. W., 1931, The equilibrium between cerebrospinal fluid and plasma, *Arch. Neurol. Psychiatry* **25:**1271–1289.

Friedman, S. B., Austen, W. G., Rieselbach, R. E., Block, J. B., and Rall, D. P., 1963, Effect of hypochloremia on cerebrospinal fluid chloride concentration in a patient with anorexia nervosa and in dogs, *Proc. Soc. Exp. Biol. Med.* **114:**801–805.

Frotscher, M., Misgeld, U., and Nitsch, C. 1981, Ultrastructure of mossy fibre endings in *in vitro* hippocampal slices, *Exp. Brain Res.* **41:**247–255.

Fujii, T., Baumgartl, H., and Lübbers, D. W., 1982, Limiting section thickness of guinea pig olfactory cortical slices studied from tissue pO_2 values and electrical activities, *Pflügers Arch.* **393:**83–87.

Fujii, T., Buerk, D. G., and Whalen, W. J., 1981, Activation energy in the mammalian brain slice as determined by oxygen microelectrode measurement, *Jpn. J. Physiol.* **31:**279–283.

Garthwaite, J., 1982, Excitatory amino acid receptors and guanosine 3′,5′-cyclic monophosphate in incubated slices of immature and adult rat cerebellum, *Neuroscience* **7:**2491–2497.

Garthwaite, J. and Wilkin, G. P., 1982, Kainic acid receptors and neurotoxicity in adult and immature rat cerebellar slices, *Neuroscience* **7:**2499–2514.

Garthwaite, J., Woodhams, P. L., Collins, M. J., and Balazs, R., 1979, On the preparation of brain slices: Morphology and cyclic nucleotides, *Brain Res.* **173:**373–377.

Garthwaite, J., Woodhams, P. L., Collins, M. J., and Balazs, R., 1980, A morphological study of incubated slices of rat cerebellum in relation to postnatal age, *Dev. Neurosci.* **3:**90–99.

Gerschenfeld, H. M., Wald, F., Zadunaisky, J. A., and De Robertis, E. D. P., 1959, Function of astroglia in the water-ion metabolism of the central nervous system. An electron microscope study, *Neurology* **9:**412–425.

Gibson, G. E., Pulsinelli, W., Blass, J. P., and Duffy, T. E., 1981, Brain dysfunction in mild to moderate hypoxia, *Am. J. Med.* **70:**1247–1254.

Grisar, T. and Franck, G., 1981, Effect of changing potassium ion concentrations on rat cerebral slices *in vitro*: A study during development, *J. Neurochem.* **36:**1853–1857.

Gutnick, M. J. and Prince, D. A., 1981, Dye coupling and possible electrotonic coupling in the guinea pig neocortical slice, *Science* **211:**67–70.

Haas, H. L., Schaerer, B., and Vosmansky, M., 1979, A simple perfusion chamber for the study of nervous tissue slices *in vitro, J. Neurosci. Methods* **1:**323–325.

Hablitz, J. J. and Lundervold, A., 1981, Hippocampal excitability and changes in extracellular potassium, *Exp. Neurol.* **71:**410–420.

Hackett, J. T., 1976, Calcium dependency of excitatory chemical synaptic transmission in the frog cerebellum *in vitro, Brain Res.* **114:**35–46.

Hamberger, A., Chiang, G., Nylen, E. S., Scheff, S. W., and Cotman, C. W., 1978, Stimulus evoked increase in biosynthesis of the putative neurotransmitter glutamate in the hippocampus, *Brain Res.* **143:**549–555.

Harvey, J. A., Scholfield, C. N., and Brown, D. A., 1974, Evoked surface-positive potentials in isolated mammalian olfactory cortex, *Brain Res.* **76:**235–245.

Hatton, G. I., Armstrong, W. E., and Gregory, W. A., 1978, Spontaneous and osmotically-stimulated activity in slices of rat hypothalamus, *Brain Res. Bull.* **3:**497–508.

Hatton, G. I., Doran, A. D., Salm, A. K., and Tweedle, C. D., 1980, Brain slice preparation: Hypothalamus, *Brain Res. Bull.* **5:**405–414.

Henderson, G., Pepper, C. M., and Shefner, S. A., 1982, Electrophysiological properties of neurones contained in the locus coeruleus and mesencephalic nucleus of the trigeminal nerve *in vitro, Exp. Brain Res.* **45:**29–37.

Hendry, E. B., 1962, The osmotic pressure and chemical composition of human body fluids, *Clin. Chem.* **8:**246–265.

Hounsgaard, J. and Yamamoto, C., 1979, Dendritic spikes in Purkinje cells of the guinea pig cerebellum studied *in vitro, Exp. Brain. Res.* **37:**387–398.

Hunter, G. and Smith, H. V., 1960, Calcium and magnesium in human cerebrospinal fluid, *Nature* **186:**161–162.

Jefferys, J. G. R., 1981, The Vibroslice, a new vibrating blade tissue slicer, *J. Physiol.* (*London*) **324:**2P.

Jefferys, J. G. R. and Haas, H. L., 1982, Synchronized bursting of CA1 hippocampal pyramidal cells in the absence of synaptic transmission, *Nature* (*London*) **300:**448–450.

Kaas, I. S. and Lipton, P., 1982, Mechanisms involved in irreversible anoxic damage to the *in vitro* rat hippocampal slice, *J. Physiol.* (*London*) **332:**459–472.

Kandel, E. R., Spencer, W. A., and Brinley, F. J., 1961, Electrophysiology of hippocampal neurons. I. Sequential invasion and synaptic organization, *J. Neurophysiol.* **24:**225–242.

Kelly, J. S., Simmonds, M. A., and Straughan, D. W., 1975, Microelectrode techniques, in: *Methods in Brain Research* (P. B. Bradley, ed.), John Wiley and Sons, New York, pp. 333–377.

Kemény, A., Boldizar, H., and Pethes, G., 1961, The distribution of cations in plasma and cerebrospinal fluid following infusion of solutions of salts of sodium, potassium, magnesium and calcium, *J. Neurochem.* **7:**218–227.

Kerkut, G. A. and Wheal, H. V., (eds.), 1981, *Electrophysiology of Isolated Mammalian CNS Preparations,* Academic Press, New York.

Kimelberg, H. K., Biddlecome, S., Narumi, S., and Bourke, R. S., 1978, ATPase and carbonic anhydrase activities of bulk-isolated neuron, glia and synaptosome fractions from rat brain, *Brain Res.* **141:**305–323.

King, G. L. and Parmentier, J. L., 1983, Oxygen toxicity of hippocampal tissue *in vitro, Brain Res.* **260:**139–142.

King, G. L. and Somjen, G. G., 1981, Effects of variations of extracellular potassium activity ($[K^+]_o$) on synaptic transmission and $[Ca^{++}]$ responses in hippocampal tissue *in vitro, Neurosci. Abstr.* **7:**439.

Langmoen, I. A. and Andersen, P., 1981, The hippocampal slice *in vitro*. A description of the technique and some examples of the opportunities it offers, in: *Electrophysiology of Isolated Mammalian CNS Preparations,* (G. A. Kerkut and H. V. Wheal, eds.), Academic Press, New York, pp. 51–105.

Lee, H. K., Dunwiddie, T., and Hoffer, B. 1980, Electrophysiological interactions of enkephalins with neuronal circuitry in the rat hippocampus. II. Effects on interneuron excitability. *Brain Res.* **184:**331–342.

Lee, H., Dunwiddie, T., Deitrich, R., Lynch, G., and Hoffer, B., 1981, Chronic ethanol consumption and hippocampal neuron dendritic spines: A morphometric and physiological analysis, *Exp. Neurol.* **71:**541–549.

Leung, L-W. S., 1979, Orthodromic activation of hippocampal CA1 region of the rat, *Brain Res.* **176:**49–63.

Li, C.-L. and McIllwain, H., 1957, Maintenance of resting membrane potentials in slices of mammalian cerebral cortex and other tissues *in vitro, J. Physiol.* (*London*) **139:**178–190.

Lighthall, J. W., Park, M. R., and Kitai, S. T., 1981, Inhibition in slices of rat neostriatum, *Brain Res.* **212:**182–187.

Lipton, P., 1973, Effects of membrane depolarization on light scattering by cerebral cortical slices, *J. Physiol.* (*London*) **231:**365–383.

Lipton, P. and Heimbach, C. J., 1977, The effect of extracellular potassium concentrations on protein synthesis in guinea-pig hippocampal slices, *J. Neurochem.* **28:**1347–1354.

Lipton, P. and Heimbach, C. J., 1978, Mechanism of extracellular potassium stimulation of protein synthesis in the *in vitro* hippocampus, *J. Neurochem.* **31:**1299–1307.

Lipton, P. and Korol, D., 1981, Evidence that decreases in intracellular pH rapidly inhibit transmission in the guinea-pig hippocampal slice, *Neurosci. Abstr.* **7:**440.

Lipton, P. and Robacker, K. M., Cerebral metabolism: Glycolysis is required for K_o activation of protein synthesis and K uptake, *Fed. Proc.*, in press.

Lipton, P. and Whittingham, T. S., 1979, The effect of hypoxia on evoked potentials in the *in vitro* hippocampus, *J. Physiol.* (*London*) **287:**427–438.

Lipton, P. and Whittingham, T. S., 1982, Reduced ATP concentration as a basis for synaptic transmission failure during hypoxia in the *in vitro* guinea pig hippocampus, *J. Physiol. (London)* **325:**51–65.

Llinás, R. and Sugimori, M., 1980, Electrophysiological properties of *in vitro* Purkinje cell somata in mammalian cerebellar slices, *J. Physiol. (London)* **305:**171–195.

Lynch, G. and Schubert, P., 1980, The use of *in vitro* brain slices for multidisciplinary studies of synaptic function, *Annu. Rev. Neurosci.* **3:**1–22.

Marshall, K. C. and Engberg, I., 1980, The effects of hydrogen ion on spinal neurons. *Can. J. Physiol. Pharmacol.* **58:**650–655.

McGale, E. H. F., Pye, I. F., Stonier, C., Mutchinson, E. C., and Aber, G. M., 1977, Studies of the inter-relationship between cerebrospinal fluid and plasma amino acid concentrations in normal individuals, *J. Neurochem.* **29:**291–297.

McIlwain, H., 1975, Preparing neural tissues for metabolic study in isolation, in: *Practical Neurochemistry* (H. McIlwain, ed.), Churchill Livingstone, London, pp. 105–132.

Mishra, O. P. and Kovachich, G. B., 1982, Elevation of K/Na ratio and depression of water uptake in incubated brain slices obtained from ascorbate-treated rats, *Fed. Proc.* **41:**1319 (abstr.)

Nicoll, R. A. and Alger, B. E., 1981, A simple chamber for recording from submerged brain slices, *J. Neurosci. Methods* **4:**153–156.

Nicoll, R. A., Eccles, J. C., Oshima, T., and Rubia, F., 1975, Prolongation of hippocampal inhibitory postsynaptic potentials by barbiturates, *Nature (London)* **258:**625–627.

Nicholson, C., 1980, Dynamics of the brain cell microenvironment, *NRP Bull.* **18:**177–322.

Nicholson, C. and Kraig, R. P., 1981, The behavior of extracellular ions during spreading depression, in: *The Application of Ion-Selective Microelectrodes* (G. Zeuthen, ed.), Elsevier/North Holland, Amsterdam, pp. 217–238.

Ogata, N., 1979, Substance P causes direct depolarization of neurones of guinea pig interpeduncular nucleus *in vitro, Nature (London)* **277:**480–481.

Okamoto, K. and Quastel, J. H., 1973, Spontaneous action potentials in isolated guinea-pig cerebellar slices: Effects of amino acids and conditions affecting sodium and water uptake, *Proc. R. Soc. London, Ser. B.* **184:**83–90.

Oppelt, W. W., MacIntyre, I., and Rall, D. P., 1963a, Magnesium exchange between blood and cerebrospinal fluid, *Am. J. Physiol.* **205:**959–962.

Oppelt, W. W., Owens, E. S., and Rall, D. P., 1963b, Calcium exchange between blood and cerebrospinal fluid, *Life Sci.* **2:**599–605.

Pepper, C. M. and Henderson, G., 1980, Opiates and opioid peptides hyperpolarize locus coeruleus neurons *in vitro, Science* **209:**394–396.

Perry, T. L., Hansen, S., and Kennedy, J., 1975, CSF amino acids and plasma-CSF amino acid ratios in adults, *J. Neurochem.* **24:**587–589.

Pittman, Q. J., Hatton, J. D., and Bloom, F. E., 1981, Spontaneous activity in perfused hypothalamic slices: Dependence on calcium content of perfusate, *Exp. Brain Res.* **42:**49–52.

Purves, R. D., 1981, *Microelectrode Methods for Intracellular Recording and Iontophoresis*, Academic Press, New York.

Rawlins, J. N. P. and Green, K. F., 1977, Lamellar organization in the rat hippocampus, *Exp. Brain Res.* **28:**335–344.

Richards, C. D. and Sercombe, R., 1968, Electrical activity observed in guinea pig olfactory cortex maintained *in vitro, J. Physiol. (London)* **197:**667–683.

Richards, C. D. and Sercombe, R., 1970, Calcium, magnesium and the electrical activity of guinea-pig olfactory cortex *in vitro, J. Physiol. (London)* **211:**571–584.

Salminen, S. and Luomanmäki, K., 1962, Distribution of sodium and potassium in serum, cerebrospinal fluid, and serum ultrafiltrate in some diseases, *Scand. J. Clin. Lab. Invest.* **14:**425–429.

Schaer, H., 1974, Decrease in ionized calcium by bicarbonate in physiological solutions, *Pflügers Arch.* **347:**249–254.

Scholfield, C. N., 1978, Electrical properties of neurones in the olfactory cortex slice *in vitro, J. Physiol. (London)* **275:**535–546.

Schwartzkroin, P. A., 1975, Characteristics of CA1 neurons recorded intracellularly in the hippocampal *in vitro* slice preparation, *Brain Res.* **85:**423–436.

Schwartzkroin, P. A., 1977, Further characteristics of hippocampal CA1 cells *in vitro, Brain Res.* **128:**53–68.

Schwob, J. E., Fuller, T., Price, J. L., and Olney, J. W., 1980, Widespread patterns of neuronal damage following systemic or intracerebral injections of kainic acid: A histological study, *Neuroscience* **5:**991–1014.

Segal, M., 1981, The action of norepinephrine in the hippocampus: Intracellular studies in the slice preparation, *Brain Res.* **206:**107–128.

Siesjo, B. K., 1978, *Brain Energy Metabolism,* John Wiley and Sons, New York.

Somjen, G., Dingledine, R., Connors, B., and Allen, B., 1981, Extracellular potassium and calcium activities in the mammalian spinal cord, and the effect of changing ion levels on mammalian neural tissues, in: *Ion-sensitive Microelectrodes and Their Uses in Excitable Tissue* (E. Sykova, P. Hnik, and F. Vyklicky, eds.), Plenum Press, New York, pp. 159–180.

Spray, D. C., Harris, A. L., and Bennett, M. V. L., 1981, Gap junctional conductance is a simple and sensitive function of intracellular pH, *Science* **211:**712–715.

Struble, R. G., Desmond, N. L., and Levy, W. B., 1978, Anatomical evidence for interlamellar inhibition in the fascia dentata, *Brain Res.* **152:**580–585.

Teyler, T. J., 1980, Brain slice preparation: Hippocampus, *Brain Res. Bull.* **5:**391–403.

Teyler, T. J., Vardaris, R. M., Lewis, D., and Rawitch, A. B., 1980, Gonadal steroids: Effect on excitability of hippocampal pyramidal cells, *Science* **209:**1017–1019.

van Harreveld, A., 1966, *Brain Tissue Electrolytes,* Butterworth, Washington, D.C.

van Herreveld, A. and Stamm, J. S., 1953, Effect of pentobarbital and ether on the spreading cortical depression, *Am. J. Physiol.* **173:**164–170.

Voskuyl, R. A. and ter Keurs, H. E. D. J., 1981, Modification of neuronal activity in olfactory cortex slices by extracellular K^+, *Brain Res.* **230:**372–377.

Vyskocil, F., Kriz, N., and Bureš, J., 1972, Potassium selective microelectrodes used for measuring the extracellular brain potassium during spreading depression and anoxic depolarization in rats, *Brain Res.* **39:**255–259.

Wanko, T. and Tower, D. B., 1964, Combined morphological and biochemical studies of incubated slices of cerebral cortex, in: *Morphological and Biochemical Correlates of Neural Activity* (M. M. Cohen and R. S. Snider, eds.), Harper and Row, New York, pp. 75–97.

White, W. F., Nadler, J. V., and Cotman, C. W., 1978, A perfusion chamber for the study of CNS physiology and pharmacology *in vitro, Brain Res.* **152:**591–596.

Yamamoto, C., 1973, Propagation of afterdischarges elicited in thin brain sections in artificial media, *Exp. Neurol.* **40:**183–188.

Index